XIII International Workshop on Neutrino Factories, Super beams and Beta beams

(NUFACT11)

Journal of Physics: Conference Series Volume 408

Geneva, Switzerland
1-6 August 2011

Editors:

Alain Blondel
Gersende Prior
Ilias Efthymiopoulos

ISBN: 978-1-62276-986-5
ISSN: 1742-6588

Printed from e-media with permission by:

Curran Associates, Inc.
57 Morehouse Lane
Red Hook, NY 12571

Some format issues inherent in the e-media version may also appear in this print version.

Printed by Curran Associates, Inc. (2013)

For permission requests, please contact the Institute of Physics
at the address below.

Institute of Physics
Dirac House, Temple Back
Bristol BS1 6BE UK

Phone: 44 1 17 929 7481
Fax: 44 1 17 920 0979

techtracking@iop.org

Additional copies of this publication are available from:

Curran Associates, Inc.
57 Morehouse Lane
Red Hook, NY 12571 USA
Phone: 845-758-0400
Fax: 845-758-2634
Email: curran@proceedings.com
Web: www.proceedings.com

TABLE OF CONTENTS

PLENARY

WORKING GROUP 1 (WG1): NEUTRINO OSCILLATION PHYSICS

WORKING GROUP 2 (WG2): NEUTRINO SCATTERING PHYSICS

WORKING GROUP 3 (WG3): ACCELERATOR PHYSICS

WORKING GROUP 4 (WG4): MUON PHYSICS

WORKING GROUP 1 AND 2 JOINT SESSION

WORKING GROUP 3 AND 4 JOINT SESSION

POSTERS

ROUND TABLE

Journal of Physics: Conference Series **408** (2013) 011001 doi:10.1088/1742-6596/408/1/011001

Preface

These proceedings present the written contributions from participants of the International Workshop on Neutrino Factories, Superbeams and Betabeams (NuFact11), which was held in Geneva, Switzerland, on 1-6 August 2011. This was the 13thof the series of NuFact annual meetings, which started in 1999. The six-day scientific programme of NuFact11 consisted of 42 plenary and 111 parallel talks and various poster sessions, with the participation of 191 researchers from 19 countries. In addition the International Neutrino Summer School 2011 was held in the two weeks preceding the workshop. It was intended for young scientists starting their research in neutrino physics in such a way that they would be able to participate and contribute to the workshop.

The objective of the NuFact workshops is to build up, time after time, the scientific knowledge on accelerator facilities aimed at the study of neutrino oscillation. The organization of the workshop is tailored to ensure continuity and growth of knowledge: working group conveners have a mandate of three consecutive meetings and maintain an updated list of questions. In order to emphasize further the 'work' (by opposition to 'shop') part of the meeting, we organized a 'round table discussion' at the beginning of the workshop, in which the invitees were asked not only to provide their answers to a preset list of questions, but to propose their own questions to the participants. This was quite successful and we have compiled the answers in this book. Topics addressed at NuFact included neutrino oscillation physics, neutrino cross-sections and detectors, accelerator physics, muon physics and high-intensity applications of muons beams. This variety of subjects is reflected in these proceedings. The editors hope to provide more than enjoyable and fruitful reading for those interested in neutrino physics, a stepping stone along the path to always more insightful research on these mysterious neutrinos, which certainly contain the solution to several essential questions.

It is a pleasure to thank CERN, University of Geneva, the Swiss Institute for Particle Physics (CHIPP) and the European FP7 mandate EuCARD that sponsored the meeting and the Summer School. The Editors wish to thank the Scientific Program Committee and the Working group conveners for their engagement in making NUFACT11 a success, and all participants for the high quality of their contributions. The Local organizing committee members worked beyond the call of duty to ensure a great and lively working atmosphere during the meeting.

The Editors: **Alain Blondel, Ilias Efthymiopoulos and Gersende Prior**

Participants of the 13th International Workshop on Neutrino Factories, Superbeams and Betabeams, 01-06 August 2011, CERN & Geneva University (Switzerland)

Journal of Physics: Conference Series **408** (2013) 011002 doi:10.1088/1742-6596/408/1/011002

Scientific Programme Committee

A. Blondel , University of Geneva
A. Cervera, IFIC
M. Dracos, IN2P3
I. Efhymiopoulos, CERN
J. Ellis , CERN
S. Geer, FNAL
R. Garoby, CERN
M. Goodman, ANL
D. Harris, FNAL
T. Hasegawa, KEK
P. Huber, Virginia Tech.
D. Kaplan, IIT
Y.D. Kim, Sejong University
H. Kirk, BNL
Y. Kuno, Osaka University
K. Long, Imperial College
N.K. Mondal, TIFR
J. Morfin, FNAL
Y. Mori, Kyoto University
K. Nishikawa, KEK
V. Palladino, University of Napoli
C. Prior, RAL
F.J.P. Soler, University of Glasgow
J. Strait, FNAL
R. Svoboda, University of California Davis
F. Terranova, LN Frascati
M. Zisman, LBNL

Local Organizing Committee

E. Benedetto, CERN/NTUA
C. Blanchard, University of Geneva
A. Blondel, University of Geneva (co-chair)
I. Efthymiopoulos, CERN (co-chair)
F. Dufour, University of Geneva
F. Girard-Madoux, CERN
E. Gschwendtner, CERN
A. Korzenev, University of Geneva
M. Morer-Olafsen, CERN
S. Murphy, University of Geneva
G. Prior, CERN
G. Wikström, University of Geneva
E. Wildner, CERN

Sponsors

EuCARD
European Organization for Nuclear Research (CERN)
Swiss Institute for Particle Physics (CHIPP)
University of Geneva

Journal of Physics: Conference Series **408** (2013) 011003 doi:10.1088/1742-6596/408/1/011003

List of Participants

ABGRALL, Nicolas	University of Geneva
ABRAMS, Robert	Muons, Inc.
ADAMSON, Phil	Fermilab
ADEY, David	University of Warwick
AGARWALLA, Sanjib Kumar	IFIC (University of Valencia/CSIC)
ALBRIGHT, Carl	Northern Illinois University & Fermilab
ALEKOU, Androula	Imperial College London
AMEMIYA, Naouyki	Kyoto University
ANKOWSKI, Artur	University of Wrocław & University of Roma Sapienza & INFN Roma
AOKI, Masaharu	Osaka University
BALLETT, Peter	Durham University
BAUSSAN, Eric	CNRS IPHC
BAYES, Ryan	University of Glasgow
BEARD, Kevin	Muons, Inc.
BENEDETTO, Elena	CERN & NTUA
BENHAR, Omar	INFN
BERGER, Niklaus	University of Heidelberg
BERNSTEIN, Robert	Fermilab
BLONDEL, Alain	University of Geneva
BOGACZ, Alex	Jefferson Laboratory
BONESINI, Maurizio	INFN Milano & University of Milano-Bicocca
BROOK-ROBERGE, Daniel	University of British Columbia
BROSS, Alan	Fermilab
CAVANNA, Flavio	INFN LNGS
CERUTTI, Francesco	CERN
CHILDRESS, Sam	Fermilab
COLOMA, Pilar	Autonomous University of Madrid
CONEY, Linda	University of California, Riverside
COOK, Sam	University College London
DAVIDSON, Sacha	CNRS IPNL
D'ARCY, Richard	University College London
DE GOUVEA, Andre	Northwestern University
DE MELO MENDONCA, Tania	CERN
DENSHAM, Chris	STFC Rutherford Appleton Laboratory
DEVAN, Joshua	College of William & Mary
DI LODOVICO, Francesca	Queen Mary University of London
DI LUISE, Silvestro	ETH
DJURCIC, Zelimir	Argonne National Laboratory
DORNAN, Peter	Imperial College London
DRACOS, Marcos	CNRS IPHC
DUFOUR, Fanny	University of Geneva
EDGECOCK, Rob	STFC Rutherford Appleton Laboratory
EFTHYMIOPOULOS, Ilias	CERN
EREDITATO, Antonio	University of Bern
FERNANDEZ MARTINEZ, Enrique	CERN
FORREST, David	University of Glasgow

Journal of Physics: Conference Series **408** (2013) 011003 doi:10.1088/1742-6596/408/1/011003

GALYMOV, Vyacheslav	York University
GAROBY, Roland	CERN
GHOSH, Tapasi	IFIC (University of Valencia/CSIC)
GILARDONI, Simone	CERN
GIUNTI, Carlo	INFN
GNINENKO, Sergei	INR
GOODMAN, Maury	Argonne National Laboratory
GRANGE, Joe	University of Florida
GSCHWENDTNER, Edda	CERN
HALZEN, Francis	University of Wisconsin-Madison
HANSEN, Christian	CERN
HANSON, Gail	University of California, Riverside
HARTNELL, Jeff	University of Sussex
HASEGAWA, Takuya	KEK
HASLER, Alexis	University of Geneva
HAYASAKA, Kiyoshi	Nagoya University
HAYATO, Yoshinari	Kamioka Observatory & ICRR & University of Tokyo
HERNANDEZ, Pilar	IFIC (University of Valencia/CSIC)
HERRERO-GARCIA, Juan	IFIC (University of Valencia/CSIC)
IBARRA, Alejandro	TUM
IKEDA, Motoyasu	Kyoto University
JENNER, Leo	Imperial College & Fermilab
JONES, Benjamin	MIT
KAKUNO, Hidekazu	University of Tokyo
KAPLAN, Daniel	Illinois Institute of Technology
KAWAMURA, Naritoshi	KEK
KELLIHER, David	STFC Rutherford Appleton Laboratory
KIELCZEWSKA, Danuta	University of Warsaw
KIKAWA, Tatsuya	Kyoto University
KORDOSKY, Michael	College of William and Mary
KORZENEV, Alexandre	University of Geneva
KUNO, Yoshitaka	Osaka University
LAGRANGE, Jean-Baptiste	Kyoto University
LALAKULICH, Olga	Giessen University
LAMY, Thierry	CNRS LPSC
LEE, Hye-Sung	Brookhaven National Laboratory
LIU, Jianglai	Shanghai Jiao Tong University
LONG, Kenneth	Imperial College London
LOPEZ PAVON, Jacobo	Durham University
LOUIS, William	LANL
MACHADO, Pedro	CEA IRFU
MACHIDA, Shinji	STFC Rutherford Appleton Laboratory
MARCHIONNI, Alberto	ETH
MARIANI, Camillo	Columbia University
MARTINI, Marco	CEA DAM
MATEV, Rosen	University of Sofia
MATSUDA, Yasuyuki	University of Tokyo
MAVROMATOS , Nikolaos	CERN
MCCAULEY, Neil	University of Liverpool

Journal of Physics: Conference Series **408** (2013) 011003 doi:10.1088/1742-6596/408/1/011003

MCDONALD, Kirk T.	Princeton University
MELONI, Davide	University of Roma Tre
MENA, Olga	IFIC (University of Valencia/CSIC)
MENTION, Guillaume	CEA IRFU
MEZZETTO, Mauro	INFN
MIHARA, Satoshi	KEK
MINAKATA, Hisakazu	Tokyo Metropolitan University
MOHAPATRA, Rabindra	University of Maryland
MORFIN, Jorge G.	Fermilab
MORI, Takaaki	Okayama University
MORI, Yoshiharu	Kyoto University
MUETHER, Mathew	Fermilab
MURPHY, Sebastien	University of Geneva
NAKADAIRA, Takeshi	KEK
NAKAMURA, Satoshi	Jefferson Lab
NELSON, Jeff	College of William and Mary
NEMEVSEK, Miha	ICTP Trieste
NEUFFER, David	Fermilab
NICOLO, Donato	Pisa University & INFN Pisa
NIEVES, Juan M.	IFIC (University of Valencia/CSIC)
NOLEN, Jerry	Argonne National Laboratory
NUNOKAWA, Hiroshi	Pontifical Catholic University of Rio de Janeiro
OLIVEIRA, Roberto	UNICAMP
OTA, Toshihiko	MPI Munich
PALAMARA, Ornella	INFN LNGS
PALLADINO, Vittorio	INFN Napoli & University of Napoli Federico II
PAPADIMITRIOU, Vaia	FNAL
PARDONS, Ans	CERN
PARKE, Stephen	Fermilab
PASCOLI, Silvia	Durham University
PASTERNAK, Jaroslaw	Imperial College London & STFC Rutherford Appleton Laboratory
PAYET, Jacques	CEA IRFU
PERDUE, Gabriel	University of Rochester
PIETROPAOLO, Francesco	INFN Padova
POSIADALA, Magdalena Zofia	University of Warsaw
PRAKASH, Suprabh	IIT Mumbai
PRIOR, Gersende	CERN
RAVONEL, Melody	University of Geneva
RICCIARDI, Stefania	STFC Rutherford Appleton Laboratory
RUBBIA, Andre	ETH
SAITO, Naohito	KEK
SASAKI, Kenichi	KEK
SCHMITZ, David	Fermilab
SCHWETZ, Thomas	MPI Heidelberg
SENJANOVIC, Goran	ICTP Trieste
SHAPOSHNIKOV , Mikhail	EPFL
SHILTSEV, Vladimir	Fermilab
SHIMOMURA, Koichiro	KEK
SINHA, Nita	IMSc Chennai Institute of Technology

Journal of Physics: Conference Series **408** (2013) 011003 doi:10.1088/1742-6596/408/1/011003

SMIRNOV, Alexei	ICTP Trieste
SOBCZYK, Jan	University of Wrocław
SOLER, Paul	University of Glasgow
SOREL, Michel	IFIC (University of Valencia/CSIC)
SOUSA, Alexandre	Harvard University
STEERENBERG, Rende	CERN
STEFAN, Dorota	INFN LNGS
STORA, Thierry	CERN
STRAIT, James	Fermilab
SULAK, Larry	Boston University & CERN
SVOBODA, Robert	University of California, Davis
TERRANOVA, Francesco	INFN LNF
TORTORA, Ludovico	INFN Roma Tre
TORUN, Yagmur	Illinois Institute of Technology
TOYODA, Akihisa	KEK
TSENOV, Roumen	University of Sofia
TUNNELL, Christopher	Oxford University
VAHLE, Patricia	College of William and Mary
VANCRAEYVELD, Pieter	Ghent University
VASSILOPOULOS, Nikolaos	CNRS IPHC
WANG, Zhimin	University of Beijing
WARK, David	STFC Rutherford Appleton Laboratory
WIKSTROM, Gustav	University of Geneva
WILDNER, Elena	CERN
WINTER, Walter	Wurzburg University
WOMERSLEY, William John	STFC
WONG, Chan Fai (Steven)	Durham University
YANG, Un-Ki	University of Manchester
YASUDA, Osamu	Tokyo Metropolitan University
YOKOYAMA, Masashi	University of Tokyo
YONEHARA, Katsuya	Fermilab
YOSHIDA, Makoto	KEK
ZHANG, He	MPI Heidelberg
ZIEMER, Benjamin	University of California, Irvine
ZISMAN, Michael	Lawrence Berkeley National Laboratory
ZITO, Marco	CEA IRFU
ZUKANOVICH FUNCHAL, Renata	University of São Paulo

Peer review statement

All papers published in this volume of *Journal of Physics: Conference Series* have been peer reviewed through processes administered by the proceedings Editors. Reviews were conducted by expert referees to the professional and scientific standards expected of a proceedings journal published by IOP Publishing.

XIII International Workshop on Neutrino Factories, Super beams and Beta beams (NUFACT11) IOP Publishing
Journal of Physics: Conference Series **408** (2013) 012001 doi:10.1088/1742-6596/408/1/012001

Experimental status of neutrino physics

Fanny Dufour[1] and David Wark[2]

[1]Section de Physique, Université de Genève, 1205 Genève, Switzerland
[2]Imperial College London, Department of Physics, London, United Kingdom

E-mail: [1]`fanny.dufour@unige.ch` and [2]`david.wark@stfc.ac.uk`

Abstract. After a fascinating phase of discoveries, neutrino physics still has a few mysteries such as the absolute mass scale, the mass hierarchy, the existence of CP violation in the lepton sector and the existence of right-handed neutrinos. It is also entering a phase of precision measurements. This is what motivates the NUFACT 11 conference which prepares the future of long baseline neutrino experiments. In this paper, we report the status of experimental neutrino physics. We focus mainly on absolute mass measurements, oscillation parameters and future plans for oscillation experiments [1].

1. Introduction

Over the last 15 years experiments have demonstrated that neutrinos oscillate and therefore mix and have masses, however few oscillation parameters are well measured. In addition, their absolute mass scale is still only described by upper limits. Finally, we do not know the nature of neutrino masses. Therefore despite the tremendous results of the last decade, there is still a vast open field with many discoveries in the making. In this paper we present the status of the absolute mass scale measurements and oscillation parameters measurements, and describe several possible new experiments for improving these measurements. This paper has been updated to take into account the results presented at the Neutrino 2012 conference in June 2012 in Kyoto [2].

2. Measurements of the absolute neutrino mass

Absolute neutrino masses can be measured in several ways. Astrophysical neutrinos, kinematic limits and neutrinoless double beta decay are all used to measure the absolute neutrino mass scale or set a limit on it.

2.1. Supernovae and cosmological constraints

Supernovae are copious sources of neutrinos: by measuring time shifts, it is in principle possible to measure neutrino masses down to 30 eV as in the case of SN1987a [3]. Since neutrinos are so numerous in the universe, even a tiny neutrino mass can have cosmological implications. Current cosmological bounds are down around $M_\nu < 0.17 - 0.33$ eV [4, 5] and new data from the Planck observatory should be available soon [6]. These results are quite model dependent and mostly illustrate the sensitivity of structure formation to neutrinos masses. They cannot replace direct laboratory experiments.

2.2. Tritium beta decay and KATRIN

Double beta decay experiments study the end of the beta decay spectrum in order to measure $m^2_{\nu_e} = \sum_{i=1}^{3} |U_{ei}|^2 m_i^2$. The current upper electron-neutrino mass limit from beta decay experiments is $m_{\nu_e} < 2.3$ eV [7]. The next generation experiment is KATRIN at Karlsruhe. It is sensitive down to 0.2 eV [8, 9].

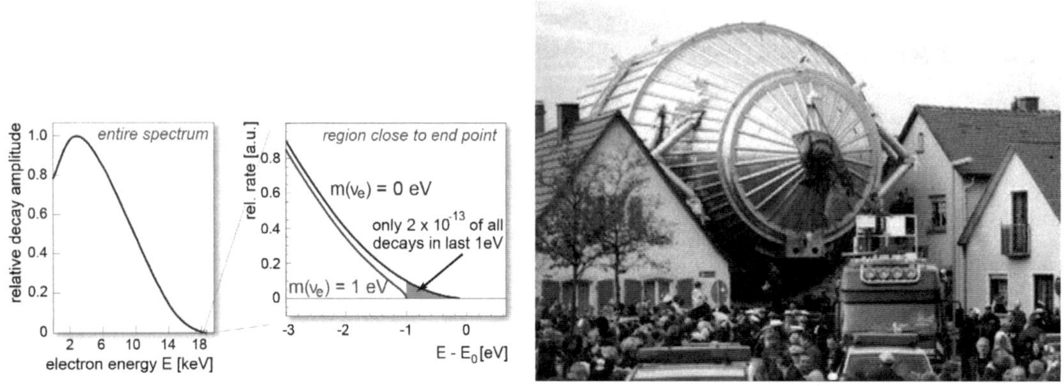

Figure 1. Left: Beta decay spectrum. **Right**: KATRIN detector [8].

2.3. Neutrinoless double beta decay

Neutrinoless double beta decay ($0\nu\beta\beta$) experiments are the main method to investigate whether neutrinos have a Majorana mass term which couples left-handed neutrinos to right-handed anti-neutrinos. It is also an excellent tool to probe the absolute mass scale and, in combination with oscillation experiments, gives a hint on the mass hierarchy (Fig. 2).

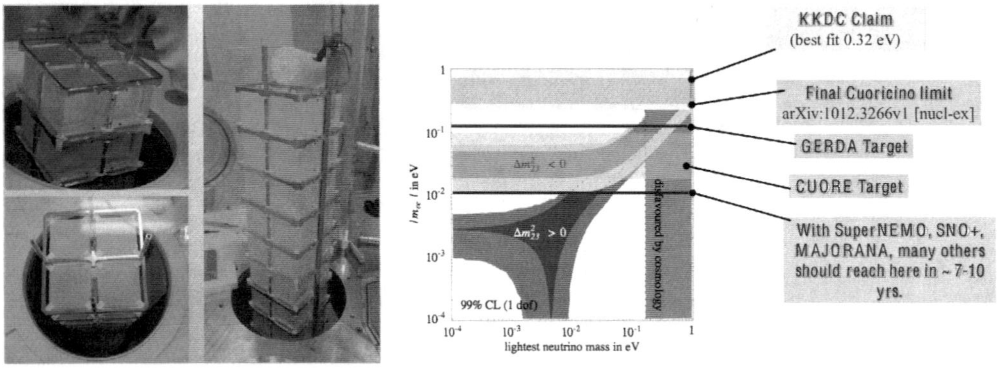

Figure 2. Left: CUORE-0 detector [10]. **Right**: Neutrinoless double beta decay. Target sensitivity of several experiments [1].

A whole set of $0\nu\beta\beta$-decay experiments are currently running or are about to start taking data: GERDA, CUORE-0, EXO and KamLAND-Zen began in 2011. SNO+, Majorana and NEXT should start in 2013, and SuperNEMO, Lucifer, EXO-gas, XMASS and CUORE are scheduled to start later. While we cannot report every result here, we report that KamLAND-Zen found that $\langle m_{\beta\beta} \rangle < 0.26 - 0.54$ eV at 90% C.L. [11] and that EXO-200 found $\langle m_{\beta\beta} \rangle < 0.14 - 0.38$ eV at the same significance [12].

3. Oscillation parameters measurements

Since solar neutrino experiments revealed the famous solar neutrino problem, we have known that neutrinos behave atypically. Since then the fact neutrinos transmute and thus have mass has been established by the Super-Kamiokande measurement of atmospheric neutrinos [13] and by solar data from SNO [14]. Neutrino oscillations is governed by a 3×3 unitary mixing matrix and by mass differences. The mixing angles (Fig. 3, left) have all been measured but even though the two mass splittings have also been measured (Fig. 3, right), the mass hierarchy and the CP violating phase are still unknown. In this section, we will present the current status of the measurement of each oscillation parameter.

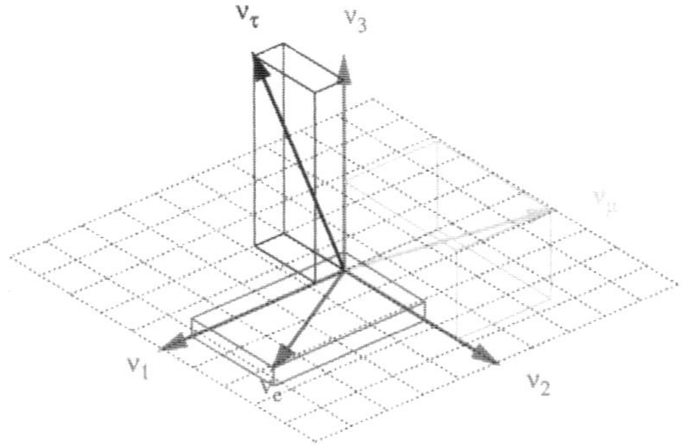

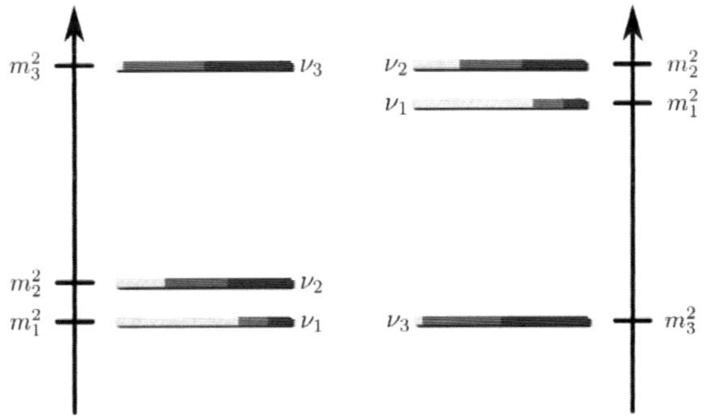

Figure 3. Left: Neutrino mixing angles. **Right**: Neutrino mass hierarchy (normal, inverted).

3.1. Measurement of θ_{13}

Until last year, θ_{13} was the sole remaining unknown mixing angle, the only knowledge of this parameter came from the Chooz experiment and was an upper limit of 0.15 at 90% C.L. [15]. A hint that θ_{13} was non-zero did however come from global fit of available oscillation data [16]. In June of 2011, T2K presented their first electron neutrino appearance measurement, six events on a predicted background of $1.5 \pm 0.3(syst.)$ indicating a non-zero value of θ_{13} at 2.5σ [17]. The analysis was not strictly speaking blind but all the cuts for the electron neutrino event selection in the Super-Kamiokande detector were decided well before T2K even started taking data. At Neutrino 2012, all the data up to May 15th 2012 were analyzed and presented [18]. Ten events are observed in the far detector and θ_{13} is excluded at 3.2σ. At 90% C.L., the T2K results is $\sin^2 2\theta_{13} = 0.104^{+0.60}_{-0.45}$ for $\delta_{CP} = 0$ and normal hierarchy. The T2K selected ν_e candidates are shown in Fig. 4. Two weeks after T2K released their results, the MINOS collaboration presented results which disfavour $\theta_{13} = 0$ at 89% C.L. with 62 events on a background of $49.6 \pm 7.0(stat) \pm 2.7(syst.)$ [19]. At Neutrino 2012, MINOS disfavors $\theta_{13} = 0$ at 96% C.L. [20].

Finally in March 2012, outstanding results came from the reactor experiments Daya Bay [21, 22] and Reno [23]. The Daya Bay collaboration excluded $\theta_{13} = 0$ at 5.2σ [21, 22], while RENO excluded it at 4.9σ [23]. Their best fit values are respectively $\sin^2 2\theta_{13} = 0.089 \pm 0.010(stat.) \pm 0.005(syst.)$ and $\sin^2 2\theta_{13} = 0.113 \pm 0.013(stat.) \pm 0.019(syst.)$ and their data is also shown in Fig. 6. Double Chooz excluded $\theta_{13} = 0$ at 3.1σ and their best fit is $\sin^2 2\theta_{13} = 0.109 \pm 0.030(stat.) \pm 0.025(syst.)$ [24]. All experiments are still statistically dominated and the clearly demonstrate that θ_{13} is greater than zero.

Reactor experiments measure the disappearance of electron anti-neutrinos to measure θ_{13} as shown in Fig. 5 and Eq. 1. Current accelerator experiments measure the appearance of electron neutrinos in a muon-neutrino beam (Eq. 3). In addition the accelerator measurement depends on

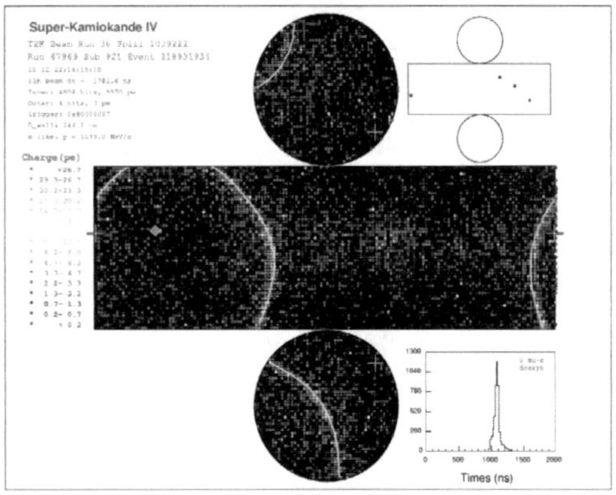

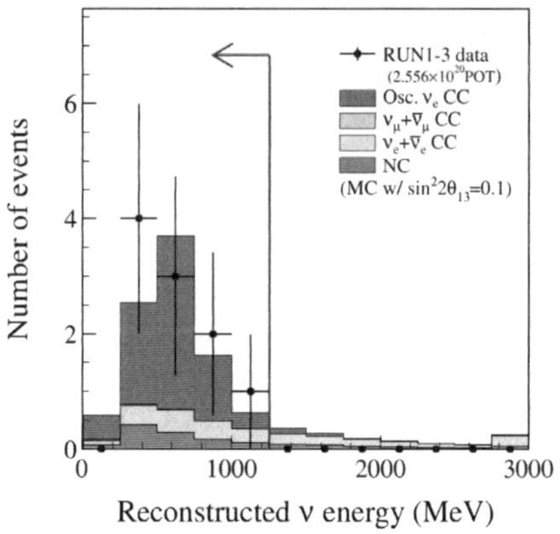

Figure 4. Left: An electron neutrino event in SK. **Right**: Reconstructed neutrino energy distribution for fully-contained, fiducial volume, single-ring e-like events with Evis > 100 MeV, no decay-e and POLfit mass less than 105 MeV/c^2 in RUN1+2+3 [18].

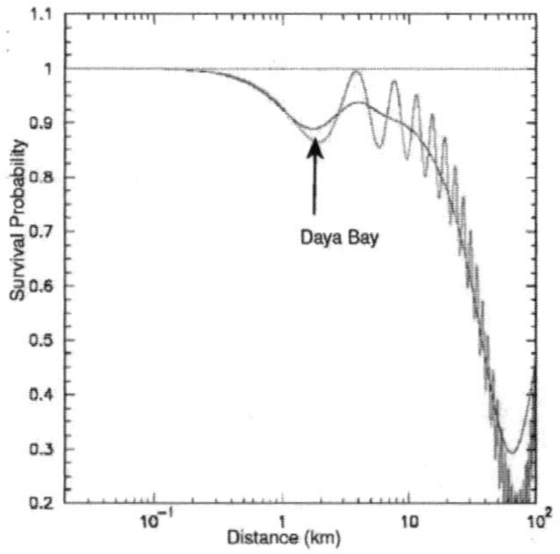

Figure 5. Left: Picture of Daya Bay detectors [22]. **Right**: Survival probability of electron neutrino as a function of the distance.

the value of the CP phase δ while the reactor measurement does not. Because these experiments are fundamentally different it is very interesting to continue both in parallel.

$$P_{\bar{\nu}_e \to \bar{\nu}_e} \approx 1 - \sin^2 2\theta_{13} \sin^2(1.267 \Delta m^2_{13} L/E) \tag{1}$$

$$\begin{aligned}
P[\nu_\mu \;\; &\to \;\; \nu_e] = \sin^2 2\theta_{13} s^2_{23} \sin^2\left(\frac{\Delta m^2_{31} L}{4E}\right) - \frac{1}{2} s^2_{12} \sin^2 2\theta_{13} s^2_{23} \left(\frac{\Delta m^2_{21} L}{2E}\right) \sin\left(\frac{\Delta m^2_{31} L}{2E}\right) \\
&+ \;\; 2J_r \cos\delta \left(\frac{\Delta m^2_{21} L}{2E}\right) \sin\left(\frac{\Delta m^2_{31} L}{2E}\right) - 4J_r \sin\delta \left(\frac{\Delta m^2_{21} L}{2E}\right) \sin^2\left(\frac{\Delta m^2_{31} L}{4E}\right) \\
&+ \;\; \cos 2\theta_{13} sin^2 2\theta_{13} s^2_{23} \left(\frac{4Ea(x)}{\Delta m^2_{31}}\right) \sin^2\left(\frac{\Delta m^2_{31} L}{4E}\right) \\
&- \;\; \frac{a(x)L}{2} \sin^2 2\theta_{13} \cos 2\theta_{13} s^2_{23} \sin\left(\frac{\Delta m^2_{31} L}{2E}\right) + c^2_{23} \sin^2 2\theta_{12} \left(\frac{\Delta m^2_{21} L}{4E}\right)^2,
\end{aligned} \tag{2}$$

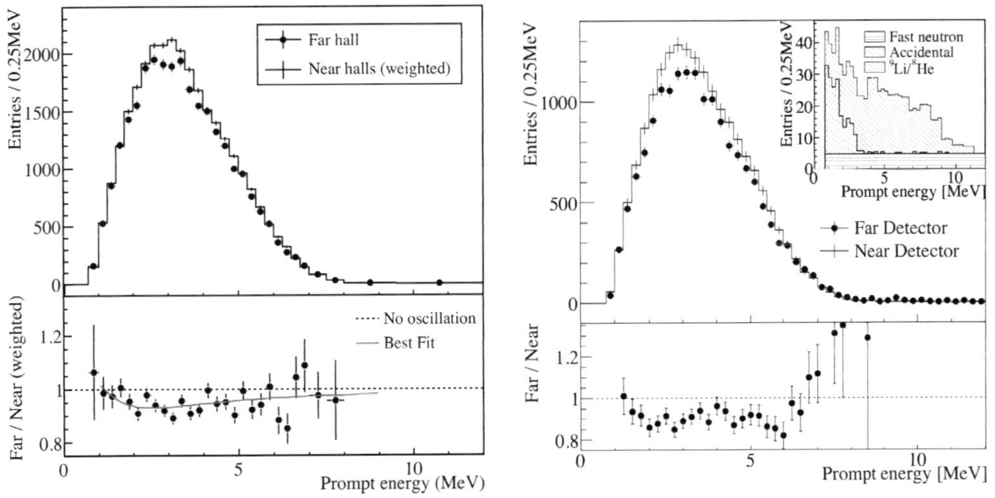

Figure 6. Left, Daya Bay data: Top: Measured prompt energy spectrum of the far hall (sum of three ADs) compared with the no-oscillation prediction from the measurements of the two near halls. Spectra were background subtracted. Uncertainties are statistical only. Bottom: The ratio of measured and predicted no-oscillation spectra. The red curve is the best- fit solution with $\sin^2 2\theta_{13} = 0.089$ obtained from the rate-only analysis. The dashed line is the no-oscillation prediction [21, 22]. **Right, RENO data:** Observed spectrum of the prompt signals in the far detector compared with the non-oscillation predictions from the measurements in the near detector. The backgrounds shown in the inset are subtracted for the far spectrum. The background fraction is 5.5% (2.7%) for far (near) detector. Errors are statistical uncertainties only. Bottom: The ratio of the measured spectrum of far detector to the non-oscillation prediction [23].

(3)

where $a(x) = \sqrt{2}G_F N_e(x)$, G_F is the Fermi constant, $N_e(x)$ is the electron number density at x in the earth, $J_r = c_{12}s_{12}c_{12}^2 s_{13}c_{23}s_{23}$ is the Jarlslog determinant.

3.2. Measurement of the solar parameters: Δm^2_{21} and θ_{12}

The solar parameters are now well constrained by both solar and reactor data, and the newly measured value of θ_{13} further reduces the uncertainty of this measurement. A very good summary of current data is presented by T. Schwetz [25] and the current values of these parameters are given below.

$$\Delta m^2_{21} = 7.59^{+0.20}_{-0.18} \times 10^{-5}\ \text{eV}^2 \quad \text{and} \quad \sin^2\theta_{12} = 0.312^{+0.017}_{-0.015}$$

The values of Δm^2_{21} and $\sin^2\theta_{12}$ have been obtained using the KamLAND neutrino data and the Super-Kamiokande and SNO solar data.

3.3. Measurement of atmospheric parameters: $|\Delta m^2_{31}|$ and θ_{23}

The atmospheric parameters have also been well measured by the Super-Kamiokande atmospheric data and by the MINOS experiment. Their results have also been summarized by T.Schwetz [25] and are given below.

$$|\Delta m^2_{31}| = \begin{cases} 2.45 \pm 0.09 & \times 10^{-3}\ \mathrm{eV}^2\ \ (\mathrm{NH}) \\ 2.34^{+0.10}_{-0.09} & \times 10^{-3}\ \mathrm{eV}^2\ \ (\mathrm{IH}) \end{cases} \quad \text{and} \quad \sin^2\theta_{23} = 0.51 \pm 0.06$$

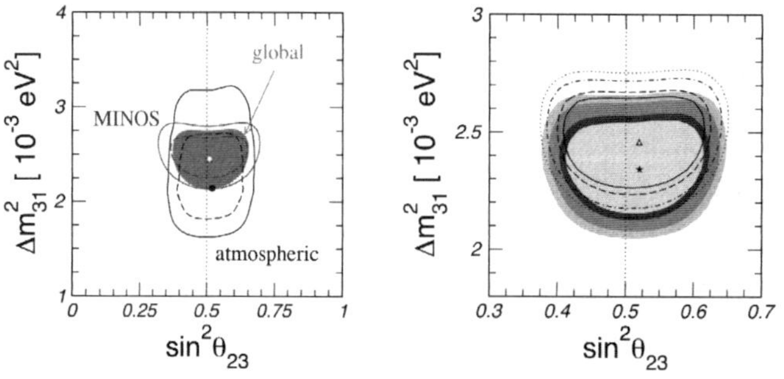

Figure 7. Determination of the atmospheric oscillation parameters. **Left**: interplay of atmospheric (black) and MINOS disappearance (blue) data and the combination (red/shaded region) for normal hierarchy at 90% C.L. (dashed) and 3σ (solid). **Right**: combined allowed regions for normal (black curves) and inverted hierarchy (colored regions) at 90%, 95%, 99%, 99.73% C.L. [25].

The $|\Delta m^2_{31}|$ value is obtained from doing a combined fit of MINOS and Super-Kamiokande atmospheric results. In addition the first measurements of $\sin^2 2\theta_{23}$ using an off-axis neutrino beam has been presented by the T2K collaboration [26]. The value of the mass splittings are well known but the ordering of the masses, (the mass hierarchy), is still unknown. Discovering the mass hierarchy is one of the goals of future experiments not only because it obscures the CP violation measurement, but if neutrino masses were inverted it would be the first time that fermions do not have increasing masses with increasing generation number.

4. Future plans

With the discovery that θ_{13} is large, the gateway for studying CP violation in the lepton sector and the mass hierarchy has been opened. Future experiments plan to do just that. Three types of proposals have emerged: high power conventional super beams, beta-beams and neutrino factories. In addition given the large value of θ_{13} Daya Bay could be able to study the mass hierarchy by placing a 20 kton detector at 60 km [27, 28].

4.1. Super beams

A super beam is a conventional beam, based on pion decays but reaching powers of the order of the mega-watt. Three such beams have been proposed. In Japan, the current T2K beam will be a super beam once it reaches its design luminosity of 750kW. This beam in combination with the proposed Hyper-Kamiokande detector [29] would be a very powerful tool to measure δ_{CP} assuming that the mass hierarchy is known. In the US the LBNE project plans to direct a neutrino beam from FermiLab to the Homestake mine where a large LAr detector would be built. And in Europe the LAGUNA-LBNO project presents three possible setups where the preferred option is a super-beam from CERN-SPS to Pyhäsalmi in Finland (baseline = 2300 km).

Today's large value of θ_{13} implies that any future experiment will quickly be systematic-limited. Therefore an effort needs to be made to reduce systematic errors. Experiments like NA61 [30], SciBoone [31], MINERνA [32] and nuSTORM [33] are extremely important to constrain hadron production cross-sections and neutrino cross-sections.

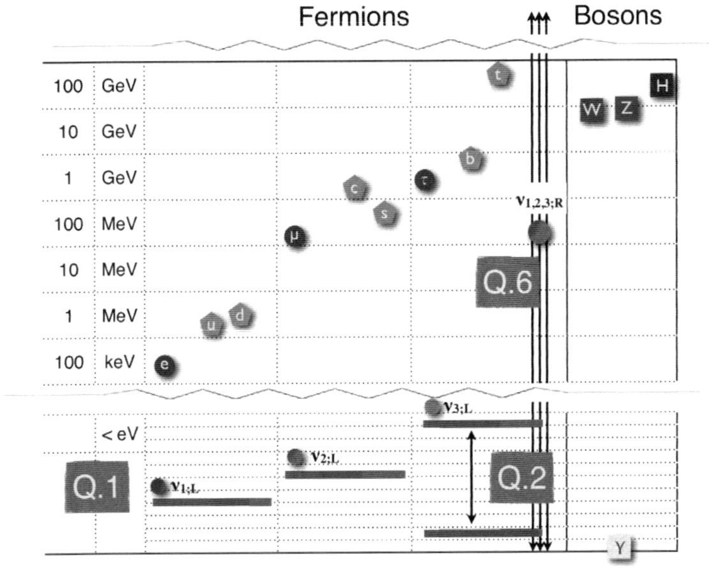

Figure 8. The remaining questions of the Standard model are: **Q.1** Determination of the absolute mass scale of neutrinos. **Q.2** Determination of the mass hierarchy of the active neutrinos. **Q.3** CP violation in neutrino oscillations. **Q.4** Violation of unitarity of the neutrino mixing matrix. **Q.5** Neutrinoless double beta decay. **Q.6** Discovery of effects implying unambiguously the existence of sterile neutrino(s).

4.2. Beta-beams and neutrino factories

The super beams can probably give us the value of δ_{CP} and the mass hierarchy. But precision similar to the precision of the CKM matrix will only be achieved with more powerful facilities. These facilities will also be needed to test the unitarity of the PMNS matrix and test whether neutrinos oscillate to become sterile neutrinos. Beta-beams [34, 35], where a radioactive ion is accelerated and then beta decays creating $\bar{\nu}_e$ beam, could be an option.

Another option is the neutrino factory. Here, μ^- and μ^+ trains simultaneously circulate in opposite direction in the storage ring and are separated in time. They are subsequently allowed to decay in flight to create pure (ν_μ, $\bar{\nu}_e$) and ($\bar{\nu}_\mu$, ν_e) beams respectively. The golden channel of the neutrino factory is $\nu_e \rightarrow \nu_\mu$, but since ν_μ are also present in the beam, measurements similar to today's (electron neutrino appearance and muon neutrino disappearance) are also possible. The presence of ν_μ, $\bar{\nu}_e$, and $\bar{\nu}_\mu$, ν_e in different trains is perfect to study CP violation since we can directly compare the oscillation of a neutrino to the oscillation of the anti-neutrino. Furthermore, because ν_μ, $\bar{\nu}_e$, $\bar{\nu}_\mu$ and ν_e are present in the beam, the near detector of a neutrino factory would be able to measure all four types of neutrinos cross-sections and reduce systematics considerably. In order to see oscillations, the far detector needs to be magnetized to differentiate $\nu_e \rightarrow \nu_\mu$ from an interacting $\bar{\nu}_\mu$. A magnetized iron detector is envisaged as far detector. It was found that an optimal neutrino factory given the current value of θ_{13} uses 10 GeV muons and a baseline of 2200 km [36, 37].

5. Conclusions

The years 2011 and 2012 have without any doubt seen tremendous progress in neutrino physics. Only a year ago, nobody knew if θ_{13} was even large enough to be measured, yet today the precision on θ_{13} is already better than the precision on θ_{23}. With this measurement the prospect of measuring the mass hierarchy and especially the CP phase δ is better than ever and the years to come will be fascinating. With the announcement on July 4th 2012 that the Higgs was probably found, the remaining questions in the Standard Model (Fig. 8) are in neutrinos physics.

We thank Alain Blondel and Mark Rayner for the careful reading of the manuscript.

References

[1] Wark D *"Experimental Status of Neutrino Physics"*, NuFACT 2011
[2] Neutrino 2012, in Kyoto URL `http://neu2012.kek.jp/neu2012/programme.html`
[3] Arnett W D and Rosner J L 1987 *Phys. Rev. Lett.* **58** 1906
[4] Giunti and Kim *"Fundamentals of Neutrino Physics and Astrophysics"*, p.616, Oxford University Press, 2007
[5] Spergel D *et al.* (WMAP Collaboration) 2003 *Astrophys. J. Suppl.* **148** 175–194 (*Preprint* `astro-ph/0302209`)
[6] *"Planck Space Observatory"*, www.rssd.esa.int/planck/
[7] Kraus C *et al.* 2005 *Eur. Phys. J.* C **40** 447–468 (*Preprint* `hep-ex/0412056`)
[8] Fraenkle F 2009 *EPS* URL `http://www.ifj.edu.pl/hep2009/`
[9] Bornschein L (KATRIN Collaboration) *"Probing the absolute neutrino mass with KATRIN"*, rencontres de Blois 2012
[10] Pedretti M (CUORE Collaboration) *"Status of the CUORE Experiment"*, neutrino 2012
[11] Inoue K (KamLAND-Zen Collaboration) *"Results from KamLAND-Zen"*, neutrino 2012
[12] Auger M *et al.* (EXO Collaboration) 2012 (*Preprint* `arXiv:1205.5608 [hep-ex]`)
[13] Ashie Y *et al.* (Super-Kamiokande Collaboration) 2005 *Phys. Rev.* D **71** 112005 (*Preprint* `hep-ex/0501064`)
[14] Ahmad Q *et al.* (SNO Collaboration) 2002 *Phys. Rev. Lett.* **89** 011301 (*Preprint* `nucl-ex/0204008`)
[15] Apollonio M *et al.* (CHOOZ Collaboration) 2003 *Eur. Phys. J.* C **27** 331–374 (*Preprint* `hep-ex/0301017`)
[16] Fogli G L *"Global Analysis of Neutrino Oscillations"*, neutrino 2012
[17] Abe K *et al.* (T2K Collaboration) 2011 *Phys. Rev. Lett.* **107** 041801 (*Preprint* `arXiv:1106.2822 [hep-ex]`)
[18] Nakaya T (T2K Collaboration) *"New Results from T2K"*, neutrino 2012
[19] Adamson P *et al.* (MINOS Collaboration) 2011 *Phys. Rev. Lett.* **107** 181802
[20] Nichol R (MINOS Collaboration) *"Final MINOS Results"*, neutrino 2012
[21] An F *et al.* (DAYA-BAY Collaboration) 2012 *Phys. Rev. Lett.* **108** 171803
[22] Dwyer D (DAYA-BAY Collaboration) *"Improved Measurement of Electron-Disappearance at Daya Bay"*, neutrino 2012
[23] Ahn J *et al.* (RENO collaboration) 2012 *Phys. Rev. Lett.* **108** 191802 (*Preprint* `arXiv:1204.0626 [hep-ex]`)
[24] Ishitsuka M (Double Chooz Collaboration) *"Double Chooz results"*, neutrino 2012
[25] Schwetz T, Tortola M and Valle J 2011 *New J. Phys.* **13** 063004 (*Preprint* `hep-ph/1103.0734`)
[26] Abe K *et al.* (T2K Collaboration) 2012 (*Preprint* `arXiv:1201.1386 [hep-ex]`)
[27] NuTURN 2012, at Gran Sasso URL `http://nuturn2012.lngs.infn.it/`
[28] Cao J (DAYA-BAY Collaboration) *"Observation of Electron Anti-neutrino Disappearance at Daya Bay"*, nuTURN 2012
[29] Abe K *et al.* 2011 (*Preprint* `arXiv:1109.3262 [hep-ex]`)
[30] Abgrall N *et al.* (NA61/SHINE Collaboration) 2011 *Phys. Rev.* C **84** 034604
[31] Mariani C (SciBooNE Collaboration) 2011 (*Preprint* `arXiv:1110.1506 [hep-ex]`)
[32] 2006 MINERvA Technical Design Report
[33] Kyberd P *et al.* (nuSTORM Collaboration) 2012 (*Preprint* `arXiv:1206.0294 [hep-ex]`)
[34] Zucchelli P 2002 *Phys.Lett.* **B 532** 166–172
[35] Wildner E 2011 *Nucl. Phys. Proc. Suppl.* **217** 199–201
[36] International Design Study for Neutrino Factory URL `https://www.ids-nf.org/wiki/FrontPage`
[37] Huber P *"Neutrino facilities comparison of performance"*, European Strategy for Neutrino Oscillation Physics, May 2012,

Accelerators for future neutrino facilities: strengths and challenges

Michael S Zisman

Accelerator & Fusion Research Division, Lawrence Berkeley National Laboratory, Berkeley, CA 94720 U.S.A.

E-mail: mszisman@lbl.gov

Abstract. Several different ideas have been put forward for future neutrino facilities, including a high-power conventional neutrino beam (a "Superbeam" facility), a Beta Beam facility based on stored beta-unstable ion beams, and a Neutrino Factory based on stored muon beams. Each of these has its advantages and disadvantages, and these are described here. After briefly describing the physics context, the various options will be described and their respective strengths and challenges indicated. The supporting R&D programs will be briefly discussed and a few specific technical questions for each facility will be outlined.

1. Introduction

Since the discovery of neutrino oscillations, there has been considerable interest in the possibility of generating intense beams of accelerator-produced neutrinos. Such a facility may be capable of observing CP violation in the lepton sector, a phenomenon that could explain the survival of matter in the universe.

Over the years, a number of schemes for producing the required neutrino beams have been proposed. These are:

- a "Superbeam" facility based on the decays of an intense pion beam
- a "Beta Beam" facility based on the decays of a stored beam of beta-unstable ions
- a "Neutrino Factory" based on the decays of a stored beam of muons

The latter facility could also serve as a precursor to an eventual Muon Collider.

All of the proposed approaches have their advantages and disadvantages. The various choices are being compared on an equal footing in the EUROnu study [1], which should be enlightening for the community. Suffice it to say here that all of the facilities would be challenging and expensive.

After briefly mentioning the physics context for these facilities, we describe each of them. We then discuss the strengths of each approach, its technical challenges, and some of the approaches to managing the challenges. Finally, we mention the main topics of the worldwide R&D programs now in place.

2. Physics Context

The Superbeam facility provides muon neutrinos or anti-neutrinos via the process $\pi \rightarrow \mu + \nu_\mu$. The polarity of the focusing horn determines the sign of the pions, and thus whether neutrinos or anti-neutrinos are delivered to the detector. Because the initial pion beam is impure, the purity of the neutrino beam is not more than ~98%.

The Beta Beam facility gives pure beams of electron neutrinos or anti-neutrinos via decays such as

$$^{6}\mathrm{He} \rightarrow {}^{6}\mathrm{Li} + e^{-} + \overline{\nu_e} \text{ or}$$
$$^{18}\mathrm{Ne} \rightarrow {}^{18}\mathrm{F} + e^{+} + \nu_e \ .$$

These decays, which correspond to the present baseline configuration of the Beta Beam facility, produce low energy neutrinos.

The Neutrino Factory beams, which are generated from the decays of muons, produce equal amounts of both muon and electron neutrinos

$$\mu^{-} \rightarrow e^{-} \overline{\nu_e} \nu_\mu \text{ or}$$
$$\mu^{+} \rightarrow e^{+} \nu_e \overline{\nu_\mu} \ .$$

In the baseline scenario, with 25 GeV muons, the neutrinos are produced at high energy, above the tau threshold.

Electron neutrinos are the most favorable for doing the science, because the oscillations from ν_e to ν_μ gives easily detectable "wrong-sign" muons. Note that the Superbeam does not produce electron neutrinos.

3. Description of Facilities

3.1. Superbeam

A Superbeam facility is basically a higher power version of today's neutrino beam facilities, that is, the approach is evolutionary rather than revolutionary. Nonetheless, increasing the beam power to 4 MW is a big step forward and does result in challenges as will be discussed below. A schematic of the EUROnu baseline Superbeam design, which would take a neutrino beam from CERN to a detector at Fréjus, is shown in Fig. 1. It is based on a 5 GeV, 4 MW proton beam from a high-power version of the CERN SPL [2].

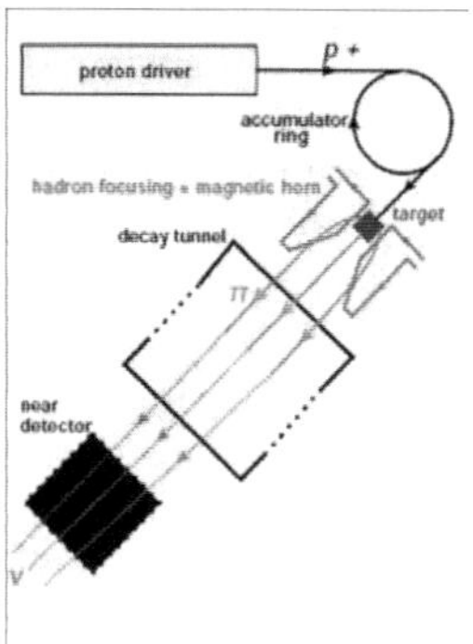

Figure 1. Schematic of CERN Superbeam layout.

3.2. Beta Beam

The baseline Beta Beam includes a number of systems—a proton driver, an ISOL target, an ion source, an acceleration system comprising a linac, a rapid cycling synchrotron (RCS), and two existing CERN accelerators (the PS and SPS), and finally a 6.9 km circumference decay ring with a 2.5 km straight section aimed at the detector. As noted earlier, the baseline scheme uses ^{6}He and ^{18}Ne ions to produce the neutrinos. This is referred to as the "low-*Q*" scenario. An alternative high-*Q* scenario, based on ^{8}Li and ^{8}B beams, is also under study. This latter concept would offer higher energy neutrino beams, but at the cost of a more complicated production scheme. To take advantage of the higher neutrino energy, a longer detector distance is required, so the high-*Q* scheme uses Gran Sasso as the detector location. Figure 2 shows the schematic layout of the Beta Beam facility illustrating both the low-*Q* and high-*Q* schemes.

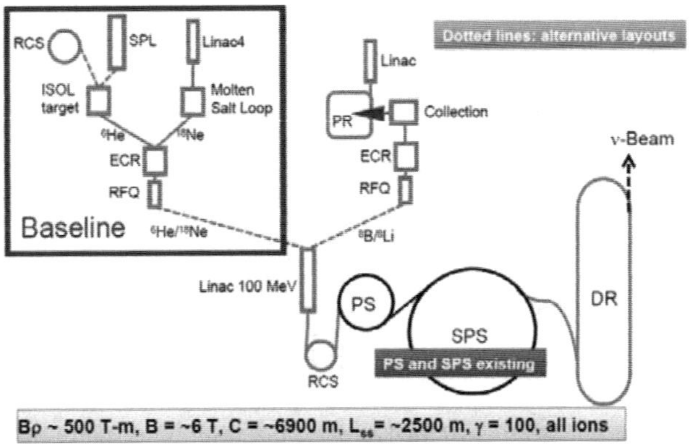

Figure 2. Beta Beam facility layout showing both the low-*Q* baseline and high-*Q* alternative configurations.

3.3. Neutrino Factory

Like the Beta Beam facility, a Neutrino Factory (Fig. 3) comprises a number of systems. These include [3]:

- a proton driver to give 4 MW of 5–15 GeV protons on a production target
- a target, capture and decay section where pions are produced, captured, and decay to muons
- a bunching and phase rotation section to reduce the energy spread of the beam
- an ionization cooling section to reduce the transverse emittance of the muon beam
- an acceleration section to increase the muon energy from 130 MeV to 25 GeV using a linac, recirculating linear accelerators (RLAs), and a non-scaling, fixed-field alternating gradient (FFAG) accelerator
- a pair of decay rings, with long straight sections aimed at the detector sites, where the muons circulate for ~1000 turns

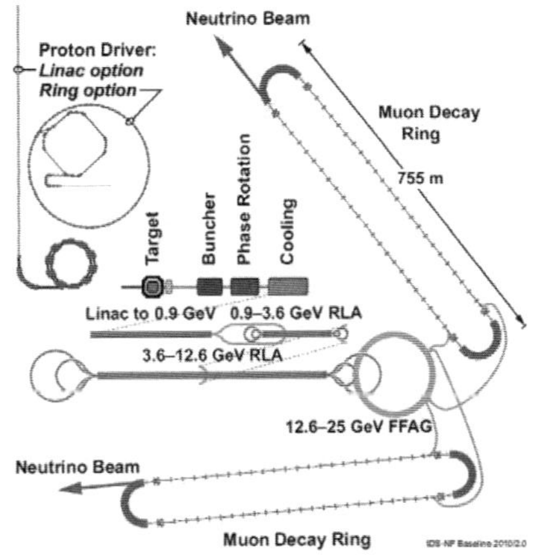

Figure 3. Neutrino Factory baseline layout.

An alternative design for a 4 GeV Neutrino Factory (Fig. 4) is also being explored [4]. This scenario is motivated by the following considerations:

- the expectation of reduced facility cost compared with the baseline design
- a neutrino energy well matched to the Fermilab to Homestake Mine distance
- a detector concept—the totally active scintillator detector (TASD)—capable of the required performance at the chosen energy

The main ingredients for the low-energy scenario are the same as those outlined for the baseline device, but there are fewer items on account of the lower energy (less acceleration, smaller decay ring, and a single detector).

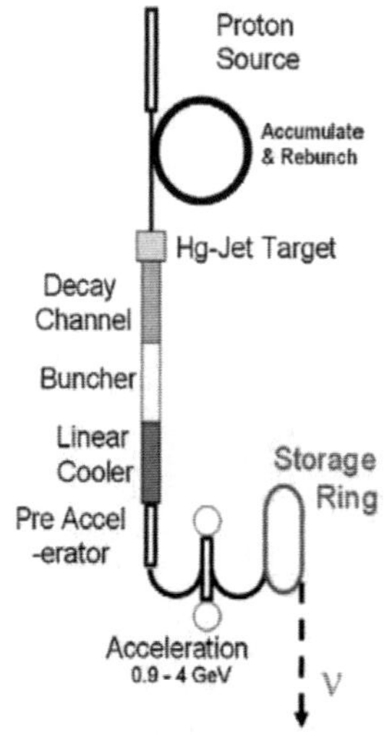

Figure 4. Schematic layout of low-energy Neutrino Factory.

3.4. Common Aspects

A common feature of all future neutrino facilities is the need for a substantially increased quantity of data. This leads to the requirements for intense particle sources and very large detectors. Both of these represent major technical challenges that extend the state-of-the-art by factors of 5–10.

All current approaches to providing the required neutrino intensities rest on producing secondary, or even tertiary, beams.

4. Strengths

For the Superbeam, its main strength compared with the other potential approaches is that it is closest to today's technology. For this reason, it is likely to be the least expensive of the three alternatives, though it nevertheless will be expensive.

For the Beta Beam, its main strengths are that it *i*) takes advantage of existing CERN infrastructure; *ii*) offers potential synergy with nuclear physics interests in isotope production; and *iii*) provides a clean beam[1] having only electron neutrinos.

The main strength of a Neutrino Factory is that it has the most experimental sensitivity, and thus the best physics reach of the three options. In this facility, both electron and muon neutrinos are produced simultaneously, and in equal proportions. The Neutrino Factory also offers a strong synergy with intense muon and/or Muon Collider programs should either or both of these be of interest to the scientific community.

[1] However, the Beta Beam facility must be used in combination with a Superbeam to fully elucidate the physics.

5. Technical Challenges and Mitigations

5.1. Superbeam

The challenges in this case are related mainly to the intensity requirement. In particular, one needs:

- a target capable of handling 4 MW of protons
- a horn capable of handling 4 MW of protons and operating at a 50 Hz repetition rate
- good charge selection (for beam purity)

A particular constraint for this facility is that the target and focusing horn must be in close proximity. Spatial constraints favor a solid, or possibly a powder-jet target. Materials compatibility issues make the Hg-jet target (favored by the Neutrino Factory community) impractical. Cooling of the horn in a crowded area is not easy, though solutions exist. The high radiation environment makes repairs challenging in this region, though it is inevitable that repairs (or component replacement) will be needed. A remote handling system is clearly mandatory.

The proposed approach to deal with these challenges is illustrated in Fig. 5. The idea is to use four horns and four targets, all sharing a common decay pipe. With this approach, each target must handle a more tractable 1 MW of beam power, and the individual horns can pulse at 12.5 Hz rather 50 Hz. Although the beam delivery system gets more complex, the benefits of relaxing the power and repetition rate make this a good trade-off. To simplify the design, each horn is designed as a single optimized device, with no reflector. The favored target concept, shown in Fig. 6, is a so-called "pebble bed" system using Ti spheres with aggressive cooling.

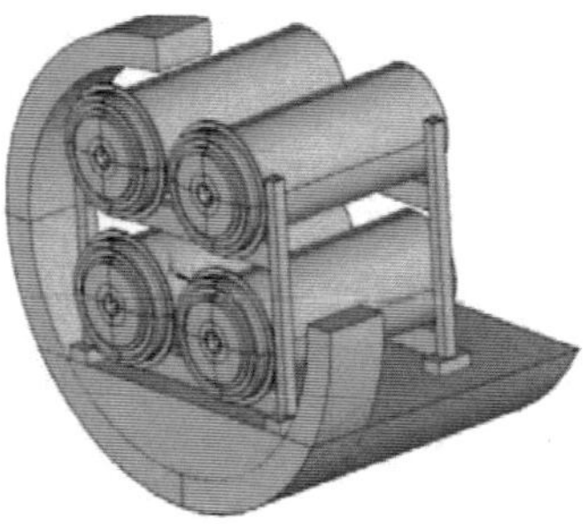

Figure 5. Four-horn target system for Superbeam facility.

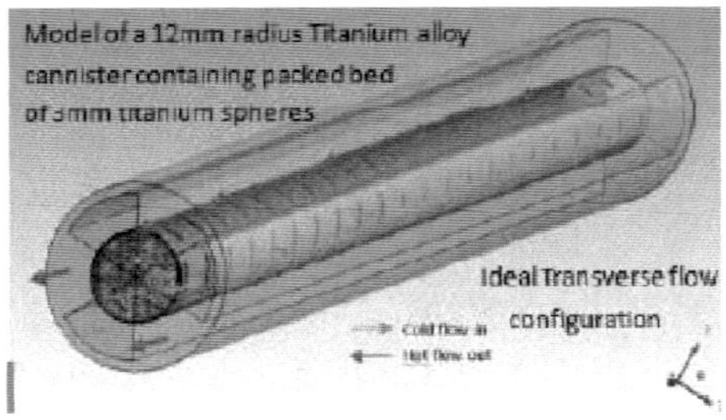

Figure 6. Schematic of pebble bed target showing cooling concept.

5.2. Beta Beam

The primary technical challenge for the Beta Beam facility is to produce the desired ion species at the required intensities. "Production" in this context includes not only the initial production of the unstable nuclei, but transporting them to an ion source, ionizing them to a useful charge state, and bunching them in a form suitable for further acceleration and transport. At present ^{6}He production looks acceptable and ^{18}Ne appears challenging, but probably possible with the ^{19}F(p,2n) reaction.

Unfortunately, the higher-Z species will typically exit the ion source in multiple charge states, reducing the overall efficiency of the process for a given charge state choice.

For the high-Q isotopes, a new production concept using a production ring and reverse kinematics has been explored [5]. Unfortunately, due to the wider decay cone for the high-Q isotopes, an additional factor of 5 in intensity is needed. Present estimates suggest that the required gas target thickness in the production ring would be about 10^4 times that of any existing target, which does not seem promising. A possible workaround involving forward kinematics and a liquid-lithium thin film target is being considered.

Another technical issue that is now being assessed is that of collective effects. Initial estimates [6] indicate that transverse mode coupling will limit intensities in the decay ring (and possibly in the SPS), exacerbated by the relatively low duty factor (0.5%) required. Work to understand this issue is in progress. It may be possible to modify the decay ring parameters to avoid this limitation.

Finally, the Beta Beam baseline scenario, with $\gamma = 100$ and a very short ring-to-detector distance do not appear optimal for the science program. Beams with $\gamma = 350$ and a longer detector distance are preferred. This scenario would require an upgrade of the SPS, which has not been seriously studied yet, as it is not in CERN's plans at present.

5.3. Neutrino Factory

The primary challenges for a Neutrino Factory are related to the difficulties in creating an intense, accelerator quality muon beam. Muons are produced as a tertiary beam ($p \rightarrow \pi \rightarrow \mu$), a process that results in a beam with a large energy spread and large transverse phase space. Thus, some form of emittance cooling is required. The intensity requirement is high, demanding a target that can tolerate up to 4 MW of protons—and a proton driver that can provide such intensity in short (1–3 ns) bunches. Finally, the muon has a very short lifetime (2.2 μs at rest), which means that all beam manipulations must be rapid. In particular, the only cooling method deemed fast enough for muons is the presently untested technique of ionization cooling [7].

To accommodate the high-intensity proton beam, the favored concept is a free Hg-jet target in a 20-T solenoidal field. The specified jet velocity of 20 m/s is such that a new, undisturbed target is re-established in time to receive the next beam pulse at a 50 Hz rate. The shielding required to operate a 20-T hybrid magnet in such a high radiation environment is daunting, but appears manageable.

Though it came as somewhat of a surprise, the operation of high-gradient normal conducting rf cavities in a magnetic field has not turned out to be straightforward. Experiments at 805- and 201-MHz have shown substantial degradation of maximum gradient with even a few-T of axial field. It is known [8] that gas-filled cavities do not suffer a loss in gradient in a magnetic field, but their behavior with an intense beam of ionizing radiation must be studied. Such studies are just getting under way. Initial indications [9] are that the passage of the beam results in severe beam loading in the cavity, but its onset may be slow enough that the cavities will be functional in the Neutrino Factory parameter regime.

6. R&D Activities

To address the challenges described in Section 5, worldwide R&D efforts are under way. Of most interest here are the work of the EUROnu study [1] and those of the IDS-NF [10] and Muon Accelerator Program, MAP [11].

6.1. Superbeam

The main R&D activity has focused on target and horn. These represent the most difficult technical challenges. An optimized horn has been designed, but needs to be fabricated and tested with its power supply. Cooling and fatigue damage are issues to examine. Testing the device in a high radiation area is worthwhile, but may not be practical. A pebble-bed target is the favored design. A scale model of this needs to be built and tested with a heat source comparable to what the real beam would provide.

6.2. Beta Beam

The main items to study here—all of which are in progress—are ion production, collective effects, and beam-loss issues. The R&D effort here has suffered somewhat by being understaffed, and more help would be of great benefit. More detailed study of the implications of the heavy ion transport in the existing CERN machines (PS and SPS) is warranted. It is likely that some modifications to existing facilities will be needed and these need to be assessed and discussed in the context of future CERN plans.

6.3. Neutrino Factory

For the Neutrino Factory, the primary R&D topics are the target, ionization cooling, and the issue of how to improve the rf gradient in the presence of an axial magnetic field. The "target" R&D involves the entire target facility, including Hg circulation and cooling, superconducting magnet design and shielding, remote handling, and safety issues. Ionization cooling R&D must focus on completing the MICE experiment [12] at Rutherford Appleton Laboratory, including an analysis of the results in sufficient detail to confirm the basic understanding of the cooling process. Finally, the degradation in maximum gradient for cavities in an axial magnetic field must be understood and cures or at least mitigation strategies developed.

7. Questions to Consider

In the spirit of having a workshop, a few questions are listed here that could be topics of discussion for this or a future meeting.

For the Superbeam group:

- Is the layout of the proton beam transport compatible with horn repair or replacement?

For the Beta Beam group:

- Given the complications of producing and capturing ^{8}Li and ^{8}B, and the need for five times higher intensity for these isotopes compared with ^{6}He and ^{18}Ne, is the cost-benefit ratio for this option really favorable?
- Are there limitations (either technical or operational) in the baseline CERN-based scenario that are severe enough to justify consideration of a "green-field" site?

For the Neutrino Factory group:

- What combination of proton beam energy and bunch length is the best compromise for integrated muon beam intensity?
- Is the rf R&D plan sufficiently well focused or too broad?

For all of the groups:

- What is the time frame to complete the work necessary to request funding to build a facility?

8. Summary

As should be clear from this paper, substantial progress is being made toward the designs of accelerator-based neutrino facilities to study CP violation in the lepton sector. While there are technical challenges in all cases, these challenges are being understood and overcome by means of vigorous international R&D programs. Experiments play a critical role in this endeavor and must be a substantial part of the overall R&D effort.

The work being carried out extends the state-of-the-art in accelerator science. Examples include high-power targets, new beam cooling techniques, ion source development, and rapid acceleration techniques.

Acknowledgments

I would like to thank my accelerator colleagues in EUROnu, the IDS-NF and MAP for sharing both their expertise and their enthusiasm on the topics covered here.

This work was supported by the Director, Office of Science, Office of High Energy Physics, of the U.S. Department of Energy, under Contract No. DE-AC02-05CH11231.

References

[1] See http://www.euronu.org/

[2] Gerigk F (ed.), "Conceptual design of the SPL II – a high-power superconducting H^- linac at CERN," CERN 2006-006 (2006); see http://cdsweb.cern.ch/record/975366/files/CERN-2006-006.pdf

[3] The IDS-NF Collaboration Interim design report, see https://www.ids-nf.org/wiki/FrontPage/Documentation?action=AttachFile&do=get&target=IDS-NF-020-v1.0.pdf.

[4] Li T "The low energy neutrino factory," https://www.ids-nf.org/wiki/FrontPage/Documentation?action=AttachFile&do=get&target=IDS-NF-012-v1.0.pdf

[5] Benedetto E "Ionization cooling in a low-energy ion ring with internal target for beta beams production," in Proc. 2010 International Particle Accelerator Conf., Kyoto, Japan, May 23–28, 2010, p. 849

[6] Hansen C and Rumolo G "Collective effect studies of the beta beam decay ring," see http://heplnv135.pp.rl.ac.uk/joomla/index.php?option=com_docman&task=doc_download&gid=59&Itemid=5

[7] Neuffer D 1983 *Part. Accel.* **14**, 75

[8] BastaniNejad M *et al.* "Studies of breakdown in a pressurized rf cavity," EPAC'08, Genoa, June 2008, MOPP080, p. 736; see http://www.JACoW.org

[9] Yonehara K "Progress of high pressure rf cavity test," https://indico.fnal.gov/getFile.py/access?contribId=0&resId=0&materialId=slides&confId=4661

[10] See https://www.ids-nf.org/wiki/FrontPage

[11] See http://map.fnal.gov/

[12] See http://mice.iit.edu/

Neutrinos and the Universe

Nick E Mavromatos[1,2]

[1]Theoretical Particle Physics and Cosmology Group, Department of Physics, King's College London, Strand, London WC2R 2LS, UK
[2]Theory Division, Physics Department, CERN, CH 1211 Geneva 23, Switzerland

E-mail: Nikolaos.Mavromatos@kcl.ac.uk

Abstract. In this talk, I review the potential connection of neutrinos to the Physics of the Early Universe, in particular the role of (sterile) neutrinos to leptogenesis/baryogenesis and the dark sector of the Universe. The possibility of CPT Violation among active neutrinos at early times and its role in leptogeneis/baryogenesis without sterile neutrinos is also touched upon.

1. Introduction and motivation

Neutrinos are fascinating particles, still full of mysteries and surprises. Not long after the end of this meeting, we have heard the claims from OPERA Collaboration [1] on the measurements of the arrival time of neutrinos, indicating *superluminal* propagation. Although this result may not be actually due to fundamental (Lorentz violating or modifying) physics but rather due to measurement uncertainties, given the complicated nature of the measurement and the many potential systematic errors that may have entered (including the particle decays that can produce the final signal, which undoubtedly introduce statistical uncertainties in the arrival times), nevertheless neutrinos made big headlines all over the world once again, and opened up interesting avenues for research, given that superluminal propagation, that characterises already existing theoretical models, may not be incompatible with causality. In this talk I will not discuss such issues, however I will touch upon a different rôle of neutrinos (assumed throughout to respect Lorentz Invariant kinematics, thus being *subluminal*, almost light-like due to their tiny masses) that relates to the Physics of the Early Universe. In particular, one of the most important questions of fundamental Physics that is still unanswered today, which pertains to our very existence, is the reason for the observed baryon asymmetry in the Universe, or why the Universe is made up mostly of matter. According to the Big bang theory, matter and antimatter have been created at equal amounts in the Early universe. The observed charge-parity (CP) violation in particle physics [2], prompted A. Sakharov [3] to conjecture that non-equilibrium Physics in the Early Universe produce Baryon number (B), charge (C) and charge-parity (CP) violating, but CPT *conserving*, interactions/decays of anti-particles in the early universe, resulting in the observed baryon-antibaryon ($n_B - n_{\overline{B}}$) asymmetry. In fact there are two types of non-equilibrium processes in the Early universe that could produce this asymmetry: the first type concerns processes generating asymmetries between leptons and antileptons, while the second produce asymmetries between baryons and antibaryons. The almost 100% observed asymmetry today, is estimated in the Big-Bang theory [4] to be of order:

$$\Delta n(T \sim 1\ \mathrm{GeV}) = \frac{n_B - n_{\overline{B}}}{n_B + n_{\overline{B}}} \sim \frac{n_B - n_{\overline{B}}}{s} = (8.4 - 8.9) \times 10^{-11} \tag{1}$$

at the early stages of the expansion, e.g. for times $t < 10^{-6}$ s and temperatures $T > 1$ GeV. In the above formula n_B ($n_{\overline{B}}$) denotes the (anti) baryon density in the Universe, and s is the entropy density. Unfortunately, the observed CP violation within the Standard Model (SM) of Particle Physics (found to be of order $\epsilon = O(10^{-3})$ in the neutral Kaon experiments [2]) cannot reproduce (1). Let us review why [5, 6]. Although classically, the baryon (B) and Lepton ($L_{e,\mu,\tau}$) numbers are conserved in the SM, at a quantum level *anomalies* in general break these symmetries. The anomalies are associated with the non-conserved currents corresponding to the above classical symmetries in the combination $\partial^\mu J^B_\mu = \partial^\mu J^L_\mu = \frac{N_f}{32\pi^2}\mathrm{Tr}F_{\mu\nu}\tilde{F}^{\mu\nu} + \mathrm{U}(1) - $ parts, with N_f the flavour number. Since the allowed processes in the SM are those which entail a change of B by multiples of 3, that is bosons $\longleftrightarrow$ bosons + 9 quarks + 3 leptons, there is a conservation law for the three combinations $L_i - B/3$, $i = e, \mu, \tau$. However the observed neutrino oscillations among flavours, imply that only one global number may be conserved in the SM, the combination $B - L$, where $L = \sum_i L_i$ is the total Lepton number. In fact if neutrinos are Majorana, the L would be itself violated, and hence there would be no conserved numbers at all! With this in mind one may evaluate the rate of B violation in the SM [5, 6]: $\Gamma \sim (\alpha_W T)^4 \,(M_{\rm sph}/T)^7 \exp(-\frac{M_{\rm sph}}{T})$, if $T \le M_{\rm sph}$, while $\Gamma \sim (\alpha_W T)^4 \alpha_W \log(1/\alpha_W)$ if $T \ge M_{\rm sph}$, with α_W the fine structure constant of the electroweak SU(2) symmetry, and $M_{\rm sph}$ is the spaleron mass scale, in particular $M_{\rm sph}/\alpha_W$ is the height of the energy barrier separating SU(2) vacua with different topologies. Thermal equilibrium (i.e. $\Gamma > H$ (Hubble rate)) for B non conserving processes occurs only for [5] $T_{\rm sph}(m_h) < T < \alpha_W^5 M_{\rm P} \sim 10^{12}$ GeV; with the Higgs mass m_h assumed in the range (in agreement with current LHC exclusion data) $m_h \in [100, 300]$ GeV , one has $T_{\rm sph} \in [130, 190]$ GeV. Baryon Asymmetry in the Universe (BAU) could be produced only when the sphaleron interactions freeze out, that is for temperatures $T \simeq T_{\rm sph}$. One should compute, within the SM, the CP violation effects at such a regime of parameters and temperature, using the Cabbibo-Kobayashi-Maskawa (CKM) matrix. The lowest SM CP violating structures are encoded in the quantity $\delta_{CP} = \sin(\theta_{12})\sin(\theta_{23})\sin(\theta_{13})\sin\delta_{CP}(m_t^2 - m_c^2)(m_t^2 - m_u^2)(m_c^2 - m_u^2)(m_b^2 - m_s^2)(m_b^2 - m_d^2)(m_s^2 - m_d^2)$, where $\delta_{CP} = D/T^{12}$ is the Kobayashi-Maskawa CP Violating phase, and D is the Jarlskog determinant, related to the appropriate quark mass matrices as [6]: $D = \mathrm{Tr}(\mathcal{M}_u^2 \mathcal{M}_d^2 \mathcal{M}_u \mathcal{M}_d)$. Computing the parameter δ_{CP} at $T \simeq T_{\rm sph}$ in the above range, one obtains $\delta_{CP} \sim 10^{-20}$ which yields a baryon asymmetry ten orders of magnitude smaller than the observed one (1). Hence the SM CP violation cannot be the source for the observed BAU. There are several ideas that go beyond the SM (e.g. GUT models, Supersymmetry, extra dimensional models etc.) in an attempt to find extra sources for CP violation that could generate the observed BAU. Since massive neutrinos (as evidenced by the observed flavour oscillations) constitute the simplest extension of SM, it is reasonable to seek for a possible rôle of neutrinos in providing us with the required amount of CP violation to explain the observed BAU. Right-handed supermassive (Majorana) neutrinos (*sterile)* may do the job and more than that. Namely, as we shall discuss below, such extensions of the SM can provide sufficient extra sources for CP Violation to explain the Origin of Universe's matter-antimatter asymmetry due to the relevant neutrino masses, without the need for Supersymmetry or extra dimensions. Moreover, as we shall discuss below, such models can also incorporate a natural Dark Matter candidate (the lightest of the sterile neutrinos), in agreement with observations [6].

2. Sterile neutrino SM extensions and Leptogenesis/Baryogenesis

Several authors have suggested the use of right-handed, supermassive sterile neutrinos as possible extensions beyond the SM, with relevance to the physics of the Early universe. In this talk, I concentrate on the simplest of such extensions, a non supersymmetric SM augmented with N generations of right-handed massive, with masses M_I, Majorana fermions, termed νMSM, a terminology I will use from now on: [6, 7] $L_{\nu{\rm MSM}} = L_{\rm SM} + \overline{N}_I i\gamma^\mu \partial_\mu N_I - F_{\alpha I}\overline{L}_\alpha N_I \tilde{\phi} - \frac{M_I}{2}\overline{N}_I^c N_I +$ h.c , where the suffix SM denotes the SM part of the Lagrangian, N_I, $I = 1, \ldots N$ denote the Majorana sterile neutrinos, the superscript c denotes charge conjugate, and L_α, $\alpha = e, \mu, \tau$ are the leptons. The field $\tilde{\phi}$ is the SU(2) dual of the Higgs scalar $\tilde{\phi} = \epsilon_{ij}\phi_j^\star$, i, j SU(2) indices, while $F_{\alpha I}$ are matrix valued Yukawa couplings involving majorana phases and mixing angles [6, 7]. The model with N=1 sterile neutrino is excluded by the current data [8], while the models with N=2,3 work well in reproducing BAU and are consistent with the current experimental data on neutrino oscillations. Of particular interest to us will be the Model with N=3 singlet neutrinos, which in fact allows one of the Majorana fermions to almost decouple from the rest of the SM fields, thus providing candidates for light (keV region of mass) sterile neutrino Dark Matter [7]. Moreover, these models can also be consistent with inflationary scenarios, through, e.g., non-minimal couplings of the Higgs scalars to the Einstein curvature tensor [9], $L_G \in \zeta\phi^\dagger\phi R + \ldots$, with flat effective potential for large values of ϕ, thus being compatible with the inflation slow roll conditions. We shall not discuss such important issues here though. The light (active) neutrino ν masses in the model are generated through appropriate see-saw mechanisms, $m_\nu = -M^D\frac{1}{M_I}[M^D]^T$, with T indicating matrix transposition, $M^D = F_{\alpha I}v$, $v =< \phi >$ is the standard model Higgs v.e.v., assumed in [7] to be of order 175 GeV thereby yielding $M_D \ll M_I$.

For the connection with Leptogenesis/Baryogenesis we need to know the thermal properties of the model. A detailed analysis [7, 10], which we shall omit here, indicates that the relevant decay processes in the early Universe, which – when out of equilibrium could lead to matter-antimatter asymmetries, that is $Nt \longleftrightarrow \nu h$, $h \longleftrightarrow N\nu$, $N \longleftrightarrow h\nu$, where h is the Higgs field, have a rate $\Gamma \sim 9F^2 f_t^2 T/(64\pi^3)$, whereby f_t denotes the top quark Yukawa coupling, and the amplitude of the sterile Neutrinos Yukawa couplings $|F|^2 \sim v^{-2}\, m_{\rm atm} M_I \sim 2 \times 10^{-15}\frac{M_I}{\rm GeV}$, $m^2_{\rm atm} \equiv |\Delta m_{\rm atm}|^2 = (2.40 + 0.12 - 0.11) \times 10^{-3}$ eV^2, is the measured neutrino mass difference form atmospheric experiments. From this rate the conditions for thermal equilibrium (i.e. $\Gamma >$ H(Hubble)) can be computed and the equilibrium temperature is $T_{\rm eq} \simeq \frac{9 f_t^2 m_{\rm atm} M_0}{64\pi^3 v^2} M_I \simeq 5M_I$, with $M_0 = M_P/(1.66\sqrt{g_{\rm eff}})$, with M_P the (four dimensional) Planck mass and $g_{\rm eff}$ the effective degrees of freedom in the radiation era of the Universe.

There are two physically distinct cases we consider: $M_I \ge M_W$ and $M_I < M_W$, where M_W the electroweak breaking scale ($M_W \sim 100$ GeV).

(I) Case where $M_I \ge M_W$: In the first case, the decay of the right-handed fermions occurs at temperatures [7, 6] $T_{\rm decay} = \left(\frac{m_{\rm atm} M_0}{24\pi^2 v^2}\right)^{1/3} M_I \simeq 3M_I$. Such processes are out of equilibrium for $T > T_{\rm eq} \sim 5M_I$ or $T < T_{\rm decay} \sim 3M_I$. If $T_{\rm eq} > T_{\rm sph}$, then the decays of the right-handed fermions occur during the period for which the sphaleron processes are active, i.e. $T_{\rm decay} > T_{\rm sph}$, this leads to thermal leptogenesis [11] as follows: the Heavy Right-handed Majorana neutrinos enter equilibrium at $T = T_{\rm eq} > T_{\rm decay}$, which in fact is independent of the initial conditions at $T \gg T_{\rm eq}$. Subsequently, as the Universe continues to expand and cools down, the out of equilibrium *Lepton Number violating* decays of the heavy neutrinos at $T \simeq T_{\rm decay} > T_{\rm sph}$: $N_I \to \nu\, h$, $\overline{\nu}\,\overline{h}$ produce a Lepton-antilepton *asymmetry* , which is then communicated (through the induced effective low-energy Lepton - number violating interactions in the SM Lagrangian of the form $L \ni \frac{2}{M_I}\overline{L}_L L_L \phi\phi$ + h.c., where L_L denote the SM left-handed lepton doublets) to the baryon sector through equilibrated B+L violating sphaleron interactions, independent of the initial conditions, to produce the observed Baryon Asymmetry. At this point we make some important remarks: to calculate the precise contributions to the BAU and compare it with the observed one, one needs to estimate the (thermal) relic abundance η_{N_i} of the heavy neutrinos by solving the appropriate Boltzmann equations, $\frac{d}{dt}\eta_{N_i} + 3H\eta_{N_i} = C[f]$, $\eta_{N_i} \equiv \int d^3p f_N(p,x,t)$. The results are sensitive to the underlying theoretical model used, of course [1] (*cf.* fig. 1). If in the νMSM model, we concentrate here for definiteness, with N=3 heavy Majorana neutrinos, all neutrino flavours are non degenerate in mass, $|M_I - M_J| \sim M_K$, then unfortunately, in order to reproduce the correct BAU (with the ratio $n_B/s \sim F^2 10^{-3} \sim 10^{-10}$ (*cf.* (1)) one needs

[1] Although the Boltzmann equations are classical equations for the time evolution of phase space distribution functions f(p,x.t), however the collision terms $C[f]$ on the r.h.s. are quantum effects involving loop corrected cross sections. Since the Leptogenesis processes are out of (thermal) equilibrum, the correct way to incorporate the quantum effects would be to consider the non-equlibrium field theory Schwinger-Keldysh formalism and the Kadanoff-Baym (KB) equations based on Green's functions, not on densities [12]. The conclusion of such more elaborate treatments is that the conventional Boltzmann equations for Lepton asymmetry give pretty good results when SM gauge interactions are taken into account.

Figure 1. *Left:* relevant graphs contributing to the BAU in the νMSM model with three sterile neutrinos. The lower two graphs are responsible for enhancement of CP violation if the two heavy sterile neutrinos $N_{2,3}$ are degenerate in mass. *Right:* An example of heavy N_1 neutrino abundance: thermal relic density η_{N_1} vs m_{N_1}/T in the scenario of [13] of resonant Leptogenesis.

Yukawa couplings of order $F^2 \sim 10^{-7}$ [6, 7], which yield masses for the heavy neutrinos of order $M_{N_I} = 10^{11}$ GeV. Such a scenario is plagued by instabilities of the Higgs mass under quantum corrections, e.g. the one-loop corrections are found to be of order $F^2 M_I^2/(4\pi) \sim 10^{14}$ GeV^2. A resolution to this problem is provided by considering models in which two of the heavy neutrinos are degenerate in mass, say $N_{2,3}$ [13, 14, 7, 6]. In such a case, much smaller Yukawa couplings are allowed, $f^2 \sim \frac{M_2 - M_3}{M_2} \sim \frac{m_\nu M_W}{v^2} \sim 10^{-13}$, with $M_I \sim M_W$. There is an enhancement of the induced CP violation in this case (due to the lower two graphs in the left picture of fig. 1). This enhanced CP violation due to the existence of degenerate in mass quantum states is familiar from the case of neutral Kaons [2]. Such models yield $n_B/s \sim 10^{-3} f^2 \frac{M_2 \Gamma_{\rm tot}}{(M_2-M_3)^2 + \Gamma_{\rm tot}^2}$, with $|M_2 - M_3| \sim \Gamma_{\rm tot}$, and $\Gamma_{\rm tot}$ the total sterile neutrinos decay width. For $M_I < 10^7$ GeV there is no problem with Higgs stability in such models [6]. A specific example has been considered in [13] in which only one of the heavy neutrinos, N_1, decays out of equilibrium, while the heavier ones $N_{2,3}$ stay in thermal equilibrium. One lepton number (τ) is resonantly produced by the out-of-equilibrium decays of N_1. To avoid excess of L_τ - number violation, the decay rates of $N_{2,3}$ are suppressed. The predicted BAU in such a class of models (termed resonant Leptogenesis for obvious reasons) can be naturally made to agree with the observed one, as following from a calculation of the thermal relic abundances of the N_1 neutrino, by means of solving the pertinent Boltzmann equations (cf. right picture in fig. 1). Leptogenesis is possible for these models if M_I is in the range $M_I \in [M_W - \text{TeV}]$. The predicted Lepton-number violating process $\mu \to e\,\gamma$ has branching ratio of order $B(\mu \to e\gamma) = 6 \times 10^{-4} a^2 b^2 v^4 / M_I^4$, where a, b are parameters entering the appropriate Yukawa coupling matrix of the model [13]. For natural values $a, b \sim 10^{-3}$, required by Leptogenesis, one has $B(\mu \to e\gamma) \leq 1.2. \times 10^{-11}$ for the above range of M_I. This is just one order of magnitude larger than the current sensitivity of the MEG Experiment [15] to the Branching ratios for $\mu^+ \to e^+\gamma$, $B(\mu^+ \to e^+ + \gamma) \leq 2.4 \times 10^{-12}$. In [13] there were also examined possible effects of such degenerate models in linear e^+e^- colliders, where one should study the production of electroweak scale $N_{2,3}$ via their decays to e, μ but not τ.

(II) Case where $M_I < M_W$: consider, for instance, the case where $M_I = O(1)$ GeV. In such a case, in order to guarantee the smallness of the light neutrino masses, the relevant Yukawa couplings should be of order [6]: $F_{\alpha I} \sim \frac{\sqrt{m_{\rm atm} M_I}}{v} \sim 4 \times 10^{-8}$. One may further assume that two of the heavy Majorana neutrinos are degenerate in mass, say $N_{2,3}$ as before, which enhances CP violation. In such a case, the scenario of [10], on Baryogenesis through coherent oscillations of the two degenerate in mass right-handed fermions, may be realised. The Heavy Majorana fermions N_I thermalize only for $T < M_W$, so their decays at for $T > M_W$ are out of

equilibrium. One may worry in such a case that the induced BAU would depend on the initial conditions. However, inflation may set the initial concentration of N_I to practically zero value at the end of inflation. The relevant Majorana masses are small compared to the Sphaleron freeze-out Temperature, hence the total lepton number is conserved (the total Lepton number is zero but unevenly distributed between active and sterile neutrinos), this leads to "apparent" lepton number violation, so that the Lepton number of active left-handed ν is transferred to Baryons due to equilibrated sphaleron processes. The coherent oscillations between the mass-degenerate singlet fermions have a frequency [10, 6]: $\omega \sim |M_2^2 - M_3^2|/E_I \sim M_2 \Delta M/T$, for $\Delta M \equiv |M_2 - M_3| \ll M_2 \simeq M_3, E_I \sim T$. For CP violation to occur (so that one can achieve the (observed) BAU) one must have that the oscillation rate is larger than the Hubble expansion rate of the Universe. Under the (important and rather delicate) assumption that the interactions with the plasma of SM particles in the early universe do not destroy quantum mechanical coherence of oscillations, one may have in this model baryogenesis occurring at 100 GeV, and a maximal baryon asymmetry $\Delta = n_B - n_{\overline{B}}/(n_B + n_{\overline{B}}) \sim 1$ when $T_B \simeq T_{\rm sph} \simeq T_{\rm eq}$. Thus this mechanism for producing BAU seems quite effective, if in operation. For the case $T_B \gg T_{\rm sph} > T_{\rm eq}$ the predicted n_B/s in this version of the νMSM model reads [6]: $n_B/s \sim 1.7 \times 10^{-10} \delta_{CP} \left(\frac{10^{-5} M_2}{\Delta M(T)}\right)^{2/3} \left(\frac{M_2}{10 \text{ GeV}}\right)^{5/3}$, with the CP violating factor expressed through the various mixing angles and CP-violating phases. This can be of O(1), according to the recent experimental data of neutrino oscillations. The observable value of the BAU, then, can be obtained for a wide range of parameters of the νMSM [6]. In particular, one may have Mass N_2 (N_3) / (Mass N_1) = O(10^5), implying that the lightest sterile neutrino may have masses in the keV range, if $M_{2,3}$ are of O(1-10) GeV. As we shall argue in the next section, this Lightest Sterile neutrino of this version of $N = 3$ νMSM is a natural Dark Matter (DM) candidate [6, 7].

Before closing this section we should mention that there are several theoretical scenarios that can explain the required mass hierarchy $M_{2,3} \gg M_1$ among the sterile neutrinos. These scenarios range from: (i) the imposition of flavour symmetries [6, 16, 17] (one starts with $M_1 = 0$ and $M_2 \simeq M_3 >$ GeV if symmetry is unbroken, while Breaking of global Lepton symmetry generate singlet fermion mass hierarchy, $M_1 = O(\text{keV}) \ll M_{2,3}$), to (ii) brane world scenarios [18] exploiting the associated exponential factor in a Randall-Sundrum-like framework [19] to obtain a large mass splitting (one sterile neutrino at the keV scale, while the other two could have masses of around 10^{11} GeV or heavier, and at the same time ensuring that the seesaw mechanism for active neutrinos works) from very moderately tuned parameters, and (iii) to Froggatt-Nielsen type mechanisms [20], whereby one fermion acquires mass via Higgs mechanism, while the rest via higher order multiple see-saw.

3. Neutrinos and the dark sector of the Universe

3.1 Sterile neutrinos in νMSM and DM

To be DM, the lightest massive sterile neutrino of the $N = 3$ νMSM, N_1, must have a life time larger than that of the Universe. For this to happen, its coupling with SM matter must be superweak, $\theta_1 G_F = \sum_{\alpha=e,\mu,\tau} \frac{v^2 |F_{\alpha I}|^2}{M_1^2}$, with G_F denoting the Fermi coupling of the weak interactions. The width of N_1 for the decay $N_1 \to \gamma\nu$ is expressed in terms of θ_1 mixing angle as follows [6]: $\Gamma_{N_1 \to \gamma\nu} = \frac{9\alpha G_F^2}{1024\pi^4} \sin^2(2\theta_1) M_1^5 \simeq 5.5. \times 10^{-22} \theta_1^2 \left(\frac{M_1}{\text{keV}}\right)^5 \text{ s}^{-1}$, which implies that, in order for the life time of N_1 to be larger than the Universe age, one must have $\theta_1^2 \leq 1.8 \times 10^{-5} \left(\frac{M_1}{\text{keV}}\right)^{-5}$. The contributions to the mass matrix of the active neutrinos from this light sterile is estimated to be of order $\delta m_\nu \sim \theta_1^2 M_1$, which can be within the experimental error for the solar mass differences of active neutrinos if $M_1 \geq 2$ keV. The estimation of the production of the N_1 sterile neutrino in the Early universe is essential in order to have an idea of whether this candidate for DM satisfies the current astrophysical constraints [7, 6]. Such an estimation requires taking into account the interactions of N_1 with the heavy degenerate

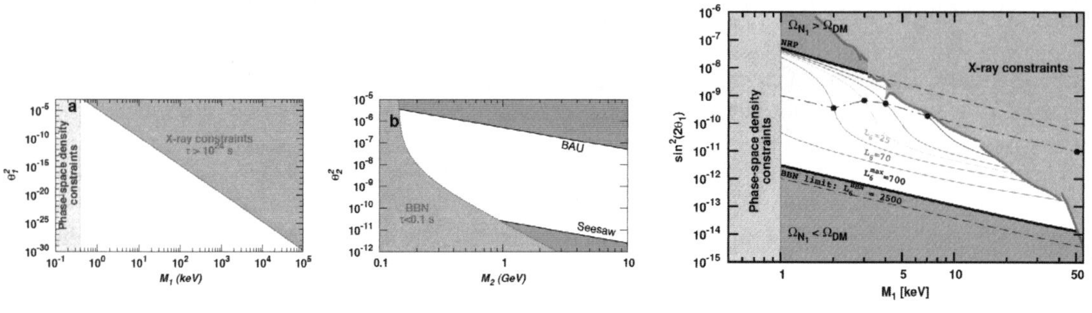

Figure 2. Astrophysical constraints for the νMSM model [7]. *Left:* Plot of the mixing angle θ_1 of the lightest sterile neutrino (LSN) with SM matter vs its mass M_1 . *Middle:* Plot of the mixing angle θ_2 vs mass of the other sterile neutrinos (assumed degenerate). *Right:* Plot of the $\sin^2(2\theta_1)$ vs Mass M_1 of the LSN and the current BBN, x-ray and DM density (Ω_{N_1}) constraints.

$N_{2,3}$ neutrinos. Because the decaying N_1 produces a narrow spectral line in the spectra of DM dominated astrophysical objects (e.g. galaxies), an important astrophysical constraint comes from the x-rays from such objects. The model is found consistent with all current astrophysical data, including x-ray constraints, Big Bang Nucleosynthesis (BBN) and Structure Formation data [7] (*cf.* fig. 2).

3.2 Neutrino condensates and dark energy in the Universe

Formation of fermion condensates dynamically in the Early Universe as in Nambu-Jona-Lasinio model, has been considered by several authors. In this spirit one may consider models of (effective) four-fermion interactions of sterile Majorana neutrino in the early Universe. The respective condensates may be formed, e.g. through a heavy scalar exchange. In the model of [21] it is argued that one light sterile neutrino (of mass of $O(10^{-3}$ eV)) may form the condensate at a late era of the early universe, and be responsible for the observed acceleration (and dark energy component of the energy budget) of the Universe. The model is argued to be consistent with the solar neutrino data. Such ideas of neutrino forming condensates through self interactions and their relation to cosmology appear in other contexts, also related to baryogenesis, in a variety of works [22].

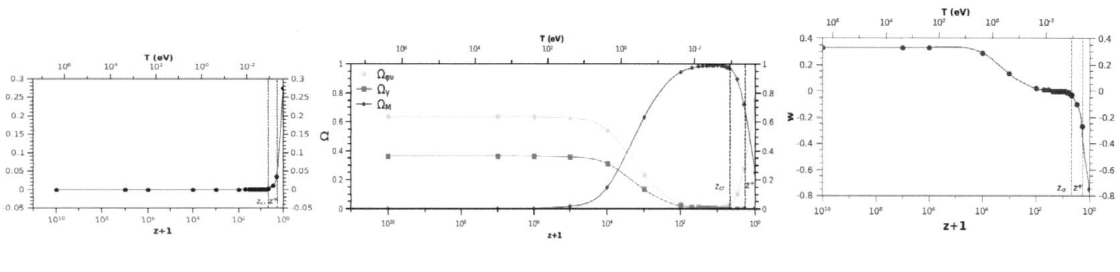

Figure 3. *Left:* The evolution of the mass of the neutrino vs T and the redshift z. *Middle:* Neutrino Dark Energy evolution vs that of Dark Matter . *Right:* The relevant equation of state, for a potential $U(\varphi) = M^{\alpha+4}/\varphi^{\alpha}$, with $M = 2.39 \times 10^{-3}$ eV, $\alpha = 0.01$. From [24].

3.3. Dark energy from mass varying neutrinos

This attractive idea [23] can be formulated simply as follows: one couple scalar cosmic fields with potential U(φ, T), where T is the temperature of the Universe at a given era, and massless fermions ψ through Yukawa couplings $g \int_0^{\beta=(k_B T)^{-1}} dt \int a^3(t) d^3x \, \varphi\overline{\psi}\psi$. The (T-dependent) fermion mass $m = g\varphi_c(T)$ is acquired through minimization w.r.t. to φ (i.e. $\varphi = \varphi_c(T) =<\varphi>$ at the minimum) of the effective potential density $\Omega(\varphi) = U(\varphi) - \frac{1}{\beta a^3 V}\log Z_D(\varphi)$, where Z_D is the partition function at temperature T. The neutrino mass (which can be significantly higher at early stages of the Universe evolution) decreases with the temperature T in an expanding Universe, and this has as a consequence the presence of a dark Energy fluid, whose equation of state resembles that of a cosmological constant at late eras [24] (*cf.* fig. 3). The models can be made consistent with current cosmologies for some choices of the potential $U(\varphi)$. However, a fundamental microscopic origin of such potentials is still lacking.

4. CPT Violation in the early Universe, active neutrinos and baryon asymmetry

So far we have assumed that the CPT symmetry holds in the Early Universe, and this produces matter and antimatter in equal amounts. An interesting idea is that during the Big Bang, one or more of the assumptions for the CPT theorem (Lorentz Invariance, unitarity and/or locality of interactions) breakdown (*e.g.* due to quantum gravity influences that may be strong at such early times), which results in CPT Violations (CPTV) and a naturally induced matter-antimatter asymmetry, without the need for extra sources of CP violation, such as sterile neutrinos. The simplest possibility [25] is through particle-antiparticle mass diffferences $m \neq \overline{m}$. These would affect the (anti) particle distributions $f(E,\mu) = [\exp(E-\mu)/T) \pm 1]^{-1}$, $E^2 = p^2 + m^2$ and similarly for antimatter $m \to \overline{m}$, and thus generate a matter antimatter asymmetry in the relevant densities $n - \overline{n} = g_{d.o.f.} \int \frac{d^3p}{(2\pi)^3} [f(E,\mu) - f(\overline{E},\mu)]$. Assuming [25] quite reasonably that dominant contributions to Baryon asymmetry come from quarks-antiquarks, and that their masses are increasing, say, linearly with temperature $m \sim gT$, one estimates the induced baryon asymmetry by the fact that the maximum quark-antiquark mass difference is bounded by the current experimental bound on proton-antiproton mass difference, which is known to be less than $2 \cdot 10^{-9}$ GeV. This produces, unfortunately, too small BAU compared to the observed one. However, active neutrino-antineutrino mass differences alone may reproduce BAU; some phenomenological models in this direction have been considered in [26], considering for instance particle-antiparticle mass differences for active neutrinos compatible with current oscillation data. But particle-antiparticle mass difference may not be the only way by which CPT is violated. As discussed in [27], quantum gravity fluctuating effects that may be strong in the early unvierse, may act as an environment inducing decoherence for the (anti) neutrino , but with couplings between the particles and the environment that are different between the neutrino and antineutrino sectors. In [27] simple models of Lindblad decoherence were considered, with zero decoherence parameters in the particle sector, and non trivial only in the antiparticle sector, and such that there was a mixed energy dependence (some of the coefficients (with dimension of energy) were proportional to the antineutrino energies, $\overline{\gamma}_i = (T/M_P)E$, while others were inversely proportional to them (and subdominant) $\overline{\gamma}_j = 10^{-24}\frac{1}{E}$, $j \neq i$). The model is phenomenological and its choice was originally motivated by fitting the LSND "anomalous data" in the antineutrino sector with the rest of the neutrino data. In this way one can derive an active (light) $\nu - \overline{\nu}$ asymmetry of order $\mathcal{A} = (n_\nu - n_{\overline{\nu}})/(n_\nu + n_{\overline{\nu}}) = \widehat{\gamma}_1 = \frac{T}{M_P} \cdot \frac{E}{\sqrt{\Delta m^2}}$. This Lepton number violation is communicated to the Baryon sector by means of B+L violating sphaleron processes, as usual, and one can thus reproduce the observed BAU without the need for extra sources of CP violation and thus sterile neutrinos. Unfortunately, at present such models lack microscopic understanding, but we think they are worth pursuing. For other scenarios of neutrino-antineutrino CPT violation induced by local curvature effects in geometries of the early universe, and its connection to baryogenesis, see [28].

Acknowledgments

I thank the organisers of Nufact2011 for the invitation. This work was supported in part by the London Centre for Terauniverse Studies (LCTS), using funding from the European Research

Council via the Advanced Investigator Grant 267352.

References

[1] Adam T *et al.* (OPERA Collaboration) 2011 Measurement of the neutrino velocity with the OPERA detector in the CNGS beam, *Preprint* arXiv:1109.4897 [hep-ex].
[2] Christenson J H, Cronin J W, Fitch V L and Turlay R 1964 Evidence for the 2 pi Decay of the K(2)0 Meson, *Phys. Rev. Lett.* **13** 138.
[3] Sakharov A D 1967 Violation of CP Invariance, C Asymmetry, and Baryon Asymmetry of the Universe, *Pisma Zh. Eksp. Teor. Fiz.* **5** 32.
[4] Gamow G 1946 Expanding universe and the origin of elements, *Phys. Rev.* **70** 572.
[5] Kuzmin V A, Rubakov V A and Shaposhnikov M E 1985 On the Anomalous Electroweak Baryon Number Nonconservation in the Early Universe, *Phys. Lett.* B **155** 36.
[6] Shaposhnikov M E 2009 Baryon asymmetry of the universe and neutrinos, *Prog. Theor. Phys.* **122** 185 and references therein.
[7] Shaposhnikov M and Tkachev I 2006 The nuMSM, inflation, and dark matter, *Phys. Lett.* B **639** 414.
Boyarsky A, Ruchayskiy O and Shaposhnikov M 2009 The Role of sterile neutrinos in cosmology and astrophysics, *Ann. Rev. Nucl. Part. Sci.* **59** 191 and references therein.
[8] Giunti C and Hernandez P these proceedings and references therein.
[9] Bezrukov F L and Shaposhnikov M, 2008 The Standard Model Higgs boson as the inflaton, *Phys. Lett.* B **659** 703.
[10] Akhmedov E K, Rubakov V A and Smirnov A Y 1998 Baryogenesis via neutrino oscillations, *Phys. Rev. Lett.* **81** 1359.
[11] Fukugita M and Yanagida T 1986 Baryogenesis Without Grand Unification, *Phys. Lett.* B **174** 45.
Buchmuller W, Peccei R D and Yanagida T 2005 Leptogenesis as the origin of matter, *Ann. Rev. Nucl. Part. Sci.* **55** 311, and references therein.
[12] Anisimov A, Buchmuller W, Drewes M and Mendizabal S 2011 Quantum Leptogenesis I, *Annals Phys.* **326** 1998.
[13] Pilaftsis A 2005 Resonant tau-leptogenesis with observable lepton number violation, *Phys. Rev. Lett.* **95** 081602.
Pilaftsis A and Underwood T E J 2005 Electroweak-scale resonant leptogenesis, *Phys. Rev.* D **72** 113001.
[14] De Simone A and Riotto A 2007 On Resonant Leptogenesis, JCAP 0708 013.
[15] Adam J *et al.* (MEG Collaboration) 2011, New limit on the lepton-flavour violating decay $\mu^+ \to e^+\gamma$, *Phys. Rev. Lett.* **107**, 171801.
[16] Mohapatra R these proceedings and references therein.
[17] Lindner M and Merle A and Niro V 2011 Soft $L_e - L_\mu - L_\tau$ flavour symmetry breaking and sterile neutrino keV Dark Matter, JCAP 1101 034.
[18] Kusenko A, Takahashi F and Yanagida T T 2010 Dark Matter from Split Seesaw, *Phys. Lett.* B **693** 144.
[19] Randall L and Sundrum R 1999 A Large mass hierarchy from a small extra dimension, *Phys. Rev. Lett.* **83** 3370.
[20] Merle A, Niro V 2011 Deriving Models for keV sterile Neutrino Dark Matter with the Froggatt-Nielsen mechanism, JCAP 1107 023.
Barry J, Rodejohann W and Zhang H 2011 Light Sterile Neutrinos: Models and Phenomenology, JHEP 1107 091.
[21] Bhatt J R, Desai B R , Ma E, Rajasekaran G and Sarkar U 2010 Neutrino Condensate as Origin of Dark Energy, *Phys. Lett.* B **687** 75.
[22] Barenboim G and Rasero J, 2011 Baryogenesis from a right-handed neutrino condensate, JHEP 1103 097.
Mavromatos N E 2011 Quantum Gravity, Flavour Vacua and Supersymmetry, *Fortschr. Phys.* **59** No. 11 12, 1113.
[23] Fardon R, Nelson A E and Weiner N 2004 Dark energy from mass varying neutrinos, JCAP 0410 005.
[24] Chitov G Y, August T, Natarajan A and Kahniashvili T, 2011 Mass Varying Neutrinos, Quintessence, and the Accelerating Expansion of the Universe, *Phys. Rev.* D **83** 045033.
[25] Dolgov A D 2010 CPT violation and particle-antiparticle asymmetry in cosmology, *Phys. Atom. Nucl.* **73** 588 and references therein.
[26] Barenboim G, Borissov L, Lykken J D and Smirnov, A Y 2002 Neutrinos as the messengers of CPT violation, JHEP 0210 001.
[27] Barenboim G and Mavromatos N E 2004 Decoherent neutrino mixing, dark energy and matter-antimatter asymmetry, *Phys. Rev.* D **70** 093015.
[28] Debnath U, Mukhopadhyay B and Dadhich N 2006 Space-time curvature coupling of spinors in early universe: Neutrino asymmetry and a possible source of baryogenesis, *Mod. Phys. Lett.* A **21** 399.

T2K results and future plans

Francesca Di Lodovico [1], for the T2K Collaboration

School of Physics and Astronomy, Queen Mary University of London, London E1 4NS, UK

E-mail: f.di.lodovico@qmul.ac.uk

Abstract. We present the $\nu_\mu \rightarrow \nu_e$ appearance and the ν_μ disappearance results, using a total of 1.43×10^{20} protons on target collected with the T2K experiment. T2K is long baseline neutrino experiment in Japan with detectors located at J-PARC, Tokai, and at Kamioka in the Gifu Prefecture, situated 295 km away from J-PARC. The muon neutrino beam is produced and measured at the near detectors at J–PARC whilst the neutrino rates after oscillation are measured with the Super-Kamiokande detector, at Kamioka. A total of six events pass all the selection criteria for $\nu_\mu \rightarrow \nu_e$ oscillations at the far detector Super-Kamiokande, leading to $0.03(0.04) < \sin^2 2\theta_{13} < 0.28(0.34)$ for $\delta_{CP}= 0$ and normal (inverted) hierarchy at 90% C.L. The ν_μ disappearance analysis excludes no oscillations at 4.3σ. At 90% C.L., the best fit values are $\sin^2 2\theta_{23} > 0.84$ and $2.1 \times 10^{-3} < \Delta m^2_{23}(\mathrm{eV}^2) < 3.1 \times 10^{-3}$. Finally, we present an overview of the T2K plans from 2011 onwards.

1. Introduction

The main goal of the T2K (Tokai-to-Kamioka) long baseline neutrino experiment [1] is to measure the appearance of electron neutrinos from a beam of muon neutrinos, thereby measuring θ_{13}, the last unknown mixing angle in the lepton sector. The results on the muon into electron neutrino appearance [2] and the muon neutrino disappearance presented in these proceedings are based on the data collected in the the first two periods of data taking from January 2010 to March 2011. This is followed by an overview of the future goals for data collections at T2K.

2. The T2K experiment

The T2K experiment's main components are the beam source, the near detector complex and the far detector. Details of the T2K experimental setup are described in Ref. [1]. Here we provide a brief review of the main components. The T2K muon neutrino beam is produced at J-PARC[2] and detected with Super-Kamiokande, the far detector, located in Kamioka in the Gifu Prefecture 295 km away from J-PARC. T2K adopts the off–axis method [3] to generate a narrow–band neutrino beam using the new proton synchrotron at J-PARC. The neutrino beam is thus directed at an angle of 2.5° with respect to the baseline connecting the proton target and Super-Kamiokande, corresponding to a peak energy of about 0.6 GeV, which maximizes the effect of the neutrino oscillation at 295 km and minimizes the background to electron-neutrino appearance detection. The angle can be reduced down to 2.0°, allowing variation of the peak neutrino energy, if necessary.

[1] Partially funded by the European Research Council, Grant agreement no. 207282-T2KQMUL.

[2] Japan Proton Accelerator Research Complex jointly constructed and operated by KEK and JAEA.

The near detector site at about 280 m from the production target houses on–axis and off–axis detectors with respect to the direction of the neutrino beam. The on–axis detector INGRID (Interactive Neutrino GRID) [4] is composed of an array of iron/scintillator sandwiches and its main goal is to measure the neutrino beam direction and profile. INGRID can measure the beam centre with a precision better than 28 cm corresponding to 1 mrad or a shift of around 2% in the peak of the neutrino energy spectrum. The off–axis detector (ND280), see Ref. [5] of this conference proceedings, is a multipurpose magnetised detector which measures the muon neutrino flux and energy spectrum and intrinsic electron neutrino contamination in the direction of Super-Kamiokande. It also measures the rates for exclusive neutrino reactions, relevant to the disappearance and appearance searches at Super-Kamiokande. These measurements are essential in order to characterize signals and backgrounds that are observed in the Super-Kamiokande far–detector. The off–axis detector measures both charged and neutral particles. It consists of a water-scintillator detector optimized to identify the π^0's (the PØD); a tracker system consisting of time projection chambers (TPCs) [6] and fine grained detectors (FGDs) optimized to study charged current interactions; and finally an electromagnetic calorimeter (ECal) that surrounds the PØD and the tracker. The whole off–axis detector is placed in a 0.2 T magnetic field provided by the recycled UA1 magnet, which also serves as part of a side muon range detector (SMRD).

The far detector, Super-Kamiokande, is located in the Mozumi mine, near the village of Higashi-Mozumi, Gifu, Japan. A detailed description of the detector can be found elsewhere [7]. Here we provide a summary. The detector cavity lies under the peak of Mt. Ikenoyama, with 1000 m of rock equivalent to 2700 meters–water–equivalent (m.w.e.) which reduces the cosmic ray background by around five orders of magnitude compared to at the Earth's surface.

Super-Kamiokande is a water Cherenkov detector with a fiducial volume (FV) of 22.5 kton within its cylindrical inner detector (ID). Surrounding the ID is the 2 m-wide outer detector (OD). The inner detector (ID) is a cylindrical space 33.8 m in diameter and 36.2 m in height which currently houses along its inner walls 11,129 inward-facing 50 cm diameter PMTs. The OD contains along its inner walls 1,885 outward-facing 20 cm diameter PMTs. The ID and OD boundaries are defined by a cylindrical structure about 50 cm wide. The front-end readout electronics allows for a zero–deadtime software trigger. Spill timing information, synchronized by the Global Positioning System (GPS) with < 150 ns precision, is transferred online to Super-Kamiokande and triggers the recording of photomultiplier hits within ± 500 μs of the expected arrival time of the neutrinos. Super-Kamiokande has been running since 1996 and has had four distinct running periods. The latest period, SK-IV, is running stably and features upgraded PMT readout electronics.

Construction of the neutrino beamline started in April 2004. The complete chain of accelerator and neutrino beamline was successfully commissioned during 2009, and T2K began accumulating neutrino beam data for physics analysis in January 2010.

Construction of the majority of the ND280 detectors was completed in 2009 with the full installation of INGRID, the central ND280 off-axis sub-detectors (PØD, FGD, TPC and downstream ECal) and the SMRD. The ND280 detectors began stable operation in February 2010. The rest of the ND280 detector (the ECals) was completed in the fall of 2010.

The results presented in these proceedings are based on the first two physics runs: run 1 (Jan–Jun 2010) and run 2 (Nov 2010–Mar 2011) whereas the proton beam power was continually increased and reached 145 kW with 9×10^{13} protons per pulse by March 2011. A total of 1.43×10^{20} protons on target (p.o.t.) were collected for analysis, corresponding to about 2% of the final approved goal for T2K.

3. Indication of electron neutrino appearance

The $\nu_\mu \to \nu_e$ analysis at Super-Kamiokande requests only a single electron-like (e-like) ring, thus selecting a sample enhanced in ν_e charged-current quasi-elastic interactions (CCQE). The

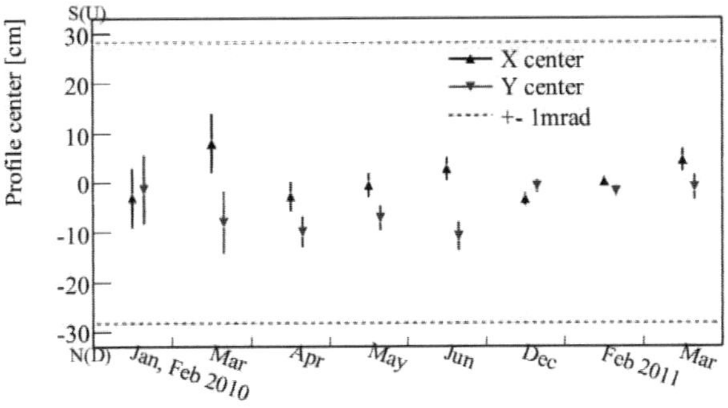

Figure 1. Beam centering stability in horizontal (x, South–North) and vertical (y, Down–Up) directions as a function of time, as measured by INGRID. Errors shown are only statistical. The dashed lines correspond to a change of beam direction by ± 1 mrad.

selection criteria for this analysis were fixed from Monte Carlo (MC) studies before analysing the data and were optimized for the initial running conditions.

The main backgrounds to the $\nu_\mu \rightarrow \nu_e$ oscillated events are due to the intrinsic ν_e contamination in the beam and to the neutral current (NC) interactions with a misidentified π^0.

The observed number of events is compared to expectations based on neutrino flux and cross-section predictions for signal and all sources of backgrounds, which are corrected using an inclusive ν_μ charged-current (CC) measurement in the off–axis near detector.

We compute the neutrino beam fluxes starting from models and tune them to experimental data. A detailed description can be found in Ref. [8] of this conference proceedings. The estimated uncertainties of the intrinsic ν_μ and ν_e fluxes below 1 GeV are around 14%. Above 1 GeV, the intrinsic ν_e flux error is dominated by the uncertainty on the kaon production rate with resulting errors of 20–50%.

The neutrino beam profile and its absolute rate (1.5 events/10^{14} p.o.t.) as measured by INGRID were stable and consistent with expectations. The beam profile center (Fig. 1) indicates that beam steering was better than ± 1 mrad. The correlated systematic error is $\pm 0.33(0.37)$ mrad for the horizontal(vertical) direction. The error on the Super-Kamiokande position relative to the beamline elements was obtained from a dedicated GPS survey and is negligible.

The NEUT MC event generator [9], which has been tuned with recent neutrino interaction data in an energy region compatible with T2K, is used to simulate neutrino interactions in the near and far detectors. The GENIE [10] generator provides a separate cross-check of the assumed cross-sections and uncertainties, and yields consistent results.

A list of reactions and their uncertainties relative to the CCQE total cross-section is shown in Table 1. An energy–dependent error on CCQE is assigned to account for the uncertainty in the low energy cross–section, especially for the different target materials between the near and far detectors.

An inclusive ν_μ CC measurement in the off-axis near detector is used to constrain the expected event rate at the far detector. The results are described in Ref. [11] of this conference proceedings. The measured data/MC ratio is

$$R_{ND}^{\mu,Data}/R_{ND}^{\mu,MC} = 1.036 \pm 0.028(\text{stat.})^{+0.044}_{-0.037}(\text{det.syst.}) \pm 0.038(\text{phys.syst.}), \quad (1)$$

where $R_{ND}^{\mu,Data}$ and $R_{ND}^{\mu,MC}$ are the p.o.t. normalized rates of ν_μ CC interactions in data and MC. The detector systematic errors mainly come from tracking and particle identification efficiencies, and physics uncertainties are related to the interaction modeling. Uncertainties that effectively cancel between near and far detectors were omitted.

At the far detector, we extract a fully-contained fiducial volume (FCFV) sample by requiring no event activity in either the OD or in the 100 μs before the event trigger time, at least 30

Table 1. Summary of systematic uncertainties for the relative rate of different charged-current (CC) and neutral-current (NC) reactions to the rate for CCQE.

Process	Systematic error
CCQE	energy-dependent (7% at 500 MeV)
CC 1π	30% ($E_\nu < 2$ GeV) – 20% ($E_\nu > 2$ GeV)
CC coherent $\pi^\pm$	100%
CC other	30% ($E_\nu < 2$ GeV) – 25% ($E_\nu > 2$ GeV)
NC $1\pi^0$	30% ($E_\nu < 1$ GeV) – 20% ($E_\nu > 1$ GeV)
NC coherent π	30%
NC other π	30%
FSI	energy-dependent (10% at 500 MeV)

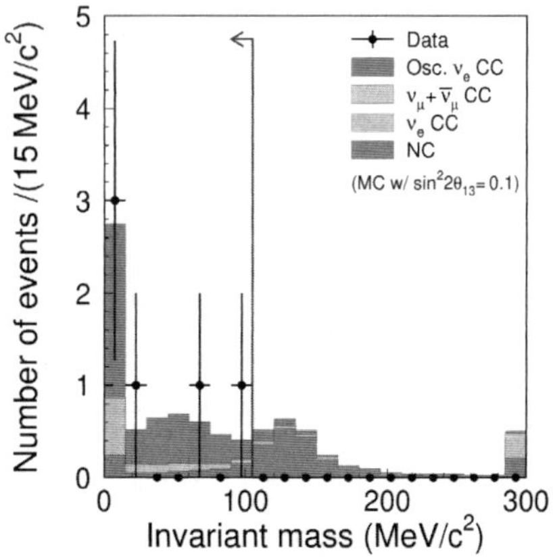

Figure 2. Invariant mass distribution reconstructing each event as two rings. The data are shown using points with statistical error bars and the MC predictions are the histograms. The last bin shows overflow entries and the vertical line the applied cut at 105 MeV.

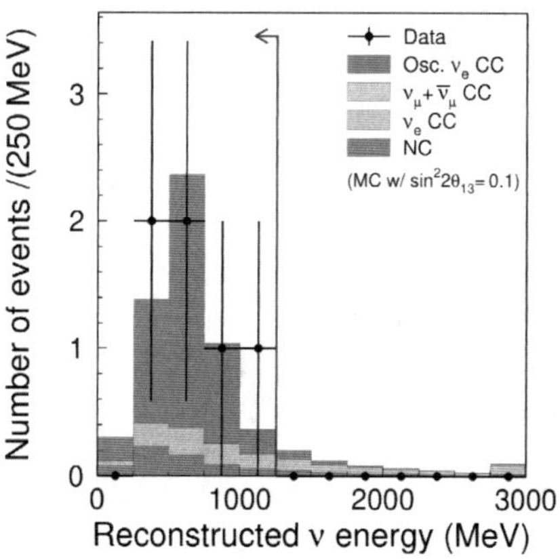

Figure 3. Same as Fig. 2 for the reconstructed neutrino energy spectrum of the events which pass all ν_e appearance signal selection criteria with the exception of the energy cut. The vertical line shows the applied cut at 1250 MeV.

MeV electron-equivalent energy deposited in the ID (defined as visible energy E_{vis}), and the reconstructed vertex in the fiducial region. The data have 88 such FCFV events that are within the timing range from -2 to 10 μs around the beam trigger time. The accidental contamination from non–beam related events is determined from the sidebands to be 0.003 events. A Kolmogorov-Smirnov (KS) test of the observed number of FCFV events as a function of accumulated p.o.t. is compatible with the normalized event rate being constant (p-value $=$ 0.32). Forty-one events are reconstructed with a single ring, and eight of those are e-like. Six of these events have $E_{vis} > 100$ MeV and no delayed-electron signal. To suppress misidentified π^0 mesons, the reconstruction of two rings is forced by comparison of the observed and expected light patterns calculated under the assumption of two showers, and a cut on the two-ring invariant mass $M_{inv} < 105$ MeV$/c^2$ is imposed. No events are rejected (Fig. 2). Finally, the neutrino energy E_ν^{rec} is computed using the reconstructed momentum and direction of the ring, by assuming quasi–elastic kinematics and neglecting Fermi motion. No events are rejected

Table 2. Event reduction for the ν_e appearance search at the Super-Kamiokande. The numbers of data and ν_μ CC, intrinsic ν_eCC, NC, and ν_e CC signal MC expected events are presented after each selection criterion. All MC CC samples include three-flavor oscillations for $\sin^2 2\theta_{13}$ = 0.1 and δ_{CP}= 0.

	Data	ν_μCC	ν_eCC	NC	$\nu_\mu \to \nu_e$CC
(0) interaction in FV	n/a	67.2	3.1	71.0	6.2
(1) fully-contained FV	88	52.4	2.9	18.3	6.0
(2) single ring	41	30.8	1.8	5.7	5.2
(3) e-like	8	1.0	1.8	3.7	5.2
(4) $E_{vis} > 100$ MeV	7	0.7	1.8	3.2	5.1
(5) no delayed electron	6	0.1	1.5	2.8	4.6
(6) non-π^0-like	6	0.04	1.1	0.8	4.2
(7) $E_\nu^{rec} < 1250$ MeV	6	0.03	0.7	0.6	4.1

by requiring $E_\nu^{rec} < 1250$ MeV, whose purpose was to suppress events from the intrinsic ν_e component arising primarily from kaon decays (Fig. 3). The data and MC reductions after each selection criterion are shown in Table 2. The ν_e appearance signal efficiency is estimated from MC to be 66% while rejection for $\nu_\mu + \bar{\nu}_\mu$ CC, intrinsic ν_eCC, and NC are > 99%, 77%, and 99%, respectively. Of the surviving background NC interactions constitute 46%, of which 74% are due to π^0 mesons and 6% originate from single gamma production.

Examination of the six data events shows properties consistent with ν_e CC interactions. The distribution of the cosine of the opening angle between the ring and the incoming beam direction is consistent with CCQE events. The event vertices in cylindrical coordinates (R,ϕ,z) show that these events are clustered at large R, near the edge of the FV in the upstream beam direction. A KS test on the R^2 distribution of our final events yields a p-value of 0.03. If this was related to contamination from penetrating particles produced in upstream neutrino interactions, then the ID region outside the FV should show evidence for such events, however this is not observed. In addition, an analysis of the neutrino interactions occurring in the OD volume is consistent with expectations.

To compute the expected number of events at the far detector N_{SK}^{exp}, we use the near detector ν_μ CC interaction rate measurement as normalization, and the ratio of expected events in the near and far detectors, where common systematic errors cancel. Using Eq. 1, this can be expressed as:

$$N_{SK}^{exp} = \left(R_{ND}^{\mu,Data}/R_{ND}^{\mu,MC}\right) \cdot N_{SK}^{MC}, \tag{2}$$

where N_{SK}^{MC} is the MC number of events expected in the far detector. Due to the correlation of systematic errors in the near and far detector samples, Eq. 2 reduces the uncertainty on the expected number of events. Event rates are computed incorporating three-flavor oscillation probabilities and matter effects [12] with $\Delta m_{12}^2 = 7.6 \times 10^{-5}$ eV^2, $\Delta m_{23}^2 = +2.4 \times 10^{-3}$ eV^2, $\sin^2 2\theta_{12} = 0.8704$, $\sin^2 2\theta_{23} = 1.0$, an average Earth density $\rho = 3.2$ g/cm^3 and δ_{CP}= 0 unless otherwise noted. The expectations are 0.03(0.03) $\nu_\mu + \bar{\nu}_\mu$CC, 0.8(0.7) intrinsic ν_eCC, and 0.1(4.1) $\nu_\mu \to \nu_e$oscillation events for $\sin^2 2\theta_{13} = 0(0.1)$, and 0.6 NC events. As shown in Table 3, the total systematic uncertainty on N_{SK}^{exp} depends on θ_{13}. Combining the above uncertainties the total far detector systematic error contribution to $\delta N_{SK}^{exp}/N_{SK}^{exp}$ is 14.7%(9.4%) for $\sin^2 2\theta_{13}$ = 0(0.1).

Our oscillation result is based entirely on comparing the number of ν_e candidate events with predictions, varying $\sin^2 2\theta_{13}$ for each δ_{CP}value. Including systematic uncertainties the expectation is 1.5±0.3(5.5±1.0) events for $\sin^2 2\theta_{13} = 0(0.1)$. The probability to observe six or more candidate events is 7×10^{-3} for $\sin^2 2\theta_{13} = 0$. Thus, we conclude that our

Table 3. Contributions from various sources and the total relative uncertainty for $\sin^2 2\theta_{13} = 0$ and 0.1, and δ_{CP}= 0.

Source	$\sin^2 2\theta_{13} = 0$	$\sin^2 2\theta_{13} = 0.1$
(1) neutrino flux	$\pm$ 8.5%	$\pm$ 8.5%
(2) near detector	$^{+5.6}_{-5.2}\%$	$^{+5.6}_{-5.2}\%$
(3) near det. statistics	$\pm$ 2.7%	$\pm$ 2.7%
(4) cross section	$\pm$ 14.0%	$\pm$ 10.5%
(5) far detector	$\pm$ 14.7%	$\pm$ 9.4%
Total $\delta N_{SK}^{exp}/N_{SK}^{exp}$	$^{+22.8}_{-22.7}\%$	$^{+17.6}_{-17.5}\%$

data indicate ν_e appearance from a ν_μ neutrino beam. At each oscillation parameter point, a probability distribution for the expected number of events is constructed, incorporating systematic errors [13], which is used to make the confidence interval (Fig. 4), following the unified ordering prescription of Feldman and Cousins [14].

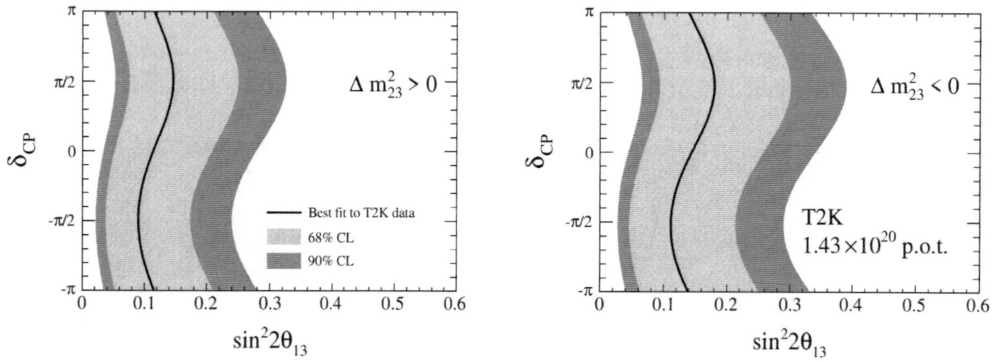

Figure 4. The 68% and 90% C.L. regions for $\sin^2 2\theta_{13}$ for each value of δ_{CP} consistent with the observed number of events in the three-flavor oscillation case for normal (left) and inverted (right) mass hierarchy. The other oscillation parameters are fixed (see text). The best fit values are shown with solid lines.

This result converted into a confidence interval yields $0.03(0.04) < \sin^2 2\theta_{13} < 0.28(0.34)$ at 90% C.L. for $\sin^2 2\theta_{23} = 1.0$, $|\Delta^2 m^2_{23}| = 2.4\times10^{-3}$ eV^2, δ_{CP}= 0 and for normal (inverted) neutrino mass hierarchy. Under the same assumptions, the best fit points are 0.11(0.14), respectively. For non–maximal $\sin^2 2\theta_{23}$, the confidence intervals remain unchanged to first order by replacing $\sin^2 2\theta_{13}$ by $2\sin^2\theta_{23}\sin^2 2\theta_{13}$. More data are required to firmly establish ν_e appearance and to better determine the angle θ_{13}. These results have been published in Ref. [2] and are consistent with the results published subsequently by the MINOS collaboration [15] that measures $2\sin^2\theta_{23}\sin^2 2\theta_{13} < 0.12(0.20)$ at 90% C.L. for δ_{CP}= 0 and the normal (inverted) mass hierarchy. More details are given in Ref. [16] in this conference proceedings.

4. Muon neutrino disappearance

We present preliminary results on the ν_μ disappearance analysis in the following. The ν_μ analysis follows the same steps of the $\nu_\mu \rightarrow \nu_e$ appearance analysis up to the single ring selection criteria, that corresponds to the event reduction criteria in Table 2 number (2). From the selected events, that are predominantly CCQE, thirty–three ν_μ–like events are indentified. By adding a further cut on the minimum muon momentum and on the observed delayed electrons, the CCQE purity

Table 4. Event reduction for the ν_μ disappearance search at Super-Kamiokande. The selection criteria (0) to (2) are shown in Table 2. The numbers of data and ν_μ CCQE, ν_μCC non–CCQE, intrinsic ν_eCC and NC MC expected events are presented after each selection criterion. All MC CC samples include three-flavor oscillations for $\sin^2 2\theta_{23} = 1$ and $\Delta m^2_{23} = 0.0024$ eV^2.

	Data	ν_μCCQE	ν_μCC non–CCQE	ν_eCC	NC
(3) μ-like	33	17.6	12.4	<0.1	1.9
(4) $P_\mu > 200$ MeV	33	17.5	12.4	<0.1	1.9
(5) 0 or 1 delayed electron	31	17.3	9.2	<0.1	1.8

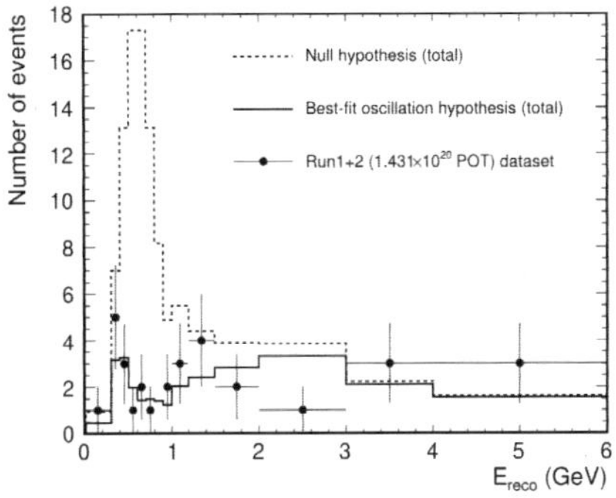

Figure 5. Reconstructed neutrino energy spectrum of the thirthy–one single μ–like ring events (points with statistical error bars), with no oscillations and best–fit MC prediction (histograms).

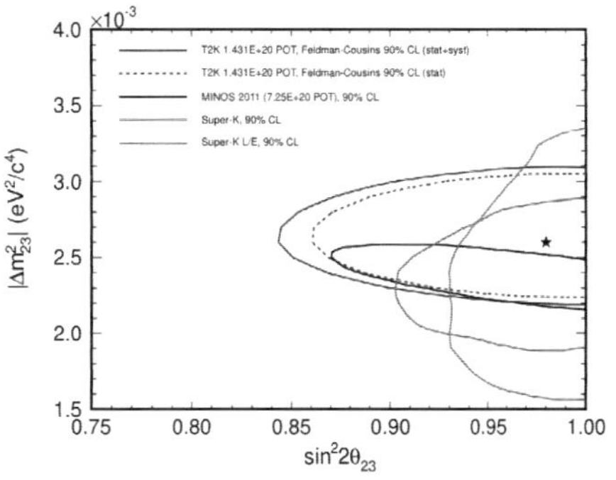

Figure 6. The 90% ν_μ–disappearance confidence regions (with and without systematics) compared with Super-Kamiokande and MINOS results. The star indicates the best fit point.

is increased with a negligible decrease in efficiency. A total of thirty–one single μ–ring events are observed with an expected number of events of about 104 without oscillations.

The same strategy as for the $\nu_\mu \to \nu_e$ analysis with regard to the beam simulation, the neutrino interactions and the treatment of the ND280 results is adopted (see Section 3). The data and MC reductions after each selection criterion are shown in Table 4.

The reconstructed energy spectrum of the thirty–one single μ–like ring events in the combined run 1 and 2 dataset, with prediction for no oscillations, and best–fit prediction is presented in Figure 5. The statistical significance of the observed deficit without oscillations is 4.3σ. The energy spectrum in Figure 5 is shown in a variable binning scheme for the purpose of illustration only; the actual analysis used equal–sized 50 MeV bins. The best fit parameters are $\sin^2 2\theta_{23} = 0.98$, $|\Delta m^2_{23}| = 2.6\times10^{-3}$ eV^2, and the corresponding 90% Feldman-Cousins confidence regions (with and without systematic errors) for the 2–flavour ν_μ –disappearance fit are shown in Figure 6. We also perform an independent oscillation analysis that gives very similar results: $\sin^2 2\theta_{23} = 0.99$, $|\Delta m^2_{23}| = 2.6\times10^{-3}$ eV^2 with a statistical significance of the observed deficit without oscillations of 4.4σ. Our results are consistent with the ones from Super-Kamiokande[17] and MINOS [18].

5. Future prospects

The T2K data taking stopped abruptly on March 11 2011 due to the Great East Japan Earthquake. No serious damages were found in the accelerator complex, the neutrino beamline or the near detectors. The Super-Kamiokande detector was not affected by the earthquake. We are on schedule to resume operations at J-PARC in December 2011 with the T2K experiment expected to start data taking as soon as possible early 2012. The current goal is to collect 1×10^{21} p.o.t. by summer 2013. Such a data sample will allow the confirmation of a non-zero value of θ_{13} with 5 sigma statistical significance at the present best fit value.

6. Conclusions

The T2K experiment, which comprises a new neutrino beamline and new near detector complex at J-PARC and the upgraded Super-Kamiokande detector, has been constructed successfully and collected data from January 2010 to March 2011 for a total of 1.43×10^{20} p.o.t. We published our first results on the ν_e appearance analysis [2] where we report the observation of six single ring e-like events. The probability to observe six or more events is $7{\times}10^{-3}$ for $\sin^2 2\theta_{13} = 0$. This result yields $0.03(0.04) < \sin^2 2\theta_{13} < 0.28(0.34)$ at 90% C.L. for $\sin^2 2\theta_{23} = 1.0$, $|\Delta m^2_{23}| = 2.4{\times}10^{-3}$ eV^2, $\delta_{CP}= 0$ and for normal (inverted) neutrino mass hierarchy. Under the same assumptions, the best fit points are 0.11(0.14), respectively. These results are consistent with the results published subsequently by MINOS.

We presented preliminary results on the ν_μ disappearance analysis using the same data set. A total of thirty-one single ring μ-like events are observed at the far detector, yielding to the best fit parameters of $\sin^2 2\theta_{23} = 0.98$, $|\Delta m^2_{23}| = 2.6{\times}10^{-3}$ eV^2, consistent with the results from Super-Kamiokande and MINOS.

T2K has collected 2% of its projected data and is on schedule to restart data taking early 2012. J-PARC will start operation in December 2011.

References

[1] Abe K *et al.* (T2K Collaboration) 2011 *Nuclear Instruments and Methods in Physics Research Section A: Accelerators, Spectrometers, Detectors and Associated Equipment* **659** 106
[2] Abe K *et al.* (T2K Collaboration) 2011 *Physical Review Letters* **107** 041801
[3] Beavis D *et al.* (E889 Collaboration) 1995 *Physics Design Report BNL* 52459
[4] Otani M *et al.* 2010 *Nuclear Instruments and Methods in Physics Research Section A* **623** 368
[5] McCauley N 2012 "Performance of the T2K Near Detectors" these proceedings
[6] Abgrall N *et al.* 2011 *Nuclear Inst. and Methods in Physics Research Section A* **637** 25
[7] Fukuda Y *et al.* (Super-Kamiokande Collaboration) 2003 *Nuclear Instruments Methods A* **501** 418
[8] Galymov V 2012 "Predicting Neutrino Flux for the T2K Experiment" these proceedings
[9] Hayato Y 2002 *Nuclear Physics (Proceedings Supplements) B* **112** 171
[10] Andreopoulos C *et al.* 2012 *Nuclear Instruments Methods A* **614** 87
[11] Brook-Roberge D G 2012 "Neutrino Interaction Measurements Using the T2K Near Detector" these proceedings
[12] Barger V *et al.* 1980 *Physical Review D* **22** 2718
[13] Conrad J *et al.* 2003 *Physical Review D* **67** 012002
[14] Feldman G J and Cousins R C 1998 *Physical Review D* **57** 3873
[15] Adamson P *et al.* (MINOS Collaboration) 2011 *Physical Review Letters* **107** 181802
[16] Hartnell J 2012 "MINOS, NOVA, MINOS+", and J. Nelson "Improved search for muon-neutrino to electron-neutrino oscillations in MINOS" these proceedings
[17] Ashie Y *et al.* (Super-Kamiokande Collaboration) 2004 *Physical Review Letters* **93** 101801; Ashie Y *et al.* (Super-Kamiokande Collaboration) 2005 *Physical Review D* **71** 112005; Hosaka J *et al.* (Super-Kamiokande Collaboration) 2006 *Physical Review D* **74** 032002
[18] Adamson P *et al.* (MINOS Collaboration) 2011 *Physical Review Letters* **106** 181801

Journal of Physics: Conference Series **408** (2013) 012005 doi:10.1088/1742-6596/408/1/012005

Neutrino mass and grand unification

Rabindra N Mohapatra

Maryland Center for Fundamental Physics and Department of Physics, University of Maryland, College Park, MD 20742, USA

E-mail: rmohapat@umd.edu

Abstract. Simple grand unified theories (GUT) of matter and forces provide a natural home for the seesaw paradigm for understanding small neutrino masses within a framework that also connects quark flavor to that of leptons. This provides a promising possibility for a unified approach to the flavor puzzle. To see what recent indications of a "large" value for the neutrino mixing parameter θ_{13} mean for this approach, I report on investigations based on supersymmetric SO(10) GUTs with type II seesaw for neutrino masses and find that it seem to support this point of view, with other predictions testable in near future.

1. Introduction

Understanding neutrino masses and mixings is an integral part of attempts to unravel the flavor puzzle in particle physics. During the past decade, the large amount of information on neutrino masses and mixings gained from the study of accelerator, reactor, solar, and cosmic ray neutrino observations have provided additional information that may have moved the frontier of understanding in this field forward. Several crucial pieces of the puzzle must still be found before we can begin to have a complete picture of neutrino masse. They include Dirac vs Majorana nature of the neutrino masses, normal vs inverted nature of the mass hierarchy, as well as the values of the mixing angle θ_{13}, and CP phases. One major step in this direction is the announcement that the T2K experiment has possible indications for a non-zero value for θ_{13} [1]. The T2K lower limit, if correct, is not far below the current experimental upper bound obtained several years ago from the CHOOZ experiment [2] and the fact that it is so "large" has important theoretical implications. The MINOS experiment has also seen an excess of electron events which could be indicating a non-zero θ_{13} [3], and their allowed range for θ_{13} overlaps with the T2K one. There have also been analyses of existing oscillation data suggesting a non-zero θ_{13} [4].

Before we specifically discuss the implications of a "large" θ_{13} for our understanding of flavor, let us review in brief the already existing information to see what possible conclusions may be drawn about the nature of neutrinos and physics beyond the standard model from them. First, the fact the neutrino masses are all in the sub-eV range can be understood within a simple paradigm of the seesaw mechanism, which we will use throughout this talk. Depending on what kind of dimension five operator [5] generates neutrino masses i.e. $LHLH$ or $L^TC^{-1}\tau_2\vec{\tau}L \cdot H^T\tau_2\vec{\tau}H$, we call it type I [6] or type II seesaw mechanism [7]. These operators are scaled by one power of inverse mass and naively taking generic neutrino mass to be 0.1 eV implies that the scale $M \sim 10^{14}$ GeV. This is close to the familiar scale of supersymmteric grand unification (GUT) of 2×10^{16} GeV. It is therefore very suggestive that the seesaw scale is indeed related to the grand unification scale just as the quark or charged lepton masses are related to the standard model weak scale. This naive argument becomes more compelling when one realizes that without any specific framework like that of grand unified theories (GUT), the type I or type II seesaw scales could be anywhere from a TeV to the GUT scale by simply adjusting the neutrino Yukawa coupling parameters of the underlying theory while in GUT theories, neutrino Yukawa couplings are related to quark Yukawas and are therefore determined thereby fixing the seesaw scale. The minimal GUT group is the $SO(10)$ group which unifies all fermions of each generation including the right handed neutrinos into a single **16**-dim. spinor representation. A lot of interesting results follow if the SO(10) GUT is supplemented by the assumption of the type II seesaw mechanism for neutrino masses. Additional support for the GUT idea comes from the approximate equality of the bottom and tau masses when extrapolated to high scales via supersymmetric renormalization group equations. Also the approximate GUT scale relations like $m_\mu \simeq 3m_s$ are suggestive of specific details of GUT models and how appealing a GUT model is, is determined by how naturally they explain these relations.

2. From data to physics via the neutrino mass matrix

Before discussing the GUT models, we note some features of the neutrino observations and their broad implications for theory. First, the maximal atmospheric neutrino mixing ($\tan\theta_{23} = 1$) suggests a form for the $2-3$ sector of the neutrino mass matrix that exhibits an underlying discrete $\mu - \tau$ symmetry (denoted by $Z_{2,\mu-\tau}$). When this symmetry is exact for the whole mass matrix, it leads to vanishing θ_{13} [8]. Depending on how this symmetry is broken (e.g. in the μ-sector or e-sector), the resulting value of θ_{13} can either be very small or not so small (or "large"). The neutrino mass matrix has also been suspected to have a larger symmetry beyond this from the observation that the value of the solar mixing angle seems to have a geometric value ($\tan\theta_{12} = 1/\sqrt{2}$). The resulting lepton mixing (PMNS) matrix known as the tri-bi-maximal mixing matrix [9] (TBM for short) arises from a mass matrix that has a larger symmetry given by the larger symmetry Z_2 symmetries $Z_{2,S} \times Z_{2,\mu-\tau}$ [10]. This full symmetry of course also leads to zero θ_{13}. The form of the TBM neutrino mass matrix is given by

$$\mathcal{M}_\nu = \begin{pmatrix} b & c & c \\ c & a+b & c-a \\ c & c-a & a+b \end{pmatrix}, \qquad (1)$$

which has only three parameters given by the three neutrino masses. In fact, in the above matrix, one could set $b = 0$, without changing the TBM mixing matrix. It only affects the masses of the neutrinos. This matrix is very different from the known mass matrices in the quark sector and could be a possible clue to a unified understanding of the quark-lepton flavor puzzle. A non-zero θ_{13} suggests that the TBM PMNS mixing is not the right form, and that "large" corrections to both the $\mu - \tau$ symmetry and TBM matrix must be present. Thus the value of θ_{13} could eventually guide us not only towards a complete determination of the neutrino mass matrix but also give indications for what kind of physics lies behind neutrino masse . For instance if observations require that the corrections to TBM mass matrix are "large", it would not be too implausible to think that the symmetry approach may only be illusory and some other mechanisms is at work. We will take this point of view and explore whether the framework of SO(10) grand unification provides such an alternative to symmetry approach and may provide answers to such questions as: do the large atmospheric and solar angles as well as the curious feature of masses that $\frac{m_{solar}}{m_{atmos}} \sim \theta_{Cabibbo}$ get an easy explanation in GUT models without invoking any symmetries. The post-SO(10) theory may of course contain symmetries as a way to understand the parameters of the theory in the same way as the Yukawa couplings of SM may be reflection of some underlying symmetries.

3. Neutrino mixings in renormalizable SO(10) GUT

As noted $SO(10)$ is the minimal GUT that embeds the seesaw paradigm for neutrino masses and among many SO(10) models, the most predictive class [11] without extra symmetries appears to be the ones with renormalizable Yukawa couplings which use the Higgs fields **10**, **126** and possibly **120** or another **10**. This is a generalization of a minimal model with only three Higgs fields **10+126+210** discussed early on in connection with coupling unification [12]. Many of the attractive features of the model relating to fermion masses can be illustrated with two **10**'s plus **126**. Denoting the **10** Higgs fields (by H, H') and the $\mathbf{126} + \overline{\mathbf{126}}$ fields by $\Delta + \bar{\Delta}$, the $SO(10)$ invariant Yukawa couplings of the model can be written as:

$$\mathcal{L}_Y \;=\; h_0 \psi\psi H \;+\; f_0 \psi\psi\bar{\Delta} \tag{2}$$

where ψ's denote the **16** dimensional spinors of $SO(10)$. The Yukawa couplings, h and f are 3×3 matrices in generation space. After symmetry breaking, one obtains the following formulae for charged fermion as well as Dirac mass for neutrinos:

$$\begin{aligned} M_u &= h \;+\; r_2 f, \\ M_d &= \frac{r_1}{\tan\beta}(h+f), labeleq:2 \qquad (3)\\ M_\ell &= \frac{r_1}{\tan\beta}(h-3f), \\ M_{\nu D} &= h \;-\; 3r_2 f. \qquad (4) \end{aligned}$$

where we have absorbed the vev of Higgs doublets into h, f. The neutrino mass formula is a combination of a type I and type II term:

$$M_\nu \;=\; f v_L \;-\; M_{\nu D}\frac{1}{f v_{BL}}M_{\nu D} \tag{5}$$

If in the above neutrino mass formula, the type II term dominates, there is a relation between neutrino and charged fermion masses of the form [13]

$$M_\nu \;=\; c(M_d \;-\; M_\ell) \tag{6}$$

It was pointed out in [14], this formula combined with $b-\tau$ mass equality at GUT scale from RGER extrapolation, provides a natural explanation of large atmospheric neutrino mixing angle. Soon afterwards, it was noted in [15] that once this idea is extended to three generations, it not only explains the large solar angle but it also explains the relation $\frac{m_{solar}}{m_{atmos}} \sim \theta_{Cabibbo}$. Furthermore, it leads to a value for $\theta_{13} \sim \theta_{Cabibbo} \sim 9-10^0$ (for the case when $h' = 0$). These results are extremely interesting since without any symmetries, the simple fact of $b-\tau$ unification seems to qualitatively reproduce all observations. To see how this comes about, observe that M_d and M_ℓ are both hierarchical mass matrices- the former due to small CKM angles and latter due to the constraint of $SO(10)$ grand unification in Eq. 3. Thus one can write

$$M_\nu \;=\; cm_b(M_U)\begin{pmatrix} \sim\lambda^4 & \sim\lambda^3 & \sim\lambda^3 \\ \sim\lambda^3 & \sim\lambda^2 & \sim\lambda^2 \\ \sim\lambda^3 & \sim\lambda^2 & 1 \end{pmatrix} \;-\; cm_\tau(M_U)\begin{pmatrix} \sim\lambda^4 & \sim\lambda^3 & \sim\lambda^3 \\ \sim\lambda^3 & \sim\lambda^2 & \sim\lambda^2 \\ \sim\lambda^3 & \sim\lambda^2 & 1 \end{pmatrix} \tag{7}$$

where $\lambda \sim 0.22$, the Cabibbo angle. Note that at GUT scale if $m_b(M_U) \sim m_\tau(M_U)(1+\lambda^2)$, the 3×3 entries cancel out in the above difference and one naturally gets for the neutrino mass matrix:

$$M_\nu \;=\; cm_\tau(M_U)\lambda^2\begin{pmatrix} \sim\lambda^2 & \sim\lambda & \sim\lambda \\ \sim\lambda & 1+\lambda & 1 \\ \sim\lambda & 1 & 1 \end{pmatrix} \tag{8}$$

where we have omitted order one coefficients in the matrix elements. This mass matrix for neutrinos clearly implies that both the solar and the atmospheric mixing angles are large and $\frac{m_{solar}}{m_{atmos}} \sim \lambda$ and also $\theta_{13} \sim \lambda$ [15]. Furthermore note that without CP violation, the model has only eleven parameters and yet it fits quark and lepton masses and mixings as well as neutrino mixings and masses so well.

This model was subsequently reanalyzed after including CP violation to see if the right CKM phase can be accomodated while at the same time explaining neutrino mixings [16]. It was found that this was possible only in a small parameter space of the model where it predicted $\theta_{13} \sim 6^0$ and the Dirac phase $\delta \sim 4^0$. Extensions of this model by the addition of a **120** Higgs multiplet [17] was then considered and shown to considerably broaden the parameter space while still preserving the "large" θ_{13} prediction. Recall that **10**, **126** and **120** are the only three irreducible representations that appear in the product of two spinors of SO(10); so in some sense it was a natural thing to consider. A particularly interesting version of the 10+126+120 model is one that assumes that CP is a good symmetry prior to spontaneous breaking by the vev of the **120** field [18], which is assumed to be odd under under CP. Since **120** gives anti-symmetric Yukawa couplings, its CP odd property after symmetry breaking hermitean fermion mass matrices, responsible for CP violation both in the quark and lepton sectors. As a result, CKM phase and the Dirac phase of the neutrinos get connected. The model has a total of 16 parameters in the fermion mass sector (12 from Yukawa couplings and four from the vevs) keeping the model still predictive. Three inputs from the neutrino sector e.g. solar and atmospheric mass differences and the atmospheric mixing angle in addition to thirteen from the charged fermion sector allow one to predict θ_{12}, θ_{13} and the neutrino CP phases.

4. SO(10) with type II seesaw and connection to TBM mass matrix

Recently, two interesting connections to the standard TBM model discussion has been noted in $SO(10)$ models with type II seesaw. In [19], it was observed that the type II formula for neutrino masses allows a TBM form for the neutrino mass matrix by simply a choice of fermion basis, with no additional symmetries; corrections to the TBM form then arise from the form of the charged lepton matrix, which, in our case, is determined by the $SO(10)$ constraints from quark masses and mixings. Strictly speaking, no bottom-tau unification is invoked explicitly in this approach but clearly, consistency requires that bottom-tau unification must hold for this basis choice to eventually lead to fermion mass fits. Detailed numerical analyses of these models have been carried out and lead to excellent fits for models with **10**, **126** and **120** Higgs fields [19, 20], and, yet again, a large θ_{13} is predicted. The near-tri-bi-maximal PMNS form in this class of GUT theories is related to the dynamics of the model rather than to any symmetry. Of course, to understand the particular Yukawa textures, one may need to invoke some symmetries broken above the GUT scale, but still those symmetries are not directly related to the θ_{13} value.

It was shown in a separate work [21] that the GUT scale mass relations relations $m_b \simeq m_\tau$ and $m_\mu \simeq -3m_s$ emerge out of a post-GUT S_4 flavor symmetry that determines the Yukawa texture of the $SO(10)$ GUT with $\mathbf{10} + \mathbf{126} + \mathbf{10'}$ Higgses while successfully giving TBM form of the neutrino mass matrix (or **126** Yukawa coupling matrix). As a consequence, in this model, no basis choice is required to get TBM form for the neutrino matrix. Since the same **126**-Yukawa coupling which gives the TBM neutrino mass matrix also contributes to charged fermion mass matrix, the model is tightly constrained and charged fermion mass fits lead to a prediction for θ_{13}. A detailed quantitative analysis of the model was carried out in [22] by supplementing the minimal model [21] by additional effective GUT scale Yukawa couplings that could follow from S_4 with a total of 12 parameters. Below we summarize the basic elements and results of this.

5. Details of the model

The effective Yukawa couplings f_0, h_0, h_0' corresponding to $H, \bar{\Delta}$ and H' Higgs, are assumed to have descended from a higher scale theory which has an S_4 symmetry broken by flavon fields with particularly aligned vacuum expectation values (vevs) (see e.g. [21]). We do not need to know the detailed form of these flavon interactions for our analysis, and we will simply write down the effective form of the h_0, h_0', f_0 that follow from it. The effective GUT scale Yukawa superpotential in this case is

$$\mathcal{L}_Y = h_0 \psi\psi H + f_0 \psi\psi\bar{\Delta} + h_0'\psi\psi\Sigma \tag{9}$$

which after symmetry breaking leads to the fermion mass formulae

$$\begin{aligned} M_u &= h + r_2 f + r_3 h', \\ M_d &= \frac{r_1}{\tan\beta}(h + f + h'), \\ M_\ell &= \frac{r_1}{\tan\beta}(h - 3f + h'), \end{aligned} \tag{10}$$

(11)

where we have absorbed the Higgs vevs into h, f, h'. The neutrino mass matrix is unaffected by this extension. Turning to the flavor structure of the various couplings, we note that S_4 symmetry [21] puts the neutrino mass matrix in a form that, upon diagonalization, leads to tri-bi-maximal mixing prior to charged lepton corrections i.e. an f_0 coupling in the form:

$$f_0 \propto \mathcal{M}_\nu = \kappa \begin{pmatrix} 0 & m_1 & m_1 \\ m_1 & m_0 & m_1 - m_0 \\ m_1 & m_1 - m_0 & m_0 \end{pmatrix}, \tag{12}$$

In [21], the S_4 symmetry constrains h_0 to be a rank one matrix of the form:

$$h = \begin{pmatrix} 0 & 0 & 0 \\ 0 & 0 & 0 \\ 0 & 0 & M \end{pmatrix}, \tag{13}$$

and we choose h' of the form [22]

$$h_0' = \begin{pmatrix} \delta'' & \delta & -\delta + \delta' \\ \delta & 0 & d \\ -\delta + \delta' & d & 0 \end{pmatrix}, \tag{14}$$

which can be also generated by choice of flavon fields and the alignment of their vevs. The model has eleven parameters if we choose all except δ' real (twelve parameters when we allow δ complex to study the allowed range of Dirac *CP* phase). Recall that the model with **10**, **126** and **120** has a total of seventeen parameters [18, 19]. The best fit values for the input parameters that leads to the minimum χ^2 is given in Table 1, and the resulting mass and mixing parameter values are given in Table 1. Note that while we get a higher value for m_b and slightly lower values for m_s and m_c, all are within reasonable statistical deviation from extrapolated values in [23]. Similarly, our predictions for m_d and m_u are somewhat higher than those obtained in [23], but we believe there could easily be instanton corrections to the light quark masses, which could change these extrapolated values. It is nevertheless remarkable that we are able to reproduce all other parameters in the charged fermion sector so well. In the figures 1 and 2 the

Table 1. Best fit values for the model parameters. Note that adding a small imaginary part to δ will give us a non-negligible Dirac *CP* phase [22].

	$\tan\beta = 10$	$\tan\beta = 55$
M (GeV)	88.2	106.2
m_0 (GeV)	1.435	1.382
m_1 (GeV)	0.275	0.275
δ (GeV)	0.285	0.2605
δ' (GeV)	0.4632 - 0.279$\imath$	0.529 - 0.335$\imath$
δ'' (GeV)	-0.0652	-0.0767
d (GeV)	3.78	4.31
$r_1/\tan\beta$	0.0153	0.01586
r_2	0.13	0.129
r_3	-0.06	-0.07

Table 2. The best fit values of the quark and charged lepton masses and the most relevant quark mixing parameters. The 1σ experimental values extrapolated by MSSM renormalization group (RG) equations to the GUT scale [23] are also shown for comparison.

	best fit	RG extrapolated	best fit	RG extrapolated value
m_e (MeV)	0.3587	$0.3585^{+0.0003}_{-0.0003}$	0.3563	$0.3565^{+0.0002}_{-0.0010}$
m_μ (MeV)	75.6865	$75.6715^{+0.0578}_{-0.0501}$	75.3359	$75.2938^{+0.0515}_{-0.1912}$
m_τ (GeV)	1.2927	$1.2922^{+0.0013}_{-0.0012}$	1.6272	$1.6292^{+0.0443}_{-0.0294}$
m_d (MeV)	3.8587	$1.5036^{+0.4235}_{-0.2304}$	4.0202	$1.4967^{+0.4157}_{-0.2278}$
m_s (MeV)	23.6026	$29.9454^{+4.3001}_{-4.5444}$	23.1619	$29.8135^{+4.1795}_{-4.4967}$
m_b (GeV)	1.3726	$1.0636^{+0.1414}_{-0.0865}$	1.7078	$1.4167^{+0.4803}_{-0.1944}$
m_u (MeV)	1.9772	$0.7238^{+0.1365}_{-0.1467}$	3.6311	$0.7244^{+0.1219}_{-0.1466}$
m_c (MeV)	177.3862	$210.3273^{+19.0036}_{-21.2264}$	177.6719	$210.5049^{+15.1077}_{-21.1538}$
m_t (GeV)	88.3886	$82.4333^{+30.2676}_{-14.7686}$	106.3806	$95.1486^{+69.2836}_{-20.6590}$
V_{us}	0.2230	0.2243 ± 0.0016	0.2233	0.2243 ± 0.0016
V_{ub}	0.0032	0.0032 ± 0.0005	0.0032	0.0032 ± 0.0005
V_{cb}	0.3489	0.0351 ± 0.0013	0.0352	0.0351 ± 0.0013
δ_{CKM}	-64.35°	$(-60 \pm 14)^\circ$	-61.84°	$(-60 \pm 14)^\circ$

correlation between the various θ_{13} predictions and other mixing angles such as θ_{12} or θ_3 are displayed. This is a way to test the model.

The exact numerical results for neutrino mixing corresponding to the quark sector fit from above are given in Figures 1-2 which show the relationships between $(\theta_{23}, \theta_{13})$ and $(\theta_{12}, \theta_{13})$, respectively. Note that the value of θ_{13} is large, though not so large as the 6°-8° central value of T2K result gives. Also note that the atmospheric mixing angle θ_{23} is always larger than the maximal value of 45°. These correlations between different mixing parameters could be used to test the model once the current uncertainties in both θ_{12} and θ_{23} are reduced and a more precise value for θ_{13} has been determined. Our model also predicts small Majorana phases ($\sim 1^\circ$). Other predictions are the normal mass hierarchy for neutrinos which is also a generic features of most GUT models and an effective mass in neutrinoless double beta decay in the few meV

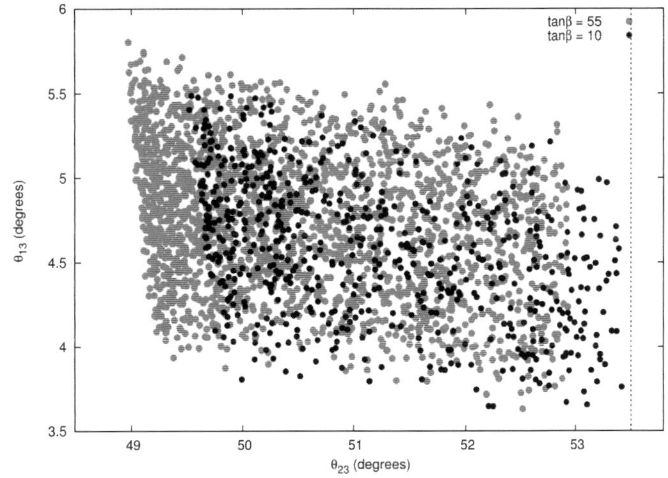

Figure 1. Correlation between θ_{13} and θ_{23} predicted in our model satisfying all the charged fermion sector constraints.

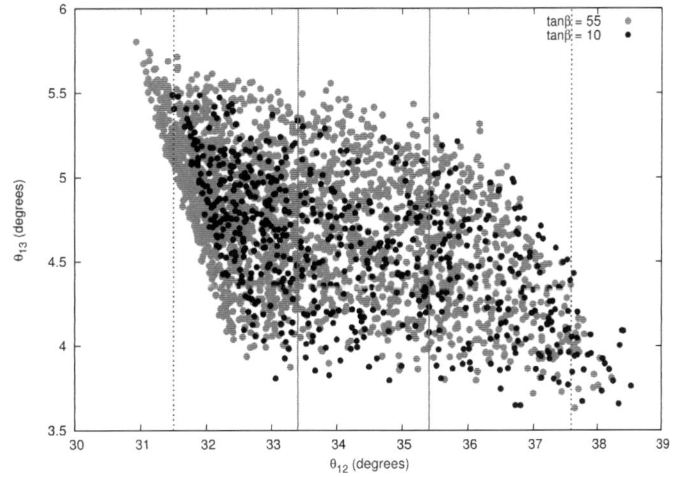

Figure 2. Correlation between θ_{13} and θ_{12} predicted in our model. The solid and dotted vertical lines are the current 1σ and 3σ limits respectively for the solar mixing angle.

range.

6. Summary

In this brief overview, an alternative to the symmetry approach to understanding the current neutrino observations is discussed based on the hypothesis of grand unification of forces and matter. A particular class of GUT models based on SO(10) with type II seesaw seems to explain several observations in a simple way with testable predictions. Furthermore, the type II dominance advocated here can also be demonstrated in SO(10) models, both supersymmetric [24] and non-supersymmetric [25]. Finally, all this discussion ignores the sterile neutrino possibility and will require drastic modification if they are confirmed. Mirror duplication may be one possibility which can accomodate the sterile neutrinos with least "damage" to the above predictions.

Acknowledgments
I thank Bhupal Dev, B. Dutta, Y. Mimura and M. Severson for collaboration and discussions. The present work is supported by the National Science Foundation grant number PHY-0968854.

References

[1] Abe K *et al.* (T2K Collaboration) *Preprint* arXiv:1106.2822 [hep-ex]
[2] Apollonio M *et al.* (CHOOZ Collaboration) 2003 *Eur. Phys. J.* C **27** 331-374 (*Preprint* hep-ex/0301017)
[3] MINOS website: `http://www-numi.fnal.gov/pr_plots/nue2011.pdf` June 2011
[4] Fogli G L, Lisi E, Marrone A, Palazzo A and Rotunno A M 2009 *Nucl. Phys. Proc. Suppl.* **188** 27-30 ; Gonzalez-Garcia M C, Maltoni M and Salvado J 2010 *JHEP* **1004** 056 (*Preprint* arXiv:1001.4524 [hep-ph])
[5] Weinberg S 1979 *Phys. Rev. Lett.* **43** 1566-1570
[6] Minkowski P 1977 *Phys. Lett.* B **77** 421; Gell-Mann M, Ramond P and Slansky R 1979 Supergravity ed Van Nieuwenhuizen F and Freedman D Z (Amsterdam: North Holland); Yanagida T 1979 *Proc. KEK workshop (unpublished)*; Glashow S L 1979 Cargese lectures; Mohapatra R N and Senjanović G 1980 *Phys. Rev. Lett.* **44** 912
[7] Lazarides G, Shafi Q and Wetterich C 1981 *Nucl. Phys.* B **181** 287; Mohapatra R N and Senjanović G 1981 *Phys. Rev.* D **23** 165; Schechter J and Valle J W F 1980 *Phys. Rev.* D **22** 2227
[8] Fukuyama T and Nishiura H *Preprint* hep-ph/9702253; Mohapatra R N and Nussinov S 1999 *Phys. Rev.* D **60** 013002 (*Preprint* hep-ph/9809415); Lam C S 2001 *Phys. Lett.* B **507** 214 (*Preprint* hep-ph/0104116); Grimus W *et. al.* 2005 *Nucl. Phys.* B **713** 151 (*Preprint* hep-ph/0408123); Mohapatra R N 2004 *JHEP* **0410** 027 (*Preprint* hep-ph/0408187); Lam C S 2005 *Phys. Rev.* D **71** 093001 (*Preprint* hep-ph/0503159); Kitabayashi T, Yasue M 2005 *Phys. Lett.* B **621** 133 (*Preprint* hep-ph/0504212); Mohapatra R N and Rodejohann W 2005 *Phys. Rev.* D **72** 053001 (*Preprint* hep-ph/0507312); Ge S F, He H J and Yin F R 2010 *JCAP* **1005** 017 (*Preprint* arXiv:1001.0940 [hep-ph])
[9] Harrison P F, Perkins D H and Scott W G 2002 *Phys. Lett.* B **530** 167 (*Preprint* hep-ph/0202074); Xing Z 2002 *Phys. Lett.* B **533** 85 (*Preprint* hep-ph/0204049); He X G and Zee A 2003 *Phys. Lett.* B **560** 87 (*Preprint* hep-ph/0302201); Wolfenstein L 1978 *Phys. Rev.* D **18** 958; Yamanaka Y, Sugawara H and Pakvasa S 1982 *Phys. Rev.* D **25** 1895 *Erratum.* 1984 *Phys. Rev.* D **29** 2135
[10] Lam C S 2007 *Phys. Lett.* B **656** 193-198 (*Preprint* arXiv:0708.3665 [hep-ph]);
Grimus W, Lavoura L and Ludl P O 2009 *J. Phys.* G **36** 115007 (*Preprint* arXiv:0906.2689 [hep-ph])
[11] Babu K S and Mohapatra R N 1993 *Phys. Rev. Lett.* **70** 2845 (*Preprint* hep-ph/9209215)
[12] Aulakh C S, Mohapatra R N 1983 *Phys. Rev.* D **28** 217;
Clark T E, Kuo T K and Nakagawa N 1982 *Phys. Lett.* B **115** 26
[13] Brahmachari B and Mohapatra R N 1998 *Phys. Rev.* D **58** 015001 (*Preprint* hep-ph/9710371)
[14] Bajc B, Senjanovic G and Vissani F 2003 *Phys. Rev. Lett.* **90** 051802 (*Preprint* hep-ph/0210207)
[15] Goh H S, Mohapatra R N and Ng S 2003 *Phys. Lett.* B **570** 215 (*Preprint* hep-ph/0303055);
Goh H S, Mohapatra R N and Ng S 2003 *Phys. Rev.* D **68** 115008 (*Preprint* hep-ph/0308197)
[16] Babu K S and Macesanu C 2005 *Phys. Rev.* D **72** 115003 (*Preprint* hep-ph/0505200);
Bertolini S, Frigerio M and Malinsky M 2004 *Phys. Rev.* D **70** 095002 (*Preprint* hep-ph/0406117);
Bertolini S, Schwetz T and Malinsky M 2006 *Phys. Rev.* D **73** 115012 (*Preprint* hep-ph/0605006);
Bertolini S, Malinsky M 2005 *Phys. Rev.* D **72** 055021 (*Preprint* hep-ph/0504241)
[17] Oshimo N 2002 *Phys. Rev.* D **66** 095010 (*Preprint* hep-ph/0206239);
Dutta B, Mimura Y and Mohapatra R N 2005 *Phys. Rev.* D **72** 075009 (*Preprint* hep-ph/0507319);
Yang W M and Wang Z G 2005 *Nucl. Phys.* B **707** 87 (*Preprint* hep-ph/0406221);
Aulakh C S and Garg S K *Preprint* hep-ph/0612021;
W. Grimus W and Kuhbock H 2007 *Eur. Phys. J.* C **51** 721-729 (*Preprint* hep-ph/0612132);
Joshipura A S, Kodrani B P and Patel K M 2009 *Phys. Rev.* D **79** 115017 (*Preprint* arXiv:0903.2161 [hep-ph])
[18] Dutta B, Mimura Y and Mohapatra R N 2005 *Phys. Rev.* D **72** 075009 (*Preprint* hep-ph/0507319)
[19] Altarelli G and Blankenburg G 2011 *JHEP* **1103** 133 (*Preprint* arXiv:1012.2697 [hep-ph])
[20] Joshipura A S and Patel K M *Preprint* arXiv:1105.5943 [hep-ph]
[21] Dutta B, Y. Mimura Y and Mohapatra R N 2010 *JHEP* **1005** 034 (*Preprint* arXiv:0911.2242 [hep-ph])
[22] Bhupal Dev P S, Mohapatra R N and Severson M 2011 *Phys. Rev.* D **84** 053005
[23] Das C R and Parida M K 2001 *Eur. Phys. J.* C **20** 121-137 (*Preprint* hep-ph/0010004)
[24] Goh H S, Mohapatra R N and Nasri S 2004 *Phys. Rev.* D **70** 075022 (*Preprint* hep-ph/0408139)
[25] Mohapatra R N and Parida M K *Preprint* arXiv:1109.2188 [hep-ph]

LAGUNA-LBNO: design of an underground neutrino observatory coupled to long baseline neutrino beams from CERN

André Rubbia
ETH Zurich, 101 Rämistrasse, CH-8092 Zurich, Switzerland
E-mail: andre.rubbia@cern.ch

Abstract. A new design study called LAGUNA-LBNO has been recently funded by the European Commission to further develop the design of a deep underground neutrino observatory for the study of neutrino oscillations at long baselines, the investigation of the Grand Unification of elementary forces and for the detection of known and unknown astrophysical sources of neutrinos. Building on the successful format and on the findings of the LAGUNA design study, LAGUNA-LBNO is wider in scope but also more focused, and is specifically considering Long Baseline Neutrino Oscillations (LBNO) with beams from CERN. Two far sites, Fréjus and Pyhäsalmi, will be considered with main priority.

1. Main physics goals and motivation

Multi-purpose underground neutrino detectors, like for instance SuperKamiokande [1], have been very successful in particle and astroparticle physics (see e.g. [2, 3] for recent updates), finding first evidence of physics beyond the Standard Model (SM). The feasibility in Europe of an underground neutrino observatory of even larger scale, able to significantly extend present bounds and explore presently inaccessible domains, has been thoroughly studied since several years in a design study called LAGUNA [4] (Large Apparatus studying Grand Unification and Neutrino Astrophysics). LAGUNA is presently one of the top priorities of the ASPERA European astroparticle roadmap [5].

If built, the LAGUNA observatory will provide a very rich physics program, bringing new scientific opportunities and discoveries, in particular:

(i) Coupled to long baseline neutrino beams, it will explore neutrino flavor oscillations with unprecedented sensitivities, complete the understanding of the mixing matrix, determine the neutrino mass hierarchy and explore the existence of CP violation in the leptonic sector *"to help solve the puzzle of the observed matter-antimatter asymmetry in the Universe"*.

(ii) It will test Grand Unification with the search for proton decay and n-$\bar{n}$ oscillations, pursuing one of the key ideas towards a more fundamental understanding of elementary particles. Grand Unified Theories (GUTs) [6] attempt to answer questions like *"what is the origin of the three forces between the known elementary particles described by the SM"* or *"are they unified at high energies, i.e. can they be derived from a single unified force"*? GUTs are supported by the quantized values of the electric charges of the SM particles, by the automatic cancellation of gauge anomalies, and by the running of the gauge couplings, which apparently cross at high energy, but experimental evidence for proton decay has not been observed.

(iii) Taking advantage of the low background environment, it will detect rare events *induced by known and unknown "neutrino messengers" from astrophysical objects to give information on processes happening in the Universe*, which cannot be studied otherwise. In particular, it would sense a large number of neutrinos emitted by exploding galactic and extragalactic type-II supernovae, allowing an accurate study of the mechanisms driving the explosion. It will also enable precision studies of other astrophysical or terrestrial sources of neutrinos like solar and atmospheric ones, and search for new sources of astrophysical neutrinos, like for example the diffuse neutrino background from relic supernovae or those produced in Dark Matter (WIMP) annihilation in the centre of the Sun or the Earth.

2. An underground observatory optimized for large liquid-target detectors

LAGUNA was organized as a 3-year long project funded by the European Commission (EC) to carry out underground sites investigations and develop a concept for a facility able to host the new underground neutrino observatory. It was primarily motivated by the fact that, although Europe currently has four national underground laboratories (at Boulby (UK), Canfranc (Spain), Gran Sasso (Italy), and Modane (France)), none of them is large enough to host a next-generation observatory. LAGUNA selected seven potential underground sites (Boulby (UK), Canfranc (Spain), Fréjus (France), Pyhäsalmi (Finland), Sieroczowice (Poland), Slanic (Romania), Umbria (Italy)) to study, and compared them in order to identify the scientifically and technically most appropriate and cost-effective strategy towards a large scale European neutrino observatory. Three detector options, following the concepts of GLACIER [7, 8], LENA [9], and MEMPHYS [10], were considered at each site. They are based on large surface-instrumented volume of liquids, respectively liquid argon (LAr), liquid scintillator (LSc) and water (WCD).

One of the key conclusion of LAGUNA is that all of the seven considered underground sites are in principle technically feasible, and able to host the desired types of detectors. The rock engineering work was performed by industrial partners with years of experience in various excavation projects completed in geological conditions nearly identical to those of the considered sites. The present engineering know-how combined with the possibilities offered by the modern technology allows to construct the required caverns in each of the chosen locations. While there are site- and detector specific differences in the excavation time and price, the cavern construction is not the dominant cost of the project. It will be only approximately 10–30% of the final price tag. An important conclusion in favor of all sites is that seismology is not an obstacle for the construction of large underground caverns and tanks. The support of the local governments is equally enthusiastic at all LAGUNA sites and also the accessibility is quite comparable. No environmental show-stoppers were found. With the reduced impact of all the other factors, site selection should be based on the physics arguments.

3. The new LAGUNA-LBNO design study

A new design study called LAGUNA-LBNO [11] has been recently funded by the EC. It is based on the successful concept and format of LAGUNA. It will further evaluate the findings of LAGUNA, and in particular, it will assess the underground construction of the large detectors, their commissioning, and the long-term operation of the facility. LAGUNA-LBNO will in addition specifically consider long baseline beams from CERN [12]. The scope of LAGUNA-LBNO is hence wider than the one of LAGUNA, and more funds have been allocated. In addition, the size of the consortium has also nearly doubled.

At the same time, LAGUNA-LBNO is more focused since two sites (instead of the seven of

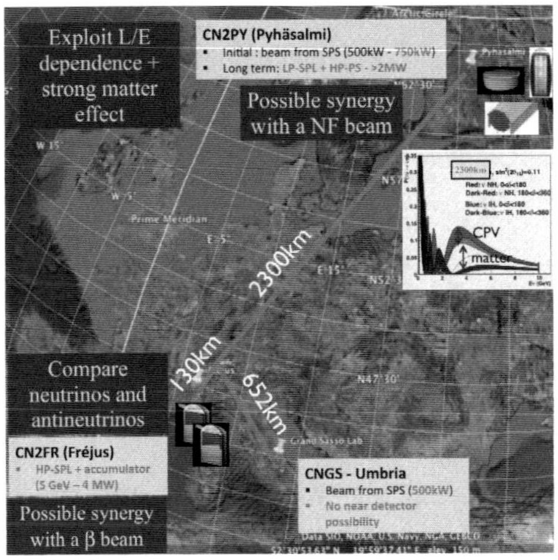

Figure 1. Thee two options considered as main focus in the LAGUNA-LBNO design studies: (1) the "shortest" CERN-Fréjus CN2FR baseline (2) the "longest" CERN-Pyhäsalmi CN2PY baseline. The CNGS baseline with a new "off-axis" site in Umbria is presently disfavored.

LAGUNA) will be developed[1]. From the seven pre-selected LAGUNA sites, the two deepest, Fréjus (4800 m.w.e.) and Pyhäsalmi (4000 m.w.e.), were found particularly attractive and deserved further attention. The study of long baseline neutrino oscillations is one of the main scientific goals, and LAGUNA-LBNO will develop an incremental path towards neutrino mass ordering determination and CP-violation discovery. As can be seen in Figure 1, the far sites provide two complementary baselines from CERN and the following scenarios are being considered:

(i) the CERN-Fréjus (CN2FR) baseline of 130 km with a large WCD detector coupled to a high-power low-energy superbeam from a 4 MW, 5 GeV HP-SPL proton driver;

(ii) the CERN-Pyhäsalmi (CN2PY) baseline of 2300 km, with a large LAr detector coupled to a high-energy conventional beam from protons accelerated with the SPS accelerator upgraded for LHC high luminosity and possibly eventually from a high-power 2 MW, 50 GeV HP-PS proton driver [12].

Both far locations offer excellent opportunities to include a large LSc detector, to enhance the physics program at the lowest energy range, in particular for solar, geo-neutrinos detection and short baseline oscillometry studies with artificial low energy neutrino sources.

4. Focusing on neutrino mass hierarchy and CP-violation at long baselines

From the broad particle and astroparticle physics program, neutrino oscillations at long baseline play a very important role, further exploring the flavor content of the SM. Especially attractive is the ultimate question of a new source of CP violation from the leptonic sector. Linear upgrades of existing long baseline oscillation facilities can find evidence for CP violation only in a limited part of the parameter space [14]. Therefore, new facilities like LAGUNA-LBNO are essential for a definitive exploration of CP violation and the determination of the neutrino mass hierarchy.

A first phase with a conventional neutrino beam is regarded as a mandatory step, especially to establish the far detector site. At conventional neutrino beams, searches for CP-violating effects are based on the electron appearance channels $\nu_\mu \to \nu_e$ and $\bar{\nu}_\mu \to \bar{\nu}_e$. Including higher order terms and matter effects, the $\nu_\mu \to \nu_e$ (resp. $(\nu_\mu \to \bar{\nu}_e)$) probabilities can be written

[1] All LAGUNA sites remain in principle available regardless of the LAGUNA-LBNO prioritization. In particular, a third solution is the continuation of CNGS (CNGS+) with a new "off-axis" site in an Italian region [13]. However, this option is disfavored because of the difficulties in locating a near detector and of the intrinsic power limitation of the facility.

as [15]: $P(\nu_\mu \to \nu_e)(\text{resp.}P(\bar{\nu}_\mu \to \bar{\nu}_e))) = P_0 \mp P_{\sin\delta} + P_{\cos\delta} + P_3$ where

$$
\begin{aligned}
P_0 &= \sin^2\theta_{23}\frac{\sin^2 2\theta_{13}}{(\hat{A}-1)^2}\sin^2((\hat{A}-1)\hat{\Delta}_{31}) \\
P_{\sin\delta} &= \alpha\,\frac{\sin\delta_{CP}\cos\theta_{13}\sin 2\theta_{12}\sin 2\theta_{13}\sin 2\theta_{23}}{\hat{A}(1-\hat{A})}\sin(\hat{\Delta}_{31})\sin(\hat{A}\hat{\Delta}_{31})\sin((1-\hat{A})\hat{\Delta}_{31}) \\
P_{\cos\delta} &= \alpha\,\frac{\cos\delta_{CP}\cos\theta_{13}\sin 2\theta_{12}\sin 2\theta_{13}\sin 2\theta_{23}}{\hat{A}(1-\hat{A})}\cos(\hat{\Delta}_{31})\sin(\hat{A}\hat{\Delta}_{31})\sin((1-\hat{A})\hat{\Delta}_{31}) \\
P_3 &= \alpha^2\frac{\cos^2\theta_{23}\sin^2 2\theta_{12}}{\hat{A}^2}\sin^2(\hat{A}\hat{\Delta}_{31})
\end{aligned}
$$

with $\Delta m^2_{21} = \alpha\Delta m^2_{31}$ and $\hat{A} = \pm A/\Delta m^2_{31} = \pm 2VE_\nu/\Delta m^2_{31}$. The phenomenology of CP is quite complex given its dependence of the oscillation probability on all oscillation parameters including solar and atmospheric sectors. Two main modes of investigations of CP-violation emerge when considering a conventional neutrino beam:

(i) to measure the CP asymmetry by comparing the appearance rates of neutrinos and antineutrinos; it requires both beam polarities with similar statistics which is more challenging for antineutrinos due to lower fluxes and cross-sections; it can be performed in a narrow band beam and experimentally requires a good control of systematic errors between neutrino and antineutrino runs;

(ii) to look for $\nu_e(\bar{\nu}_e)$ appearance and to measure its energy spectrum shape in a wide-band beam (WBB) at a given baseline L. Experimentally this measurement requires good energy resolution and low background systematics. The fit of the CP-phase is basically determined by the peak positions and height of the 1st and 2nd maximum and minimum. This method is sensitive to all values of δ_{CP} and allows determining the value of δ_{CP} using neutrinos only. However, degeneracies require the neutrino mass hierarchy to be determined beforehand. In a wide band beam and at very large distance where matter effects are significant, the hierarchy is determined unambiguously with an essentially equal sharing of neutrino and antineutrino running modes with much less statistics than needed to explore CP, so this can be achieved as an initial phase of the experiment.

Both methods are complementary for CP-determination, although the second provides a richer phenomenology.

In the LAGUNA context, the CN2FR option would allow the study of CP-violation with a precise comparison of neutrino and antineutrino oscillation rates. The CN2PY option would exploit the L/E dependence of the neutrino and antineutrino oscillations and would benefit from strong matter effect to rapidly determine the neutrino mass hierarchy.

It is important to note that the required size of the detector and integrated beam power, that are needed to discover CP-violation in the leptonic sector, are to first order independent of θ_{13} for large values of θ_{13}, but strongly depend on the true value of δ_{CP}. The physics reach of this kind of setup can be advantageously expressed as a function of the "exposure", which scales like the total active mass of the far detector and the integrated power (=number of protons times proton energy) on target. If Nature has chosen δ_{CP} close to 90° or 270°, there is a chance that significant evidence could be found with setups which are smaller in size than the ultimate detectors considered in LAGUNA. Therefore, the option to increase the scope of the project as a function of progressive findings on CP – the *incremental approach* – was strongly advocated in Ref. [16]. The LAGUNA-LBNO study will consider the physics reach achievable with "exposures" reachable in 5 to 10 years with existing accelerators at CERN as well as significantly larger "exposures" reachable with new accelerators such as the CERN HP-SPL [17] or HP-PS [12]. Conventional beams will eventually become limited by systematic uncertainties, assumed at present to be at the level of 5% or so. If the CP effect is small, e.g. if $\delta_{CP} < 15°$, evidence for CP violation could be beyond the reach of conventional beams.

In this case, the chosen LAGUNA-LBNO baselines are already optimal for upgrades towards the beta-beam [18] for the CN2FR and the neutrino factory [19] for the CN2PY, and upgrades towards one of these facilities could be envisaged. New ways to further explore CP violating effects and with higher precision would be potentially possible. The "megaton-scale" Water Cerenkov detector at Fréjus would already be adequate for the beta beam. In the case of a neutrino factory from CERN to Pyhäsalmi, magnetization of the far detector or a hybrid solution, expanding the fine grained and the magnetized muon detector, should be further considered.

5. Beam flux optimization and expected event rates

High-intensity neutrino fluxes as required for LAGUNA-LBNO require high-power proton sources. Prospects for such sources at CERN and their relevance for high intensity neutrino beams were discussed in Ref. [12]. The existing 400 GeV SPS accelerator, profiting from the LHC luminosity upgrades (LIU project), might be able to deliver up to 600–700 kW in a near future. Reaching megawatt beam power at the SPS would likely require significant replacement of its injector chains and a big stretch of its performance. A more appropriate solution which was advocated in Ref. [12] is to consider a new ring accelerator (HP-PS), with a proton energy in the range 30 to 50 GeV, designed for high power with, for instance, an intensity of 2.5×10^{14} ppp and a fast cycle of 1.2 seconds, yielding an average proton beam power of 1.6 MW. A conceptual design is presently being developed within LAGUNA-LBNO.

In LAGUNA, conventional beams to each of the seven considered far sites were analyzed, spanning baselines from 130 km up to 2300 km. In order to optimize the physics sensitivity of long baseline oscillations, the energy spectrum of the neutrino beams must be optimized for each distance to cover well the first and second maximum of the oscillation. For all possible LAGUNA baselines, this implies neutrinos in an energy range $0 \div 7$ GeV. The actual optimization depends on the chosen site and preliminary work was performed based on a full simulation of the neutrino beam line [20]. To produce the beam to shortest distance (130 km), the 4 MW, 5 GeV HP-SPL proton driver was assumed. For the other baselines, the 1.6 MW, 50 GeV HP-PS was considered. If we adopt a 200 days per calendar year, the HP-SPL delivers 56×10^{21} pot and the HP-PS yields integrated 3×10^{21} pot per year. See Table 1. The resulting fluxes are shown in Figure 2. The energies of the oscillation maximum for each baseline are indicated with vertical lines having the same color as the corresponding spectrum for $\Delta m^2_{31} = 2.45 \times 10^{-3}\,\mathrm{eV}^2$. The expected charged current event rates are listed in Table 1. They are normalized to a detector mass of 100 kton and a running time of one year. We note that contrary to naive expectation the event rates for ν_μ CC do not scale like $1/L^2$ since the neutrino focusing increases with the boost and the cross-section increases linearly with energy. The number of tau charged current events increases as well with distance, in particular for the 2300 km baseline the first oscillation maximum is above the tau production threshold.

6. Opportunities with the CERN-Pyhäsalmi (CN2PY) baseline

The Pyhäsalmi mine offers a unique infrastructure for underground access in an excellent condition, and could be readily transformed into a dedicated science facility once the mine closes, currently foreseen towards the end of this decade. The CN2PY baseline offers a very compelling long baseline physics program which can be implemented in a truly *incremental approach* as was shown in Ref. [16]. The initial phase is focused at determining the neutrino mass hierarchy and testing the maximal CP scenario. Owing to the very strong matter effects, the mass hierarchy can be determined with 100% certainty of the unknown δ_{CP} value. The

Table 1. Charged current event rates with optics optimized for the seven LAGUNA far sites (see text). Figures are normalized to a detector mass of 100 kton and a running time of one year corresponding to 3×10^{21} p.o.t. for the 50 GeV LP-SPL+HP-PS proton driver [12] and 56×10^{21} p.o.t. for the 4.5 GeV option.

baseline	neutrino polarity			antineutrino polarity		
km	ν_μ(no osc.)	ν_e(no osc.)	$\nu_\mu \to \nu_\tau$	$\bar{\nu}_\mu$(no osc.)	$\bar{\nu}_e$(no osc.)	$\bar{\nu}_\mu \to \bar{\nu}_\tau$
130	41316	174	–	5915	15	–
630	36844	486	28	13652	157	11
665	38815	516	28	14287	158	11
950	37844	349	40	14700	107	15
1050	51787	314	148	21728	88	65
1570	26785	174	170	11184	47	73
2300	17257	110	377	7577	32	172

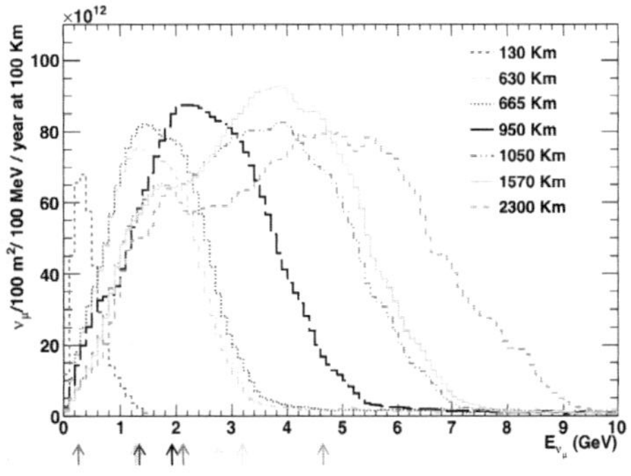

Figure 2. Neutrino fluxes at 100 km for the optimized setups. The energies of the oscillation maximum for each baseline are indicated with vertical lines having the same color as the corresponding spectrum. From Ref. [20].

CP violation parameter space will subsequently be further explored with increased "exposures", which could achieved by expanding the far detector mass and/or increasing the beam power.

Table 2 shows the assumed accelerator parameters, starting with the present SPS configuration, the expected SPS performance after the LIU upgrade, a kind of ultimate SPS++ performance, and finally the figures to be achieved with the HP-PS accelerator.

Table 2. Assumed parameters for different machine scenarios.

	SPS now	SPS+LIU	SPS++	LP-SPL+HP-PS [12]
Proton energy (GeV)	400	400	400	50
ppp	4.00×10^{13}	6.00×10^{13}	7.00×10^{13}	2.50×10^{14}
Tc (s)	6	6	6	1.2
Peak beam power (MW)	0.43	0.64	0.75	1.67
Global eff	0.85	0.85	0.85	0.85
Beam sharing	0.85	0.85	0.85	1
Running (d/year)	200	200	200	200
Npot/year	8.32×10^{19}	1.25×10^{20}	1.46×10^{20}	3.00×10^{21}
Npot equiv at 50 GeV	7.00×10^{20}	1.00×10^{21}	1.20×10^{21}	3.00×10^{21}

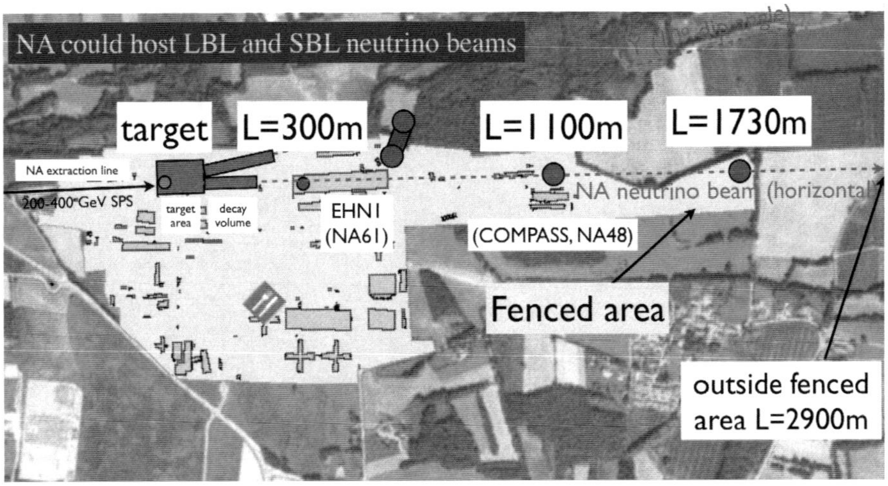

Figure 3. Neutrinos in the CERN North Area: the area could in principle host a facility with both long and short baseline beams.

7. An extension of the CERN North Area for neutrinos ?

Preliminary considerations show that the CERN North area could be the optimal location to host the CN2PY neutrino beam line facility. Protons with an energy of 400 GeV are presently slow extracted from the SPS to the North Area to drive the fixed target program. The proton beam can be extracted from the existing extraction line and horizontally bent north and downward by 10° in order to produce the beam directed at Pyhäsalmi. A preliminary layout foresees an underground near detector accessible from the North Area. See Figure 3. This configuration opens the possibility to have a synergy with a short baseline neutrino beam. The short baseline beam would require a dedicated target and an horizontal decay tunnel. Several detector locations can be identified at 300 m, 1100 m, 1730 m and possibly 2900 m (outside the CERN domain), as shown in the Figure. Further investigations are in progress.

8. Conclusions and outlook

The indications for $\sin^2 2\theta_{13} > 0.01$ have given new impulse to the design of next generation experiments with appropriate baselines and powerful conventional beams. In Europe, the LAGUNA-LBNO design study, a continuation of the LAGUNA project but with a wider scope and more focus than before, has started. The consortium has been enlarged and is aiming at proposing a "realistic plan" for a medium- and long-term European long baseline program, with large discovery potentials at each phase. The first goal is to determine the neutrino mass hierarchy (with the 2300 km baseline). Then, *incrementally* exploring the phase space, leading to CP-violation discovery. In parallel, ultimate searches for proton decay and interesting neutrino astrophysics measurements will also be possible. An expression of interest to the European Strategy Roadmap is foreseen. A graph illustrating a possible timescale for the different steps to develop the LAGUNA-LBNO CERN-Pyhäsalmi programme is shown in Figure 4.

Acknowledgements

We thank the organizers for inviting us to the very fruitful NUFACT11 workshop. We acknowledge useful discussions with I. Efthymiopoulos.

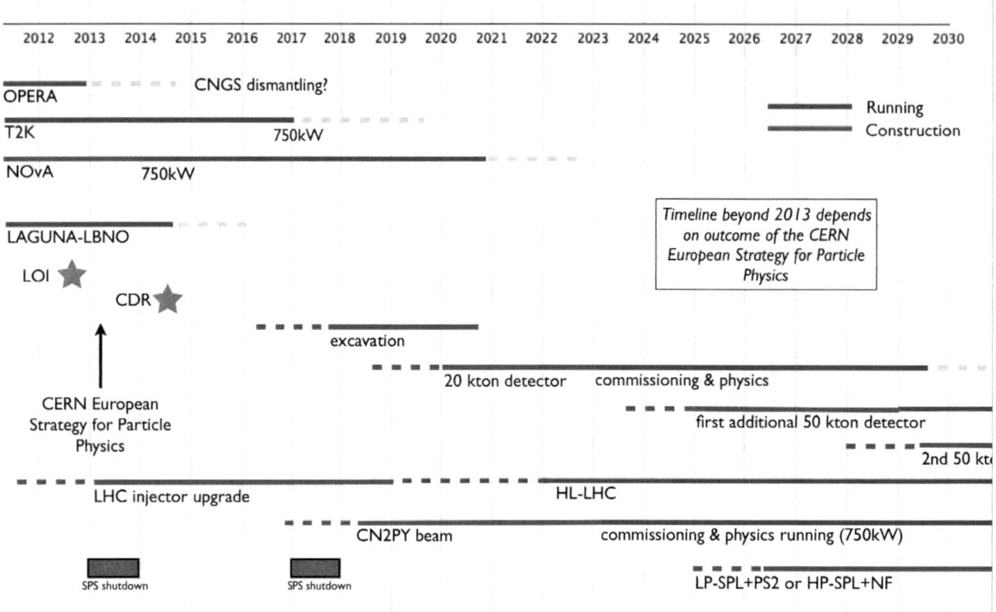

Figure 4. Timeline illustrating the milestones and phases to develop the complete LAGUNA-LBNO CERN-Pyhäsalmi programme.

References

[1] Fukuda Y *et al.* 2003 *Nucl. Instrum. Meth. A* **501** 418–462
[2] Abe K *et al.* (Super-Kamiokande Collaboration) 2011 *Phys. Rev. D* **83** 052010
[3] Kaji H (Super-Kamiokande Collaboration) 2011 *AIP Conf. Proc.* **1382** 100–102
[4] Rubbia A *et al.* (LAGUNA Collaboration) 2010 *Acta Phys. Polon. B* **41** 1727–1732
[5] URL `http://www.aspera-eu.org`
[6] Georgi H and Glashow S 1974 *Phys. Rev. Lett.* **32** 438–441
[7] Rubbia A 2004 *Experiments for CP violation: A Giant liquid argon scintillation, Cerenkov and charge imaging experiment?* (*Preprint* `hep-ph/0402110`)
[8] Rubbia A 2009 *J. Phys. Conf. Ser.* **171** 012020 (*Preprint* `arXiv:0908.1286[hep-ph]`)
[9] Oberauer L, Von Feilitzsch F and Potzel W 2005 *Nucl. Phys. Proc. Suppl.* **138** 108–111
[10] Borne J *et al.* 2011 *Nucl. Instrum. Meth. A* **639** 287–289
[11] LAGUNA-LBNO design study, FP7 Research Infrastructure "Design Studies" LAGUNA-LBNO (Grant Agreement No. 284518 FP7-INFRA-2011-1). `http://www.laguna-science.eu/`
[12] Rubbia A 2010 *A CERN-based high-intensity high-energy proton source for long baseline neutrino oscillation experiments with next-generation large underground detectors for proton decay searches and neutrino physics and astrophysics* (*Preprint* `arXiv:1003.1921[hep-ph]`)
[13] Meregaglia A and Rubbia A 2006 *JHEP* **0611** 032 (*Preprint* `hep-ph/0609106`)
[14] Huber P, Lindner M, Schwetz T and Winter W 2009 *JHEP* **0911** 044 (*Preprint* `arXiv:0907.1896[hep-ph]`)
[15] Freund M 2001 *Phys. Rev. D* **64** 053003 (*Preprint* `hep-ph/0103300`)
[16] Agarwalla S K, Li T and Rubbia A 2011 (*Preprint* `arXiv:1109.6526[hep-ph]`)
[17] Brunner O *et al.* 2009 *Phys. Rev. ST Accel. Beams* **12** 070402
[18] Benedikt M *et al.* 2011 *Eur. Phys. J. A* **47** 24
[19] Choubey S *et al.* (The IDS-NF Collaboration) 2011 Interim Design Report for the International Design Study for a Neutrino Factory
[20] Longhin A 2010 *PoS (ICHEP2010)* 325

Long baseline neutrino beams at Fermilab

S Childress and J Strait
Fermilab, Batavia, IL 60510, U.S.A.

E-mail: childress@fnal.gov, strait@fnal.gov

Abstract. Fermilab has had a very active long baseline neutrino program since initiation of the NuMI project in 1998. Commissioned in 2005, the NuMI beam with 400 kW design power has been in operation for the MINOS neutrino oscillation program since that time. Upgrade of NuMI to 700 kW for NOvA is now well advanced, with implementation of the beam upgrades to be accomplished in 2012-2013. Design development for the next generation LBNE neutrino beam is now a major ongoing effort. We report here salient features and constraints for each of these beams, as well as significant challenges both experienced and expected.

1. Introduction

The Fermilab long-baseline neutrino beams discussed here are of conventional design, but with far greater beam power than for historical neutrino beams. Accelerated protons strike a target to produce charged pions, which are magnetically focused, then decay to neutrinos in a following drift region. A hadron absorber is at the end of the decay region, backed by appropriate muon shielding prior to a near detector on the Fermilab site. Neutrinos then continue on a direction between the near detector and a far detector at a distance of hundreds of kilometres. To achieve the needed far detector neutrino event rates we must have very intense proton beams, challenging target and focusing systems, plus massive neutrino detectors. These very powerful conventional design neutrino beams at megawatt levels have become known as 'Super beams'.

Table 1. Fermilab long baseline neutrino beams.

	NuMI/MINOS	NuMI/NOvA	LBNE
Proton beam power	0.4 MW	0.7 MW	0.7 to 2.3 MW
Proton energy	120 GeV	120 GeV	60 to 120 GeV
Repetition rate	1.87 sec (design)	1.33 sec	1.33 sec
Protons per spill	$4.0x10^{13}$	$4.9x10^{13}$	$4.9x10^{13}$ to $1.6x10^{14}$
Baseline	735 km	810 km	1300 km

2. NuMI – Neutrinos at the Main Injector

The 400 kW NuMI beam was designed to enable precision measurement of muon neutrino oscillations by the MINOS experiment. Beam design parameters were determined before oscillation parameters were known, with tunability of the neutrino beam spectrum a key requirement.

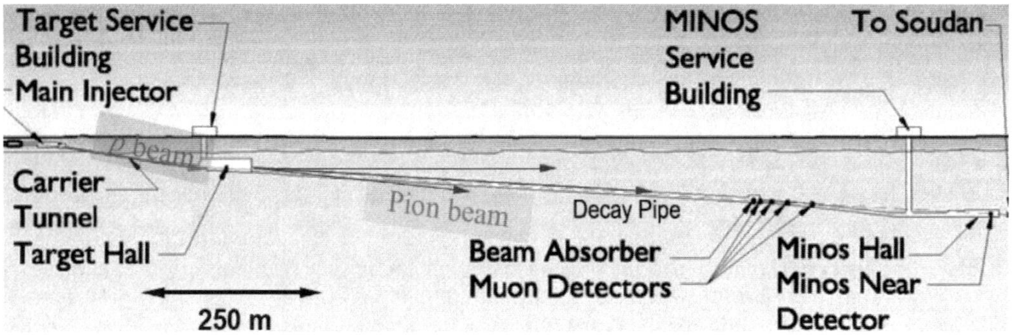

Figure 1. Profile view of NuMI beam layout.

A schematic view of the NuMI target, focusing horns and decay region is shown in figure 2. The 675 meter long decay allowed the possibility of a high energy neutrino beam when the horn separation was also increased by moving the second horn (reflector) downstream. With better knowledge of oscillation parameters prior to NuMI operation, the chosen operational configuration shown in the figure was for lower neutrino energy. Within this configuration a more limited range of neutrino energies peaked from 3 to 9 GeV is still feasible by moving the target longitudinally relative to the first horn, with the target significantly inside the horn for the normal low energy operation. This range of neutrino energies is shown in figure 3.

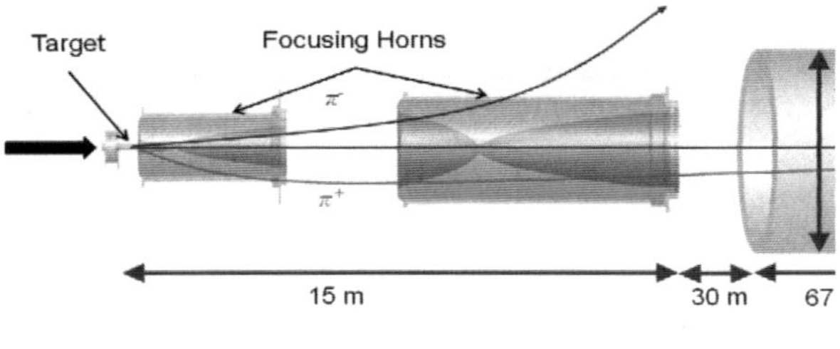

Figure 2. Focusing horns in neutrino mode collect positive pions while removing negatives. The resultant muon neutrino beam is 91.7% pure. An antineutrino beam is created by reversing the horn polarity.

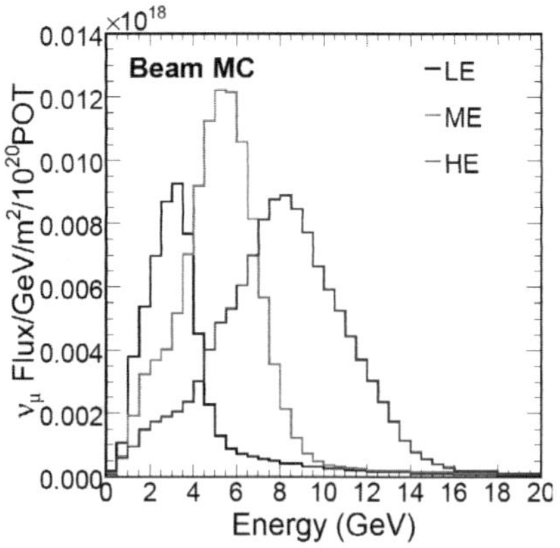

Figure 3. Range of neutrino energies achievable by moving target into (LE) or away from the first horn (ME, HE). Normal MINOS operation is in LE position, with average neutrino energy of ~ 3 GeV. ME and HE target position data is used for tuning beam flux models.

2.1 NuMI beam operation

The NuMI beam has been in operation for six years, with a considerable amount of experience gained during this process. Over this time, average beam power during operation has been ~ 250 kW, with a continuing increase in power. Recent operation has been at ~ 330 kW, with shorter periods near the design peak of 400 kW. A total of 1.3×10^{21} protons on target at 120 GeV have been accumulated with more than 50 million extracted beam pulses. Discussed below is our experience for some key NuMI systems.

2.2 Proton beam

A significant early concern with intense NuMI beam operation was for radiological protection of an aquifer region through which the unshielded primary beam tunnel passes prior to targeting in a deep target hall mined in the underlying dolomite rock. Another consideration was the need for rigorous protection against significant activation of the multi-ton magnets positioned on steep 100 to 150 milliradian slopes. Additionally, beam technical components can be damaged within a few seconds by an off trajectory proton beam.

Many improvements beyond previous beam operation were developed to address these issues. These include a comprehensive beam permit system which monitors status of more than 250 inputs prior to enabling each extracted NuMI beam pulse, open extraction and primary transport geometries to accept a range of extracted beam conditions, fully automated beam position control with no manual control of beam positions during operation, and high precision power supply regulation. In the tunnel radiation safety monitoring requires fractional beam loss for each extraction to be $< 10^{-5}$ with normal fractional loss at a few $\times 10^{-7}$ of the transported beam. All of this has worked extremely well, and going forward provides capability for precision control of megawatt proton beams.

2.3 Target hall systems

NuMI target hall components are subjected to very severe conditions, with resultant challenging learning experiences for both targets and focusing horns during our high intensity beam operation. Early failures with both systems prior to the readiness of spares necessitated the development of efficient repair capabilities for the highly activated components. These capabilities have enabled overall efficient NuMI beam operation over a period of years in spite of many challenges.

Two failure modes have been experienced with the focusing horns. Original design electrical isolators for each horn water line had a Kovar interface connecting ceramic and stainless steel. These interfaces were prone to failure after millions of horn pulsings, with several failures during beam operation. New design replacement isolators use a pressure fit seal directly between ceramic and stainless steel, which has been problem free. A second type of horn system failure was experienced at the strip-line power feed, where high-strength steel washers failed due to hydrogen embrittlement. High-strength steel is no longer used for applications inside the target chase. Experience with NuMI targets is shown in figure 4.

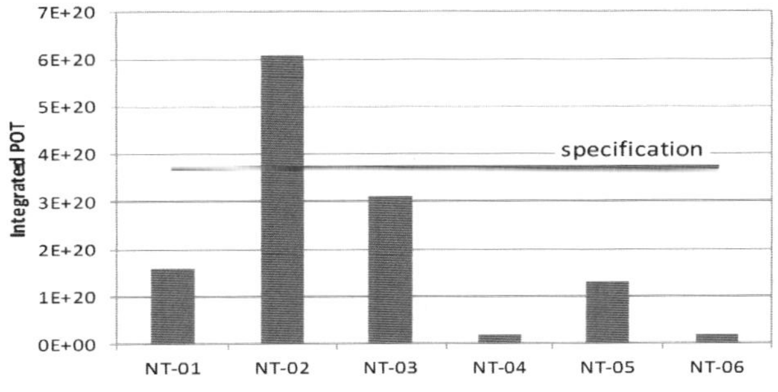

Figure 4. Integrated protons on target for the six NuMI graphite targets used between 2005 and 2011. The horizontal line indicates the target design lifetime specification of 3.7×10^{20} POT.

The second target NT-02 was problem free, and used for three years of NuMI beam operation. In later stages a progressive reduction in neutrino flux to 15% was seen, which we attribute to graphite degradation from the $6x10^{20}$ POT. The other 5 NuMI targets experienced failures related to cooling water leaks, with the worst experience for the last 3 targets produced. An extensive effort has evaluated the target weaknesses seen, with a number of improvements implemented in new production targets. Early operating experience with target NT-07 is positive.

Shown in figure 5 is a drawing of the target inserted into the first horn when in the normal low energy target position. Very tight spatial constraints exist with need for a low mass target system to minimize absorption of the low energy pions important for the desired neutrino spectrum.

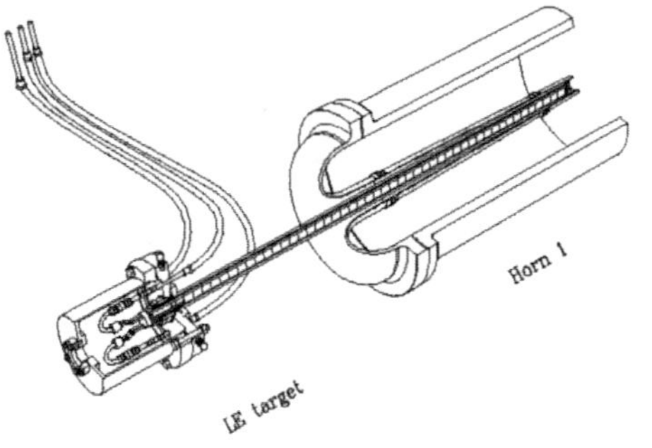

Figure 5. NuMI target inserted into horn when in normal low energy position.

2.4 Decay region

The 675 meter NuMI decay volume was designed as a vacuum system with a thin upstream aluminum window to minimize pion absorption. After corrosion was discovered on the window at the location impacted by protons not interacting in the target, a change was made to fill the decay region with helium. This reduces the neutrino flux by a few percent, but greatly lessens pressure gradient across the window.

2.5 Tritium mitigation

Along with target reliability, this has been the most significant issue for sustained NuMI beam operation. The deep NuMI beam enclosure system was designed to let some ground water enter the tunnel complex over its full length. While initial inlet flow rates were higher, the current rate of 500 liters/minute is still substantial. This water is then pumped to the surface, and into the laboratory cooling ponds system. Designing for water inflow provided a significant cost advantage for the facility construction, and also insured that water near the enclosure perimeter would be collected and released as surface water where allowable tritium levels are significantly higher than in the underlying aquifer.

NuMI produces a few hundred Curies per year of tritium in hadron showers, which is initially imbedded in the radiation shielding. But significant amounts of this tritium can become highly mobile in the presence of moist air. For the NuMI beam, a dominant source of tritium in water pumped from the tunnel was determined to have been produced in shielding surrounding the target hall chase, with the hadron absorber shield providing a smaller source. Humid air passing through these shields, in time came in contact with tunnel water outside the shields.

A significant dehumidification system has been added to intercept the majority of tritium prior to precipitation into NuMI sump waters. This system has been augmented as beam power has increased, with additional mitigation efforts being evaluated. An additional significant advantage of reduced humidity in the target hall chase is less corrosion for target hall components.

3. NOvA- Neutrino Experiment

Upgrade of the Fermilab accelerator complex and NuMI to 700 kW beam power is currently ongoing for the NOvA neutrino experiment. In parallel with the beam efforts, a new 15 kTon detector designed for precision measurement of oscillations by muon neutrinos to electron neutrinos is being constructed in northern Minnesota just south of the U.S.-Canada border. Initial NOvA beam is expected in 2013.

The NOvA far detector is off axis from the center trajectory of the NuMI beam by 14 milliradians, generating a narrower neutrino energy spectrum peaked at 2 GeV, as shown in figure 6.

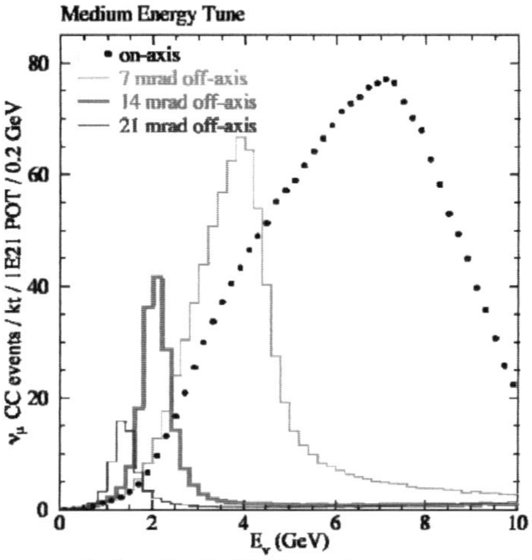

Figure 6. Effect on neutrino energy spectrum for progressively larger off axis angles. The 14 mrad spectrum peaked at 2 GeV is ideal for oscillation sensitivity at the NOvA detector baseline length of 810 km.

Accelerator improvements for NOvA take advantage of the 8 GeV Recycler storage ring, now available after completion of the Tevatron collider program. Twelve Booster accelerator batches will be injected into the Recycler, and then slip stacked into six batches for single turn extraction to the Main Injector accelerator. As this injection process occurs in parallel with the previous MI acceleration cycle, a major reduction in the cycle time between 120 GeV proton extractions to NOvA is possible. Most of the beam power increase is due to this reduction to 1.33 seconds cycle time. An additional intensity increase to 4.9×10^{13} protons per cycle is also feasible, giving 700 kW beam power to NOvA at 120 GeV.

NuMI primary beam upgrades include power system improvements to enable the 1.33 second cycle, improved precision for major dipole regulation, and more robust magnets for higher current quadrupoles. For the target hall, a significantly more robust target design is made feasible with the target no longer constrained to fit inside the first horn. Horn design is improved to reduce beam heating effects, target chase cooling is augmented, and the reflector horn is positioned further downstream enabling the optimal neutrino spectrum at the off axis NOvA detector. The on axis MINOS far detector will now see a higher energy spectrum as is also shown in figure 6. The redesigned NOvA target is shown in figure 7.

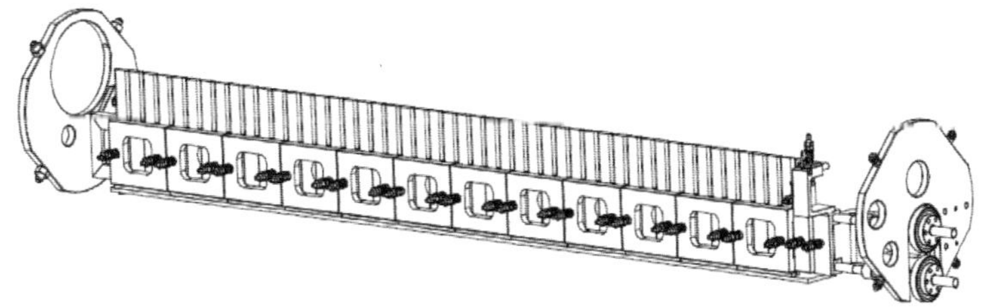

Figure 7. NOvA target. Beam passes through the graphite fins in upper part of the target canister.

4. LBNE – Long Baseline Neutrino Experiment

Physics goals for the next U.S long baseline neutrino oscillation experiment LBNE require a new beamline with capabilities not attainable with the existing NuMI/NOvA beam:

- Longer baseline to separate CP and matter effects, with 1300 km an optimal length.
- Beam power capability upgradeable to >2 MW when available with Project X beam.
- Beam optimized for lower energy and smaller electron neutrino component, obtainable with a shorter, wider decay pipe.
- Broad band beam covering 1st and 2nd oscillation maxima (2.5 and 0.8 GeV), with minimal high energy tail above ~ 5 GeV.

LBNE beam is designed to aim toward the Homestake mine in South Dakota, the site chosen by a national search process for the optimal far detector location. Beam trajectory is shown in figure 8.

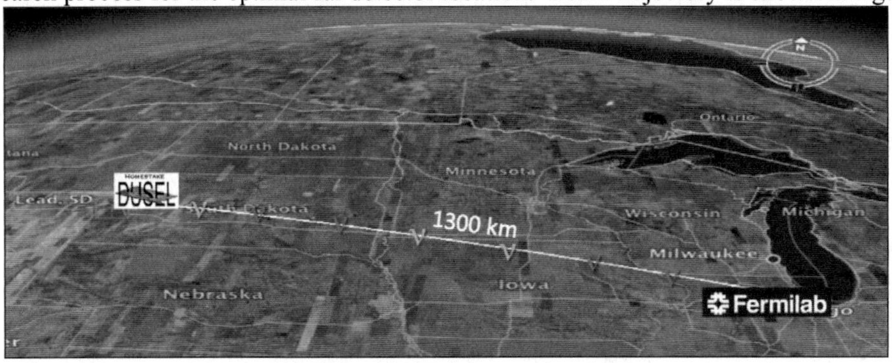

Figure 8. LBNE beam trajectory.

Other high level requirements for the LBNE beam are that initial beam power will be 700 kW available prior to Project X, with > 2 MW capability established for critical systems not readily upgradeable. Proton beam energy should be from 60–120 GeV. Beam design must meet stringent radiation safety requirements, and the LBNE design also should meet needed technical requirements while minimizing beam and facility cost.

We have evaluated two options for proton beam extraction from the Main Injector accelerator, the MI-60 northward extraction currently used for NuMI/NOvA and a westward MI-10 extraction requiring minimal additional bending for the trajectory toward South Dakota. Either option has good technical feasibility for the extraction design. MI-10 extraction is significantly more cost effective, but has more limited available space on the laboratory site for the full beam facility.

We have also evaluated two fundamentally different beam facility placements with either a deep mined beam facility as for NuMI, or with proton beam targeting above natural grade level. These two options are shown in figures 9 and 10.

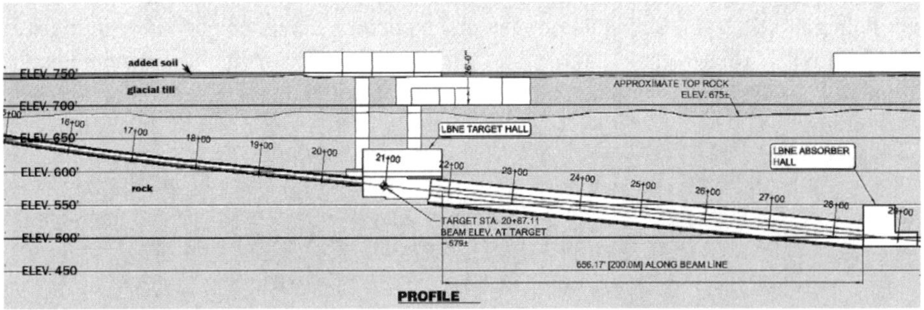

Figure 9. Profile view of a deep LBNE beam facility target hall and decay region.

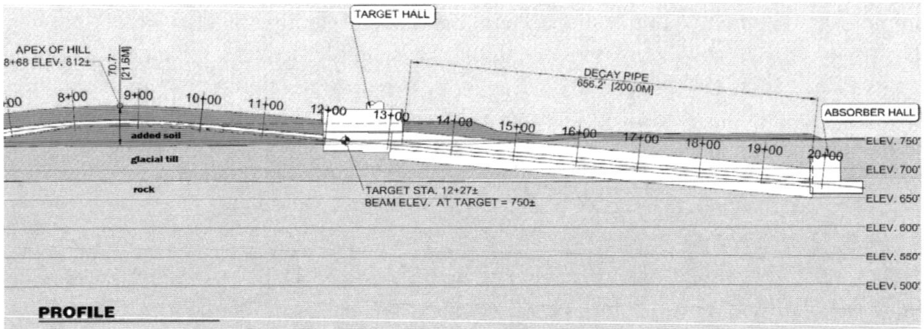

Figure 10. Profile view of a partially above grade LBNE beam facility target hall and decay region.

In each figure is shown the glacial till compacted clay region which extends ~ 20 meters below grade level, and then the underlying dolomite rock. For the above grade facility added soil above natural grade level is also shown.

For the deep facility option, placement of the target hall is chosen by reaching sufficient depth to enable a structural rock cover above the hall equal to the span of the hall. The facility is then mined in the supporting rock, as was done for NuMI construction. Required length of the proton beamline to reach this targeting depth precludes a 120 GeV proton beam with MI-10 extraction due to the limited site space, but is feasible for 60-90 GeV beams. With MI-60 beam extraction, ample site space is available. A possible advantage of the deep beam facility is that many features such as shielding and handling techniques are more readily understood due to similarity with NuMI.

For the partially above grade facility option, target hall placement is chosen to balance construction requirements for the elevated proton beam versus decay region and absorber hall located partially in rock. The only deep mined construction needed now at the Fermilab site is for the near detector hall (not shown here). The significantly shorter proton beamline now enables full energy range of 60-120 GeV with MI-10 extraction. To maintain needed component position stability, a key requirement with the above grade beam is for footing supports to underlying rock for the elevated proton beam enclosure and upstream decay region. A number of features for this beam facility design are quite different from previous site construction in the glacial till or from recent NuMI experience. Hence, there is the added requirement of a more challenging design learning curve. However, this beam facility option has two possibly compelling advantages, significantly reduced cost and major improvement for tritium mitigation.

4.1 LBNE beam technical challenges

Initial LBNE beam power of 700 kW is the same as we will have for NOvA beam operation, although some technical challenges as discussed below will be significant. Key features of LBNE beam designs are the need to enable the capability for a significant beam power upgrade when available with Project X and also to build beam systems and facility which can function with high efficiency for decades of operation. Discussed below are some challenges for key LBNE systems.

4.2 Proton beam.

Although many features for the LBNE proton beam are very different for both the two beam extraction options and also for the facility design options, the differences are ones of detail and are not fundamental. Technical solutions should be of similar difficulty regardless of selection choices, as also challenges of beam operation and control for ultimate LBNE beam power. Systems developed for NuMI, with iterative refinements underway for NOvA can be expected to work well for either LBNE proton beam.

As an example, for a deep LBNE proton beam ground water protection against loss of the very intense beam is an obvious challenge, while personnel shielding is simple with the large enclosure overburden. For an above grade beam ground water protection is easy, while significant attention is needed to insure appropriate personnel shielding. Component residual activation protection requirements are severe but similar for either design. In either case, rigorous control of the intense proton beam for every beam extraction, transport and targeting enables readily achievable solutions.

4.2 Target hall systems

The need for a relatively broad band low energy neutrino beam leads to design of an on-axis beam with target inside the first horn as for NuMI. Hence, we have the challenges of the NuMI target design, but for much higher beam power. This design challenge is well appreciated, and efforts are ongoing on several fronts including target material beam tests and design coordination with international experts. Experience gained with the NuMI target modifications and upcoming NOvA operation, as well as LBNE 700 kW operation should be productive in understanding good solutions for > 2MW LBNE beam.

A lesson from NuMI experience which is being incorporated into LBNE target hall design is the need for significantly improved remote handling capability with activated components, as well as improved time turn around for component repair or replacement. At the highest beam power, there will be many challenges with technical component design for target hall components and instrumentation in highly activated areas. Again, experience from 700 kW beam operation should aid significantly in understanding design upgrades which may be needed.

4.4 Tritium mitigation

Based on NuMI experience, the most important design features for tritium mitigation are to minimize humidity inside the target chase, decay and absorber shield, and to avoid mixing air release from these spaces with ground water sources. Addressing these issues in the LBNE facility design process is essential.

Acknowledgments

The authors wish to acknowledge inputs and assistance from many colleagues working with NuMI, NOvA and LBNE as well as the Fermilab accelerator systems

Fermilab is operated by Fermi Research Alliance, LLC under Contract No. DE-AC02-07CH11359 with the United States Department of Energy.

References

[1] The NuMI Project 2002 *NuMI Technical Design Handbook* (Fermilab)
[2] The MINOS Collaboration 1998 *MINOS Technical Design Report* (Fermilab)
[3] The NOvA Collaboration 2007 *NuMI Off-Axis ν_e Appearance Experiment – Technical Design Report* (Fermilab)
[4] Kovar D 2009 *Mission Need Statement for a Long Baseline Neutrino Experiment LBNE* (Office of High Energy Physics, Office of Science)
[5] Holmes S 2010 *Project X – Reference Design Report* (Fermilab)

Review of reactor antineutrino experiments

Zelimir Djurcic

Argonne National Laboratory, 9700 South Cass Avenue, Argonne, IL 60439, USA

E-mail: zdjurcic@hep.anl.gov

Abstract. As discussed elsewhere, the measurement of a non-zero value for θ_{13} would open up a wide range of possibilities to explore CP-violation and the mass hierarchy. Experimental methods to measure currently the unknown mixing angle θ_{13} include accelerator searches for the ν_e appearance and precise measurements of reactor antineutrino disappearance. The reactor antineutrino experiments are designed to search for a non-vanishing mixing angle θ_{13} with unprecedented sensitivity. This document describes current reactor antineutrino experiments and synergy between accelerator searches for the ν_e appearance and precise measurements of reactor antineutrino disappearance.

1. Introduction

Recent years have seen enormous progress in the physics of neutrino mixing, but several critical questions remain: What is the value of θ_{13}, the last unmeasured mixing angle in the neutrino mixing matrix? What is the mass hierarchy? Do neutrino oscillations violate CP symmetry? Why are the quark and neutrino mixing matrices so different?

Answering the questions listed above and understanding CP-violation in particular are among the most important priorities within the physics community [1]. A major motivation to search for CP-violation in neutrino oscillations is that its observation would make it more likely that the baryon-antibaryon asymmetry of the universe arose through leptogenesis [2]. The theory of leptogenesis is linked to the see-saw theory and as a consequence the light neutrinos are Majorana and have GUT-scale partners. Then the matter-antimatter asymmetry of the universe may be explained through CP-violating decays of the heavy partners, producing a state with unequal numbers of Standard Model leptons and antileptons. Furthermore, the Standard Model processes convert such a state into the world around us with an unequal number of baryons and antibaryons. It is thought that CP-violation would be very unlikely to appear in the heavy sector without happening in light neutrinos [3].

The value of θ_{13} is central to each of these questions. Measurement of non-zero value of θ_{13} would open a wide range of possibilities to explore CP-violation and the mass hierarchy. In addition, the size of θ_{13} with respect to the other mixing angles may give insights into the origin of these angles and the source of neutrino mass.

Experimental methods to measure the currently unknown mixing angle θ_{13} include accelerator searches for ν_e appearance and precise measurements of reactor antineutrino disappearance. In long-baseline appearance experiments, we search for the appearance of electron neutrinos in a fairly pure muon neutrino beam as a function of distance L from the neutrino source to the detector and the neutrino energy E. In such a case the neutrino oscillation probability may be expressed as [4]

$$
\begin{aligned}
P(\nu_\mu \to \nu_e) &= \sin^2\theta_{23}\,\sin^2 2\theta_{13}\,\frac{\sin^2(\Delta_{31}-aL)}{(\Delta_{31}-aL)^2}\,\Delta_{31}^2 \\
&+\cos\delta\,\sin 2\theta_{23}\,\sin 2\theta_{12}\,\sin 2\theta_{13}\,\cos\Delta_{32}\Big(\frac{\sin(\Delta_{31}-aL)}{(\Delta_{31}-aL)}\Delta_{31}\Big)\,\Big(\frac{\sin(aL)}{(aL)}\Delta_{21}\Big) \\
&-\sin\delta\,\sin 2\theta_{23}\,\sin 2\theta_{12}\,\sin 2\theta_{13}\,\sin\Delta_{32}\Big(\frac{\sin(\Delta_{31}-aL)}{(\Delta_{31}-aL)}\Delta_{31}\Big)\,\Big(\frac{\sin(aL)}{(aL)}\Delta_{21}\Big) \\
&+\cos^2\theta_{23}\,\sin^2 2\theta_{12}\,\frac{\sin^2(aL)}{(aL)^2}\,\Delta_{21}^2, \qquad (1)
\end{aligned}
$$

where Δ_{jk} stands for the kinematic phase, $\Delta m_{jk}^2 L/4E$. Matter effects are described with $a \equiv G_F N_e/\sqrt{2}$ which is approximately $(3500\ km)^{-1}$.

δ_{CP} is the CP-violating phase.

The first (last) term in Eq.(1) represents the atmospheric (solar) probabilities and the middle terms describe the interference between them. It is clear that θ_{13} may be probed by measuring electron neutrino appearance from accelerator produced muon neutrinos. We need the values L and E such that interference between solar and atmospheric scales can be seen. Currently there are two long-baseline experiments running: T2K [5] and MINOS [6]. Both recently reported indications of $\nu_\mu \to \nu_e$ appearance. A recent global analysis [7] with all available data including T2K and MINOS, provided >3 sigma evidence for nonzero θ_{13}. If θ_{13} is confirmed not to be too small the long-baseline oscillation experiments will have sensitivity for measuring the CP-violating phase. However, the oscillation probability is complicated and dependent not only on θ_{13} but also on

- CP-violation parameter (δ_{CP}),
- Mass hierarchy (sign of Δm_{31}^2), and
- Size of $\sin^2\theta_{23}$.

at the same time. Therefore any attempt to measure CP-violation and the mass hierarchy would be greatly simplified if θ_{13} was measured independently.

A clean measurement of θ_{13} can be performed by observing the disappearance of electron antineutrinos. In general, the electron antineutrino survival probability is given by

$$
\begin{aligned}
P(\bar\nu_e \to \bar\nu_e) &= 1-\cos^4\theta_{13}\,\sin^2 2\theta_{12}\,\sin^2\Delta_{21} \\
&- \sin^2 2\theta_{13}(\cos^2\theta_{12}\,\sin^2\Delta_{31}+\sin^2\theta_{12}\,\sin^2\Delta_{32}), \qquad (2)
\end{aligned}
$$

which further simplifies to

$$P(\bar\nu_e \to \bar\nu_e) \approx 1-\sin^2 2\theta_{13}\,\sin^2\frac{\Delta m_{31}^2 L}{4E}, \qquad (3)$$

at distances 1-2 km from a reactor source. Eq. (3) shows that the reactor experiments hold the promise of unambiguously determining the θ_{13} mixing angle. The only additional input needed is the well measured value of $\Delta m_{31}^2 \approx \Delta m_{atm}^2 = (2.32^{+0.12}_{-0.08}) \times 10^{-3}\ eV^2$ [8]. As described elsewhere [9], electron antineutrinos from large commercial nuclear reactors are playing an important role in the exploration of neutrino oscillations. The choice of distance between source and detector allows one to conveniently tune an experiment's sensitivity to either the atmospheric or solar neutrino mass splitting. KamLAND [10, 11, 12, 13] has established a connection between the MSW effect in the sun [14] and vacuum oscillations with antineutrinos, and has provided the most accurate measurement of Δm_{21}^2 to date. Earlier, CHOOZ [15] and Palo Verde [16] provided what are still the best upper limits on the mixing angle θ_{13}. Antineutrino detectors at very short distance are being explored with the purpose of detecting coherent neutrino-nucleon scattering [18], and for testing plutonium diversion at commercial reactors [19].

2. Reactor oscillation searches for θ_{13}

The reactor antineutrino experiments are designed to search for a non-vanishing mixing angle θ_{13} with unprecedented sensitivity. The measurement of θ_{13} will be conducted by searching for an apparent disappearance of electron antineutrinos in an intense flux from nuclear reactor cores available to the experiments. The current best individual constraint on the third mixing angle comes from the CHOOZ [15] reactor neutrino experiment $\sin^2 2\theta_{13} < 0.15$ at 90% C.L., with $\Delta m^2_{atm} = 2.5 \times 10^{-3}$ eV^2. The improvement of the CHOOZ result requires an increase in the statistics, a reduction of the systematic error below one percent, and a careful control of the backgrounds. The new generation of reactor oscillation searches will use two or more identical detectors to explore the range of $\sin^2 2\theta_{13}$ down to ~ 0.01.

Except for the case of coherent neutrino-nucleon scattering, which will not be addressed further here, reactor antineutrinos are usually detected using the inverse-β reaction $\bar{\nu}_e + p \to n + e^+$ whose correlated signature helps reduce backgrounds but is only sensitive to $\bar{\nu}_e$s with energy above 1.8 MeV. The kinetic energy of the positron gives a measure of the incoming $\bar{\nu}_e$ energy. Neutrino oscillation parameters may be extracted from the data by fitting the observed $\bar{\nu}_e$ spectrum $\frac{dn_\nu}{dE_\nu}$ to the following equation:

$$\frac{dn_\nu}{dE_\nu} = \sum_k^{\text{reactors}} N_p \epsilon(E_\nu)\sigma(E_\nu)\frac{P_{ee}(E_\nu, L_k)}{4\pi L_k^2} S_k(E_\nu), \quad \text{with } S(E_\nu) = \sum_i^{\text{isotopes}} f_i \left(\frac{dN_{\nu i}}{dE_\nu}\right), \tag{4}$$

Here, N_p is the number of target protons, $\epsilon(E_\nu)$ is the energy-dependent detection efficiency, $\sigma(E_\nu)$ is the detection cross-section, $P_{ee}(E_\nu, L_k)$ is the oscillation survival probability for $\bar{\nu}_e$s traveling a distance L_k from reactor k to the detector, and $S_k(E_\nu)$ is the $\bar{\nu}_e$ spectrum emitted by reactor k. $S_k(E_\nu)$ is usually expressed in form of $\frac{dN_{\nu i}}{dE_\nu}$ the $\bar{\nu}_e$ emission spectrum per fission of isotope i, and f_i the number of fissions of isotope i during the data taking period. Since >99.9% of the energy and $\bar{\nu}_e$-producing fissions are from ^{235}U, ^{238}U, ^{239}Pu, and ^{241}Pu [21], the summation in Eq.(4) is performed over just these 4 isotopes. The time variation of the flux may be included in the fit to discriminate the time-varying reactor signal from constant backgrounds. Eq.(4) should technically be multiplied by the detector resolution function and integrated over E_ν.

Nuclear power reactors operate on the principle that the fission of U and Pu isotopes and the subsequent decays of their daughter fragments release energy, generating heat. Large Q-value β-decays of unstable fission fragments are primarily responsible for the $\bar{\nu}_e$ emission of nuclear reactors. A typical 3.8 GW reactor emits $\sim 7 \times 10^{20}$ $\bar{\nu}_e/s$. Such a source would give about 1500 $\bar{\nu}_e$ interactions per year in a 1 ton detector at 1100 m distance. Decays of long-lived isotopes in the nuclear fuel and in spent fuel elements stored at the reactor site contribute at the sub-percent level to the detected $\bar{\nu}_e$ flux; this contribution has been treated elsewhere [20]. For ^{235}U, ^{239}Pu, and ^{241}Pu, the emitted $\bar{\nu}_e$ spectra $\frac{dN_{\nu_i}}{dE_\nu}$ are derived from β-spectrum measurements of the fissioning of the isotopes at ILL by thermal neutrons [22, 23, 24]. For ^{238}U, no measurements are available, so theoretical calculations of its $\bar{\nu}_e$ emission must be used [25]. Such predictions of the antineutrino flux were considered reliable because experimental observation agreed with the prediction at the 2-3% level, thus supporting nuclear reactors as "calibrated sources" of antineutrinos [9]. A recent improved derivation of antineutrino spectrum described in [26, 27], based on the ILL reference β-spectrum measurements, and updated calculation of the ^{238}U spectrum [27] resulted in a $\sim$2 σ disagreement with the data from the short-baseline neutrino measurements [28]. The data is lower than expectation and consistent with a number of hints suggesting a possible oscillation to sterile neutrinos in the $\Delta m^2 \sim 1$ eV^2 region [29].

Equally important for the understanding of the predicted antineutrino flux is a proper understanding of the systematic effects involved in the modeling of the reactor antineutrino flux and energy spectrum. The sources of systematics (thermal power, boron concentration, temperatures and densities, etc.) are usually propagated, through appropriate core simulation packages, to the fission rates needed to calculate the expected antineutrino spectrum. Moreover, as most experiments measure $\bar{\nu}_e$s from more than one reactor, it is important to understand not only the magnitude of these effects, but also the correlations between uncertainties from different sources [30]. The quality of such simulation work may be quantified through available benchmarks as shown in [31] and references therein.

3. Current reactor oscillation searches

With the exception of KamLAND, long-baseline antineutrino oscillation experiment, all other current reactor experiments are short-baseline searches. There are three reactor antineutrino disappearance experiments currently underway, constructed to perform a very sensitive measurements of the mixing angle θ_{13}. These include Double Chooz [32], RENO [33], and Daya Bay [34]. All of the current experiments use a new detector design with a gadolinium (Gd) doped liquid scintillator target surrounded by a non-doped scintillator region to detect the gammas from neutron capture on Gd. Antineutrinos will be detected through the inverse beta decay reaction: $\bar{\nu}_e + p \to n + e^+$. The $\bar{\nu}_e$ target in each detector is the liquid scintillator loaded with 0.1% Gd. Antineutrino events are identified by a coincidence between the prompt positron signal, which provides a direct measure of the neutrino energy, and the delayed signal from neutron capture on Gd. In addition, all experiments will use a combination of near and far detectors to reduce the reactor flux uncertainty. Compared to the original CHOOZ experiment, which used a single detector, the use of two detectors eliminates sensitivity to calculations of the reactor flux, and only requires a knowledge of the relative acceptance of the two detectors rather than the absolute acceptance.

The Double Chooz experiment is currently performed at the Chooz Nuclear Power Station, the site of the original CHOOZ experiment [15]. The facility has two 4.27 GWth reactors separated by 140 m. Next year the experiment will be using two identical detectors: a near detector at an average distance of about 400 m from the two reactor cores and a far detector at an average distance of 1050 m. The experiment is currently running with the single far detector. The far detector is located in the cavern with 300 m.w.e. (meters water equivalent) overburden used by the original CHOOZ experiment. For the near detector, a tunnel is under excavation and a near lab will be equipped (120 m.w.e) by the end of 2012.

The RENO experiment in Korea has both the near and far detector operational from August 2011 at the Yeong Gwang six core reactor complex. The collaboration is currently taking antineutrino data with expected performance [35]. First results are expected in 2012.

The Daya Bay experiment will eventually have four near detectors at two near sites, and four far detectors exposed to antineutrinos from six core reactor complex. Two of the four near detectors are currently operational. The four far detectors will be deployed in 2012. Data taking with the full eight detector setup is expected to start in mid 2012 [36].

Table 1 lists the parameters of the current generation of the short-baseline reactor antineutrino experiments.

An example of new detector design will be described in the case of the Double Chooz experiment. The Double Chooz detector design is shown in Fig. 1. It is optimized to reduce backgrounds to a negligible level. Each detector consists of a series of concentric cylinders.

The target vessel is an acrylic cylinder and it contains 10.3 m^3 of liquid scintillator loaded with 0.1% Gd. The next liquid volume is the Gamma Catcher, a 55 cm thick buffer of non-loaded liquid scintillator with the same optical properties (light yield, attenuation length) as the target volume. This scintillating buffer increases the efficiency for detecting the photons from the neutron capture on Gd and from the positron annihilation. A 105 cm thick region of non-scintillating liquid surrounds the Gamma Catcher and serves to decrease the level of accidental background. This buffer will reduce the singles rates in each detector by two orders

Table 1. Comparison of the parameters of the current generation of short baseline reactor antineutrino experiments.

Experiment	Thermal Power [GW_{th}]	Distance to Near/Far [m]	Shielding Near/Far [m.w.e]	Target Mass [tons]	Sensitivity $\sin^2 2\theta_{13}$ [90%C.L.]	Status
Double Chooz (France)	8.4	390/1050	115/300	8/8	0.03	Data taking started 2011 far det. only
RENO (Korea)	17.3	290/1380	120/450	16/16	0.02	Data taking from mid-2011
Daya Bay (China)	17.4	360(500)/ 1985(1615)	260/910	2x2x20 (N) 4x20(F)	0.01	Data taking started with 1 near site

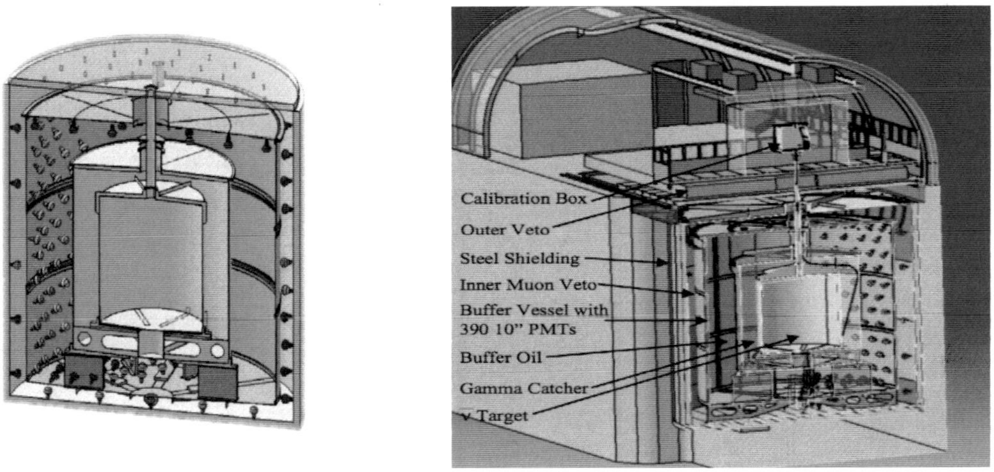

Figure 1. A cross-sectional view of the Double Chooz detector. Starting from the center of the target, the detector elements are: Target, Gamma Catcher, Non-scintillating buffer, Buffer vessel and PMT support structure, Inner veto system, Steel Shielding, and Outer veto system. The detector components are enlarged on the left.

of magnitude with respect to those in the original CHOOZ experiment, which had no such buffer. A total of 390 10" low background phototubes (PMTs) are mounted from the interior surface of the buffer vessel. Outside of the buffer vessel, an additional 50 cm thick Inner veto region is filled with liquid scintillator and instrumented with PMTs. The Inner Veto is surrounded by a 15 cm Steel Shielding, protecting the detector from rock radioactivity. As a result of the few MeV energy of the reactor antineutrinos, cosmogenic backgrounds are a serious issue for reactor experiments. Double Chooz instrumented an Inner Veto detector in combination with the Outer Veto detector to tag cosmic-ray muons which may produce backgrounds in the target region. Outer Veto is a large coverage muon system of scintillating counters to be placed above the detector. It is expected to be a powerful tool in studying the ^{9}Li background, which is

not well understood. The combined veto system reduces backgrounds from accidentals, fast neutrons, and muon capture by large factors. With about 47,000 events expected in the far detector over a 3-year run, the statistical uncertainty associated with the near to far comparison is expected to be 0.5%. In order to take advantage of this statistical precision, the experiment needs to reduce the systematic uncertainties to this level. Double Chooz will greatly reduce this systematic component by making the ratio of the antineutrino spectra in two detectors. The antineutrino rates are proportional to the number of free protons within the target volumes. The relative fiducial volume uncertainty of the detectors is studied and assumed to be 0.2%. Uncertainties associated with efficiencies and cuts may be reduced in the final analysis due to care in making the near and far detectors identical in design, materials, and target volumes. The Double Chooz detectors were designed to eliminate the problems with radioactivity encountered in CHOOZ by adding buffer volume and selecting low radioactivity materials for the detector components. Therefore, only a minimum number of analysis cuts is sufficient, resulting in an overall contribution to systematic uncertainty of about 0.5% in the two-detector run.
A comparison of the current reactor experiments is given in Table 1 with the available thermal power for each experiment, the distance of near and far detectors from the reactors, shielding of detectors, detector target mass, expected sensitivities, and current schedule.

4. Current status and future prospects

At the time of writing this article the Double Chooz collaboration presented results of the first oscillation search performed with the far-only detector [37]. The result is an indication of reactor electron antineutrino disappearance consistent with neutrino oscillations. A ratio of 0.944 $\pm$ 0.016 (stat) $\pm$ 0.040 (syst) observed to predicted events was obtained in 101 days of running at the Chooz Nuclear Power Plant in France, with two 4.25 GW_{th} reactors. The results were obtained from the far detector alone with $10m^3$ fiducial volume located 1050 m from the two reactor cores. The Bugey4 [38] measurement was used to establish expectation for antineutrino rate prediction, effectively playing the role of the near detector. The deficit can be interpreted as an indication of a non-zero value of still unmeasured neutrino mixing parameter θ_{13}. Analyzing both the rate of the prompt positrons and their energy spectrum it is found $\sin^2 2\theta_{13} = 0.086 \pm 0.041\ (stat) \pm 0.030\ (syst)$ corresponding to $0.015 < \sin^2 2\theta_{13} < 0.16$ at 90% C.L.
This is one of the important first steps in our multi-year program to establish the magnitude of θ_{13} and a valuable input to today's global fits to the three-neutrino oscillation model. Indeed, when combined with current accelerator results [5] [6] the Double Chooz result leads to $\sin^2 2\theta_{13} \neq 0$ at 3σ level [42]. Ultimately the Double Chooz will use two identical detectors to explore the range of $\sin^2 2\theta_{13}$ from 0.15 to about 0.03, within three years of data taking. RENO experiments expects first results in mid 2012 with both detectors combined. Daya Bay experiment will be producing results soon after the full setup is operational.
With respect to the hints suggesting a possible oscillation to sterile neutrinos in the $\Delta m^2 \sim 1$ eV^2 region there are various proposals to perform precise searches. Considering reactor antineutrinos one may use a small cubic meter detector near research reactors and measure the $\bar{\nu}_e$ rate as a function of distance from the core [39, 40, 41].
Aside from sterile neutrino issues and with recent results suggesting $\sin^2 2\theta_{13} \sim 0.08$ [42], the reactor experiments are in a unique position to perform a precise measurement of this quantity. If confirmed, such a large value would open a window of opportunity for long-baseline experiments with a rich program of neutrino oscillation physics associated with CP-violation and mass hierarchy measurements. The upcoming NOνA [45] experiment will be operational in 2014. It will be the first precision experiment to probe oscillations with neutrinos propagating through signifiant matter (810 km distance between neutrino production point and the far detector). Several other long(er)-baseline experiments combined with high intensity neutrino

sources are considered [43]. These would utilize very large neutrino detectors to collect sufficient event statistics at far sites. The program under development in the US is the LBNE experiment with a neutrino beam originating at Fermilab and a detector (either water Cherenkov or liquid argon) located in proposed underground laboratory in the Homestake mine in South Dakota [44]. LBNE might benefit from possible extensions such as Fermilab Project-X upgrade [46] and/or a high-intensity decay-at-rest $\bar{\nu}_e$ source proposed by the DAEδALUS collaboration [47]. Other proposals are considered worldwide including Hyper-K experiment in Japan [48], a neutrino beam from CERN to either very large water Cherenkov detector [49], liquid scintillator [50], or liquid argon [51] detector.
At the end, it will be the synergy between accelerator searches for the ν_e appearance and precise measurements of reactor antineutrino disappearance that helps us to measure the currently unknown mixing angle θ_{13} and CP-violation and mass hierarchy.

References

[1] See http://www.aps.org/policy/reports/multidivisional/neutrino/;
T. N. S. A. Group *Recommendations to the Department of Energy and the National Science Foundation on a Future U.S. Program in Neutrino Oscillations* http://www.science.doe.gov/hep/hepap_reports.shtm;
T. P. Panel *Report of the Particle Physics Project Prioritization Panel* http://www.er.doe.gov/hep/hepap_reports.shtm;
Conrad J M 2010 *Preprint* arXiv:1012.4853[hep-ex]
[2] Kayser B 2011 *Proc. 22nd Rencontres de Blois* ed L Celnikier, J Dumarchez, B Klima and J Tran Trahn Van (Vietnam: Gioi Publishers) p 91
[3] Amsler C 2008 *et al. Rev. Part. Phys.* see chapter 13 *Neutrino Mixing*;
Phys. Lett. B **667** 1
[4] Nunokawa H, Parke S J, and Valle J W F 2008 *CP Violation and Neutrino Oscillations Prog. Part. Nucl. Phys.* **60** 338D402 (*Preprint* arXiv:0710.0554[hep-ph])
[5] Abe K *et al.* 2011 *Phys. Rev. Lett.* **107** 041801 (*Preprint* arXiv:1106.2822[hep-ex])
[6] Adamson P 2011 *et al. Phys. Rev. Lett.* **107** 181802 (*Preprint* arXiv:1108.0015v1[hep-ex])
[7] Fogli G L *et al.* 2011 (*Preprint* arXiv:1106.6028v2[hep-ex])
[8] Adamson P *et al.* 2011 *Phys. Rev. Lett.* **106** 181801
[9] Bemporad C, Gratta G and Vogel P 2002 *Rev. Mod. Phys.* **74** 297;
McKeown R D and Vogel P 2004 *Phys. Rept.* **394** 315;
Fogli G L *et al.* 2006 *Prog. Part. Nucl. Phys.* **57** 742;
Gonzalez-Garcia M C and Maltoni M 2008 *Phys. Rept.* **460** 1
[10] Eguchi K *et al.* 2003 *Phys. Rev. Lett.* **90** 021802
[11] Eguchi K *et al.* 2005 *Phys. Rev. Lett.* **94** 081801
[12] Djurcic Z 2004 *Ph.D. Thesis* University of Alabama UMI-31-63548
[13] Abe S *et al.* 2008 *Phys. Rev. Lett.* **100** 221803
[14] Wolfenstein L 1978 *Phys. Rev.* D **17** 2369;
Mikheyev S P and Smirnov A Y 1985 *Sov. J. Nucl. Phys.* **42** 13
[15] Apollonio M *et al.* 2003 *Eur. Phys. J.* C **27** 331
[16] Boehm F *et al.* 2001 *Phys. Rev.* D **64** 112001
[17] Anderson K *et al.* 2004 *Preprint* hep-ex/0402041v1
[18] Barbeau P S, Collar J I and Tench O 2007 *JCAP* **09** 009
[19] Bernstein A *et al.* 2002 *J. Appl. Phys.* **91** 4672;
Bernstein A *et al.* 2008 *J. Appl. Phys.* **103** 074905
[20] Kopeikin V I *et al.* 2001 *Phys. Atom. Nuclei* **64** 849
[21] Miller L 2001 *Ph.D. Thesis* Stanford University
[22] Schreckenbach K *et al.* 1985 *Phys. Lett.* B **160** 325
[23] Von Feilitzsch A F and Schreckenbach K 1982 *Phys. Lett.* B **118** 325
[24] Hahn A A *et al.* 1989 *Phys. Lett.* B **218** 365
[25] Vogel P *et al.* 1981 *Phys. Rev.* C **24** 1543
[26] Huber P 2011 *Phys. Rev.* C **84** 024617 (*Preprint* arXiv:1106.0687[hep-ph])
[27] Mueller T A *et al.* 2011 *Phys. Rev.* C **83** 054615 (*Preprint* arXiv:1101.2663[hep-ex])
[28] Mention G *et al.* 2011 *Phys. Rev.* D **83** 073006 (*Preprint* arXiv:1101.2755[hep-ex])
[29] Louis W C *Future Short-baseline Physics at FermiLab* these proceedings;
Djurcic Z *MiniBooNE Oscillation report* these proceedings (*Preprint* arXiv:1201.1519[hep-ex]);

Djurcic Z 2011 *AIP Conf. Proc.* 1382:91-93 FERMILAB-CONF-11-654;
Karagiorgi G *et al.* 2009 *Phys. Rev.* D **80** 073001 (*Preprint* arXiv:0906.1997[hep-ph])
[30] Djurcic Z *et al.* 2009 *J. Phys.* G **36** 045002 (*Preprint* arXiv:0808.0747[hep-ex])
[31] Jones C L *et al.* 2011 *Preprint* arXiv:1109.5379v1[nucl-ex]
[32] Ardellier F *et al.* 2006 *Preprint* hep-ex/0606025v4
[33] Ahn J K *et al.* 2010 *Preprint* arXiv:1003.1391[hep-ex]
[34] Guo X *et al.* 2007 *Preprint* hep-ex/0701029v1
[35] Talk given by Joo K K at LowNu 2011 http:// workshop.kias.re.kr/lownu11/
[36] Talk given by Lu H at LowNu 2011 http:// workshop.kias.re.kr/lownu11/
[37] Abe Y *et al.* 2011 *Preprint* arXiv:1112.6353[hep-ex]
[38] Declais Y *et al.* 1994 *Phys. Lett.* B **338** 383
[39] Porta A *et al.* 2010 *J. Phys. Conf. Ser.* 203 012092
[40] Talk given by Bowden N at LowNu 2011 http:// workshop.kias.re.kr/lownu11/
[41] Talk given by Egorov V at LowNu 2011 http:// workshop.kias.re.kr/lownu11/
[42] Machado P A N *et al.* 2011 *Preprint* arXiv:1111.3330[hep-ph]
[43] Talk given by Shaevitz M at PANIC 2011 web.mit.edu/panic11/
[44] Long-Baseline Neutrino Experiment http://lbne.fnal.gov/;
Akiri T *et. al.* 2011 *Preprint* arXiv:1110.6249[hep-ex]
[45] Ayres D *et al.* 2007 FERMILAB-DESIGN-2007-01;
Djurcic Z 2011 *AIP Conf. Proc.* 1382:132-134 FERMILAB-CONF-11-655;
Davies G S 2011 *Preprint* arXiv:1110.0112[hep-ex];
Betancourt M 2011 *Preprint* arXiv:1109.6692[hep-ex]
[46] Project X http://projectx.fnal.gov/
[47] Alonso J *et al.* 2010 *Preprint* arXiv:1006.0260[physics.ins-det];
Alonso J *et al.* 2010 *Preprint* arXiv:1008.4967[hep-ex]
[48] Abe K *et al.* 2011 *Preprint* arXiv:1109.3262[hep-ex]
[49] De Bellefon A *et al.* 2006 *Preprint* hep-ph/0603172
[50] Wurm M *et. al.* 2011 *Preprint* arXiv:1104.5620[hep-ex]
[51] Rubbia A *et al.* *Preprint* arXiv:0908.1286[hep-ph]

Journal of Physics: Conference Series **408** (2013) 012009 doi:10.1088/1742-6596/408/1/012009

Phenomenology of sterile neutrinos

Carlo Giunti

INFN, Sezione di Torino, Via P. Giuria 1, I–10125 Torino, Italy

E-mail: giunti@to.infn.it

Abstract. The indications in favor of short-baseline neutrino oscillations, which require the existence of one or more sterile neutrinos, are reviewed. In the framework of 3+1 neutrino mixing, which is the simplest extension of the standard three-neutrino mixing which can partially explain the data, there is a strong tension in the interpretation of the data, mainly due to an incompatibility of the results of appearance and disappearance experiments. In the framework of 3+2 neutrino mixing, CP violation in short-baseline experiments can explain the difference between MiniBooNE neutrino and antineutrino data, but the tension between the data of appearance and disappearance experiments persists because the short-baseline disappearance of electron antineutrinos and muon neutrinos compatible with the LSND and MiniBooNE antineutrino appearance signal has not been observed.

1. Introduction

From the results of solar, atmospheric and long-baseline neutrino oscillation experiments we know that neutrinos are massive and mixed particles (see [1]). There are two groups of experiments which measured two independent squared-mass differences (Δm^2) in two different neutrino flavor transition channels:

- Solar neutrino experiments (Homestake, Kamiokande, GALLEX/GNO, SAGE, Super-Kamiokande, SNO, BOREXino) measured $\nu_e \to \nu_\mu, \nu_\tau$ oscillations generated by $\Delta m^2_{\text{SOL}} = 6.2^{+1.1}_{-1.9} \times 10^{-5}\,\text{eV}^2$ and a mixing angle $\tan^2\vartheta_{\text{SOL}} = 0.42^{+0.04}_{-0.02}$ [2]. The KamLAND experiment confirmed these oscillations by observing the disappearance of reactor $\bar{\nu}_e$ at an average distance of about 180 km. The combined fit of solar and KamLAND data leads to $\Delta m^2_{\text{SOL}} = (7.6 \pm 0.2) \times 10^{-5}\,\text{eV}^2$ and a mixing angle $\tan^2\vartheta_{\text{SOL}} = 0.44 \pm 0.03$ [2].
- Atmospheric neutrino experiments (Kamiokande, IMB, Super-Kamiokande, MACRO, Soudan-2, MINOS) measured ν_μ and $\bar{\nu}_\mu$ disappearance through oscillations generated by $\Delta m^2_{\text{ATM}} \simeq 2.3 \times 10^{-3}\,\text{eV}^2$ and a mixing angle $\sin^2 2\vartheta_{\text{ATM}} \simeq 1$ [3]. The K2K and MINOS long-baseline experiments confirmed these oscillations by observing the disappearance of accelerator ν_μ at distances of about 250 km and 730 km, respectively. The MINOS data give $\Delta m^2_{\text{ATM}} = 2.32^{+0.12}_{-0.08} \times 10^{-3}\,\text{eV}^2$ and $\sin^2 2\vartheta_{\text{ATM}} > 0.90$ at 90% C.L. [4].

These measurements led to the current three-neutrino mixing paradigm, in which the three active neutrinos ν_e, ν_μ, ν_τ are superpositions of three massive neutrinos ν_1, ν_2, ν_3 with respective masses m_1, m_2, m_3. The two measured squared-mass differences can be interpreted as

$$\Delta m^2_{\text{SOL}} = \Delta m^2_{21}\,, \qquad \Delta m^2_{\text{ATM}} = |\Delta m^2_{31}| \simeq |\Delta m^2_{32}|\,, \tag{1}$$

Journal of Physics: Conference Series **408** (2013) 012009 doi:10.1088/1742-6596/408/1/012009

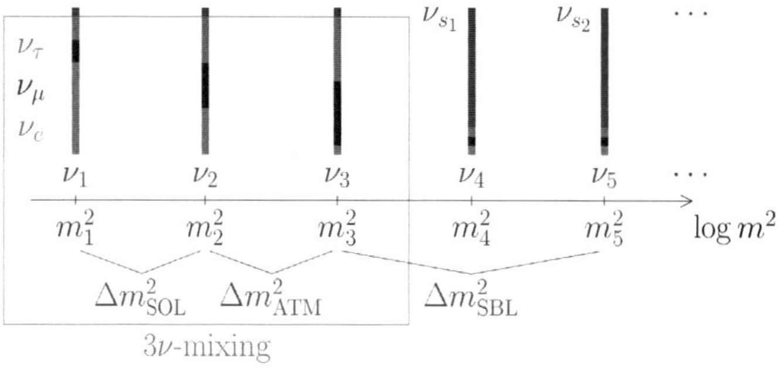

Figure 1. Schematic description of the neutrino spectrum beyond three-neutrino mixing with one or more relatively heavy additional massive neutrinos which are mainly sterile.

with $\Delta m^2_{kj} = m^2_k - m^2_j$. In the standard parameterization of the 3×3 unitary mixing matrix (see [1]) $\vartheta_{\text{SOL}} \simeq \vartheta_{12}$, $\vartheta_{\text{ATM}} \simeq \vartheta_{23}$ and $\sin^2\vartheta_{13} < 0.035$ at 90% C.L. [5].

The completeness of the three-neutrino mixing paradigm was challenged in 1995 by the observation of a signal of short-baseline $\bar{\nu}_\mu \to \bar{\nu}_e$ oscillations in the LSND experiment [6, 7], which would imply the existence of one or more squared-mass differences much larger than Δm^2_{SOL} and Δm^2_{ATM}. The MiniBooNE experiment was made in order to check the LSND signal with about one order of magnitude larger distance (L) and energy (E), but the same order of magnitude for the ratio L/E from which neutrino oscillations depend. The first results of the MiniBooNE experiment in neutrino mode did not show a signal compatible with that of LSND [8], but the results in antineutrino mode, presented in the summer of 2010 [9], show an excess of events over the background at approximately at the same L/E of LSND. This result revived the interest in the possibility of existence of one or more neutrinos with masses at the eV scale which can generate squared-mass differences for short-baseline oscillations.

Figure 1 illustrates schematically the neutrino spectrum beyond three-neutrino mixing with one or more additional massive neutrinos heavier than the three standard massive neutrinos. In the flavor basis the additional massive neutrinos correspond to sterile neutrinos, which do not have standard weak interactions and do not contribute to the number of active neutrinos determined by LEP experiments through the measurement of the invisible width of the Z boson, $N_a = 2.9840 \pm 0.0082$ [10]. The existence of sterile neutrinos which have been thermalized in the early Universe is compatible with Big-Bang Nucleosynthesis data [11, 12], with the indication however that schemes with more than one sterile neutrino are disfavored [13]. It is also compatible with cosmological measurements of the Cosmic Microwave Background and Large-Scale Structures if the neutrino masses are limited below about 1 eV [14–18]. Therefore, we consider neutrino mixing schemes in which the three standard neutrinos have masses much smaller that 1 eV and the additional neutrinos have masses at the eV scale. However, other sterile neutrinos may exist and the cosmological constraints can be avoided by suppressing the thermalization of sterile neutrinos in the early Universe and/or by considering non-standard cosmological theories. For example, a sterile neutrino with a mass scale much smaller than 1 eV could have important implications for solar neutrino oscillations (see [19]) and a sterile neutrino with a mass at the keV scale could constitute Warm Dark Matter (see [20, 21]).

A further indication in favor of short-baseline oscillations came from a new calculation of the reactor $\bar{\nu}_e$ flux presented in January 2011 [22] (see also [23]), which obtained an increase of about 3% with respect to the previous value adopted in the analysis of the data of reactor neutrino oscillation experiments. As illustrated in Fig. 2 of [24], the measured reactor rates are in agreement with those derived from the old $\bar{\nu}_e$ flux, but show a deficit of about 2.2σ with respect to the rates derived from the new $\bar{\nu}_e$ flux. This is the "reactor antineutrino anomaly" [25],

Table 1. Values of χ^2, number of degrees of freedom (NDF), goodness-of-fit (GoF) and best-fit values of the mixing parameters obtained in our 3+1 and 3+2 fits of short-baseline oscillation data. The last three lines give the results of the parameter goodness-of-fit test [43]: $\Delta\chi^2_{\text{PG}}$, number of degrees of freedom (NDF_{PG}) and parameter goodness-of-fit (PGoF).

	3+1	3+2
χ^2_{min}	100.2	91.6
NDF	104	100
GoF	59%	71%
$\Delta m^2_{41}[eV^2]$	0.89	0.90
$\lvert U_{e4}\rvert^2$	0.025	0.017
$\lvert U_{\mu 4}\rvert^2$	0.023	0.019
$\Delta m^2_{51}[eV^2]$		1.61
$\lvert U_{e5}\rvert^2$		0.017
$\lvert U_{\mu 5}\rvert^2$		0.0061
η		1.51π
$\Delta\chi^2_{PG}$	24.1	22.2
NDF_{PG}	2	5
PGoF	6×10^{-6}	5×10^{-4}

which may be an indication in the $\bar{\nu}_e \to \bar{\nu}_e$ channel of a signal corresponding to the $\bar{\nu}_\mu \to \bar{\nu}_e$ signal observed in the LSND and MiniBooNE experiments. Finally, there is a "Gallium neutrino anomaly" [26–31], consisting in a short-baseline disappearance of electron neutrinos measured in the Gallium radioactive source experiments GALLEX [32] and SAGE [33].

In the following, I consider in Sections 2 and 3 the cases of 3+1 [34–37] and 3+2 [38–41] neutrino mixing, respectively, following the discussion in [42]. Conclusions are drawn in Section 4.

2. 3+1 neutrino mixing

In this Section I consider the simplest extension of three-neutrino mixing with the addition of one massive neutrino. In 3+1 neutrino mixing [34–37] the effective flavor transition and survival probabilities in short-baseline (SBL) experiments are given by

$$P^{\text{SBL}}_{\overset{(-)}{\nu}_\alpha \to \overset{(-)}{\nu}_\beta} = \sin^2 2\vartheta_{\alpha\beta} \sin^2\left(\frac{\Delta m^2_{41} L}{4E}\right), \qquad P^{\text{SBL}}_{\overset{(-)}{\nu}_\alpha \to \overset{(-)}{\nu}_\alpha} = 1 - \sin^2 2\vartheta_{\alpha\alpha} \sin^2\left(\frac{\Delta m^2_{41} L}{4E}\right), \tag{2}$$

for $\alpha, \beta = e, \mu, \tau, s$ and $\alpha \neq \beta$, with

$$\sin^2 2\vartheta_{\alpha\beta} = 4|U_{\alpha 4}|^2 |U_{\beta 4}|^2, \qquad \sin^2 2\vartheta_{\alpha\alpha} = 4|U_{\alpha 4}|^2 \left(1 - |U_{\alpha 4}|^2\right). \tag{3}$$

Therefore:

(1) All effective SBL oscillation probabilities depend only on the absolute value of the largest squared-mass difference $\Delta m^2_{41} = m^2_4 - m^2_1$.

(2) All oscillation channels are open, each one with its own oscillation amplitude.

(3) The oscillation amplitudes depend only on the absolute values of the elements in the fourth column of the mixing matrix, i.e. on three real numbers with sum less than unity, since the unitarity of the mixing matrix implies that $\sum_\alpha |U_{\alpha 4}|^2 = 1$.

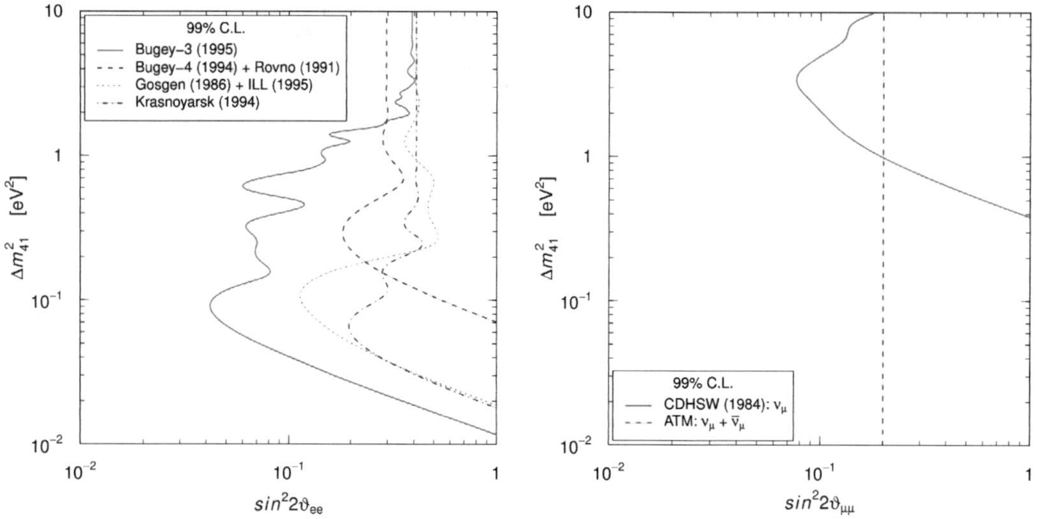

Figure 2. Left: exclusion curves obtained from the data of reactor $\bar{\nu}_e$ disappearance experiments (see [25]). Right: exclusion curves obtained from the data of the CDHSW ν_μ disappearance experiment [44], and from atmospheric neutrino data (extracted from the analysis in [40]).

(4) CP violation cannot be observed in SBL oscillation experiments, even if the mixing matrix contains CP-violation phases, because neutrinos and antineutrinos have the same effective SBL oscillation probabilities. Hence, 3+1 neutrino mixing cannot explain the difference between neutrino [8] and antineutrino [9] oscillations observed in the MiniBooNE.

The dependence of the oscillation amplitudes in Eq. (3) on three independent absolute values of the elements in the fourth column of the mixing matrix implies that the amplitude of $\overset{(-)}{\nu}_\mu \to \overset{(-)}{\nu}_e$ transitions is limited by the absence of large SBL disappearance of $\overset{(-)}{\nu}_e$ and $\overset{(-)}{\nu}_\mu$ observed in several experiments.

The results of reactor neutrino experiments constrain the value $|U_{e4}|^2$ through the measurement of $\sin^2 2\vartheta_{ee}$. Even taking into account the reactor antineutrino anomaly [25] discussed in the Introduction, the $\bar{\nu}_e$ disappearance is small and large values of $\sin^2 2\vartheta_{ee}$ are constrained by the exclusion curves in the left panel of Fig. 2. Since values of $|U_{e4}|^2$ close to unity are excluded by solar neutrino oscillations (which require large $|U_{e1}|^2 + |U_{e2}|^2$), for small $\sin^2 2\vartheta_{ee}$ we have

$$\sin^2 2\vartheta_{ee} \simeq 4|U_{e4}|^2 \,. \tag{4}$$

The value of $\sin^2 2\vartheta_{\mu\mu}$ is constrained by the curves in the right panel of Fig. 2, which have been obtained from the lack of ν_μ disappearance in the CDHSW ν_μ experiment [44] and from the requirement of large $|U_{\mu1}|^2+|U_{\mu2}|^2+|U_{\mu3}|^2$ for atmospheric neutrino oscillations [40]. Hence, $|U_{\mu4}|^2$ is small and

$$\sin^2 2\vartheta_{\mu\mu} \simeq 4|U_{\mu4}|^2 \,. \tag{5}$$

From Eqs. (3), (4) and (5), for the amplitude of $\overset{(-)}{\nu}_\mu \to \overset{(-)}{\nu}_e$ transitions we obtain

$$\sin^2 2\vartheta_{e\mu} \simeq \frac{1}{4} \sin^2 2\vartheta_{ee} \sin^2 2\vartheta_{\mu\mu} \,. \tag{6}$$

Therefore, if $\sin^2 2\vartheta_{ee}$ and $\sin^2 2\vartheta_{\mu\mu}$ are small, $\sin^2 2\vartheta_{e\mu}$ is quadratically suppressed [34, 35]. This is illustrated in the left panel of Fig. 3, where one can see that the separate effects of the constraints on $\sin^2 2\vartheta_{ee}$ and $\sin^2 2\vartheta_{\mu\mu}$ exclude only the large-$\sin^2 2\vartheta_{e\mu}$ part of the region

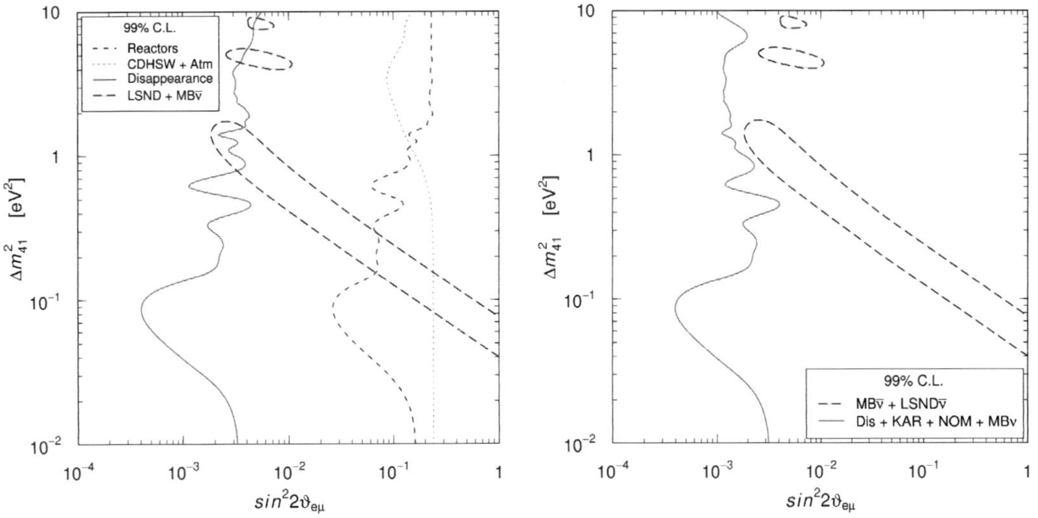

Figure 3. Left: exclusion curves in the $\sin^2 2\vartheta_{e\mu}$–Δm^2_{41} plane obtained from the separate constraints in Fig. 2 (blue and green lines) and the combined constraint given by Eq. (6) (red line) from disappearance experiments (Dis). Right: exclusion curve obtained with the addition of KARMEN [45] (KAR), NOMAD [46] (NOM) and MiniBooNE neutrino [8] (MBν) data (red line). In both figures the region enclosed by the dark-red lines is allowed by LSND and MiniBooNE antineutrino data.

allowed by LSND and MiniBooNE antineutrino data, whereas most of this region is excluded by the combined constraint in Eq. (6). As shown in the right panel of Fig. 3, the constraint becomes stronger by including the data of the KARMEN [45], NOMAD [46] and MiniBooNE neutrino [8] experiments, which did not observe a short-baseline $\overset{(-)}{\nu}_{\mu} \to \overset{(-)}{\nu}_{e}$ signal. Since the parameter goodness-of-fit [43] is 6×10^{-6} [42], 3+1 neutrino mixing is disfavored by the data. This conclusion has been reached recently in [42,47–49] and confirms the pre-MiniBooNE antineutrino results in [34–37,40,41,50].

However, in spite of the low value of the parameter goodness-of-fit it is not inconceivable to refuse to reject 3+1 neutrino mixing for the following reasons:

(A) It is the simplest scheme beyond the standard three-neutrino mixing which can partially explain the data.

(B) It corresponds to the natural addition of one new entity (a sterile neutrino) to explain a new effect (short-baseline oscillations). Better fits of the data require the addition of at least another new entity (in any case at least one sterile neutrino is needed to generate short-baseline oscillations).

(C) The minimum value of the global χ^2 is rather good: $\chi^2_{\text{min}} = 100.2$ for 104 degrees of freedom.

(D) There is a marginal appearance–disappearance compatibility: $\Delta\chi^2_{\text{PG}} = 9.2$ with 2 degrees of freedom, corresponding to PGoF = 1.0%.

(E) 3+1 mixing is favored with respect to 3+2 mixing by the Big-Bang Nucleosynthesis limit $N_{\text{eff}} \leq 4$ at 95% C.L. obtained in [13].

Therefore, we consider the global fit of all data in the framework of 3+1 neutrino mixing, which yields the best-fit values of the oscillation parameters listed in Tab. 1.

Figure 4 shows the allowed regions in the $\sin^2 2\vartheta_{e\mu}$–Δm^2_{41}, $\sin^2 2\vartheta_{ee}$–Δm^2_{41} and $\sin^2 2\vartheta_{\mu\mu}$–$\Delta m^2_{41}$ planes and the marginal $\Delta\chi^2$'s for Δm^2_{41}, $\sin^2 2\vartheta_{e\mu}$, $\sin^2 2\vartheta_{ee}$ and $\sin^2 2\vartheta_{\mu\mu}$.

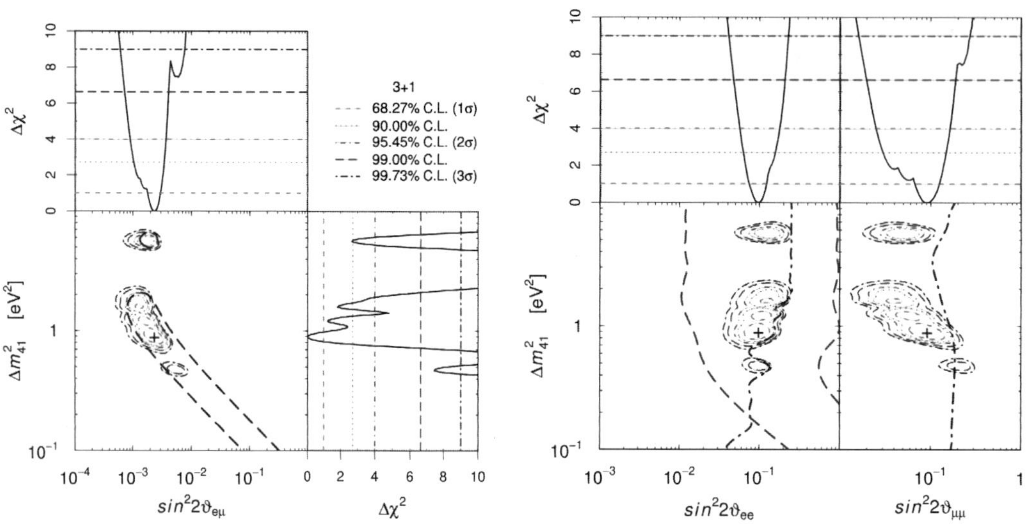

Figure 4. Allowed regions in the $\sin^2 2\vartheta_{e\mu}$–Δm^2_{41}, $\sin^2 2\vartheta_{ee}$–Δm^2_{41} and $\sin^2 2\vartheta_{\mu\mu}$–$\Delta m^2_{41}$ planes and marginal $\Delta\chi^2$'s obtained from the global fit in 3+1 neutrino mixing. The best-fit point is indicated by a cross. Left: the isolated dark-blue long-dashed contours enclose the regions allowed at 3σ by the analysis of appearance data (LSND [7], KARMEN [45], NOMAD [46], MiniBooNE [8, 9]). Right: the isolated dark-blue dash-dotted lines are the 3σ exclusion curves obtained from reactor neutrino data and from CDHSW and atmospheric neutrino data. The isolated dark-red long-dashed lines delimit the region allowed at 99% C.L. by the Gallium anomaly [31].

3. 3+2 neutrino mixing

In 3+2 neutrino mixing [38–42, 47, 49] the relevant effective oscillation probabilities in short-baseline experiments are given by

$$P^{\rm SBL}_{\overset{(-)}{\nu}_\mu\to\overset{(-)}{\nu}_e} = 4|U_{\mu4}|^2|U_{e4}|^2\sin^2\phi_{41} + 4|U_{\mu5}|^2|U_{e5}|^2\sin^2\phi_{51} + 8\Omega\sin\phi_{41}\sin\phi_{51}\cos(\phi_{54} \overset{(+)}{-} \eta)\,, \quad (7)$$

$$P^{\rm SBL}_{\overset{(-)}{\nu}_\alpha\to\overset{(-)}{\nu}_\alpha} = 1 - 4(1-|U_{\alpha4}|^2-|U_{\alpha5}|^2)(|U_{\alpha4}|^2\sin^2\phi_{41} + |U_{\alpha5}|^2\sin^2\phi_{51}) - 4|U_{\alpha4}|^2|U_{\alpha5}|^2\sin^2\phi_{54}\,, \quad (8)$$

for $\alpha, \beta = e, \mu$, with

$$\phi_{kj} = \Delta m^2_{kj}L/4E\,, \quad \Omega = |U_{\mu4}U_{e4}U_{\mu5}U_{e5}|\,, \quad \eta = \arg[U^*_{e4}U_{\mu4}U_{e5}U^*_{\mu5}]\,. \quad (9)$$

Note the change in sign of the contribution of the CP-violating phase η going from neutrinos to antineutrinos, which allows us to explain the CP-violating difference between MiniBooNE neutrino and antineutrino data.

Figure 5 shows the marginal allowed allowed regions in the Δm^2_{41}–Δm^2_{51} plane obtained in our 3+2 global fit. The best-fit values of the mixing parameters are shown in Tab. 1.

The parameter goodness-of-fit obtained with the comparison of the fit of LSND and MiniBooNE antineutrino data and the fit of all other data is 5×10^{-4}. This is an improvement with respect to the 6×10^{-6} parameter goodness-of-fit obtained in 3+1 mixing. However, the value of the parameter goodness-of-fit remains low and the improvement is mainly due to the increased number of degrees of freedom, as one can see from Tab. 1. The persistence of a bad parameter goodness-of-fit is a consequence of the fact that the $\bar{\nu}_\mu \to \bar{\nu}_e$ transitions observed in LSND and MiniBooNE must correspond in any neutrino mixing scheme to enough short-baseline disappearance of $\overset{(-)}{\nu}_e$ and $\overset{(-)}{\nu}_\mu$ which has not been observed and there is an irreducible

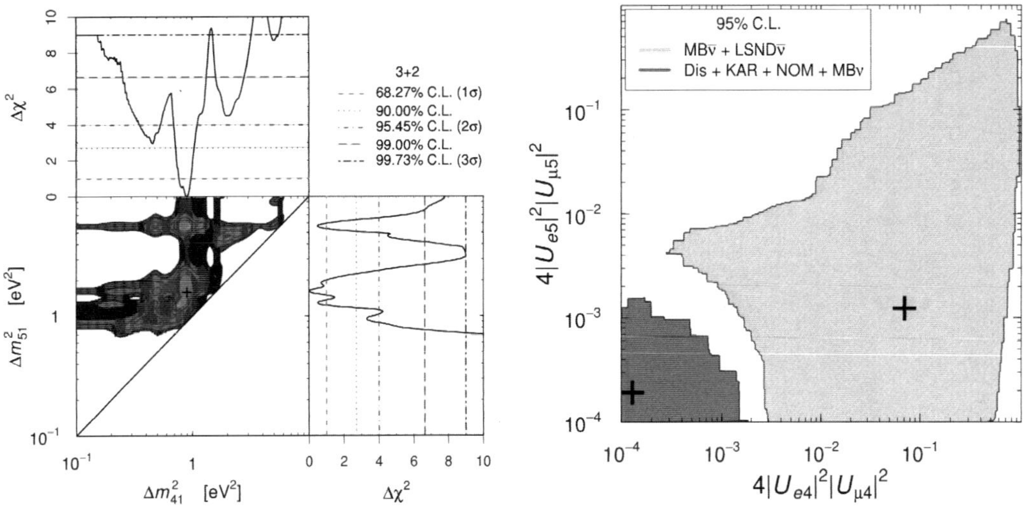

Figure 5. Left: allowed regions in the Δm^2_{41}–Δm^2_{51} plane and corresponding marginal $\Delta\chi^2$'s obtained from the global fit of all the considered data in 3+2 mixing. Right: comparison of the 95% C.L. allowed regions in the $4|U_{e4}|^2|U_{\mu4}|^2$–$4|U_{e5}|^2|U_{\mu5}|^2$ plane obtained from LSND and MiniBooNE antineutrino data on the right (green area) and disappearance, KARMEN, NOMAD and MiniBooNE neutrino data on the left (red area). Best-fit points are indicated by crosses.

tension between the LSND and MiniBooNE antineutrino data and the KARMEN antineutrino data. The only benefit of 3+2 mixing with respect to 3+1 mixing is that they allow to explain the difference between MiniBooNE neutrino and antineutrino data through CP violation. In fact, neglecting the MiniBooNE neutrino data we obtain $\Delta\chi^2_{\rm PG} = 16.6$ with PGoF $= 3 \times 10^{-4}$ in 3+1 mixing and $\Delta\chi^2_{\rm PG} = 20.4$ with PGoF $= 1 \times 10^{-3}$ in 3+2 mixing. In this case $\Delta\chi^2_{\rm PG}$ is even lower in 3+1 mixing than in 3+2 mixing!

The tension between LSND and MiniBooNE antineutrino data and disappearance, KARMEN, NOMAD and MiniBooNE neutrino data is illustrated in the right panel of Fig. 5, which is the analogue for 3+2 mixing of the right panel in Fig. 3 in 3+1 mixing. In practice, in order to show the tension in a two-dimensional figure we have marginalized the χ^2 over all the other mixing parameters, including the two Δm^2's.

4. Conclusions

In the framework of 3+1 neutrino mixing, there is a strong tension between LSND and MiniBooNE antineutrino data and disappearance, KARMEN, NOMAD and MiniBooNE neutrino data [42, 47–49]. Since however the minimum value of the global χ^2 is rather good, one may choose to consider as possible 3+1 neutrino mixing, which can partially explain the data, taking into account its simplicity and the natural correspondence of one new entity (a sterile neutrino) with a new effect (short-baseline oscillations).

In the framework of 3+2 neutrino mixing the tension between LSND and MiniBooNE antineutrino data and disappearance, KARMEN, NOMAD and MiniBooNE neutrino data is reduced with respect to the 3+1 fit, but it is not eliminated (see the right panel of Fig. 5). Moreover, the improvement of the parameter goodness of fit with respect to that obtained in the 3+1 fit is mainly due to the increase of the number of oscillation parameters, as one can see from Tab. 1. Hence it seems mainly a statistical effect.

In conclusion, I think that the interpretation of the indications in favor of short-baseline oscillations is uncertain and new experiments are needed in order to clarify the reasons of the tensions in the data and for leading us to the correct interpretation.

References

[1] Giunti C and Kim C W 2007 *Fundamentals of Neutrino Physics and Astrophysics* (Oxford: Oxford University Press) ISBN 978-0-19-850871-7
[2] Abe K *et al.* (Super-Kamiokande Collaboration) 2011 *Phys. Rev.* D **83** 052010
[3] Ashie Y *et al.* (Super-Kamiokande Collaboration) 2005 *Phys. Rev.* D **71** 112005
[4] Adamson P *et al.* (MINOS Collaboration) 2011 *Phys. Rev. Lett.* **106** 181801
[5] Schwetz T, Tortola M and Valle J W F 2008 *New J. Phys.* **10** 113011 (*Preprint* arXiv:0808.2016 [hep-ph])
[6] Athanassopoulos C *et al.* (LSND Collaboration) 1995 *Phys. Rev. Lett.* **75** 2650–2653
[7] Aguilar A *et al.* (LSND Collaboration) 2001 *Phys. Rev.* D **64** 112007 (*Preprint* hep-ex/0104049)
[8] Aguilar-Arevalo A A (MiniBooNE Collaboration) 2009 *Phys. Rev. Lett.* **102** 101802
[9] Aguilar-Arevalo A A *et al.* (MiniBooNE Collaboration) 2010 *Phys. Rev. Lett.* **105** 181801
[10] Schael S *et al.* (ALEPH, DELPHI, L3, OPAL, SLD, LEP Electroweak Working Group, SLD Electroweak Group, SLD Heavy Flavour Group) 2006 *Phys. Rept.* **427** 257 (*Preprint* hep-ex/0509008)
[11] Cyburt R H, Fields B D, Olive K A and Skillman E 2005 *Astropart. Phys.* **23** 313–323
[12] Izotov Y I and Thuan T X 2010 *Astrophys. J.* **710** L67–L71 (*Preprint* arXiv:1001.4440 [astro-ph.CO])
[13] Mangano G and Serpico P D 2011 *Phys. Lett.* B **701** 296–299 (*Preprint* arXiv:1103.1261 [astro-ph.CO])
[14] Hamann J, Hannestad S, Raffelt G G, Tamborra I and Wong Y Y 2010 *Phys. Rev. Lett.* **105** 181301
[15] Giusarma E *et al.* 2011 *Phys. Rev.* D **83** 115023 (*Preprint* arXiv:1102.4774 [astro-ph.CO])
[16] Kristiansen J R and Elgaroy O 2011 (*Preprint* arXiv:1104.0704 [astro-ph.CO])
[17] Hou Z, Keisler R, Knox L, Millea M and Reichardt C 2011 (*Preprint* arXiv:1104.2333 [astro-ph.CO])
[18] Gonzalez-Morales A X, Poltis R, Sherwin B D and Verde L 2011 (*Preprint* arXiv:1106.5052 [astro-ph.CO])
[19] de Holanda P C and Smirnov A Y 2011 *Phys. Rev.* D **83** 113011 (*Preprint* arXiv:1012.5627 [hep-ph])
[20] Kusenko A 2009 *Phys. Rept.* **481** 1–28 (*Preprint* arXiv:0906.2968 [hep-ph])
[21] Boyarsky A, Ruchayskiy O and Shaposhnikov M 2009 *Ann. Rev. Nucl. Part. Sci.* **59** 191–214
[22] Mueller T A *et al.* 2011 *Phys. Rev.* C **83** 054615 (*Preprint* arXiv:1101.2663 [hep-ex])
[23] Huber P 2011 *Phys. Rev.* C **84** 024617 (*Preprint* arXiv:1106.0687 [hep-ph])
[24] Giunti C 2011 La Thuile 2011, NeuTel 2011 and IFAE 2011 *Preprint* arXiv:1106.4479 [hep-ph]
[25] Mention G *et al.* 2011 *Phys. Rev.* D **83** 073006 (*Preprint* arXiv:1101.2755 [hep-ex])
[26] Giunti C and Laveder M 2007 *Mod. Phys. Lett.* A **22** 2499–2509 (*Preprint* hep-ph/0610352)
[27] Giunti C and Laveder M 2008 *Phys. Rev.* D **77** 093002 (*Preprint* arXiv:0707.4593 [hep-ph])
[28] Acero M A, Giunti C and Laveder M 2008 *Phys. Rev.* D **78** 073009 (*Preprint* arXiv:0711.4222 [hep-ph])
[29] Giunti C and Laveder M 2009 *Phys. Rev.* D **80** 013005 (*Preprint* arXiv:0902.1992 [hep-ph])
[30] Giunti C and Laveder M 2010 *Phys. Rev.* D **82** 053005 (*Preprint* arXiv:1005.4599 [hep-ph])
[31] Giunti C and Laveder M 2011 *Phys. Rev.* C **83** 065504 (*Preprint* arXiv:1006.3244 [hep-ph])
[32] Kaether F, Hampel W, Heusser G, Kiko J and Kirsten T 2010 *Phys. Lett.* B **685** 47–54
[33] Abdurashitov J N *et al.* (SAGE Collaboration) 2009 *Phys. Rev.* C **80** 015807
[34] Okada N and Yasuda O 1997 *Int. J. Mod. Phys.* A **12** 3669–3694 (*Preprint* hep-ph/9606411)
[35] Bilenky S M, Giunti C and Grimus W 1998 *Eur. Phys. J.* C **1** 247–253 (*Preprint* hep-ph/9607372)
[36] Bilenky S M, Giunti C, Grimus W and Schwetz T 1999 *Phys. Rev.* D **60** 073007 (*Preprint* hep-ph/9903454)
[37] Maltoni M, Schwetz T, Tortola M and Valle J 2004 *New J. Phys.* **6** 122 (*Preprint* hep-ph/0405172)
[38] Sorel M, Conrad J and Shaevitz M 2004 *Phys. Rev.* D **70** 073004 (*Preprint* hep-ph/0305255)
[39] Karagiorgi G *et al.* 2007 *Phys. Rev.* D **75** 013011 (*Preprint* hep-ph/0609177)
[40] Maltoni M and Schwetz T 2007 *Phys. Rev.* D **76** 093005 (*Preprint* arXiv:0705.0107 [hep-ph])
[41] Karagiorgi G, Djurcic Z, Conrad J, Shaevitz M H and Sorel M 2009 *Phys. Rev.* D **80** 073001
[42] Giunti C and Laveder M 2011 *Phys.Rev.* D **84** 073008 (*Preprint* arXiv:1107.1452 [hep-ph])
[43] Maltoni M and Schwetz T 2003 *Phys. Rev.* D **68** 033020 (*Preprint* hep-ph/0304176)
[44] Dydak F *et al.* (CDHSW Collaboration) 1984 *Phys. Lett.* B **134** 281
[45] Armbruster B *et al.* (KARMEN Collaboration) 2002 *Phys. Rev.* D **65** 112001 (*Preprint* hep-ex/0203021)
[46] Astier P *et al.* (NOMAD Collaboration) 2003 *Phys. Lett.* B **570** 19–31 (*Preprint* hep-ex/0306037)
[47] Akhmedov E and Schwetz T 2010 *JHEP* **10** 115 (*Preprint* arXiv:1007.4171 [hep-ph])
[48] Giunti C and Laveder M 2011 *Phys. Rev.* D **83** 053006 (*Preprint* arXiv:1012.0267 [hep-ph])
[49] Kopp J, Maltoni M and Schwetz T 2011 *Phys. Rev. Lett.* **107** 091801 (*Preprint* arXiv:1103.4570 [hep-ph])
[50] Maltoni M, Schwetz T, Tortola M A and Valle J W F 2002 *Nucl. Phys.* B **643** 321–338

Journal of Physics: Conference Series **408** (2013) 012010 doi:10.1088/1742-6596/408/1/012010

A proposed search for sterile neutrinos with the ICARUS detector at the CERN-PS

Francesco Pietropaolo

Istituto Nazionale di Fisica Nucleare Sezione di Padova, via Marzolo 8, 35131 Padova, Italy

E-mail: Francesco.Pietropaolo@pd.infn.it

Abstract. An experiment to search for sterile neutrinos beyond the Standard Model with the CERN-PS 19.2 GeV beam and the technology of imaging in ultra-pure cryogenic liquid Argon is proposed. The superior quality of the LAr-TPC, now widely demonstrated experimentally and in particular its unique e/π^0 discrimination allows a full rejection of backgrounds, offers the possibility to explore the neutrino conversion into sterile components as recently suggested by the analysis of different observed neutrino anomalies.
The experiment is based on two identical LAr-TPC detectors observing the electron-neutrino signal in the Near and Far positions, the first of about 150 tons at 127 m from the proton target, the second one of about 600 tons placed 850 m away. This project will exploit the largest LAr-TPC ever built, ICARUS T600 now running underground at LNGS exposed to CNGS neutrino beam, by moving it from Gran Sasso to the CERN Far position. An additional 150 t detector will be constructed in the Near position. The intrinsic ν_e beam spectra are expected practically identical in the two positions; in addition all cross-sections and experimental biases cancel out in the comparison of the two detector signals allowing to perform a sensitive search for neutrino oscillations in the presence of sterile neutrinos.

1. Introduction

Neutrino oscillations have established a coherent picture for a large number of experiments consistent with the mixing of three physical neutrino ν_e, ν_μ and ν_τ with three mass eigenstates. In particular the mass differences turn out to be relatively small, $|\Delta m^2_{23}| \sim 2.4\ 10^{-3}$ eV^2 , and $|\Delta m^2_{23}| \sim 8\ 10^{-5}$ eV^2 [1]. There are however a number of anomalies which, if experimentally confirmed, could hint to the presence of an additional, large squared mass difference in the framework of a four-neutrinos mixing. If more than the two oscillation signals were to be observed, additional Physics beyond the Standard Model in the neutrino sector will be necessary. The possible presence of sterile neutrinos was originally proposed by B. Pontecorvo [2], and has been discussed since a very long time, but so far without conclusive evidence. Two distinct classes of phenomena have been analyzed, namely a) the apparent reduction in the $\overline{\nu}_e$ low energy neutrinos from nuclear reactors [3] and from the signal from Mega-Curie sources in the Gallium experiments [4, 5] originally designed to detect solar neutrino deficit, and b) evidence for a $\overline{\nu}_e$ excess signal of in interactions coming from neutrinos from particle accelerators [6, 7, 8]. These experiments may all point out the possible existence of the fourth non standard neutrino state driving neutrino oscillations at a small distance, with typically $|\Delta m^2_{new}| \geq 1$ eV^2 and relatively large mixing angle $\sin^2\theta_{new} \simeq 0.1$ [9].
The class a) of phenomena hint at a significant fast disappearance rate in the initial $\overline{\nu}_e$ production

and the class b) predicts an anomalous $\nu_\mu \rightarrow \nu_e$ oscillation with similar, large $|\Delta m^2_{new}|$ values, much greater than the ones of the current three neutrino mixing model. More recently the existence of a fourth neutrino state may be also hinted or at least not excluded by cosmological data mainly coming from WMAP and other observations [10]. It is only because the masses of known neutrino species are so small, that their contribution to the Dark Matter of the Universe can be excluded. The situation could be altered by the additional presence of sterile neutrinos, provided that they are massive enough. Therefore, the presence of massive sterile neutrinos, will also contribute to clarify the Dark Matter problem.

2. A more direct approach to neutrino oscillations

In order to definitely clarify the existence of a fourth non standard neutrino state in the above mentioned parameter range, a new experiment at CERN PS beam is proposed (Fig. 1). It's based on two strictly identical LAr-TPC detectors observing the electron-neutrino signal in the Near and Far positions, the first one of about 150 tons at 127 m from the proton target, the second one of about 600 tons placed 850 m away [11]. In this way, all cross sections and experimental biases cancel out. In absence of oscillations the two experimentally observed event distributions must be identical, the background electron neutrino radial and energy distributions being extremely similar in the two positions. Any difference in the event distributions between the two detectors should be attributed to the possible existence of oscillations, and both the mixing angle $\sin^2 \theta_{new}$ and the mass difference $|\Delta m^2_{new}|$ can be separately measured.

Figure 1. Experimental set-up proposed at the CERN PS neutrino beam with two LAr-TPC detectors.

The CERN-PS ν_e spectra, expected at the level of 0.5 % of ν_μ, are very closely identical in the Near and Far positions. The physical reason of this effect has to be identified in the fact that while the ν_μ spectrum is dominated by the two body $\pi \rightarrow \mu\nu_\mu$ decays, where the neutrino directions are narrowly distributed along the axis, the ν_e contamination is dominated by the three body decays of K and μ where there is a much wider neutrino angular spread. The similarity is further enhanced by the fact that the Near and Far detectors will be designed with identical internal configurations. In absence of oscillations, after some beam related small spatial corrections, the two energy spectra should be a precise copy of each other, independently of the specific experimental event signatures. Therefore, an exact observed proportionality between the two ν_e spectra, would imply directly the absence of neutrino oscillations over the measured interval of L/E.

A key issue of the proposed experiment will be the LAr-TPC detection capability of all the genuine ν_e events, allowing the reconstruction of the totality of neutrino events without restricting to the QE interactions, and the very high level of rejection of the associated background events. The LAr-TPC technique appears very well suited for this purpose, because of its excellent imaging and calorimetric capabilities, which allow very efficient electron - π^0 separation, together with unambiguous electron identification.
The ICARUS T600 detector [12], presently in operation at LNGS, consists of a large cryostat, with a size of about 500 t of active mass, split into two identical, adjacent half-modules with internal dimensions 3.6×3.9×19.6 m^3 and filled with ultra-pure liquid Argon. Each half-module houses two TPCs separated by a common cathode, with a drift length of 1.5 m.
Ionization electrons, produced by charged particles along their path are drifted under uniform electric field (E_D = 500 V/cm) towards the TPC anode made of three parallel wire planes, facing the drift volume. A total of ~54000 wires are deployed, with a 3 mm pitch, oriented on each plane at a different angle (0^0, $+60^0$, -60^0) with respect to the horizontal direction.The first two planes provide signals in non-destructive way by appropriate voltage biasing; the ionization charge is collected and measured on the last collection plane. The relative time of each ionization signal, combined with the electron drift velocity information (v_D ~1.6 mm/μs), provides the position of the track along the drift coordinate. Combining the wire coordinate on each plane at a given drift time, a three-dimensional image of the ionizing event can be reconstructed with a remarkable resolution of about 1 mm^3 (Fig. 2). The absolute time of the ionizing event is provided by the prompt VUV (128 nm) scintillation light emitted in LAr (~5000 photons/mm for minimum ionizing particles) and measured through arrays of Photo Multiplier Tubes (PMTs), installed in LAr behind the wire planes.

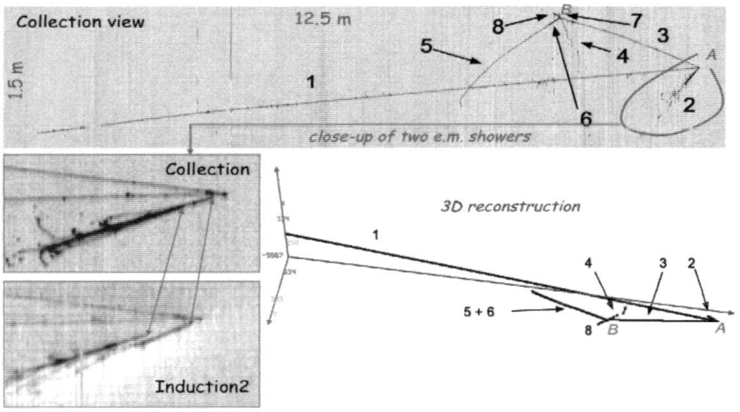

Figure 2. A fully reconstructed CNGS ν_μ-CC event. Primary vertex (A): very long leading muon (1), e.m. cascade reconstructed as a neutral pion (2), charged pion (3). Secondary vertex (B): the longest track (5) is a muon coming from stopping K (6). The muon decay is also identified. All particle momenta are measured for a complete kinematic reconstruction of the event.

According to the present proposal the ICARUS T600 detector will be moved from the LNGS laboratory into the Far position of the CERN-PS neutrino beam (Hall B-191) after the completion of the presently running CNGS2 experiment. An additional LAr-TPC using the same design of the T600 will be constructed for the Near site.

3. The CERN PS neutrino beam

The proposed experiment will exploit the CERN-PS neutrino beam-line, originally used by the BEBC-PS180 Collaboration [13] and successively re-considered by the I216/P311 Collaboration [14]. The neutrino beam will be a low energy ν_μ beam, centered at about 1 GeV, produced by 19.2 GeV protons, of intensity 1.25×10^{20} pot/yr. The CERN-PS can deliver routinely on a refurbished TT7 transfer line 3×10^{13} protons per cycle at 20 GeV/c. More than 1.25×10^{20} pot/year can be collected by the experiment in one year of data taking for an optimal machine time availability. The proton transfer line, the new target and the horn focusing system must be reconstructed. Some preliminary studies about the realization of the TT7 PS neutrino beam line [15] have been already completed at CERN.

As reference case the neutrino beam set-up used by the BEBC-PS180 experiment has been chosen. An optimized design of new and possibly improved beam optics will be the subject of further studies. The 19.2 GeV/c proton beam was extracted from the PS and impinged on a 80 cm long, 6 mm diameter beryllium target. This was followed by a pulsed magnetic horn designed to focus positive particles of momentum around 2 GeV/c into a decay tunnel of about 50 m length. Most of the muons and remaining hadrons were then absorbed by a 4 m thick iron shield and 65 m of earth.

The total accumulated number of accelerated protons from the PS in 180 days of operation is now about $\sim 3.6 \times 10^{20}$ pot/year. With one third of the protons dedicated to the neutrino beam line, the expected intensity will be about 1.25×10^{20} pot/year. Such a target value has also been already assumed for the previously proposed I216/P311 experiment. The positively and negatively focused CERN-PS neutrino spectra, originally used by the BEBC-PS180 Collaboration, are shown in Fig. 3 for the most relevant Far position at 850 m from the PS target.

Event rates in the CERN-PS beam for the un-oscillated spectra and a few parameter values of $\nu_\mu \to \nu_e$ oscillations are reported in Table 1.

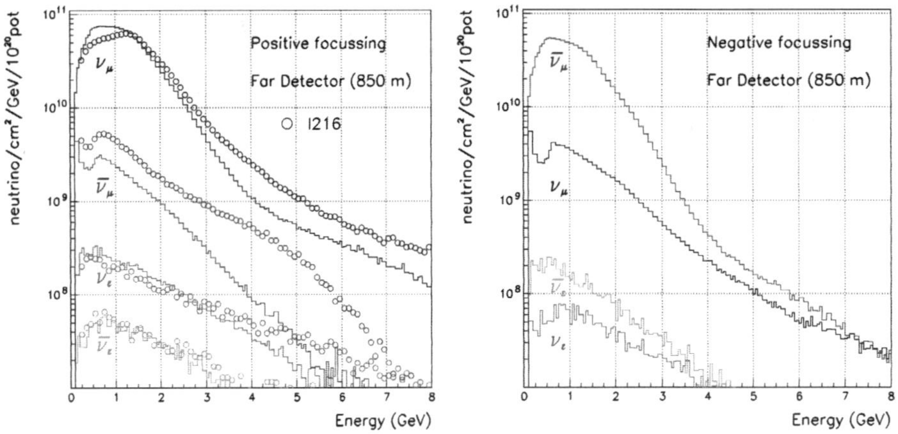

Figure 3. Expected composition of neutrino beam at Far detector position for both positive (left) and negative (right) focusing.

4. The Near and Far detectors

The two closely similar LAr-TPCs (Fig. 4) will be located respectively at 850 m and 127 from the PS target in the existing locations B-191 and B-181 respectively. The size of the buildings

Table 1. Event rates for the Far and Near detectors given for 2.5×10^{20} pot for $E_\nu < 8$ GeV. Neutrino fluxes are taken from [14]. The oscillated signals are clustered below 3 GeV of visible energy.

	Neutrino focus		Antineutrino focus	
	FAR	NEAR	FAR	NEAR
Fiducial mass	500 t	150 t	500 t	150 t
Distance from target	850 m	127 m	850 m	127 m
ν_μ interactions	1.2×10^6	18×10^6	2.0×10^5	2.3×10^6
QE ν_μ interactions	4.5×10^5	66×10^5	87000	10^6
Intrinsic $\nu_e + \overline{\nu}_e$	9000	120000	2000	29000
Intrinsic $\nu_e + \overline{\nu}_e$ $E_\nu < 3$ GeV	3900	54000	880	13000
ν_e oscillation events:				
$\Delta m^2 = 2.0$ eV^2; $\sin^2 2\theta = 0.002$	1194	1050	230	58
$\Delta m^2 = 0.4$ eV^2; $\sin^2 2\theta = 0.02$	2083	2340	330	115
$\Delta m^2 = 0.064$ eV^2; $\sin^2 2\theta = 0.96$	3350	1250	465	140
$\Delta m^2 = 4.42$ eV^2; $\sin^2 2\theta = 0.0066$	2980	25050	490	3220

is perfectly adequate and no major civil engineering is needed. The T300 test in Pavia has demonstrated that the LAr TPC detectors can safely operate on surface.

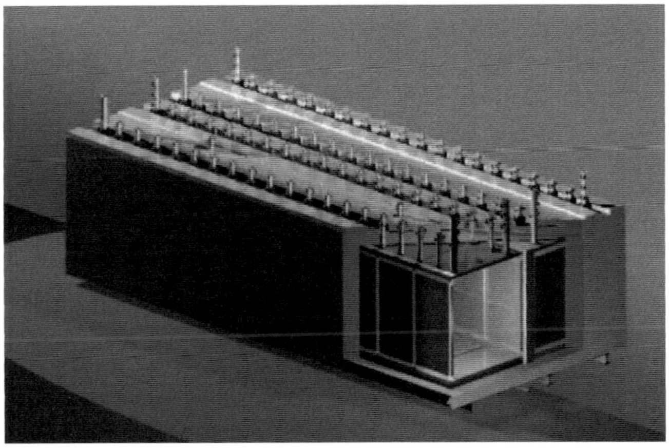

Figure 4. Schematics of T600 detector to be placed in Hall B-191.

While the Far detector is relying on the already existing T600, the Near detector must be constructed anew. In order to ensure the maximum similarity between the two detectors, required in order to ensure identical behaviors in absence of oscillations, the Near detector design will be identical to the T600 Far detector except the total mass which has been conservatively chosen to be of 150 t, namely a clone of a single T300 half-module with the length reduced by a factor 2 (about 12 m). This allows keeping untouched the inner detector layout for a total of 14200 wires. The already existing basement pit of hall B-181, previously used for neutrino oscillation experiments, fits perfectly the Near detector. Electronics and cryogenic equipments will be installed close to the detector pit. All the other component of the detector, cryogenic elements, internal photomultipliers, front-end electronics and read-out system, DAQ and ancillary equipments, will be replicated accordingly to the downscaled detector mass. Some obvious improvements and some simplifications over the by now 10 years old T600 may be implemented with the new detector.

5. Signal selection and background rejection

In the LAr-TPC all reaction channels with electron production can be analyzed without the need to restrict the search to the quasi-elastic channel, which accounts for about one half of the events. Moreover, events due to neutral currents are also very well identified and can be rejected to a negligible level.

The energy resolution and detector granularity are largely adequate for the 1-3 Gev energy range relevant for the proposed experiment. The performance of the LAr-TPC has been extensively studied in the 2001 T300 technical run with cosmic rays and now at LNGS [12]. Electromagnetic energy resolution is $\sqrt{E}/E = 0.03 \times E(GeV) + 0.01$, in agreement with π^0 invariant mass measurements in the sub-GeV energy range. At higher energies the estimated resolution for hadronic showers is $\sqrt{E}/E = 0.30 \times E(GeV)$. However the LAr-TPC detector allows to identify and measure, track by track, each hadron produced in 1 GeV neutrino interactions, through ionization and range, leading to a much better energy resolution. Indeed most of the particles, generated in the neutrino interactions, come to rest within the detector, including muons. Moreover the momentum of escaping muons will be extracted from the deflection angle along the track due to multiple scattering with a resolution which can be as good as 10 % depending mainly on track length. The measured dE/dx energy allows a clear pion to proton separation as obtained from the last part of the residual range. Moreover $\pi^0 \rightarrow 2\gamma$ events are easily recognized by detecting the two γ conversion, the measurement of dE/dx and of the two gammas invariant mass.

Quasi-elastic neutrino events in the 50 litre ICARUS LAr-TPC exposed to the CERN-WANF beam in coincidence with the NOMAD experiment [16] have been readily reconstructed in 3D with particle identification, momentum balance and π^0 rejection in a energy range which is relevant for the proposed experiment [17].

Monte Carlo simulations show that the energy reconstruction of electron neutrino interaction events at around 1.5 GeV is not affecting the signal/background ratio if a minimal cut of 50 cm in the longitudinal direction and a 10 cm cut on the sides of the sensitive volume are performed, corresponding to a fiducial volume of about 90 % of the active one. Electron identification is also ensured under these geometrical cuts. Due to the directionality of the neutrino beam, the probability that an electron escapes from the instrumented volume before initiating a shower is extremely small: only 2 % of the electrons travel through a LAr-TPC thickness smaller than $3X_0$ and 0.3 % travel less than $1X_0$ in the instrumented volume. With the previously described fiducial volume cuts, the expected average neutrino energy resolution is about 14%.

In view of the excellent imaging capability of LAr-TPC, π_0 from ν_μ NC events cannot be misidentified as electrons. In fact π^0 are almost fully separated by electrons by requiring both photon converting at least 2 cm from the ν_μ vertex, a π^0 mass reconstructed within 10 % accuracy and on the basis of dE/dX signal analysis. This method provides 90 % electron neutrino identification efficiency with a 0.1% π^0 mis-interpretation probability within the fiducial volume.

6. Sensitivity to ν_e and ν_μ disappearance signals

The intensity of the ν_e signal, much smaller than the one of the dominant ν_μ's (about 0.5%), is largely sufficient to collect an adequate number of events.

In Figure 5 the 90% confidence levels for the actual oscillation mechanism in the $\sin^2(2\theta_{new} - \Delta m^2_{new}$ plane are shown for the presently proposed experiment with an integrated intensity corresponding to (a) 2.5×10^{20} pot (protons on target) for the original beam intensity of about 30 kW of the previous CERN/PS experiments, (b) to 7.5×10^{20} pot for the newly planned 90 kW neutrino beam at CERN/PS and (c) a 270 kW curve. They are also compared with the "anomalies" from the combination of the published reactor neutrino experiments, GALLEX and Sage calibration sources experiments. The disappearance signal in the same $\sin^2(2\theta_{new}) - \Delta m^2_{new}$ range may also be studied independently with the dominant ν_μ and $\overline{\nu}_\mu$.

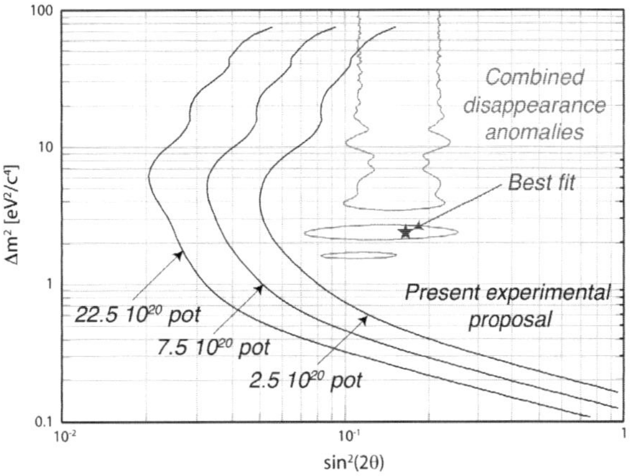

Figure 5. Sensitivities (90% C.L.) in the $(\sin^2(2\theta), |\Delta m^2|)$ region for 30, 90 and 270 kW beam power.

The $\nu_\mu \to \nu_{new}$ process has a different and independent signature than the one due to explicit $\nu_\mu \to \nu_e$ oscillations related to LSND/MiniBooNE anomalies. However the ν_μ and $\overline{\nu_\mu}$ spectral shapes, hereby primarily due to pion decays, are significantly different in the "Near"and "Far" positions. In the energy range below 3 GeV, where the effect is expected, the relative differences amount to about 30% and they may be predicted to about 1%, which is larger than what expected from the huge available statistics but quite significant for a test. The presence of the two alternate phenomena presumably with different values of $\sin^2(2\theta_{new})$, if confirmed, will hint at the presence of an appropriate fourth neutrino coupling $U_{4,k}$ with $k = \mu$ and e.

7. Sensitivity to $\nu_\mu \to \nu_e$ oscillations and other related physics

A sensitivity of $\sin^2 2\theta < 3 \times 10^{-4}$ (for $\Delta m^2 < 2$ eV^2) and $\Delta m^2 < 0.02$ eV^2 (for $\sin^2 2\theta = 1$) at 90 % C.L. is expected with a two year exposure at the CERN-PS ν_μ beam (Fig. 6). The parameter space region allowed by the LSND experiment is fully covered, except for the highest Δm^2 region. The sensitivity has been computed assuming a 3% systematic uncertainty in the "Far" to "Near" ratio prediction. Similarly in anti-neutrino focusing, twice as much exposure (5.0×10^{20} pot) allows to fully cover the LSND parameter region. Very different and clearly distinguishable patterns are indeed possible depending on the actual values in the $(\Delta m^2 - \sin^2 2\theta)$ plane.

As an additional bonus, the large statistics of excellent quality data will also profit to the knowledge of neutrino cross-sections. Precise measurements in the 0-5 GeV energy range are required by present and future neutrino oscillation experiments. Existing data on charged current quasi-elastic, deep inelastic, and single pion production are affected by a large uncertainty, especially at the lower energies [18]. Owing to the very low detection threshold of the Liquid Argon technique, the exposure at the CERN-PS neutrino beams could significantly improve the neutrino cross-section knowledge. Approximately 1.2×10^6 and 1.8×10^7 charged current events per 2.5×10^{20} pot will be recorded from the CERN-PS in the "Far" and "Near" detectors respectively, with a neutrino spectrum peaked at around 1 GeV. Neutral current cross-sections are also measured, with approximately 4.4×10^5 and 6.0×10^6 events in the "Far" and "Near" detectors respectively.

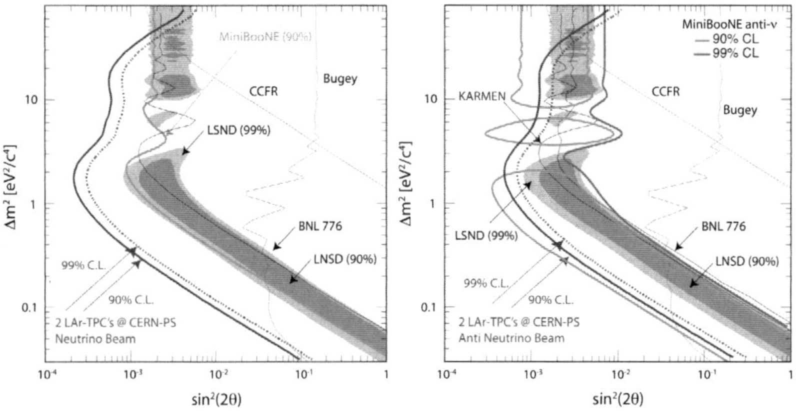

Figure 6. Expected $\nu_\mu \rightarrow \nu_e$90% C.L. sensitivity for the proposed experiment for both neutrino (left) and antineutrino (right) beams. MiniBooNe results are for 6.6×10^{20} pot (neutrino mode) and 8.58×10^{20} pot (antineutrino mode).

References

[1] Nakamura K *et al.* (Particle Data Group) 2010 *J. Phys.* G **37** 075021
[2] Pontecorvo B 1967 *Zh. Eksp. Teor. Fiz.* **53** 1717 [*Sov. Phys. JETP* 1968 **26** 984]
[3] Mention G *et al.* *Preprint* arXiv:1101.2755v1 [hep-ex] and previous references therein
[4] Abdurashitov J N *et al.* (SAGE Collaboration) 2009 *Phys. Rev.* C **80** 015807 and references therein
[5] Kaether F, Hampel W, Heusser G, Kiko J and Kirsten T 2010 *Phys. Lett.* B **685** 47 and references therein
[6] Aguilar A *et al.* (LSND Collaboration) 2001 *Phys. Rev.* D **64** 112007
[7] Aguilar-Arevalo A A (MiniBooNE Collaboration) 2009 *Phys. Rev. Lett.* **102** 101802;
Aguilar-Arevalo A A *et al.* (MiniBooNE Collaboration) *Preprint* arXiv:1007.1150 [hep-ex]
[8] Zimmerman E 2011 "Short-Baseline Neutrino Physics at MiniBooNE" talk at PANIC11 Conference
[9] Giunti C and Laveder M 2010 *Phys. Rev.* D **82** 053005 and references therein
[10] Komatsu E *et al.* *Preprint* arXiv:1001.4538 [astro-ph.CO] and references therein
[11] Baibussinov B *et al.* *Preprint* arXiv:0909.0355v3 [hep-ex];
Rubbia C *et al.* 2011 "A search for anomalous neutrino disappearance and oscillatory phenomena at the CERN-PS" CERN SPSC-M-773
Rubbia C *et al.* 2011 "A comprehensive search for anomalies from neutrino and anti-neutrino oscillations at large mass differences ($\Delta m^2 \simeq 1\ eV^2$) with two LAr-TPC imaging detectors at different distances from the CERN-PS" CERN-SPSC-P-345
[12] Rubbia C 1977 CERN-EP/77-08;
Amerio S *et al.* (ICARUS Collaboration) 2004 *Nucl. Instr. And Meth.* A **527** 329;
Kisiel J "ICARUS T600 results" these proceedings
[13] Baldo-Ceolin M *et al.* 1980 CERN/PSCC/80-130, PSCC/P33;
Angelini C *et al.* (PS180 Collaboration) 1986 *Phys. Lett.* B **179** 307
[14] Guler M *et al.* 1999 CERN-SPSC/99-26, SPSC/P311
[15] Information on CERN-PS neutrino beam refurbishing project available at: http://info-psnf.web.cern.ch/info-PSNF/Presentations.htm
[16] Astier P *et al.* 2003 *Nucl. Instr. And Meth.* A **515** 800 and references therein
[17] Arneodo F *et al.* (ICARUS Collaboration) 2006 *Phys. Rev.* D **74** 112001
[18] Zeller G P "Low Energy Neutrino Cross Sections: Comparison of various Monte Carlo Predictions to Experimental Data" talk at Nu-INt02 Irvine 2002 (*Preprint* hep-ph/0312061v1)

Future R&D experiments for Super-Beams, Neutrino Factories and Beta-Beams

R Edgecock

University of Huddersfield, Queensgate, Huddersfield, HD1 3DH, UK; and
STFC Rutherford Appleton Laboratory, Didcot, OX11 0QX, UK.

E-mail: rob.edgecock@stfc.ac.uk

Abstract. There are three principle options for future beam based neutrino oscillation facilities that could discover and measure CP-violation in the lepton sector. These are conventional Super-Beams, Neutrino Factories and Beta Beams. Several projects have been taking place world-wide to study examples of these facilities. In Europe, one of the most important of these is the Framework Programme 7 supported project, EUROnu. These projects are in the process of identifying experimental R&D work that must be done before a technical design of the facility can be finalised and construction started. This paper will summarise this work for each type of facility, based largely on what has been done in EUROnu.

1. Introduction

The EUROnu project [1] started on 1st September 2008 and is due to run for four years. It is studying the following three facilities:

- The CERN to Fréjus Super-Beam [2].
- A Neutrino Factory, in close collaboration with the International Design Study for a Neutrino Factory (IDS-NF) [3].
- Two versions of the Beta Beam [4].

Each of these is discussed in more detail below, along with the experimental R&D already identified. In addition to the accelerator facilities, EUROnu is also studying the performance of the baseline detectors for each and determining the physics performance.

2. Super-Beam

In a conventional Super-Beam, the neutrino beam is created by impinging a proton beam onto a target to make pions, focusing the pions into a decay channel using a pulsed magnetic horn and producing the neutrino beam from the pion decay (see figure 1). Future, high intensity, facilities plan to use proton beam powers up to around 4 MW to produce a higher flux of neutrinos.

In the case of EUROnu, the facility being studied is the CERN to Fréjus Super-Beam, employing the high power Superconducting Proton Linac (HP-SPL) to produce a 4 MW proton beam of 5 GeV, pulsed at 50 Hz. The linac is followed by an accumulator ring for reducing the number of proton bunches, before delivering them to the target. To minimize neutron production and hence the heat load and radiation damage to the target surroundings, it is planned to use a low atomic number material for the target and to base both the target and the horn design as closely as possible on existing technology to reduce the amount of R&D that needs to be done.

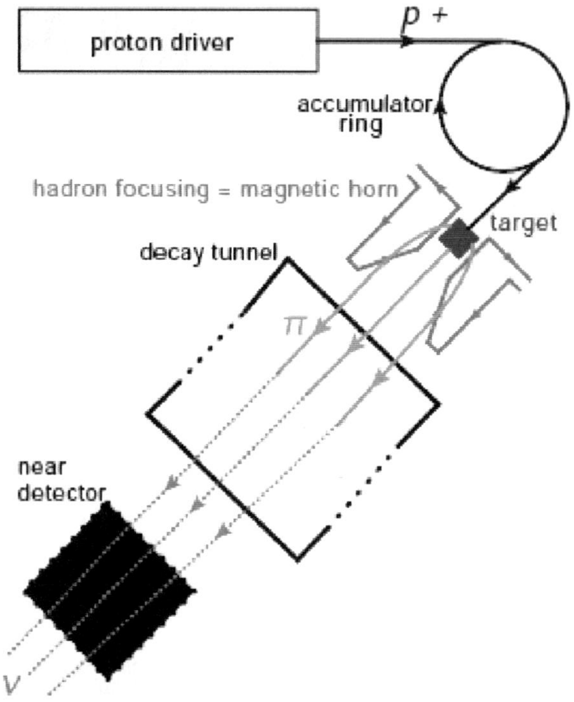

Figure 1. The layout of the CERN to Fréjus Super-Beam

If constructed, the HP-SPL will consist of a number of sections. The first of these, Linac 4 [5] is currently under construction at CERN, though an upgrade will be required to deliver sufficient current for 4 MW operation. This will accelerate the beam to 160 MeV and use entirely normally conducting technology. This will be followed by a number of superconducting sections, accelerating the beam first to 2.5 GeV, at which point it can be extracted for other uses, and then to 5 GeV.

R&D is required on all sections of the linac, in addition to that already envisaged for the low power SPL, though a number of projects are already underway. Work on an H^- ion source with a large enough duty cycle, delivering sufficient current with a high reliability needs to continue and significant R&D is required on the superconducting RF system. The latter includes the upgrade of an existing test stand at CERN, prototyping and testing of the superconducting cavities, the design and test of the high power RF couplers, higher order mode dampers and tuners required, the prototyping of the high power klystron modulators and the cryomodules. Further, injection into the accumulator ring will employ charge exchange and this will require the beam to be stripped at full power. This system will also require careful design and R&D.

To minimize the extrapolation required for the target and horn compared to existing facilities such as T2K [6] and MINOS [7], it is planned to employ four of each and scan the beam across them (see figure 2). As a result, each will need to run at 12.5 Hz, rather than 50, and each will see about 1 MW, rather than 4. To do this, a system will need to be devised which will deliver the beam to each target in turn, while keeping beam losses within allowed levels, ensuring the remnant proton beam is correctly dumped and maximizing the flux of captured pions. The baseline target design is a so-called pebble bed target consisting of 3mm diameter spheres of either beryllium or titanium tightly packed within a canister. This has slots in it which allows helium gas at a pressure of around 10 bar to flow through, taking away the heat deposited in the bar. Modelling suggests that this will keep the temperature of the spheres at such a level that their strength will be more than sufficient compared to the thermal shock from the beam. Nevertheless, a prototype of such a target will need to be constructed and the cooling tested by using a coil to heat the spheres. Furthermore, the target should be subjected to a beam of the correct energy density, while at the operating temperature, using, for example, the HiRadMat facility at CERN, to verify the effects of thermal shock and to further test the cooling. The lifetime of the target due to radiation damage and fatigue will also need to be determined. Finally, although each target will only see 1 MW, a 4 MW target station will need to be designed, located at CERN and a license to operate it obtained. Due to the location and nature of the laboratory, this will not be entirely straight-forward.

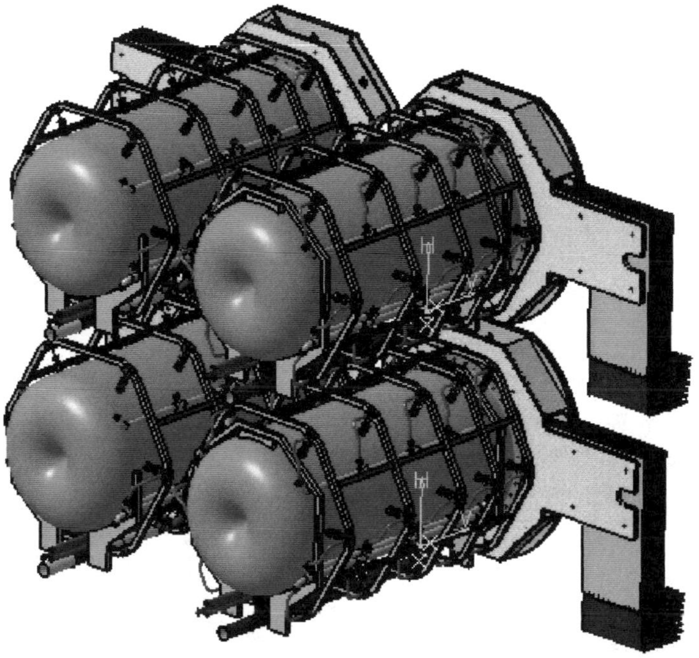

Figure 2. The arrangement of four targets and horns planned for the Super-Beam project. The targets are located within each horn.

The horn design is based on that of MiniBooNE. Each horn will need to be pulsed at 12.5 Hz using a current between 300 and 450 kA, depending on the physics requirements. It will subject to significant heating due to the combined effects of secondary particles and the current, corresponding to a maximum of 12 kW in the section closest to the target. A number of cooling schemes have been studied and modelling suggests these can keep the maximum temperature at the 300 K level. Nevertheless, a prototype will need to be constructed and tested to verify this. Furthermore, the pulsing will result in a maximum stress on the horn of at least 18 MPa and the fatigue lifetime resulting from this will need to be assessed. The horn will also be subject to significant radiation damage and the implications of this on the lifetime will need to be determined. Finally, the pulsing circuit will need to deliver a repeatable current pulse with the correct characteristics to the horn. A prototype circuit will need to be built and tested.

3. Neutrino Factory

The design of the Neutrino Factory is being studied by the IDS-NF, of which EUROnu is an important part, forming the bulk of the European contribution to the study. The baseline layout for the accelerator is shown in figure 3. The neutrino beams are produced by the decay of stored μ^+ and μ^- beams in two storage rings. To meet the physics aims, the muons are created by firing a 4 MW proton beam into a heavy metal target and focusing as many of the pions produce as possible into a decay channel using a 20 T solenoid built from combined normally and superconducting coils. The muons from the decay are captured, bunched, phase rotated and cooled, before being accelerated as quickly as possible using a combination of a linac, two re-circulating linear accelerators and a non-scaling Fixed Field Alternating Gradient (FFAG) accelerator. They are then injected into the storage ring, the neutrino beams being produced from the muon decays in the straight sections pointed towards the far detectors.

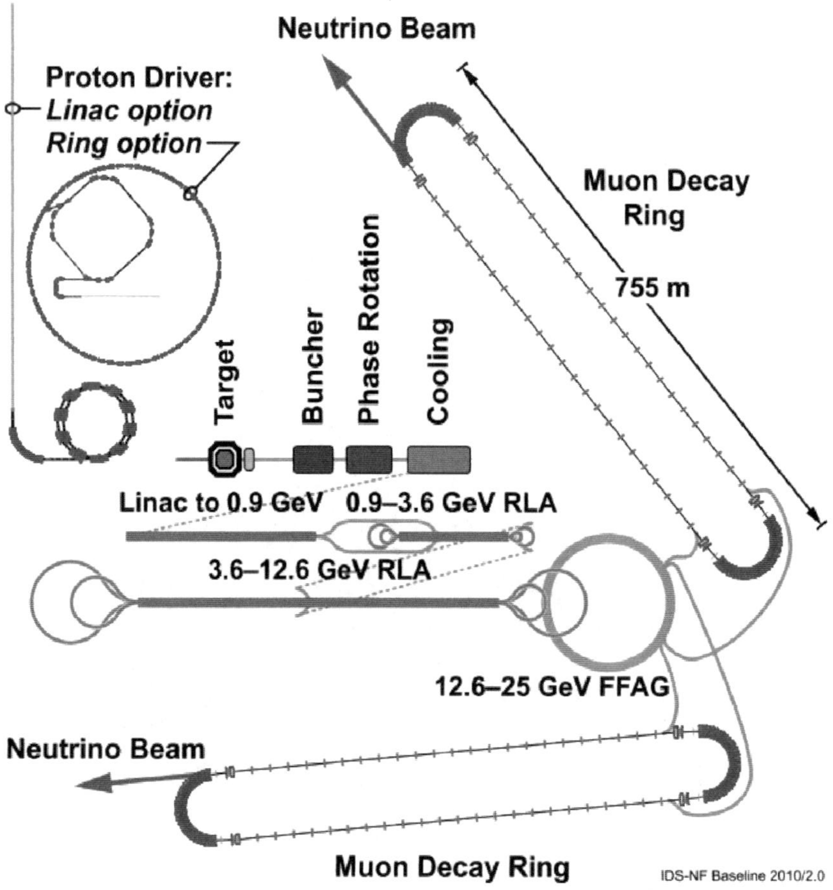

Figure 3. Current layout for the Neutrino Factory.

All parts of the Neutrino Factory are well beyond the state of the art and hence a significant amount of R&D is required. For the proton driver, two options are considered: a linac option or a ring option. As far as EUROnu is concerned, the linac option is also the HP-SPL, so much of the R&D is the same as described in section 2. However, an additional ring is required to compress the proton bunch lengths to the required 3 ns. For this, and for the ring option, R&D is required to ensure this can be achieved, to study injection and extraction and to ensure beam losses are kept under control.

Much R&D is also still required for the target and the target station. Of particular concern is the heat load in the target station from the secondaries produced and the remnant proton beam. With the original baseline, the heat load in the first two superconducting solenoid coils far exceeds a manageable level and various methods are being considered to reduce this. Adding more shielding will allow the heat load to be brought to tolerable levels, but increases the radius of the coils by a factor of two. The design of these was already difficult due to the magnetic forces and the stored energy. In addition, more than a MW is deposited in the shielding and this must be removed. Recent experience in existing high power target stations has shown that water cooling can be problematic due to tritium production. Another alternative being considered is the use of lower atomic number elements for the target as these should produce a smaller thermal neutron flux and hence reduce the heat load. This work is at an early stage.

Two major R&D projects are already underway for the next section of the machine, the muon front-end. MuCool [8] is prototyping the main components of the muon cooling channel and, in particular, testing what accelerating gradient can be achieved in the RF cavities, which will be in the strong magnetic field of the focusing solenoids of the channel [9]. The MICE experiment (see figure 4) plans to do a detailed study of the cooling process, ionization cooling. In addition to these, a number of other R&D studies are underway. Modelling suggests that as well as the desired large flux of both signs of muons, a number of other particle species, including protons, pions and electrons make it a significant way down the front-end. Although these are lost throughout the front-end, there are certain regions in which the losses are particularly high. The result is that the beam power lost would far exceed the conical 1 W/m required for hands-on maintenance. As a result, a chicane is being studied at the start of the front-end to ensure these particles are lost in a controlled way.

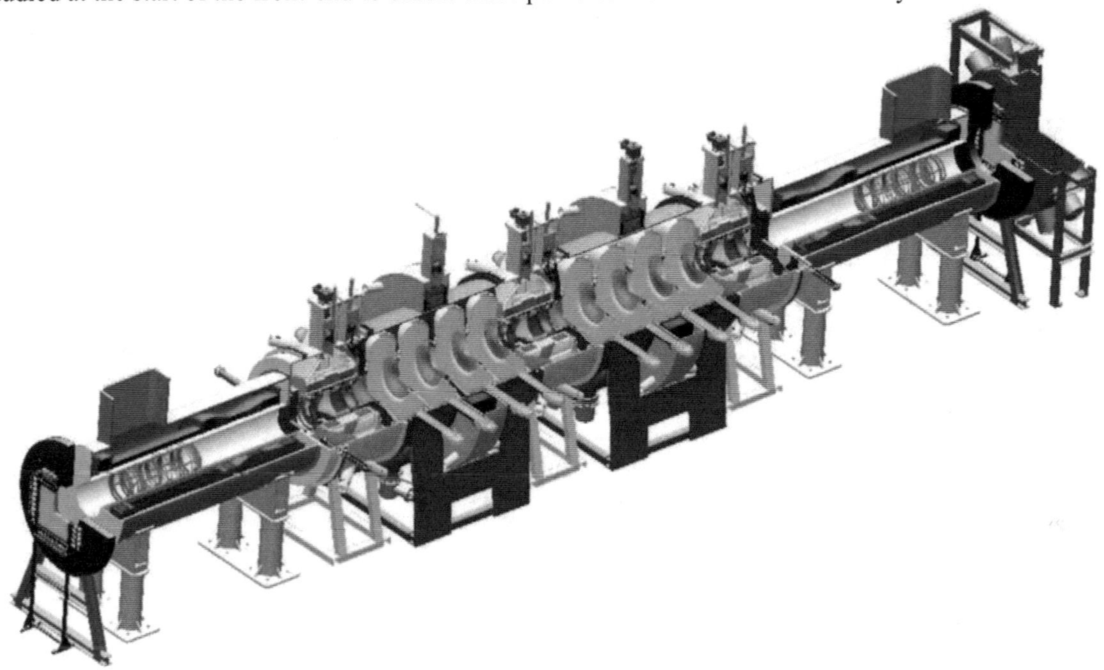

Figure 4: The MICE experiment. The two cooling cells in the centre are surrounded by instrumentation to identify muons and measure their parameters precisely.

In case the solenoid magnetic fields have too big of an impact on the RF gradient in the cavities, alternative cooling lattices which reduce this field are also under study. Promising results have already been achieved, but further work is required. Once the conceptual design of the front-end has been finalized, a full engineering design and layout will be required.

The muon acceleration systems consists of a linear accelerator to increase the beam energy to 0.9 GeV from 250 MeV, two re-circulating linear accelerators (RLA) going to 3.6 and 12.6 GeV, respectively, and a non-scaling FFAG accelerating to 25 GeV. For the linacs and RLAs, a full lattice design and physical layout still need to be made. Once this is done, R&D will be required on the main lattice components, in particular the "spreader" magnets used to steer the beam into the correct arcs of the RLAs, the arc magnets and the RF system. For the FFAG, a major R&D project has already taken its first main step. This is the EMMA non-scaling FFAG ring, see figure 5. As well as being a proof-of-principle project for this type of accelerator, EMMA is also testing out many of the features of the muon FFAG for a Neutrino Factory. For example, it has recently demonstrated that both serpentine acceleration and multiple integer resonance crossings work [10]. Nevertheless, there are still a number of issues that need to be studied, including matching the FFAG to the other Neutrino Factory systems, 6D tracking, prototyping the main ring magnets and prototyping the injection and extraction systems from the ring.

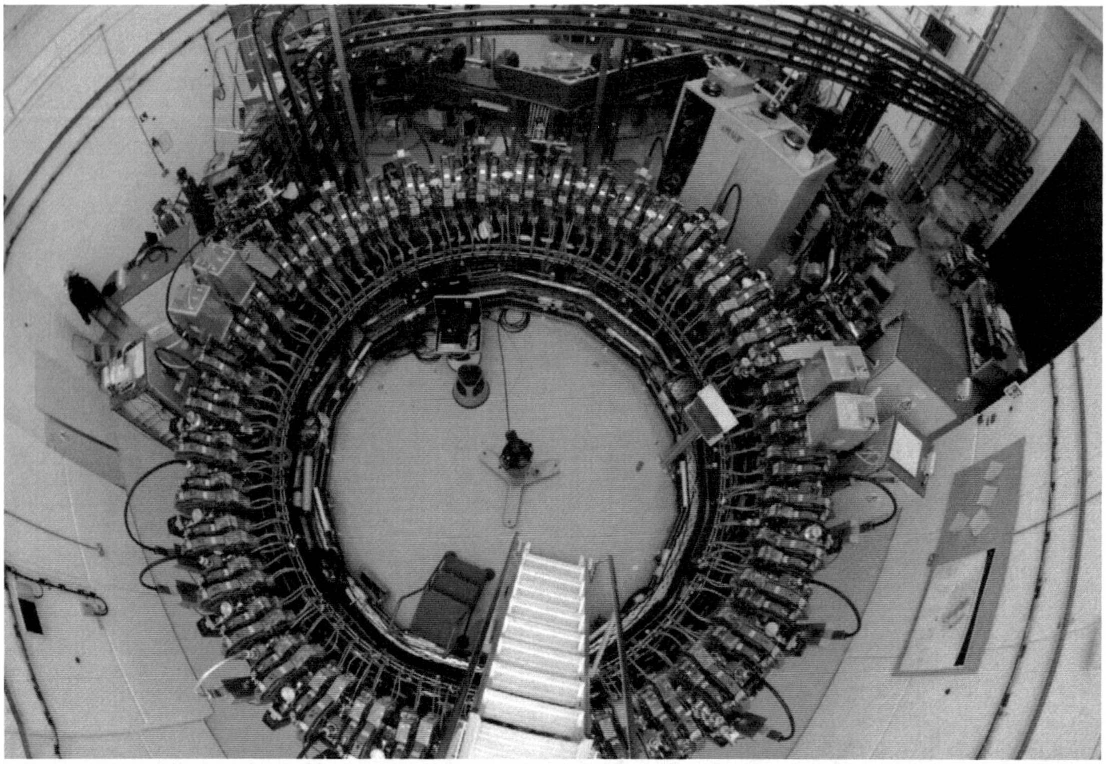

Figure 5. The EMMA FFAG, showing the ring, the injection line (to the right), a diagnostics beamline (to the left) and part of the ALICE accelerator [11], which is used as the injector (in the background).

4. Beta Beam

In a Beta Beam facility, the neutrino beams are produced by the decay of stored beta emitting ions. Two ion species are required, one decaying to electron neutrinos and the other to electron anti-neutrinos. The facility, shown in figure 6, is split into three sections: (1) ion production, capture, bunching and initial acceleration; (2) further acceleration in existing CERN machines; and (3) the ion decay ring (DR). The main challenge for the project is to create, accelerate and store a sufficient flux of both ion species to meet the physics goals and this has formed the focus of activities so far and will continue to do so for the near future.

Two pairs of ion species are being studied, 6He/18Ne and 8Li/8B. The first of these was originally studied by the EURISOL project [12] and this determined that a sufficient flux of 6He could be produced by existing techniques, but 18Ne was more problematic. As a result, EUROnu has studied the production of 8Li and 8B via a production ring and more recently, 18Ne via another technique compared to EURISOL. It should be noted that the Li and B produce higher energy neutrinos, requiring the neutrino far detector to be further away than for He and Ne. The net result is that five times the flux of these ions is required for the same physics sensitivity.

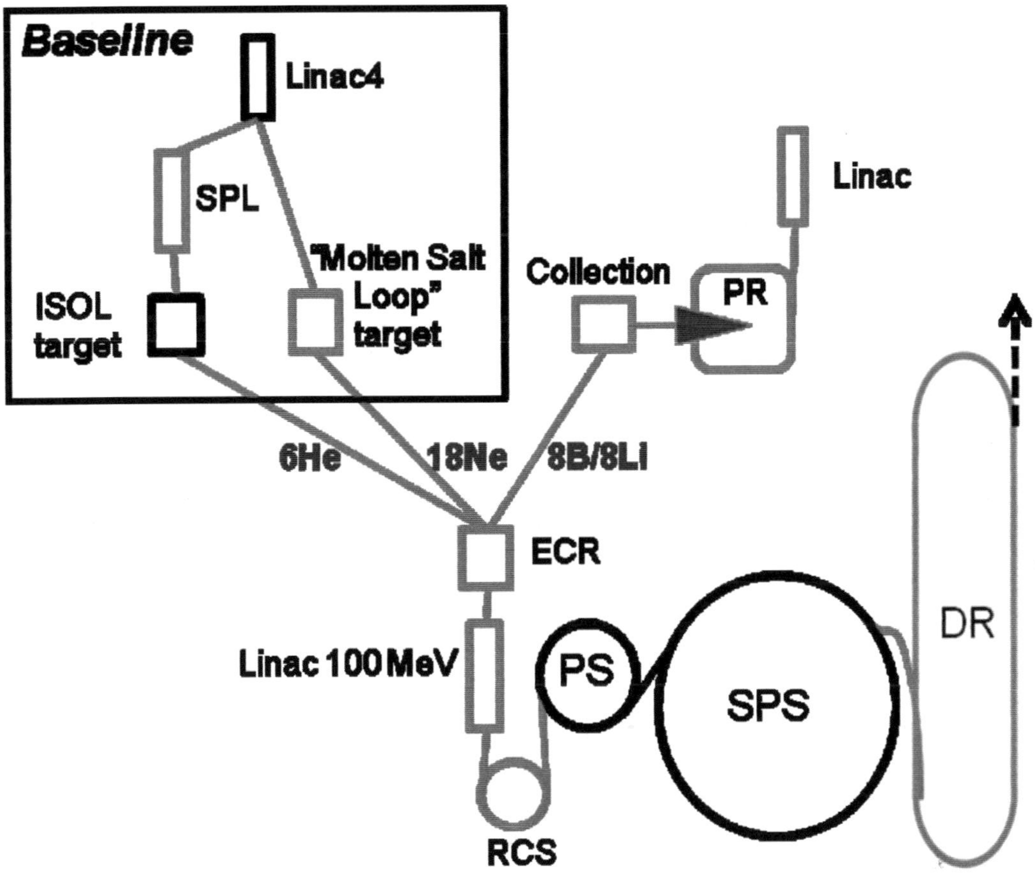

Figure 6. The layout of a Beta Beam facility. The proposed production mechanisms for both pairs of ions are shown. New installations are shown in red, while those that already exist at CERN are shown in black.

Initial studies of 8Li and 8B production via a gas jet target in a production ring suggest that it may be very difficult to achieve the required rate as the target density would need to be too high and it could lead to vacuum problems. As a result, an alternative direct production method using liquid film targets has been proposed. To determine what is possible, the cross-sections for both production processes are being measured in Legnaro (see figure 7). Meanwhile, studies of 18Ne direct production using a Linac 4 upgraded to 6 mA and a molten salt loop target have proved encouraging. It is now planned to undertake an experiment at ISOLDE to verify that the predicted production rate can be achieved. For now, the 6He/18Ne pair is considered to be the Beta Beam baseline.

In addition to the ion production, R&D is also underway on the ion collection devices and a 60 GHz ECR ion source. A prototype collection device has been built and is now being tested using 8Li. A prototype ECR has also been built and has been tested to ensure it achieves the required magnetic fields. It is now planned to modify it and test its performance with an ion beam.

Another major potential problem for the Beta Beam facility is collective effects in the accelerators and DR due to the very high ion intensity. Detailed studies of this are on-going and it is planned to make measurements wherever possible, in particular in the existing accelerators. Another significant problem is decay losses. These have been studied for the baselines ions up to the PS and no insurmountable problems have been identified. This work now needs to be extended to the SPS. The DR uses superconducting magnets and there is a problem with the heat load in these. The original plan of using open mid-plane magnets helps, but does not solve this problem. Further, a study adding more shielding in front of the magnets has been inconclusive. More work is required on this problem.

Figure 7. The RipeN detector used to measure the cross-section for 8B production.

References

[1] See http://www.euronu.org
[2] Longhin A 2011 A new design for the CERN-Fréjus neutrino Super Beam arXiv:1106.1096 [physics.acc-ph]
[3] Choubey S *et al* 2011 International Design Study for the Neutrino Factory, Interim Design Report IDS-NF-20
[4] Lindroos M and Mezzetto M 2010 Beta Beams *Ann. Rev. Nucl. Part. Sci.* **60** 299-325
[5] Arnaudon L *et al* 2006 Linac4 technical design report CERN-AB-2006-084
[6] Nakadaira T *et al* 2008 T2K Target *AIP Conf. Proc.* **981** 290-292
[7] Abramov A G *et al* 2002 Beam Optics and Target Conceptual Designs for the NuMI Project *Nucl. Instrum. Methods Phys. Res.* ***A* 485** 209-227
[8] Li D *et al* 2008 MUCOOL/MICE RF Cavity R&D Programs *AIP Conf. Proc.* **981** 299-302
[9] Palmer R *et al* 2008 RF Breakdown with and without External Magnetic Fields arXiv:0809.1633 [physics.acc-ph]
[10] Machida S *et al* 2011 Acceleration in the linear non-scaling fixed field alternating gradient accelerator EMMA, Electron Model for Many Applications, Submitted to *Nature Physics*
[11] Alexander J *et al* 2009 Progress on the Commissioning of ALICE, the Energy Recovery Linac-Based Light Source at Daresbury *Proc. Part. Acc. Conf. 2009 (Vancouver, Canada, 4-8 May 2009)* pp 1281-1283
[12] Blumenfeld Y 2007 The EURISOL Project *AIP Conf. Proc.* **891** 147-154

Journal of Physics: Conference Series **408** (2013) 012012 doi:10.1088/1742-6596/408/1/012012

Beam monitoring and near detector requirements for a Neutrino factory or long-baseline beams

Roumen Tsenov and Rosen Matev

Department of Atomic Physics, St. Kliment Ohridski University of Sofia, 5 James Bourchier Boulevard, Sofia 1164, Bulgaria

E-mail: tsenov@phys.uni-sofia.bg

Abstract. Neutrino Factory is a facility for future precision studies of neutrino oscillations. A so called near detector is essential for reaching the aimed precision of neutrino oscillation analysis. Main task of the near detector is to measure the flux of the neutrino beam. Such brilliant neutrino source like Neutrino Factory provides also opportunity for precision studies of various neutrino interaction processes in the near detector. We discuss design concepts of such a detector. Results of simulations of a high resolution scintillating fiber tracker show that it is capable to measure the neutrino flux through pure leptonic interactions with an uncertainty of less than 1%. A full set-up of the near detector consisting of high granularity vertex detector, high resolution tracker and muon catcher is also presented.

1. Neutrino Factory Near Detector(s) baseline

A future neutrino facility [1] will need near detectors in order to perform oscillation measurements with required sensitivity. It is necessary to have one near detector for each of the straight sections of the storage ring at each of the two polarities, so four near detectors designed to carry out measurements essential to the oscillation-physics programme are required. The near detector tasks include measurement of neutrino flux through the measurement of neutrino-electron scattering; measurement of neutrino beam properties needed for the flux to be extrapolated to the far detector; measurement of charm production cross sections (charm production in far detectors is one of the principal backgrounds to the oscillation signal). In addition, the brilliant Neutrino Factory beam allows for unique neutrino physics non-oscillation studies, such as measurement of cross sections, structure functions, nuclear effects, $\sin^2\theta_W$ *etc.* at neutrino energies in the 0-25 GeV range. The near detector must also be capable of searching for new physics, for example by detecting τ-leptons which are particularly sensitive probes of non-standard interactions at source and at detection. ν_τ detection is also important in a search for sterile neutrino.

Design requirements for the near detector(s) can be formulated as follows: low Z high resolution tracker for flux and cross-section measurement (ν_μ and ν_e); magnetic field for better than in MIND [2] muon momentum measurement; muon catcher and capability for e^+/e^- identification; vertex detector for charmed hadrons and τ-leptons detection (for non-standard interactions and sterile neutrino searches); good resolution on neutrino energy (much better than in the Far Detector) for flux extrapolation.

Current near detector design anticipates three subdetectors (see Fig. 1): high granularity detector for charm/τ measurement; high resolution tracker for precise measurement of the event close to the vertex and Mini-MIND detector for muon measurement.

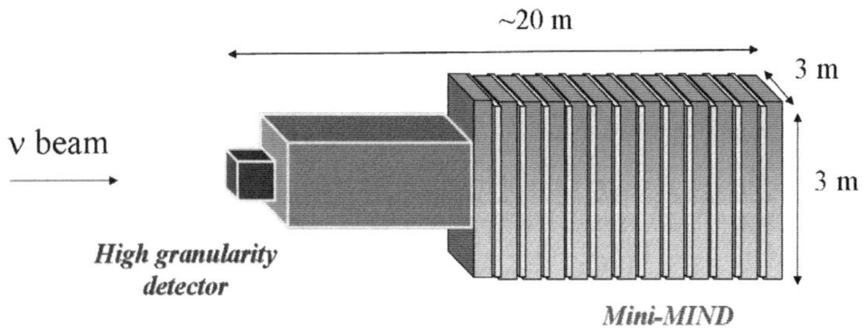

Figure 1. Block diagram design of the Near Detector

2. Measurement of the neutrino flux by neutrino-electron scattering

Neutrino-electron interaction cross sections are straightforwardly calculated in the Standard Model framework. Any small uncertainties could come only from (well measured) Standard Model parameters. Therefore, such processes are suitable for measurement of neutrino beam fluxes, provided that beams are intense enough.

There are two pure leptonic interactions of the Neutrino Factory beam producing a muon in the final state:

$$\nu_\mu + \mathrm{e}^- \rightarrow \mu^- + \nu_\mathrm{e} \text{ and } \overline{\nu}_\mathrm{e} + \mathrm{e}^- \rightarrow \mu^- + \overline{\nu}_\mu \quad (IMD). \tag{1}$$

The first one is known as inverse muon decay, while the second one produces a muon in the final state through annihilation. The neutrino energy threshold (for electrons at rest) for both processes is 10.9 GeV .

There are four pure leptonic reactions of interest producing energetic electron in the final state:

$$\nu_\mu + \mathrm{e}^- \rightarrow \nu_\mu + \mathrm{e}^- \text{and} \quad \overline{\nu}_\mathrm{e} + \mathrm{e}^- \rightarrow \overline{\nu}_\mathrm{e} + \mathrm{e}^- \quad (ES^-) \tag{2}$$

$$\nu_\mathrm{e} + \mathrm{e}^- \rightarrow \nu_\mathrm{e} + \mathrm{e}^- \text{and} \quad \overline{\nu}_\mu + \mathrm{e}^- \rightarrow \overline{\nu}_\mu + \mathrm{e}^- \quad (ES^+). \tag{3}$$

Despite the smallness of the total cross sections for the above processes, a massive detector placed close to the straight section end can provide a sufficient interaction rate - Fig. 2. However, inclusive CC and NC neutrino interactions with nuclei

$$\nu_\ell + N \rightarrow \ell + X \text{ and } \quad \nu_\ell + N \rightarrow \nu_\ell + X \tag{4}$$

are a few orders of magnitude more intensive. For example, at 15 GeV the muon neutrino CC total cross section is $\sim 1 \times 10^{-37}$ cm^2, compared to $\sim 2 \times 10^{-41}$ cm^2 for inverse muon decay $\nu_\mu \mathrm{e}^- \rightarrow \mu^- \nu_\mathrm{e}$. An obvious distinction between purely leptonic processes and processes (4) is the lack of hadronic system X in the former. Thus, the measured recoil energy of the hadronic system can be used as a good criterion for background suppression. Muons from quasi-elastic neutrino-electron scattering have angular distribution peaked at very forward direction. At the Neutrino Factory, the polar angle of these muons does not exceed 5 mrad. The angular spread comes mainly from the intrinsic scattering angle ~ 4 mrad in processes (1), while neutrino beam divergence (and solid angle covered by detector) has little contribution. This kinematic property can be used as another event selection criterion. The polar angle distribution of electrons from neutrino-electron elastic scattering is ten times wider and is not suitable for event selection. On the other hand, the composite variable $\theta_\ell^2 E_\ell$, proportional to Bjorken's $y = 1 - E_l/E_\nu$ in elastic scattering, provides good separation between signal and background for all neutrino-electron scattering processes provided lepton angle and energy are measured with sufficient precision.

Two options for the high-resolution tracker are being considered: *a scintillating fiber tracker* and *a straw tube tracker*.

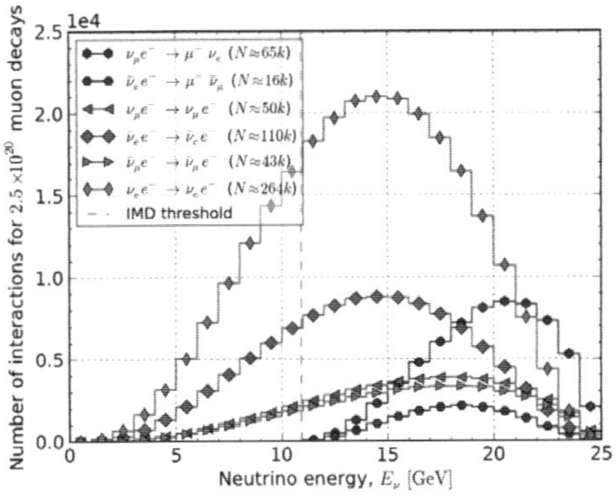

Figure 2. Number of neutrino-electron interactions for 2.5×10^{20} muon decays per muon charge and per straight section– a nominal year of Neutrino Factory operation. Rates are calculated for 2.7 t detector with 1.5×1.5 m^2 frontal cross section and average $Z/A \approx 0.54$. Detector is placed 100 m after the straight section end. Dashed vertical line indicates threshold for the inverse muon decay.

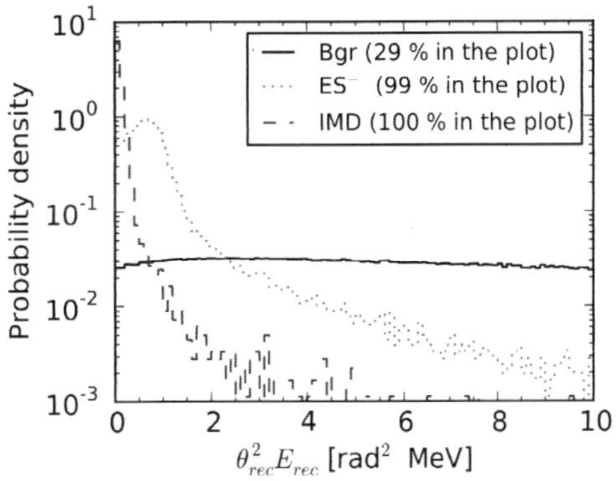

Figure 3. Distributions of reconstructed $\theta_\ell^2 E_\ell$ variable for IMD (blue), ES$^-$ (red) and background (black) events in μ^- mode. The fraction of events contained in the plot is indicated in the legend. All distributions are normalized to a unit area.

3. Scintillating fiber tracker

A schematic drawing of a scintillating fiber tracker with an incorporated calorimeter is shown on Fig. 4. The detector consists of 20 square shaped modules placed perpendicular to the beam axis. Each module has a calorimeter section and a tracker section (also called tracker station). Modules are positioned equidistantly forming gaps filled with air. With larger distance between tracker stations, X and Y displacement of hits is increased and thus angular resolution improved. The sides of the air gaps are covered with a layer of plastic scintillating bars. These layers are referred to as *side slabs*. The detector is placed in 0.5 T dipole magnetic field. Each station consists of one layer of fibers with horizontal orientation and another with vertical orientation. Each layer has four planes made of 1 mm round fibers. They form a hexagonal pattern in the layer, thus minimizing dead volume. There are 12 000 fibers per station, thus 240 000 fibers in total. Calorimeter sections consist of plastic scintillating bars perpendicular to the magnetic field and arranged in 5 planes in each section. Bars are co-extruded with a wavelength shifting (WLS) fibers inside and have 10 mm by 30 mm cross-section. Both tracker fibers and WLS fibers in bars are read from both ends by silicon photomultipliers (SiPMs). Overall dimensions of the detector are ~ 1.5 m $\times$ 1.5 m $\times$ 11 m. The detector mass is ~ 2.7 t.

3.1. Simulation of the detector response and signal extraction

The neutrino flux at the near detector has been generated by a Monte Carlo simulation of muon decays along the straight section of the Neutrino Factory decay ring [3, 4]. Neutrino interactions in the detector have been simulated by the GENIE package [8]. For the simulation of the detector response to them, the Geant4 software platform [5] was used. Simple algorithms have been developed for vertex and scattered lepton track reconstruction.

The distribution of the difference between the reconstructed and true value of the scattering angle is shown on Fig. 5. The resolution (σ parameter of the fit) is ~ 0.5 mrad for both muons

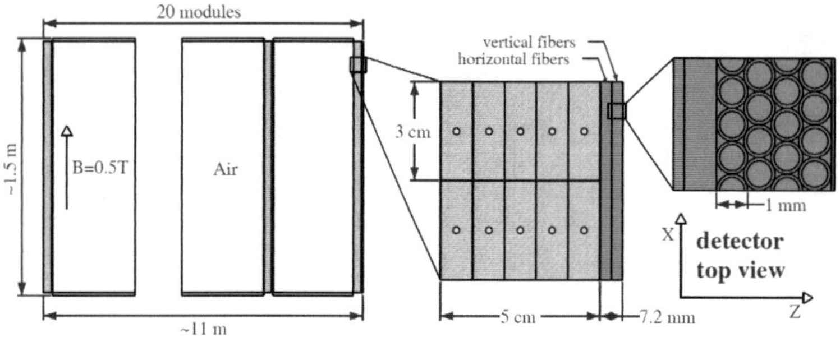

Figure 4. Schematic drawing of the detector.

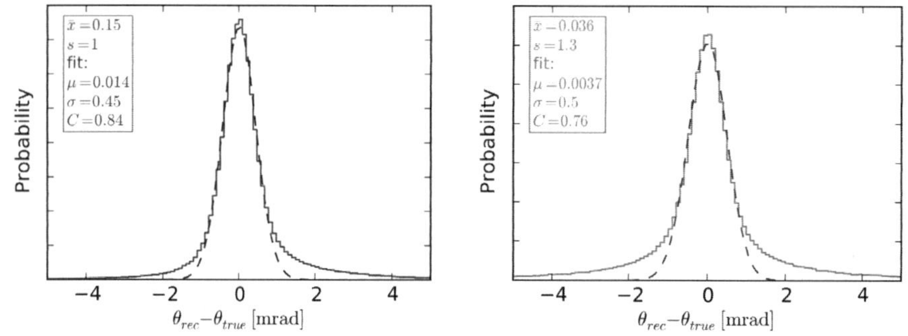

Figure 5. Obtained angular resolution for muons (left) and electrons (right). Gaussian fits are shown with dashed lines.

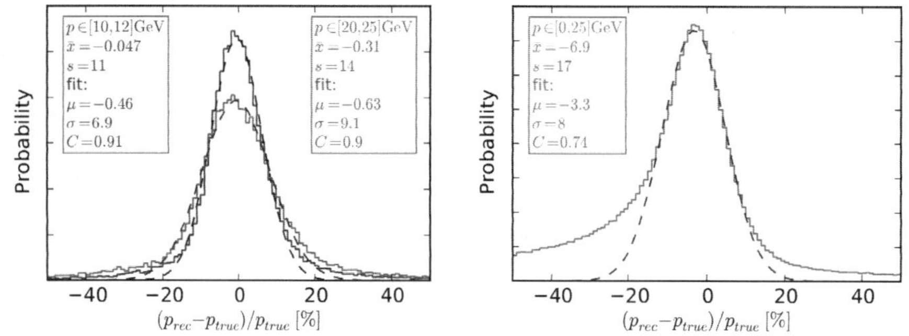

Figure 6. Obtained momentum resolution for muons (left) and electrons (right). Gaussian fits are shown with dashed lines. For muons, distribution is shown for two samples of events: one with true muon momentum in the [10-12] GeV range (blue) and one in the [20-25] GeV range (green).

and electrons. The reconstructed momentum resolution is shown on Fig. 6. For muons it goes up to ~ 9 % for the highest energy muons. For electrons, the distribution is biased towards the negative values with a heavy negative tail. The reason for this is that they loose momentum due to bremsstrahlung and ionization. If energy loss is taken into account, for instance with Kalman filter fitting [6], bias can be reduced.

Both IMD and ES events have a property of low (consistent with single particle) energy

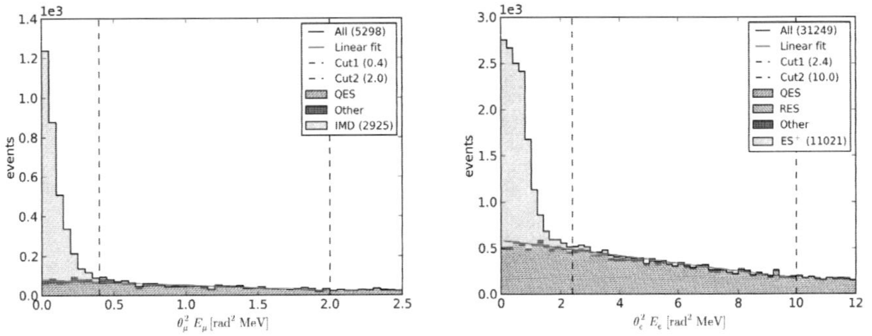

Figure 7. Distributions over $\theta_\mu^2 E_\mu$ for the IMD sample (left) and for the ES^+ sample (right). The leptonic events histogram is filled with solid gray, the hadronic events histogram is hatched and the total spectrum is in black. The two cuts bounding the fit interval are drawn with dashed line. The red line indicates the background extrapolation.

deposition near the vertex. To exploit that, a cut on energy deposit around the vertex is imposed. Some other kinematic and calorimetric cuts have also been applied in order to get a sample enriched with signal events. As a result, the signal-to-background ratio has been increased from $\sim 10^{-4}$ to $\sim$30-50 %. Further on, extrapolations of certain background distributions should be made in order to subtract background from event samples. We have chosen to do background subtraction in terms of the primary lepton (μ or e) kinematic variables – the lepton's scattering angle θ_ℓ and initial momentum p measured in the detector. In the case of IMD signal extraction, the scattering angle θ_ℓ and $\theta_\ell^2 E_\ell$ variable can be used to discriminate signal from background. In the case of ES signal, background is well separated only when one exploits the $\theta_\ell^2 E_\ell$ variable. The distributions of $\theta_\ell^2 E_\ell$ for IMD signal, ES^- signal and background are shown on Fig. 3. The background distribution is nearly flat. This fact allows for its simple parameterization.

Two methods of obtaining the number of signal events are discussed below: *the linear fit method* and *the μ^+-method.*

The linear fit method relies on the nearly flat shape of the respective background distribution. The idea is to estimate the background under the signal peak by linear extrapolation from signal-free region towards the signal one. Examples are shown in Fig. 7. Comparison between the estimated and the true number of signal events is made in Table 1. It is seen that the true values lie within the 95 % confidence intervals of the predictions. The systematic uncertainty, estimated as the difference between the fit result and the true number of signal events is less than 1 %. However, to give conclusive estimation of the systematic error, one should investigate if and how various parameters of simulation and selections influence the background shape.

IMD interactions are present only in the μ^- decay mode. The idea of the μ^+-method is to estimate the background under the IMD signal peak exploiting the distribution of positive muons detected in ($\overline{\nu}_\mu$, ν_e)-beam [7]. In the near detector, an event sample from the ($\overline{\nu}_\mu$, ν_e)-beam events is selected with the same selection cuts as the IMD sample. The $\theta_\mu^2 E_\mu$ histogram for μ^+ is normalized to the $\theta_\mu^2 E_\mu$ histogram for μ^-. An interval outside the IMD signal peak and with approximately constant ratio of μ^-- and μ^+-events is defined and the normalization factor is calculated within this interval. The μ^- histogram and the normalized μ^+ histogram are shown in Fig. 8 (left). Subtraction is made using the normalized μ^+ histogram.

Table 1 demonstrates that the number of neutrino-electron scattering events can be measured exploiting the $\theta_\ell^2 E_\ell$ distribution with a good precision. A direct comparison between the

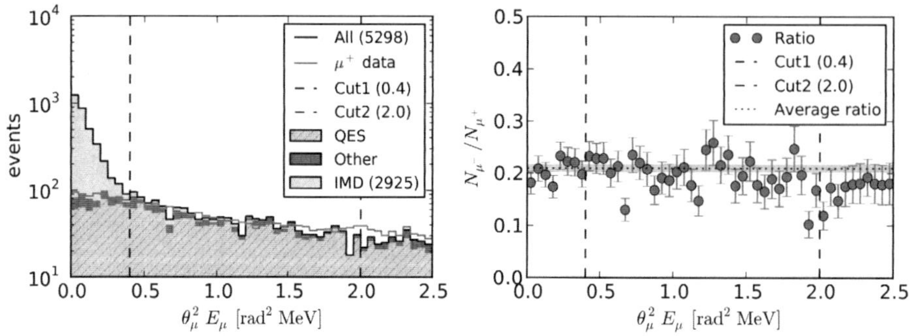

Figure 8. *Left:* distributions over $\theta_\mu^2 E_\mu$ for the IMD sample. The leptonic events histogram is filled with solid gray, the hadronic events histogram is hatched and the total spectrum is in black. The two cuts bounding the normalization interval are drawn with dashed line. The red line indicates the normalized μ^+ histogram. *Right:* ratio of the μ^- histogram and the μ^+ histogram over $\theta_\mu^2 E_\mu$. Horizontal dotted line indicates the (constant) normalization factor.

Table 1. Estimated number of signal events for the three event samples. The result in the last row was obtained using the μ^+ background subtraction method, while the other three results were obtained using linear fit background subtraction method. Statistics correspond to 2.3×10^{19} μ^- decays and 2.3×10^{19} μ^+ decays, which is approximately a tenth of the nominal year.

Event sample	Selection eff.	Overall eff.	Purity	All events	Signal events	Signal events from fit
IMD	86 %	46 %	81 %	3520	2850	2926 ± 59
ES$^-$	70 %	32 %	61 %	7355	4491	4479 ± 86
ES$^+$	83 %	37 %	63 %	16964	10607	10512 ± 131
IMD	86 %	46 %	81 %	3520	2850	2831 ± 61

measured and true number of signal events shows a deviation of no more than 1 %. It is worth noting that the MC truth was not used in reconstruction and signal extraction. Thus, with the presented design of the tracker we can achieve 1% uncertainty on the flux normalisation by exploring IMD and/or ES scattering.

4. High resolution straw tube tracker

Another option for the near detector is a high resolution straw tube tracker inspired by the HiResMν detector [9] being considered for the LBNE project at Fermilab as a near detector.

Building upon the NOMAD experience [11], this low-density tracking detector will have a fiducial mass of 7.4 tons as an active neutrino target, similar to the ATLAS Transition Radiation Tracker [12] and the COMPASS detector [13]. The tracker will be composed of straw tubes with 1 cm diameter, in the vertical (y) and horizontal (x) direction. In front of each module a plastic radiator made of many thin foils allows the identification of electrons through their transition radiation. The nominal fiducial volume is: $350 \times 350 \times 600$ cm^3, corresponding to 7.4 tons of mass with an overall density $\rho < 0.1$ g/cm^3. The straw-tube tracker will be surrounded by an electromagnetic calorimeter (sampling Pb/scintillator) covering the forward and side regions. Both sub-detectors will be installed inside a dipole magnet providing a magnetic field of $\sim$0.4 T.

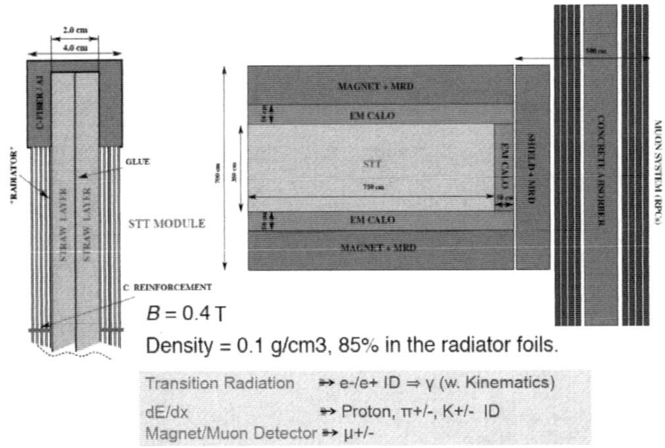

Figure 9. Sketch of the proposed HiRes detector showing the inner straw tube tracker (STT), the electromagnetic calorimeter (EM CALO) and the magnet with the muon range detector (MRD). Also shown is one module of the proposed straw tube tracker (STT). Two planes of straw tubes are glued together and held by an aluminium frame.

The detector will provide full reconstruction of charged particles and γs; identification of electrons, pions, kaons, and protons from dE/dx; electron (positron) identification from transition radiation ($\gamma > 1000$); full reconstruction and identification of recoil protons down to momenta of 250 MeV; reconstruction of electrons down to momenta of 80 MeV from curvature in the B-field.

Detailed simulations of this detector have been carried out in the context of the LBNE proposals [10]. These simulations are to be adapted to the neutrino spectra at a Neutrino Factory to derive the performance parameters of this detector in this context.

5. Charm and Tau Detector

A near detector at a neutrino factory needs to measure the charm cross-section to validate the size of the charm background in the far detector, since this is the main background to the wrong-sign muon signature. The charm cross-section and branching fractions are poorly known, especially close to threshold. For this reason, it is paramount to make an independent near detector measurement of the charm cross-section and make the error in the charm cross-section negligible in the estimation of the neutrino oscillation background.

Since events with a τ-lepton in the final state have a similar signature to charm events, any detector that can measure charm should be able to measure τ's as well. This is important to explore couplings of Non Standard Interactions (NSI) at source $\epsilon^s_{\tau\mu}$, $\epsilon^s_{\tau e}$ or detection $\epsilon^d_{\tau\mu}$, $\epsilon^s_{\tau e}$ A semiconductor vertex detector for charm and τ-lepton detection could potentially be used for this purpose. The advantage of this type of detector is that it is able to operate at a high event rate and still have very good spatial resolution. This is necessary to distinguish the primary neutrino interaction vertex from the secondary vertex due to the short lived charm hadron or the τ-lepton. The vertex detector could be similar to the NOMAD–STAR detector that was installed upstream of the first drift chamber of the NOMAD neutrino oscillation experiment [11] used to measure the impact parameter and double vertex resolution to determine the charm detection efficiency. The reconstruction of τ-leptons from an impact parameter signature with a dedicated silicon vertex detector was studied in the NAUSICAA proposal [14]. A silicon vertex detector with a B_4C target was proposed as an ideal medium to identify τ-leptons. Standard ν_μ CC interactions have an impact parameter $r.m.s$ of 28 μm, while tau decays have an impact parameter$r.m.s$ of 62 μm. By performing a cut on the impact parameter significance (σ_{IP}/IP) one can separate one prong decays of the tau from the background. For three prong decays of the tau, a double vertex signature is used to separate signal from background. The total net efficiency of the tau signal in NAUSICAA was found to be 12%.

A silicon strip vertex detector as part of the near detector could have the following dimensions [1]: 18 layers of B_4C (2.49 g/cm^3), 150x150x2 cm^3 each; total mass = 2.02 tonnes, 20 layers of silicon strip or pixel detectors, e.g. 45 m^2 of silicon; about 64 000 channels per layer, 1.28 million channels in total. At the Neutrino Factory in such a detector about 3×10^7 ν_μ CC interactions per year are expected and 10^6 charm events among them. With the efficiency of the tau detection found in NAUSICAA, one could have a sensitivity of $P_{\mu\tau} < 3 \times 10^{-6}$ at 90% C.L. on the $\nu_\mu \to \nu_\tau$ conversion probability.

6. Summary and outlook

A near detector(s) at the Neutrino factory is a valuable tool for neutrino flux measurement and standard and non-standard neutrino interactions. The envisioned set-up consists of high granularity vertex detector followed by high resolution tracker and muon catcher. Silicon vertex detector+SciFi tracker+mini-MIND set-up is most advanced with respect to simulations with Neutrino Factory beam. They show that the neutrino flux can be measured with 1% uncertainty. A second option exists for the tracker – HiResMν. Simulations with a Neutrino Factory beam are needed to confirm its ability to select and measure neutrino-electron scattering.

Further tasks include the simulation of the full set-up in order to estimate systematic errors coming from near-to-far extrapolation (determination of the so-called *migration matrices*); expectation on cross-section measurements and other physics studies, sensitivity to non-standard interactions (τ-lepton production). In a bit more distant future, serious R&D efforts would be needed to validate the technology choices for the vertex detector, high-resolution tracker, *etc.*

References

[1] Choubey S *et al.* 2011 Interim Design Report of the International Design Study for the Neutrino Factory *Preprint hep-ex/1112.2853*
[2] Laing A Ph.D. Thesis (University of Glasgow: 2010) `http://theses.gla.ac.uk/2216/`
[3] Karadzhov Y 2010 *AIP Conf. Proc.* **1222** 467–70
[4] Karadzhov Y and Tsenov R 2010 *Annual of University of Sofia, Faculty of Physics* **103** `http://www.phys.uni-sofia.bg/annual/arch/103/full/GSU-Fizika-103-06_full.pdf`
[5] Agostinelli S *et al.* (GEANT4 Collaboration) 2003 *Nucl. Instrum. and Methods* A **506** 250–303; Amako K, Apostolakis J, Araujo H and Dubois P *et al.* 2006 *IEEE Trans. Nucl. Sci.* **53** 270
[6] Kalman R E 1960 *Transactions of the ASME–Journal of Basic Engineering*, Series D, **82** 35–45
[7] Geiregat D*et al.* (CHARM-II Collaboration) 1990 *Phys. Lett.* B **247** 131–6; Vilain P *et al.* (CHARM-II Collaboration) 1995 *Phys. Lett.* B **364** 121–6
[8] 2010 Andreopoulos C *et al.*, *Nucl. Instrum. and Methods* A **614** 87–104 (*Preprint hep-ph/0905.2517*)
[9] Mishra S R, Petti R and Rosenfeld C 2008 *PoS*, **NUFACT08** 069 (*Preprint hep-ex/0812.4527*)
[10] *Long Baseline Neutrino Experiment*, `http://lbne.fnal.gov/` (2010).
[11] Altegoer J *et al.* (NOMAD collaboration) 1998 *Nucl. Instrum. and Methods* A **404** 96–128
[12] Akesson T *et al.* (ATLAS TRT collaboration) 2004 *Nucl. Instrum. and Methods*, A **522** 131–45 and 50–3; Abat E *et al.* (ATLAS TRT collaboration) 2008 *JINST*, **3** P02013.
[13] Bychkov V N *et al.* 2002 *Part. Nucl. Lett.* **111** 64–73
[14] Gomez-Cadenas J J, Hernando J A and Bueno A 1996 *Nucl. Instrum. and Methods* A **A378** 196–220

XIII International Workshop on Neutrino Factories, Super beams and Beta beams (NUFACT11) IOP Publishing
Journal of Physics: Conference Series **408** (2013) 012013 doi:10.1088/1742-6596/408/1/012013

Status of the Neutrino Factory accelerator design studies

Gersende Prior[1]
CERN, 1211 Geneva 23, Switzerland
E-mail: gersende.prior@cern.ch

Abstract. This document is a review of the present status of the Neutrino Factory design study, after the publication of the Interim Design Report and before the publication of the Reference Design Report. The different components of the accelerator as well as their current design stage and future tasks are described here.

1. Introduction

In March 2011, the Interim Design Report (IDR) [1] was published. It documents in details the Neutrino Factory design study and progresses made, since the Accelerator Working Group (AWG) of the International Scoping Study (ISS) identified in 2006, the baseline design as well as the required R&D activities and issues to address. Details of the work performed by the ISS AWG group have been published [2]. Following up this publication the International Design for a Neutrino Factory (IDS-NF) collaboration was born, whose mandate is to complete the design study by addressing the current technical issues and working on possible mitigation options for the publication of the Reference Design Report (RDR), foreseen at the beginning of 2013. At present 134 authors and 47 institutes are members of the IDS-NF collaboration. In Europe, the Neutrino Factory design study is funded by the EU FP7 contract EUROnu, which also encompasses the design study of two other accelerator options: the Super Beam and the Beta Beams.

2. The Neutrino Factory: wish-list and constraints

The next generation of neutrino facilities will perform precision measurements of the last unknown mixing angle θ_{13}, search for CP-invariance violation in neutrino oscillations, determine the sign of Δm^2_{31} and measure all the oscillation parameters with an unprecedented precision. It requires intense (4 MW, 10^{21} neutrinos/year), high-energy (> 20 GeV) neutrino and anti-neutrino beams. This puts a number of constraints on the target and accelerator systems. The most important important ones being:

- the target should be able to withstand beam-induced shocks
- the muon beam should be bunched, rotated and cooled over a small distance
- a rapid muon acceleration system able to transport the muons beam to two decay rings with minimum beam losses is needed

[1] on behalf of the EUROnu and IDS-NF collaborations.

The Neutrino Factory buncher allows both muon signs transport in different radio-frequency (RF) buckets, whereas the rotator will permit a reduction of the energy spread and the cooler a reduction of the beam emittance.
The feasibility study will determine if we can overcome its technical challenges, identify the cost driving factors and detail possible risk mitigation solutions.
In addition, if one considers building in the future a muon collider, a detailed reference design of the Neutrino Factory is necessary as it is a key component of the muon collider complex [3].

3. Proton driver and annexes

A schematic layout of the Neutrino Factory is shown on Figure 1. Three options have been identified for the proton driver which are described in more details below.

3.1. CERN SPL-based scenario

As part of the upgrade of the CERN accelerator complex, a Super Proton Linac (SPL) will be built in the future. It is an H^- linac, with a bunch frequency of 352.2 MHz and a repetition rate of 50 Hz. It comprises a high-speed chopper permitting to form bunches less than 2 ns wide, including rise and fall times. It could serve as a proton driver [4] for the Neutrino Factory and two working options have been identified which are summarized in Table 1.

Table 1. CERN SPL configuration options for the Neutrino Factory and beam energies (a) = 2.5 GeV and (b) = 5 GeV.

	Option 1	Option 2
Power (MW)	2.25 (a) or 4.5 (b)	5 MW (a) and 4 MW (b)
Number of protons/pulse	1.1×10^{14}	2.14×10^{14} (a) and 1×10^{14} (b)
Pulse current (mA)	20	40
Pulse duration (ms)	0.9	1 (a) and 0.4 (b)

The SPL is then followed by an accumulator and compressor rings, where the proton beam is transformed into three bunches aimed at the target. Figure 2 shows the transformation of the bunch structure from the SPL to the output of the compression ring. A study of the beam instabilities in the accumulator for the case of three bunches has been performed. MADX [5] simulations of the accumulator and compressor rings are available for the case of three bunches. The components of the accumulator and compressor rings are being listed for costing purposes.

3.2. Fermilab scheme and the upgrade to Project X

The proton driver based on the upgrade of the Fermilab Project X acceleration complex (see Figure 3) would provide 4 MW power at 8 GeV. By increasing the 3 GeV CW linac average current to 5 mA, its duty factor to $\sim$10% (Project X is $\sim$ 5%) and the number of particles per linac bunch, one could reach the Neutrino Factory proton driver requirements. An additional accumulator and a compressor rings (as in the CERN scenario) would be necessary. In the present design, the accumulator has a circumference of $\sim$ 250 m aiming at forming 14 $\times$ 100 ns-long bunches with 1.3×10^{13} protons/bunch. The use of a stripping foil or a novel laser-stripping technique are under consideration. The compressor would get at its entrance $\sim$ 50 ns bunches needing to be shortened into few ns long bunches. Detailed studies of the instabilities and space charge have to be completed. An optimization study of the beam size and angle at target is also under way.

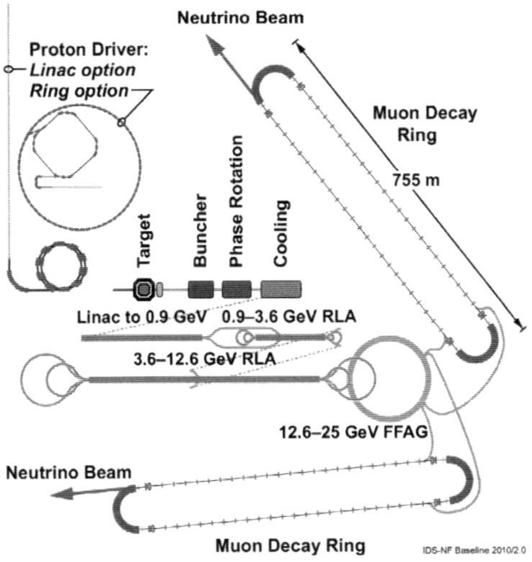

Figure 1. Schematic layout of the Neutrino Factory.

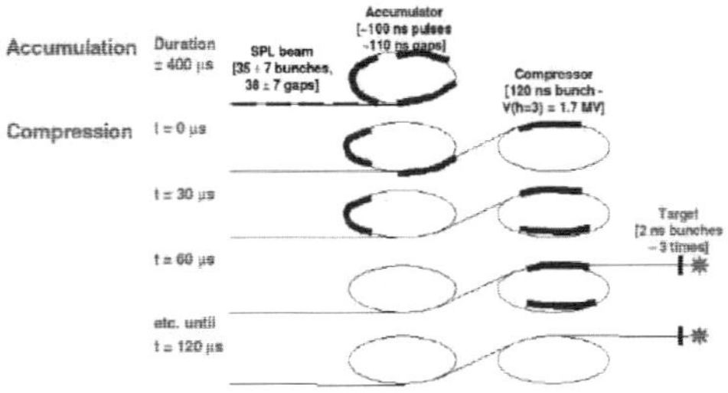

Figure 2. Bunches accumulation and compression in the CERN accumulation and compression rings.

3.3. RAL scenario and the upgrade to ISIS

The third option consists of an upgrade of the Rutherford Appleton Laboratory (RAL), neutron spallation source ISIS, to provide beam powers of 2-5 MW in the few GeV energy range. The facility could be shared between a short pulse spallation neutron source and the Neutrino Factory. A layout of the proposed accelerator complex is shown in Figure 4. It would require an additional Rapid Cycling Synchrotron (RCS) or a Fixed Field Alternating Gradient (FFAG) booster in order to bring the proton beam to the necessary energy and perform the appropriate bunch compression. Several studies are underway, including a high-intensity ~ 3.3 GeV booster synchrotron and beam lines, a 800 MeV high-intensity linac and the RCS and FFAG options for the main ring accelerator. The study of a high-power Front End Test Stand (FETS), RF systems, stripping foils options, diagnostics and kickers will be addressed in future R&D experiments.

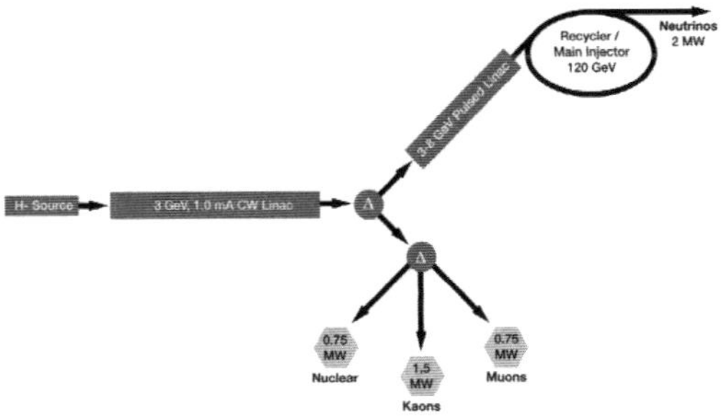

Figure 3. Project X layout.

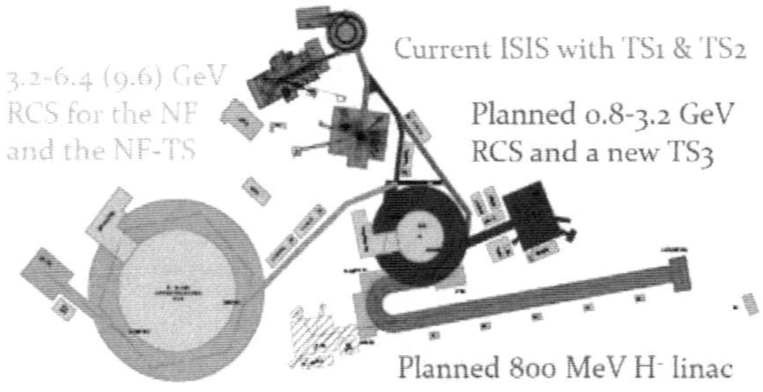

Figure 4. RAL accelerator complex layout.

4. Target system

4.1. Hg-jet target developments

The Neutrino Factory baseline target is made of a Hg-jet target surrounded by a solenoid field of 20 T followed by an adiabatic taper to 1.5 T, to allow for optimal secondaries capture. Previous simulations using MARS 15 [6] and FLUKA [7] showed high levels of energy deposition in the magnets, requiring the shielding to dissipate about 2.4 MW. Also the Hg-jet and proton beam are sent to a Hg pool and disrupting it, showing the need for a study of splash mitigation. During a redesign campaign [8] a better shielding of the Super Conducting (SC) magnets from radiation was identified. Splash mitigation options are under study [9] and the design of the mechanical support has been improved. The layout of the new target design is shown in Figure 5. The R&D experiment MERIT [10] whose analysis of data was completed in 2007, validated 4 MW proton beam operation in mercury. Further work on the target station infrastructure, including outer shielding, remote handling, mercury cooling loop, the design of the beam windows and the beam dump are required.

4.2. Alternative/mitigation options
A few alternative target systems [11] are under consideration, made either of a metal-powder jet or a system of solid tungsten bars that would be exchanged between pulses.
A test rig has been built at RAL where 100 kg of W powder with grain size < 250 μm could be operated continuously for about 20 min. A coherent free flow jet at P $\sim$ 2 bars was produced and the results were validated with simulations. A photograph of the jet is shown in Figure 6. For the solid target options, shock studies were performed using high-currents in thin W and Ta wires. The results obtained are in agreement with LS-DYNA [12] simulations. The preliminary design for the target change system is underway.

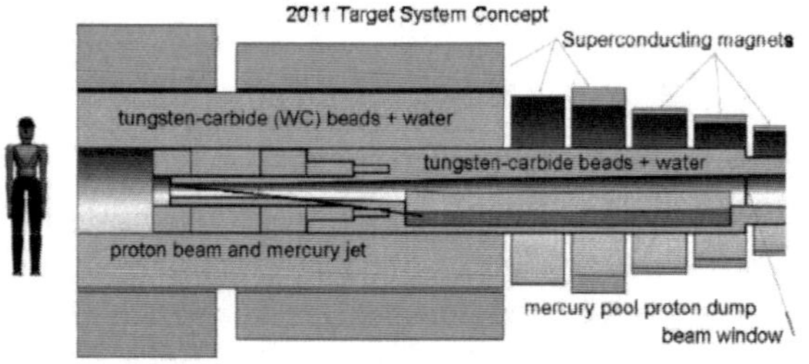

Figure 5. Hg-jet target system layout.

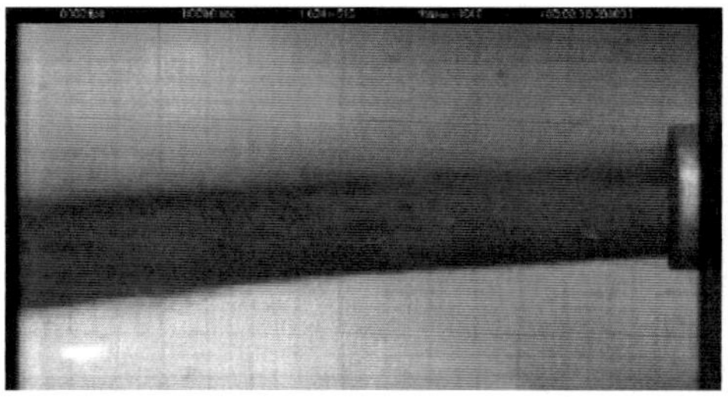

Figure 6. Photograph of the W powder jet.

Future R&D will include the study of flow improvement by mitigation of the flux breakdown or phase separation for the powder target. An irradiation experiment of tungsten powder and tungsten pebble bed will be performed at the CERN HiRadMat facility [13].

5. Front-end system
The front-end system is located after the target system. It comprises a buncher, a rotator and a cooling sections.

5.1. Front-end status
The baseline lattice described in the IDR has undergone an optimization study [14] identifying possible options to get rid of the unwanted particles in the accelerator, as early as possible. A proton absorber is under consideration in order to remove the low-momentum protons. The design of a chicane (see Figure 7) is underway to remove the high-momentum particles. Finally a transverse collimation system will take care of the remaining particles.
The reference lattice parameters have started to be listed in order to provide a detailed engineering study, where possible, and provide an accurate costing.
A realistic operational RF gradient limit has still to be determined and future R&D experiments

[15] will be performed at the Fermilab Muon Testing area (MTA). There is also a need to assess in details, and mitigate, the energy deposition from particle losses. An optimization of the lattice matching sections need also to be done. Finally a complete engineering design for the magnets, RF and absorbers is required.

5.2. Alternative/mitigation cooling options

Efficient bunching/rotation and cooling of the muons beam requires high (9-16 MV/m) RF cavity gradients in high (1-3 T) magnetic fields. This increases the risk of breakdown as suggested by experiments performed at the MTA. Three scenarios are under study, as alternative to the breakdown problem. A bucked coil lattice, a magnetically insulated lattice and a high-pressure gas RF (HPRF) lattice.

The bucked coil lattice provides a reduced magnetic field in the RF. It is made of 1.8 or 2.1 m long cells (see Figure 8). Three different solenoid current configurations for two cooling cells

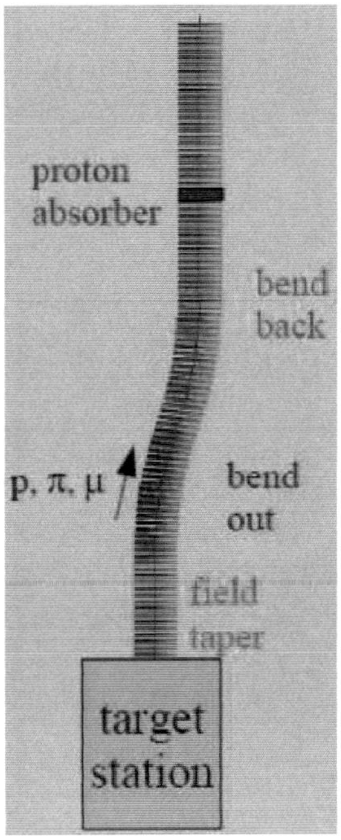

Figure 7. Layout of the front-end chicane and proton absorber.

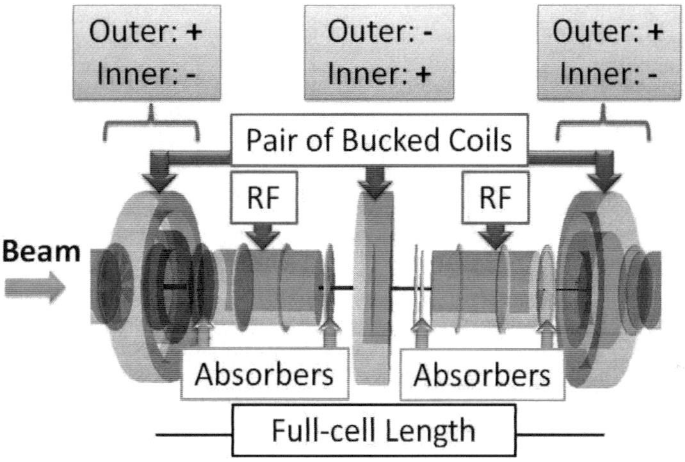

Figure 8. Schematic of a pair of bucked coils and RF cavities.

were simulated in G4MICE [17]. These configurations have been tested with both a reduced (~ 1000 muons) and full statistics. This lattice shows a good transmission in comparison with the ISS lattice.

The magnetically insulated lattice [18] has a configuration where the electric field is perpendicular to the magnetic field in the cavity. Simulations results are giving similar performance to the ISS lattice. In addition the behavior of a pillbox cavity in presence of electric field having a small angle with the magnetic field was tested at the MTA. The drawback of such a design is that the tolerance to coils misalignment is reduced to below 2 mm. Additional issues such as multipactoring and power consumption have still to be addressed.

The HPRF lattice comprises a cavity filled with high-pressure H_2 gas [19]. LiH absorber disks will be used on the windows as in the IDR configuration to provide muons cooling. Studies of windows material and windows thickness for different pressure configuration have been performed. In addition, tests with a gas-filled cavity were done at the MTA.

6. Muon acceleration

The acceleration system [20] is made of a linac for low-energies (below 0.9 GeV), followed by two Recirculating Linac Accelerators (RLA's) allowing multiple passes (to 12.6 GeV). Final acceleration to 25 GeV is provided by a FFAG (see Figure 9).

6.1. Linac and RLA's

The linac is made of short (3 m, 3.8 MV/m), medium (5 m, 5.1 MV/m) and long (8 m, 6.4 MV/m) cells containing SC RF cavities. Focusing is provided with solenoids which are better for low-energy and large-emittance beams. As the beam goes in the linac, the acceleration is increased by moving the RF phase toward the crest.

The RLA's have a dogbone shape providing greater separation at the switchyard (over a racetrack shape). They are made of SC RF and quadrupoles. The beam injection is done at the center of the linac. The muons are going for 4.5 passes per linac before beam extraction.

The validation of the switchyard design for the RLA's needs to be completed, as well as the lattice design, including the matching sections, injection system and the overall layout. A tracking through all the subsystems with realistic errors needs to be performed. The engineering design for all the components (including magnets and RF) needs also to be completed.

6.2. FFAG rings

The FFAG system [21] is a linear non-scaling ring made of a single arc with a large energy acceptance. It consists of identical FDF triplets. Almost all drifts contain SC cavities or injection/extraction hardware. For the injection and the extraction, the kickers are shared for both muons signs. The injection is made from inside and the extraction to outside. The magnet apertures are slightly bigger in the injection and extraction regions. The remaining tasks are to finalize the chromatic correction scheme, determine the optimal longitudinal phase space matching, design the matching to upstream and downstream systems. In addition a complete 6D tracking with errors need to be performed.A complete design for the main components (magnets, RF, injection and extraction) will also be provided. Finally a cost comparison with an equivalent RLA solution has to be done.

6.3. Decay rings

As the goal of the Neutrino Factory is to measure all together δ the CP violation phase, θ_{13} and the neutrino mass hierarchy, two distinct baselines working in combination would permit an optimum measurements of these parameters. At a so-called "magic" [23] distance of 7500 km, the matter effects cancel out the CP-violation effects allowing a clean measurement of θ_{13} and the sign of Δm^2_{31} with the help of a 50 kT fiducial mass Magnetised Iron Neutrino Detector (MIND) [22]. In combination with the magic baseline, a second baseline set at 4000 km, allows the measurement of δ with a 100 kT fiducial mass detector based on the MIND technology. The Neutrino Factory contains two racetrack shaped rings (see Figure 10), one per detector. Three trains of 50 bunches at 25 GeV enter the rings. The muons decay in the straight section which represents a large fraction of the total circumference. Both muons signs are stored simultaneously. The beam divergence from the lattice is at most $0.1/\gamma$. The rings circumference is 1609 m with 599 m-long (not including the matching cells) straight sections. The rings tilt angles are large, 36° for the detector located at 7500 km and 18° for the detector located at 4000 km. This places the decay ring tunnels lower points at a depth of 445 m and 234 m respectively. The β of the beam is 150 m in the straights and 13 m in the arcs.

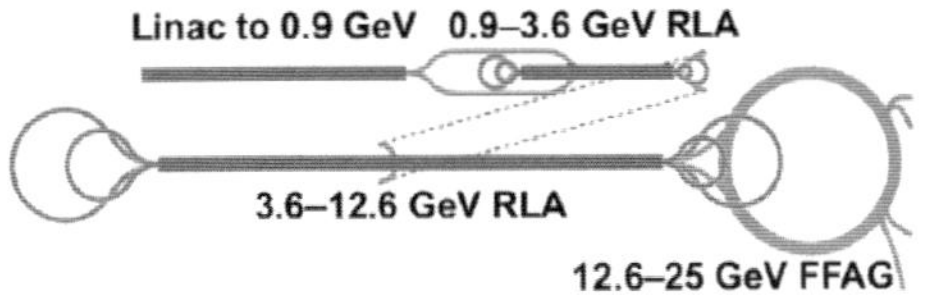

Figure 9. Layout of the muon acceleration systems linac and RLA's.

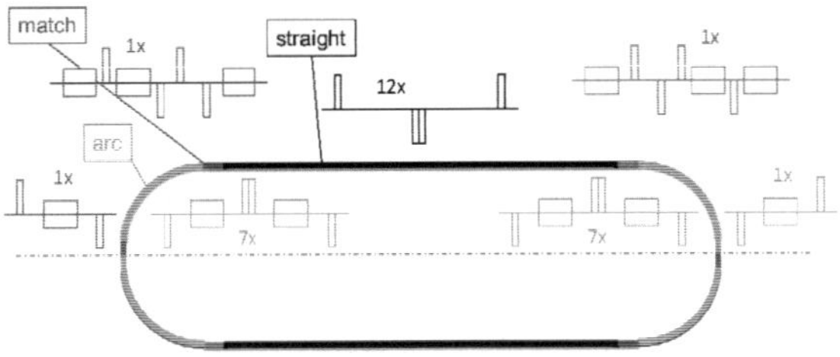

Figure 10. Layout of the muon decay ring.

Beam diagnostics [24] are necessary and the use of a polarimeter to measure the decay electrons is under consideration. It is based on measurements of the g-2 muon spin precession. It can also measure precisely the beam energy and the energy spread.
Measurements of the beam divergence can be provided by in-beam devices. Two options are under consideration, a cherenkov He-gas detector and an optical transition radiation (OTR) device. It still has to be demonstrated if the desired precision level can be reached by these devices, as the natural beam divergence is 4 mrad.
The remaining tasks include the design of the injection system, to assess the need for chromatic corrections and a beam abort scheme. The design of the diagnostics and the identification of their specifications has to be completed. Means to measure the neutrino flux spectrum at the far detectors will also have to be identified.

7. Conclusion: from the IDR to the RDR

The European Committee for Future Accelerator (ECFA) Review Panel was mandated to review the EUROnu mid-term report and the IDS-NF IDR in May 2011. The remaining steps toward the RDR include the development of a complete and technically feasible design with the required performance. An end-to-end tracking of the entire facility has be carried out to validate the performance estimate. A cost estimate of the whole facility has to be performed. After completion of these steps the reference design will be published in the RDR.

Acknowledgments

We acknowledge the financial support of the European Community under the European Commission Framework Programme 7 Design Study: EUROnu, Project Number 212372. The EC is not liable for any use that may be made of the information contained herein.

References

[1] Choubey S *et al.* (The IDS-NF collaboration) 2011 *Interim Design Report* IDS-NF-020
[2] Berg J S *et al.* (The ISS Accelerator Working Group) 2007 *Accelerator design concept for future neutrino facilities* RAL-TR-2007-23
[3] Shiltsev V *Toward a Muon Collider* these proceedings
[4] Garoby R *Proton drivers for Neutrino beams and other high-intensity applications* these proceedings
[5] Schmidt F 2002 *MADX user's guide* CERN-T5001
[6] Mokhov N and Striganov S I 2007 *MARS15 Overview* Fermilab-Conf-07-008-AD
[7] Battistoni G *et al.* 2007 *The FLUKA code: description and benchmarking* AIP Proc. Conf. 896 31
[8] McDonald K *Simulation of dynamic interaction of the Neutrino Factory mercury jet with the mercury collection pool/beam dump* these proceedings
[9] McDonald K *Multi-MW target and capture design for NF* these proceedings
[10] McDonald K *et al.* 2009 *The MERIT high-power target experiment at the CERN PS* IPAC10 Conf. Proc. WEPE078
[11] Densham C *Target option for NF* these proceedings

[12] LS-DYNA http://www.lstc.com/
[13] Efthymiopoulos I *et al. Feasibility experiment of granular target options for future neutrino facilities* these proceedings
[14] Neuffer D *Neutrino factory front-end and extensions* these proceedings
[15] Yonehara K *Commissioning and status of the MTA* these proceedings
[16] Alekou A *Performance comparison between FSIIA and bucked coils for the Neutrino Factory cooling* these proceedings
[17] G4MICE http://www.mice.iit.edu/software/
[18] Stratakis D *et al.* 2011 *Numerical study of a magnetically insulated front-end channel for a neutrino factory* PRSTAB 14 011001
[19] Zisman M *Accelerators for future neutrino facilities: strength and challenges* these proceedings
[20] Beard K *Linac & RLA design status and simulations* these proceedings
[21] Pasternak J *Recent development on the muon non-scaling FFAG for the Neutrino Factory and its subsystems* these proceedings
[22] Abe T *et al. Preprint* arXiv:0712.4129 [physics.ins-det]
[23] Huber P and Winter W 2003 *Phys. Rev.* D **68** 037301
[24] Blondel A *Neutrino flux monitoring in the Neutrino Factory* these proceedings

Journal of Physics: Conference Series **408** (2013) 012014 doi:10.1088/1742-6596/408/1/012014

EMMA

S Machida[1]
On behalf of EMMA collaboration
ASTeC, STFC Rutherford Appleton Laboratory, Harwell Oxford, Didcot, Oxfordshire OX11 0QX, UK

E-mail: shinji.machida@stfc.ac.uk

Abstract. EMMA stands for Electron Model for Many Applications and is expected to demonstrate the novel concept of a non-scaling Fixed Field Alternating Gradient (FFAG) accelerator. A non-scaling FFAG accelerator was originally invented to accelerate muons up to a few tens of GeV energy range although it is now hoped to have a wider variety of applications such as a proton driver for Accelerator Driven Sub-critical Reactor (ADSR). We will report the recent developments of the EMMA experiment, with more emphasis on acceleration of a muon beam.

1. Introduction

The accelerator for a muon beam has different requirements from that for a conventional particle beam such as protons and electrons. Acceleration has to be very quick because of its lifetime in the rest frame which is 2.2 μs. Acceptance should be large enough to accommodate the large muon beam emittance, which is true even after the ionisation cooling. A linear accelerator (or linac) could satisfy the requirements because it is basically a sequence of accelerating units and its aperture could be as large as the muon beam emittance. One drawback of a linac for muon acceleration is the cost. Using an acceleration unit only once for each beam pulse is the most inefficient way of transferring energy to the beam. In order to reach the final energy, the total sum of the gap voltage over all the acceleration units must at least be equal to the final energy.

Circular accelerators like the cyclotron and synchrotron use their acceleration units repeatedly for the same beam pulse in order to minimize the required rf voltage and reduce the total machine cost to achieve the final energy. However, the cyclotron does not work as a muon accelerator because muon beams are in the ultra-relativistic regime. The synchrotron has a problem as well because of its slow repetition of acceleration. As a hybrid type of accelerator between linear and circular ones, a re-circulating linear accelerator (RLA) had been proposed [1]. It utilizes a linear accelerator as an accelerating unit but with the several arcs of 180 degree bending at the end of the linac so that multiple passes through the linac can be realized. In practice, however, more than 4 or 5 re-circulations is not feasible because the separator/combiner between the linac and the arc becomes increasingly complicated.

A Fixed Field Alternating Gradient (FFAG) accelerator is a circular accelerator with a strong focusing lattice like a synchrotron [2]. However, as its name implies, the guiding magnetic field is

[1] To whom any correspondence should be addressed.

constant in time and the time required for acceleration is determined only by the available rf voltage. For example, if we could provide 1 GV rf voltage per turn, 10 GeV energy gain would be possible in 10 turns within a few micro seconds. Considering the lifetime of muons which is Lorentz boosted, acceleration in a few micro seconds is still fast enough. If large physical and dynamic aperture are assured, a FFAG accelerator becomes a very promising candidate for muon acceleration with relatively low capital cost.

Proposal of muon acceleration with a FFAG has been published first at Nufact01 workshop [3]. It assumed a scaling FFAG in which the lattice magnets have transverse field profile of r^k so that the transverse tune stays constant through the acceleration. Constancy of tune is essential for more conventional acceleration where a beam circulates in an accelerator for a few tens of thousands or even more turns. If tune changes during acceleration by more than a unit, a beam will cross dangerous integer and half-integer resonances which are excited by imperfections of lattice magnets and misalignments. However, the resonance is an accumulative phenomenon and it may not affect a beam if it circulates in an accelerator for only 10 turns.

If constancy of transverse tune is eliminated, the lattice magnets can be simpler and, more importantly, becomes more compact, which further enhances the advantage of a FFAG accelerator. That is a non-scaling FFAG which uses linear quadrupole magnets instead of more complicated magnets with nonlinear fields [4]. For acceleration of a muon beam within around 10 turns, particle tracking simulation shows that it is indeed possible to use the non-scaling FFAG without deterioration of beam quality [5]. A neutrino factory design with a non-scaling FFAG as the main muon accelerator becomes now a baseline [6].

2. Goals of the experiment

The idea of a non-scaling FFAG seems feasible by paper, but it certainly needs an experimental demonstration. Fortunately, a project was approved and a small-scale model using 10 to 20 MeV electron beams from ALICE (Accelerators and Lasers in Combined Experiments) facility [7] was constructed at Daresbury Laboratory in the UK. With more applications of this new type of accelerator in mind, the model accelerator is named Electron Model for Many Applications (EMMA).

Table 1. Principal parameters of EMMA.

	Value	Unit
Momentum	10.5 to 20.5	MeV/c
Circumference	16.57	m
Number of cells	42	
Focusing	focusing/defocusing quadrupole doublet	
Nominal integrated quadrupole gradient (QF/QD)	0.402/-0.367	T
Radio frequency	1.301	GHz
Number of radio frequency cavities	19	
Tune shift for the momentum range above	~0.3 to ~0.1/cell ~12 to ~4/ring	
Acceptance (normalized)	3	π mm rad

The construction of the major parts has completed by June 2010 and the beam commissioning started right after. The main parameters are listed in Table 1.

In the EMMA experiment, we set three main goals. One is the rapid acceleration with large tune shift during acceleration, which is inevitable due to lattice focusing with linear quadrupole magnets. In the accelerator terminology, the accelerator is operated with natural chromaticity over a wide momentum range (+/- 33%). Figure 1 shows the horizontal and vertical tune per ring as a function of beam momentum from 10.5 to 20.5 MeV/c. The total ring consists of 42 identical doublet focusing cells. The cell tune can be defined as the ring tune divided by the cell number of 42. Therefore, the cell tune is within 0 and 0.5, meaning that there is no systematic integer and half-integer resonance lines crossed during acceleration. However, a beam must traverse non-systematic integer and half-integer resonances. The experiment should show whether those non-systematic resonances can be safely avoided with rapid acceleration.

Second goal is to demonstrate acceleration in a serpentine channel. We need an explanation of the serpentine channel here. In order to minimize the orbit excursion in horizontal direction from the lowest to the highest momenta, the dispersion function everywhere in the ring is minimized. This leads to the minimum variation of the orbital length in the whole momentum range and the orbital length has a parabolic dependence as a function of momentum. In the momentum range of 10.5 to 20.5 MeV/c for electrons, it is already ultra-relativistic so that the orbital period also has a parabolic shape as a function of momentum. This means that the accelerator is operated almost on the isochronous condition, but not exactly. In other words, a beam goes through the transition energy at the middle of acceleration. When we choose the rf frequency such that it synchronizes at one momentum below the transition, this implies that there is a corresponding synchronous momentum above the transition. In the longitudinal phase space, two stable fixed points appear slightly below and above the transition energy whose phase position is 180 degree apart. Now if we keep increasing the rf voltage, the rf bucket size increases until its separatrix merges with the unstable fixed point associated with the other bucket. The longitudinal phase space geometry suddenly changes and the continuous channel starting in the area below the lower stable fixed point to the one above the higher stable fixed is created [8]. The beam is injected outside of the separatrix, but it gained energy at almost fixed rf phase. It was reported that similar acceleration dynamics had been observed in a cyclotron where the isochronism was not exactly satisfied [9]. Figure 2 shows the evolution of a beam in longitudinal phase space in the serpentine channel.

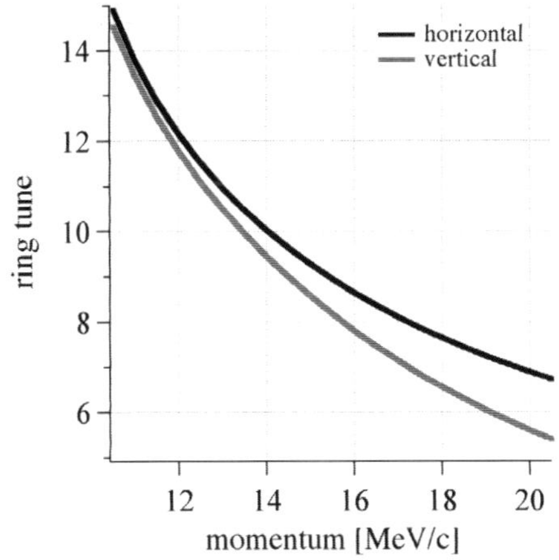

Figure 1. Tune shift during acceleration. The ring consists of 42 identical cells so that cell tune is 1/42 of ring tune.

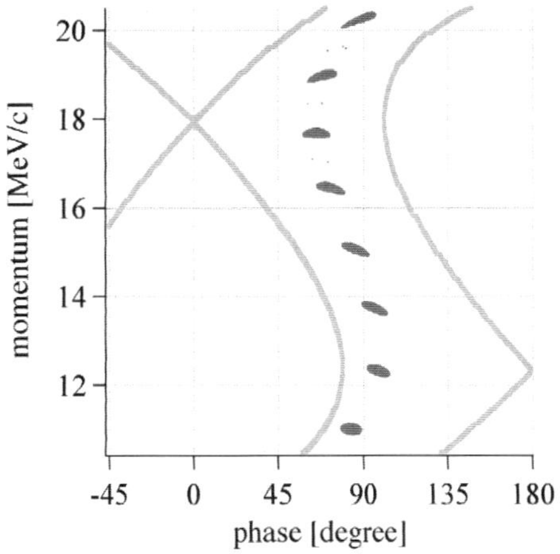

Figure 2. Beam (red bunch) is accelerated in serpentine channel. Grey curve shows separatrix.

The third goal is to show large acceptance. Since the lattice focusing magnet has no nonlinearity, dynamic aperture is expected to be large and simulation study confirms it. However, it is necessary to demonstrate that is the case experimentally.

3. Experimental observation

The beam commissioning started last summer in 2010 and it continues until next year in 2012. The recent development will be published in a conference proceedings [10] and a paper [11].

References

[1] Bogacz S A and Lebedev V A 2003 *Nucl. Instrum. Methods Phys. Res.* A **503** 306.
[2] Symon K R, Kerst D W, Jones L W, Laslett L J and Terwillinger K M 1956 *Phys. Rev.* **103** 6 1837.
[3] Machida S 2003 *Nucl. Instrum. Methods Phys. Res.* A **503** 41.
[4] Mills F and Johnstone C 1997 4th *Inter. Conf. Phys. Potential and Development of* $\mu+\mu$-*colliders* 693.
[5] Machida S and Kelliher D J 2007 *Phys. Rev. ST Accel. Beams* **10** 114001.
[6] Apollonio M *et al* 2009 *Journal of Instrumentation* **4** P07001.
[7] Saveliev Y *et al* 2010 *Inter. Particle Accelerator Conf.* 2350.
[8] Chao A W and Maury T (ed.) 1998 *Handbook of Accelerator Physics and Engineering* 92.
[9] Craddock M K 1977 *IEEE Trans. on Nucl. Sci. NS-24* **3** 1615.
[10] Machida S 2011 *Inter. Particle Accelerator Conf. 951.*
[11] Machida S *et al* 2012 *Nature Physics* March 2012 **8** 3 243.

Neutrino physics within and beyond the three flavour oscillation

Mikhail Shaposhnikov

Institut de Théorie des Phénomènes Physiques, École Polytechnique Fédérale de Lausanne, CH-1015 Lausanne, Switzerland

E-mail: `mikhail.shaposhnikov@epfl.ch`

Abstract. I will discuss several possible choices for the mass scale M of the Majorana leptons, complementing the Standard Model to describe neutrino masses and oscillations. Depending on M, they can or cannot accommodate eV neutrino anomalies, lead to baryogenesis, provide the dark matter candidate, ensure the stability of the Higgs mass against radiative corrections, and be directly searched at some experiments.

1. Introduction

The discovery of neutrino oscillations provides an undisputed signal that the Standard Model (SM) of elementary particles is not complete. However, what kind of new physics it brings to us remains still unclear - we do not know yet the properties of new particles which are believed to be behind this phenomenon. In this talk I will overview a number of possibilities. The most conservative one is the scenario of three flavour oscillations, encoded in (3×3) Majorana neutrino mass matrix for active neutrinos. The particle physics model in this case is nonrenormalisable, calling for an "integrating in" of some new degrees of freedom. The most attractive possibility is the extension of the SM by three right-handed neutrinos, making the leptonic sector similar to the quark one. The masses and couplings of new leptons remain largely unknown. The four different possibilities for the masses M of new leptons will be shortly discussed: (i) the "GUT see-saw" corresponding to the scale 10^{10-16} GeV, (ii) "electroweak see-saw" with the scale 10^{2-3} GeV, (iii) the νMSM (Neutrino Minimal SM) choice $M \sim$ keV to GeV, and (iv) the eV scale, associated with the "canonical " sterile neutrinos with this mass.

2. Three flavour oscillations scenario

The most minimal way (no extra degrees of freedom are introduced) to incorporate non-zero neutrino masses into the Standard Model is to add to its Lagrangian the SU(3)×SU(2)×U(1) invariant five-dimensional operator [1, 2]

$$O_5 = A_{\alpha\beta}\left(\bar{L}_\alpha\tilde{\phi}\right)\left(\phi^\dagger L^c_\beta\right) , \tag{1}$$

suppressed by the scale Λ. Here L_α are the leptonic doublets ($\alpha = e,\ \mu,\ \tau$), ϕ is the Higgs field, $\tilde{\phi}_i = \epsilon_{ij}\phi^*_j$, and c is the sign of charge conjugation. Then the neutrino mass matrix is given by:

$$M_\nu \sim A_{\alpha\beta}\frac{v^2}{\Lambda} . \tag{2}$$

Published under licence by IOP Publishing Ltd

where $v = 174$ GeV is the vacuum expectation value of the Higgs field. If $|A_{\alpha\beta}| \sim 1$, then the correct neutrino mass scale $\mathcal{O}(0.1)$ eV is obtained at $\Lambda \sim 3 \times 10^{14}$ GeV.

The matrix M_ν depends on 9 physical parameters which potentially can be determined in low energy neutrino experiments. They are: 3 absolute values of ν masses, (only mass square differences can be determined in neutrino oscillation experiments, and δm^2_{sol}, δm^2_{atm} are known with good accuracy); 3 mixing angles θ_{23}, θ_{12} and θ_{13}, 1 Dirac CP-violating phase and 2 Majorana phases. Four parameters out of these 9 are not known (absolute value of neutrino masses and 3 CP-violating phases). The mixing angles and Dirac phase determine the PMNS mixing matrix [3, 4].

The ultimate experimental goal within 3 family scenario is to determine all 9 parameters with as high accuracy as possible. The self-consistency of this scheme would require the verification of the unitarity of the PMNS mixing matrix. The methods include long and short baseline neutrino oscillation experiments, search for neutrinoless double beta decay, determination of the end point of electron spectra in β decays. An incomplete list of experiments include T2K, MINOS, GERDA, CUORE, NOνA, MiniBooNE, Majorana, Double Chooz, Opera, NEMO, RICE, KATRIN, SNO, RENO, IceCube, LAGUNA-LBNO, Daya Bay, LSND and Karmen. The theoretical challenges of the 3 family scenario were discussed in [5], and will not be addressed here. This scenario can be rejected if the existence of extra neutrino flavours, such as sterile neutrinos, will be firmly established.

3. Neutrino physics beyond three flavour oscillations

An effective field theory approach, described in Section 2 allows to "solve" phenomenological aspects of the problem of neutrino masses and oscillations. However, it leaves unanswered the following fundamental questions:

- What is the physics behind the non-renormalizable terms?
- What is the energy scale of physics beyond the SM, leading to neutrino masses?

The success of relativistic quantum field theory, associated with the fact that the SM agrees with most experiments, strongly indicates that the origin of neutrino masses is the existence of new unseen particles and that the complete theory should be a renormalisable extension of the Standard Model. From the SM quantum numbers of active neutrinos one can identify several possible sources for neutrino masses. If no new fermionic degrees of freedom are introduced, one needs to have a Higgs triplet with weak hypercharge 2. Another option is an introduction of singlet (with respect to the SM gauge group) Majorana fermions N_I (other names for them are sterile neutrinos or heavy neutral leptons). In this talk I will consider only the latter possibility. The dimension five operator in (1) appears then due to an exchange of the singlet fermions.

This extension of the SM is associated with the Lagrangian

$$L = L_{\rm SM} + \bar{N}_I i\partial_\mu\gamma^\mu N_I - F_{\alpha I}\,\bar{L}_\alpha N_I\tilde{\phi} - \frac{M_I}{2}\,\bar{N}_I^c N_I + h.c., \tag{3}$$

where $L_{\rm SM}$ is the Lagrangian of the SM. Since N_I are SU(3)×SU(2)×U(1) singlets, Majorana mass terms for them are consistent with the symmetries of the SM. The number of singlet fermions cannot be deduced from symmetry principles; the minimal number is 2, to get 2 different mass square differences in active neutrino sector. We take it to be 3 in analogy with the number of generations of quarks and leptons. These particles complement nicely the fermionic content of the SM, making it left-right symmetric in neutrino sector as well, see Fig. 1. This Lagrangian is usually used for the explanation of the small values of neutrino masses via the see-saw mechanism [6, 7, 8, 9], which *assumes* that the Yukawa coupling constants $F_{\alpha I}$ of the singlet fermions are of the order of the similar couplings of the charged leptons or quarks. We are not going to make such an assumption.

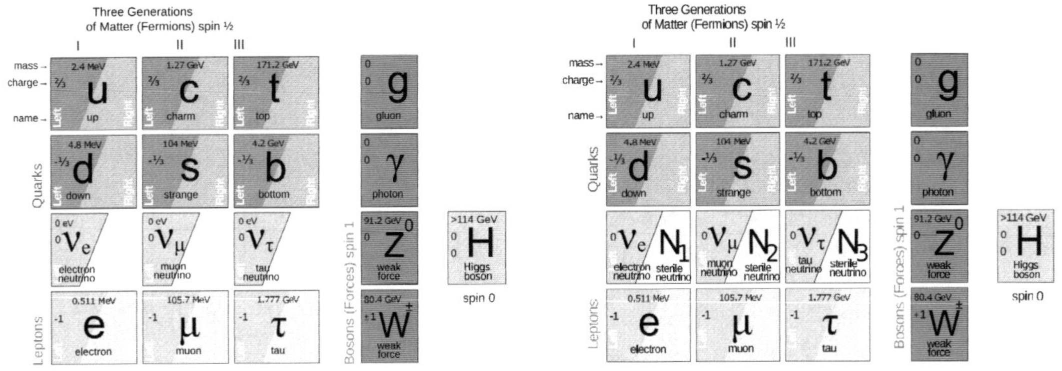

Figure 1. Particle content of the SM and its extension in neutrino sector. In the SM (left) right-handed partners of neutrinos are missing. In the νMSM (right) all fermions have both left and right-handed components.

To reduce the uncertainties in the further discussion, we will assume the validity of this theory up to the Planck scale. This may be a pretty safe bet: no supersymmetry, or extra dimensions, or technicolor are seen at the LHC at present.

In comparison with the SM, this theory contains 18 new parameters: 3 Majorana masses of new neutral fermions N_I, and 15 new Yukawa couplings in the leptonic sector, corresponding to 3 Dirac neutrino masses, 6 mixing angles and 6 CP-violating phases. The number of parameters is almost doubled in comparison with the SM; none of them can be determined theoretically within this model, in complete analogy with the SM parameters (which are all taken from experiment).

The new parameters can be divided in two different groups. The first one is the new mass scale - a generic value of the Majorana neutrino mass (denoted by M), and the second one is the typical amplitude of the Yukawa coupling constants Y, which may be defined as $Y^2 = Trace[F^\dagger F]$. We know very little about the actual values of Y and M. Basically, M can have any value between zero (corresponding to Dirac neutrinos) to 10^{16} GeV, whereas Y can vary from 10^{-13} (Dirac neutrino case) to 1 (the onset of the strong coupling). The admitted region is shown in Fig. 2 (left).

There is no preference to any value of M or Y inside the triangle shown in Fig. 2 from the point of view of particle theory described by the Lagrangian (3). The four "preferred" choices, relying on different physics and marked by "GUT see-saw", "LHC", "νMSM" and "LSND", will be discussed below.

3.1. GUT see-saw

The key *assumption* of the see-saw mechanism [6, 7, 8, 9] is that the Yukawa couplings of N to the Higgs and left-handed lepton doublets are similar to those in quark or charged lepton sector (say, $Y \sim 1$, as for the top quark). This leads to an estimate the mass scale of the singlet leptons:

$$M_I \simeq \frac{v^2}{m_{\rm atm}} \simeq 6 \times 10^{14}\ \text{GeV}\ , \tag{4}$$

where $m_{\rm atm} = \sqrt{\Delta m^2_{\rm atm}} \simeq 0.05$ eV is the atmospheric mass difference. This scale is rather close to that of Grand Unification, providing an argument that the gauge coupling unification and neutrino masses might be related [10]. In addition, the decays of superheavy N_I in the early Universe may lead to baryogenesis through leptogenesis [11] and rapid anomalous fermion number non-conservation at high temperatures [12] (see also [13]). There are several challenges of the GUT see-saw. One of them is the stabilisation of the Higgs mass hierarchy $m_H \ll M$ against radiative corrections to m_H involving the loops with superheavy Majorana leptons [14]. It requires low energy supersymmetry. If SUSY is not discovered at the LHC then so small value of m_H in comparison with M is "unnatural" and requires an enormous fine-tuning. From more phenomenological side, this choice of M does not provide a Dark Matter candidate and thus calls for a further extension of the model.

The experimental neutrino physics within GUT see-saw is identical with that within the three flavour oscillations - new particles are too heavy to be searched directly at any experiment. However, if the see-saw mechanism is embedded in some grand unified theory, one should expect to have a proton decay.

3.2. EW see-saw

The key assumption of the "EW see-saw" is that the masses of new Majorana leptons are related to (unknown) physics of electroweak symmetry breaking and are thus of the order of 100 GeV - 1 TeV. If all the Yukawa coupling are of the same order of magnitude, then they are too small to make these particles directly observable at any type of experiment (this corresponds to the lower part of the triangle in Fig. 2). However, extra symmetries can allow for cancellations [15, 16], and some of Yukawa couplings can be of the order of 1, corresponding to the upper horisontal line in Fig. 2. If true, LHC can find them [16, 17]. This question was discussed in [18]; the low energy signatures of the EW see-saw were analysed in [20, 19]. As in the GUT see-saw case, the baryogenesis through resonant leptogenesis is possible [21]. However, no dark matter candidate exists with this choice of parameters.

3.3. The νMSM

The νMSM approach [22, 23] (for reviews see [24, 25]) is driven by neutrino experiments and cosmology. It requires that three new particles - Majorana leptons - must explain simultaneously neutrino masses and oscillations, Dark Matter, and baryon asymmetry of the Universe. Once the range of the parameters of the model, leading to these facts, is determined, one can figure out how and were to search for the new particles.

3.3.1. Dark matter Though the νMSM does not have any extra stable particle in comparison with the SM, the lightest singlet fermion, N_1, may have a life-time τ_{N_1} greatly exceeding the age of the Universe and thus play a role of a dark matter particle [26, 27, 28, 29]. The following considerations determine the range of masses and couplings of the DM sterile neutrino:
(i) Cosmological production. N_1 are created in the early Universe in reactions $l\bar{l} \to \nu N_1$, $q\bar{q} \to \nu N_1$, etc. We should get the correct DM abundance.
(ii) Structure formation. If N_1 is too light it may have considerable free streaming length and erase fluctuations on small scales. This can be checked by the study of Lyman-α forest spectra of distant quasars and structure of dwarf galaxies.
(iii) X-rays. N_1 decays radiatively, $N_1 \to \gamma\nu$, producing a narrow line which can be detected by X-ray telescopes (such as Chandra or XMM-Newton). This line has not been seen yet.

The summary of these constrains is presented in Fig. 2 where the mixing angle θ is the ratio of the Dirac and Majorana masses,

$$\theta = \frac{m_D}{M_1} \; . \tag{5}$$

The interactions of N_1 with particles of the SM is weaker than the weak interactions by a factor θ (in the amplitude). So, they fall into the SuperWIMP category of the DM particle physics candidates. It is important that the DM sterile neutrino production requires the presence of large, $\Delta L/L > 2 \times 10^{-3}$ lepton asymmetry at temperature $T \sim 100$ MeV. It can only be produced in the νMSM [33] (the GUT, EW, or eV see-saw cannot make it).

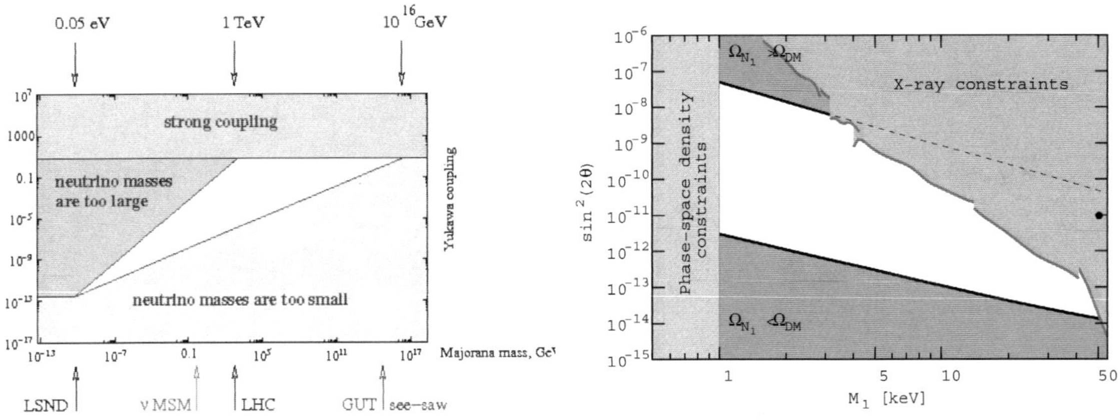

Figure 2. *Left*: The admitted values of the Yukawa couplings as a function of the Majorana fermion mass. *Right*: The allowed region of parameters for dark matter sterile neutrinos produced via mixing with active neutrinos (unshaded region). The two thick black lines bounding this region represent production curves for zero lepton asymmetry (upper line) and for the maximal lepton asymmetry attainable in the νMSM. The red shaded region in the upper right corner represents X-ray constraints. The region below 1 keV is ruled out according to the phase-space density arguments [30, 31]. The Lyman-α constraints are in general stronger but depend essentially on lepton asymmetry, see [32] for detailed discussion. For zero lepton asymmetry the lower bound on M_1 is around 8 keV, while for large asymmetries it is as small as 2 keV. From Ref. [25].

The constraints shown in Fig. 2 (right) allow to make a number of predictions for neutrino physics [22, 34]. The minimal number of sterile neutrinos, which can explain the dark matter in the Universe and neutrino oscillations, is $\mathcal{N} = 3$. Only one sterile neutrino can be the dark matter. Moreover, it practically decouples and does not contribute to active neutrino masses. Also, the absolute neutrino mass scale is fixed: the mass of the lightest active neutrino is bounded from above by $m_1 \leq 2 \cdot 10^{-3}$ eV. This leads to the following values of the masses of other active neutrinos: $m_2 = [9.05^{+0.2}_{-0.1}] \cdot 10^{-3}\text{eV} \simeq \sqrt{\Delta m^2_{solar}}$, $m_3 = [4.8^{+0.6}_{-0.5}] \cdot 10^{-2}\text{eV} \simeq \sqrt{\Delta m^2_{atm}}$ (normal hierarchy), or $m_{2,3} = [4.7^{+0.6}_{-0.5}] \cdot 10^{-2}$ eV (inverted hierarchy). Yet another prediction is the effective Majorana mass $m_{\beta\beta}$ for neutrinoless double β decay [35]: 1.3 meV $< m_{\beta\beta} <$ 3.4 meV (normal hierarchy) and 13 meV $< m_{\beta\beta} <$ 50 meV (inverted hierarchy). Moreover, knowing $m_{\beta\beta}$ experimentally will allow to fix Majorana CP-violating phases in neutrino mass matrix, provided θ_{13} and Dirac phase δ are known.

The strategy for search of DM sterile neutrino was discussed in a number of papers, for a review see [25]. In short, one should use the X-ray telescopes (such as Chandra and XMM Newton) to look for a narrow γ line against astrophysical background. The astrophysical objects leading to the best signal to background ratio are the dwarf satellite galaxies and the Milky Way. So, the answer to John Womersley question: "Do neutrinos play a role in dark matter, especially if there is no light neutralino?" is: " Yes, sterile neutrino with mass from 1 keV to 50 keV is an excellent warm or cold DM candidate. Search for it with X-ray telescopes in space!"

3.3.2. Baryon asymmetry In addition to DM sterile neutrino the νMSM contains a pair of more heavier singlet fermions, N_2 and N_3. The parameters of these particles can be constrained from the following conditions:
(i) BAU generation via singlet fermion oscillations [36, 23] requires out of equilibrium: mixing

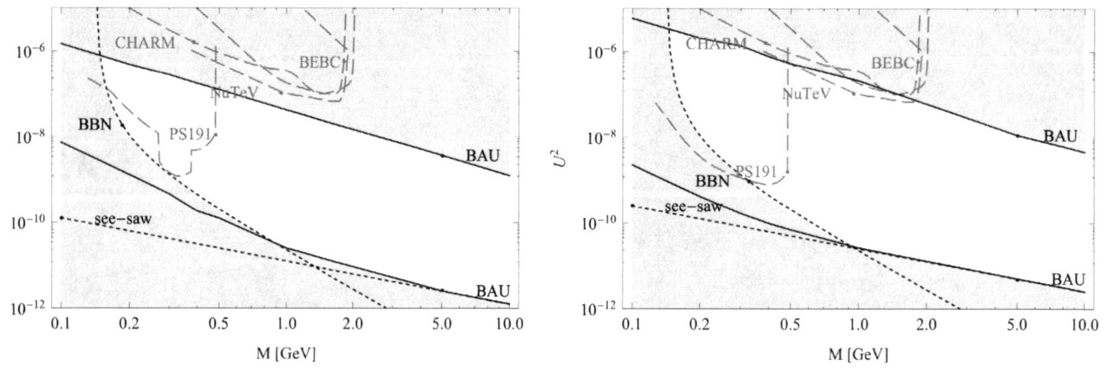

Figure 3. Constraints on U^2 coming from the baryon asymmetry of the Universe (solid lines), from the see-saw formula (dotted line) and from the big bang nucleosynthesis (dotted line) [37]. Experimental searched regions are in red - dashed lines. *Left:* normal hierarchy. *Right:* inverted hierarchy.

angle of $N_{2,3}$ to active neutrinos cannot be too large. In addition, due to the smallness of the Yukawa couplings, the asymmetry generation must have a resonant character, leading to the requirement that $N_{2,3}$ must be almost degenerate.
(ii) Neutrino masses: mixing angle of $N_{2,3}$ to active neutrinos cannot be too small.
(iii) BBN: decays of $N_{2,3}$ must not spoil Big Bang Nucleosynthesis.
(iv) Experiment: $N_{2,3}$ have not been seen yet.

The summary of these constrains is presented in Fig. 3 where the mixing angle U^2 is defined in full analogy with (5).

The answer to Sachio Komamiya question: "θ_{13} and δ are similar to V_{ub} and CP phase in the quark sector. In heavy flavours, angles are not really the interesting thing - CP violation there is found to be insufficient to generate baryogenesis. How about neutrinos? We know about lepto-genesis, is there further fundamental physics in neutrinos beyond the numerical values of the angle and phase?" is: "The knowledge of θ_{13} and δ is necessary, but not sufficient to find theoretically baryon asymmetry of the Universe. It is not zero even if $\theta_{13} = 0$ or $\delta = 0$. To compute it, one *must* find N's and determine their properties experimentally". The ways how this can be done is discussed below.

3.3.3. Experimental searches of $N_{2,3}$ Several distinct strategies can be used for the experimental search of these particles [38]. The first one is related to their production (U^2 effect). The singlet fermions participate in all the reactions the ordinary neutrinos do with a probability suppressed roughly by a factor U^2. Since they are massive, the kinematics of, say, two body decays $K^\pm \to \mu^\pm N$, $K^\pm \to e^\pm N$ or three-body decays $K_{L,S} \to \pi^\pm + e^\mp + N_{2,3}$ changes when $N_{2,3}$ is replaced by an ordinary neutrino. Therefore, the study of *kinematics* of rare K, D, and B meson decays can constrain the strength of the coupling of heavy leptons. This strategy has been used in a number of experiments for the search of neutral leptons in the past [39, 40], where the spectrum of electrons or muons originating in decays π and K mesons has been studied. The precise study of kinematics of rare meson decays is possible in Φ (like KLOE), charm, and B factories, or in experiments with kaons where their initial 4-momentum is well known (like NA48 or E787 experiments).

The second strategy is to use the proton beam dump (U^4 effect). As a first step, the proton beam heating the fixed target creates K, D or B mesons, which decay and produce $N_{2,3}$. The

	N mass	ν masses	eV ν anomalies	BAU	DM	M_H stability	direct search	experiment
GUT see-saw	10^{10-16} GeV	YES	NO	YES	NO	NO	NO	-
EWSB	10^{2-3} GeV	YES	NO	YES	NO	YES	YES	LHC
ν MSM	keV – GeV	YES	NO	YES	YES	YES	YES	a'la CHARM
ν scale	eV	YES	YES	NO	NO	YES	YES	a'la LSND

Figure 4. This table shows whether the corresponding choice of the mass for Majorana fermions may explain neutrino masses and oscillations, accomodate eV neutrino anomalies, lead to baryogenesis, provide the dark matter candidate, ensure the stability of the Higgs mass against radiative corrections, and be directly searched at some experiments.

second step is a search for decays of N in a near detector, looking for the processes "nothing" $\rightarrow$ leptons and hadrons [41, 42, 43, 44]. To this end, quite a number of already existing or planned neutrino facilities (related, e.g., to CERN SPS, MiniBooNE, MINOS or J-PARC), complemented by a near *dedicated* detector, can be used. Finally, these two strategies can be unified, so that the production and the decay occurs inside the same detector [45].

For the mass interval $M_N < M_K$, both strategies can be used. According to the estimates, an upgrade of NA62 experiment at CERN would allow the finding or exclusion of singlet fermions with the mass below that of the kaon. If $m_K < M_{2,3} < m_D$, the search for the missing energy signal, potentially possible at beauty, charm, and τ factories, is unlikely to gain the necessary statistics. Thus, the search for decays of neutral fermions is the most effective opportunity. The dedicated experiments on the basis of the SPS proton beam at CERN can touch a very interesting parameter range for $M_N < 1.8$ GeV. The sensitivity is proportional to total delivered protons on target (PoT); for 2.5×10^{20} PoT the constraints shown in Fig. 3 can be improved by one order of magnitude (without accounting for improvement of experimental technique). An upgrade of the LHCb experiment, allowing to use the combination of two strategies, could potentially enter in a cosmologically interesting region for masses and mixing angles of singlet fermions. Going above D-meson but still below B-meson thresholds is very hard if not impossible with the present or planned proton machines or B-factories. To enter into a cosmologically interesting parameter space would require the increase in the present intensity of, say, CERN SPS beam by two orders of magnitude or to produce and study the kinematics of more than 10^{10} B-mesons.

3.4. eV scale: "canonical" sterile neutrinos

The key *assumption* of the "eV see-saw" is that the new scale coincides with that found in neutrino oscillations. This leads to existence of eV sterile neutrinos which can be used to explain a number of anomalies (LSND, MiniBOONE) indicating that the three neutrino oscillation paradigm may not be complete. This possibility has been discussed in detail at this conference (for theory and phenomenology see [46, 47, 5, 48, 49], while experimental situation was discussed in [50, 51]). I would like just to add that the theory (3) with the eV mass for sterile neutrinos cannot explain baryon asymmetry of the Universe and does not provide a dark matter candidate, so that an extra extension of the theory (3) is required.

4. Conclusions

Even the simplest extension of the Standard Model by three Majorana fermions leads to rich and interesting physics. Independently on the masses of new leptons, they can explain oscillations of active neutrinos. For some ranges of parameters, they can provide the Dark Matter candidate, lead to baryogenesis, and allow for direct experimental search. The summary of different possibilities is given in Fig. 4. Hopefully, future experiments in neutrino and flavour physics will allow to clarify what is the physics beyond the three flavour oscillation.

This work was supported by the Swiss National Science Foundation.

References

[1] Weinberg S 1979 *Phys. Rev. Lett.* **43** 1566
[2] Wilczek F and Zee A 1979 *Phys. Rev. Lett.* **43** 1571
[3] Pontecorvo B 1957 *Sov. Phys. JETP* **6** 429
[4] Maki Z, Nakagawa M and Sakata S 1962 *Prog. Theor. Phys.* **28** 870
[5] Smirnov A these proceedings
[6] Minkowski P 1977 *Phys. Lett.* B **67** 421
[7] Yanagida T 1980 *Prog. Theor. Phys.* **63** 354
[8] Gell-Mann M, Ramond P and Slansky R 1979 *Supergravity* ed. by D. Freedman et al (North Holland)
[9] Mohapatra R N and Senjanovic G 1980 *Phys. Rev. Lett.* **44** 912
[10] Mohapatra these proceedings
[11] Fukugita M and Yanagida T 1986 *Phys. Lett.* B **174** 45
[12] Kuzmin V A, Rubakov V A and Shaposhnikov M E 1985 *Phys. Lett.* B **155** 36
[13] Mavromatos N these proceedings
[14] Vissani F 1998 *Phys. Rev.* D **57** 7027
[15] Shaposhnikov M 2007 *Nucl. Phys.* B **763** 49
[16] Kersten J and Smirnov A Y 2007 *Phys. Rev.* D **76** 073005
[17] Senjanovic G 2011 *Riv. Nuovo Cim.* **34** 1
[18] Senjanovic G these proceedings
[19] Ibarra A, Molinaro E and Petcov S 2011 *Phys. Rev.* D **84** 013005
[20] Ibarra A these proceedings
[21] Pilaftsis A 1997 *Phys. Rev.* D **56** 5431
[22] Asaka T, Blanchet S and Shaposhnikov M 2005 *Phys. Lett.* B **631** 151
[23] Asaka T and Shaposhnikov M 2005 *Phys. Lett.* B **620** 17
[24] Shaposhnikov M 2007 *preprint hep-th/0708.3550*
[25] Boyarsky A, Ruchayskiy O and Shaposhnikov M 2009 *Ann. Rev. Nucl. Part. Sci.* **59** 191
[26] Dodelson S and Widrow L M 1994 *Phys. Rev. Lett.* **72** 17
[27] Shi X D and Fuller G M 1999 *Phys. Rev. Lett.* **82** 2832
[28] Dolgov A D and Hansen S H 2002 *Astropart. Phys.* **16** 339
[29] Abazajian K, Fuller G M and Patel M 2001 *Phys. Rev.* D **64** 023501
[30] Boyarsky A, Ruchayskiy O and Iakubovskyi D 2009 *JCAP* **0903** 005
[31] Gorbunov D, Khmelnitsky A and Rubakov V 2008 *JCAP* **0810** 041
[32] Boyarsky A, Lesgourgues J, Ruchayskiy O and Viel M 2009 *Phys. Rev. Lett.* **102** 201304
[33] Shaposhnikov M 2008 *JHEP* **08** 008
[34] Boyarsky A, Neronov A, Ruchayskiy O and Shaposhnikov M 2006 *JETP Lett.* **83** 133
[35] Bezrukov F 2005 *Phys. Rev.* D **72** 071303
[36] Akhmedov E K, Rubakov V A and Smirnov A Y 1998 *Phys. Rev. Lett.* **81** 1359
[37] Canetti L and Shaposhnikov M 2010 *JCAP* **1009** 001
[38] Gorbunov D and Shaposhnikov M 2007 *JHEP* **10** 015
[39] Yamazaki T *et al.* 1984 *Proc. Int. Conf. on High Energy Physics (Leipzig)* Vol 1 p 262
[40] Daum M *et al.* 2000 *Phys. Rev. Lett.* **85** 1815
[41] Bernardi G *et al.* 1986 *Phys. Lett.* B **166** 479
[42] Bernardi G *et al.* 1988 *Phys. Lett.* B **203** 332
[43] Vaitaitis A *et al.* (NuTeV) 1999 *Phys. Rev. Lett.* **83** 4943
[44] Astier P *et al.* (NOMAD) 2001 *Phys. Lett.* B **506** 27
[45] Achard P *et al.* (L3) 2001 *Phys. Lett.* B **517** 75
[46] De Gouvea A these proceedings
[47] Giunti C these proceedings
[48] Hernandez P these proceedings
[49] Zhang H these proceedings
[50] Luis B these proceedings
[51] Halzen F these proceedings

Proton drivers for neutrino beams and other high intensity applications

R Garoby[1], K Gollwitzer[2], T Koseki[3], J Thomason[4]

[1]CERN, BE/HDO, 1211 Geneva 23, Switzerland
[2]Fermilab, Batavia, IL 60510, USA
[3]J-PARC, KEK and JAEA, Tokai, Ibaraki 319-1195, Japan
[4]ISIS, Rutherford Appleton Laboratory, STFC, OX11 0QX, UK

Abstract. CERN, Fermilab, J-PARC and RAL tentatively plan to have proton accelerators delivering multi-MW of beam power in view of enhancing their physics reach especially in the domain of neutrinos. These plans are described, together with their benefits for other applications.

1. Introduction
High power proton accelerators ("proton drivers") have a wide range of applications including spallation neutron sources, radioactive ion sources, rare decay experiments, muon and neutrino production, etc. Among the existing laboratories equipped with medium power (100 kW-class) proton accelerators, CERN, Fermilab, J-PARC and RAL have a declared interest in neutrino physics, including the potential implementation of high power (MW-class) proton accelerators.

2. Plans at Fermilab
2.1. Project X
Fermilab is planning to build Project X [1] with operation beginning around 2021. Project X will be a high intensity proton machine capable of supporting several different experiments at the same time that could later be used as a starting point for a proton driver for a Neutrino Factory. Project X (figure 1) will consist of a low energy (up to 3 GeV) continuous wave (CW) linac and a higher energy (3 to 8 GeV) pulsed linac. The CW linac will deliver up to 3 MW of beam power at 3 GeV, principally for the study of rare muon and kaon processes, while the pulsed linac will provide beam to the Main Injector and increase its beam power for the Long Baseline Neutrino Experiment (LBNE). At a later stage in the 2020s, these new linacs may be upgraded to serve as the proton driver for a possible future Neutrino Factory and/or Muon Collider.

The front end of Project X will consist of a negatively charged hydrogen ion source, a low energy beam transport system, which will bunch the beam at 162.5 MHz, and a programmable chopper system that will form different desired pulse trains for the different experiments. For the initial configuration of Project X, the ion source will be run at 5 mA. The CW linac will initially be able to support an average beam current of 1 mA, thus requiring the above-mentioned chopper system to intercept more than 80% of the beam. The CW linac will consist of superconducting RF systems operating at 325 (up to 160 MeV) and 650 MHz (between 160 MeV and 3 GeV).

After the CW linac, the majority of the possible 3 MW of beam power will be sent to an experimental campus. A transverse RF splitter system will direct beam pulses to different experiments.

The programmable chopper will determine whether a beam pulse is present at different phases of the RF splitter system at the desired frequency for each experimental area: the chopper system can send either pulse trains or every n^{th} pulse as long as the average CW linac beam current is less than 1 mA. A minor fraction of the beam from the CW linac will be diverted away from the 3 GeV experimental area by a pulsed dipole and sent into the pulsed linac.

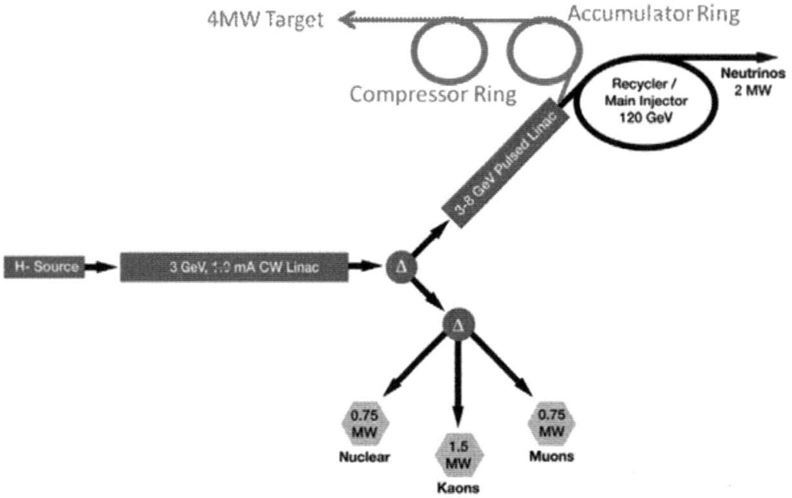

Figure 1. Schematic of Fermilab Project X with additional rings for beam formatting for a Neutrino Factory.

The 3–8 GeV pulsed linac will use 1.3 GHz superconducting RF cavities using technology developed for the ILC. Accelerating 4 ms long beam pulses every 100 ms, the pulsed linac will deliver approximately 300 kW of beam power at 8 GeV to the Main Injector/Recycler. Project X foresees upgrades to the Main Injector/Recycler rings for accumulation and subsequent acceleration of the intense beam: more than 2 MW can then be delivered to LBNE.

2.2. Future Project X upgrades

With the Neutrino Factory proton driver requirement of 4 MW of beam power between 5 and 15 GeV, it is natural to use the output of the 8 GeV pulsed linac. The necessary increase of beam power by an order of magnitude will be obtained by boosting the ion source beam current, the average CW linac beam current, and the length and repetition rate of pulse trains diverted to the pulsed linac. The Project X ion source is capable of operating at 10 mA. The associated chopper system and beam dump are being designed to handle 10 mA. The CW linac is being designed to handle an average 5mA beam current; there will have to be additional RF power purchased. The pulsed linac will require an upgrade of the power couplers and more RF power for longer pulses of nearly 7 ms at a repetition rate of 15 Hz. The cryogenic capacity will have to be increased by approximately 50% to support both linacs; the layout of utilities is being done with future expansion in mind.

An upgrade of Project X outlined above will deliver 4 MW of beam power at 8 GeV; however, the pulsed linac output does not satisfy the desired short-intense beam bunch on target as required for a Neutrino Factory. The beam will have to be accumulated and RF manipulations must be performed in new accelerator rings (see figure 1) to achieve the few-ns bunch length. A new accumulation ring (200–250 m circumference) is being designed. The programmable chopper will send beam that will only be injected at the correct RF phase for easy capture. Due to the beam power, stripping the hydrogen ions of electrons to form a proton beam is a concern and is being studied. Several bunches at a time will be transferred to a second ring for final bunch length compression. RF bunch rotation will be performed to reduce the bunch length to the desired few-ns length and then each bunch will be extracted towards the target with the desired time interval. For example, accumulation is done at 15 Hz into h = 12 buckets; three bunches are transferred to the second ring and undergo bunch rotation; then the three bunches are extracted to the target separated by 120 μs. This example results in three bunch trains of desired bunch length and repetition rate of 60 Hz. Design and simulations of these rings are in progress.

3. Plans at RAL

The Rutherford Appleton Laboratory (RAL) is home of ISIS, the world's most productive spallation neutron source. ISIS has two neutron producing target stations (TS-1 and TS-2), driven at 40 Hz and 10 Hz, respectively, by a 50 Hz, 800 MeV proton beam from a rapid cycling synchrotron (RCS), which is fed by a 70 MeV H^- drift tube linac (DTL) [2]. Potential upgrades of the ISIS accelerators to provide beam powers of 2–5 MW in the few-GeV energy range could be envisaged as the starting point for a proton driver shared between a short pulse spallation neutron source and the Neutrino Factory (NF). Although the requirements for the NF baseline proton energy and time structure are different from those for a spallation neutron source, an additional RCS or FFAG booster bridging the gap in proton energy and performing appropriate bunch compression seems feasible.

3.1. ISIS megawatt upgrades

The first stage of the upgrade path is to replace parts or all of the ISIS 70 MeV H^- injector in order to address obsolescence issues with the present linac, and ensure reliable operation for the foreseeable future. The more exciting, but more challenging, option is to install a higher energy linac (up to ≈180 MeV), with a new optimised injection system into the present ring, which could give a substantial increase in beam power (up to ≈0.5 MW) [3].

The next stage is a new ≈3.2 GeV RCS that can be employed to increase the energy of the existing ISIS beam to provide powers of ≈1 MW. This new RCS would require a new building, along with a new ≈1 MW target station. There are a number of possible candidates for the ≈3.2 GeV, 50 Hz RCS, but studies are presently focused on a 3.2 GeV doublet-triplet design with five superperiods (5SP) and a 3.2 GeV triplet design with four superperiods (4SP), both of which will include features required for fast injection directly from the existing ISIS RCS, together with the option for optimised multi-turn injection from a new 800 MeV H^- linac [4].

The final upgrade stage is to accumulate and accelerate beam in the ≈3.2 GeV RCS from a new 800 MeV linac for 2–5 MW beams [5]. It should be noted that a significant collimation section or 'achromat' would be required after the linac to provide a suitably stable beam for injection into the RCS. These upgrades to the ISIS facility are shown in figure 2.

3.2. Common proton driver

In a common proton driver for a neutron source and the NF, based on a 2–5 MW ISIS upgrade, bunches of protons are shared between the two facilities at ≈3.2 GeV, and a dedicated RCS or FFAG booster must then accelerate the NF bunches to meet the requirements for the NF baseline. Some possible bunch sharing scenarios are outlined in table 1.

It would appear that the 5SP design is most suitable, as it meets all the requirements of the NF baseline and provides more beam power to the neutron source, but its merits need to be established by thorough beam dynamics studies. In order to give some flexibility in case the total power at ≈3.2 GeV is somewhat less than 5 MW, 6.4–10.3 GeV RCS and FFAG booster designs will be considered. Figure 2 shows the conceptual layout of the common proton driver.

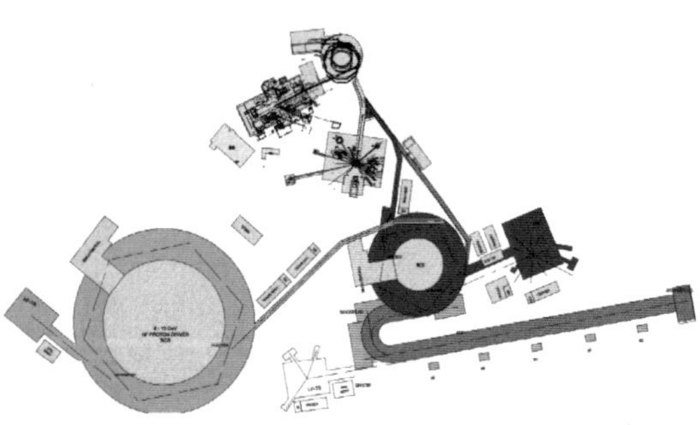

Figure 2. Conceptual layout of the NF proton driver with ISIS (green), ≈ 3.2 GeV RCS (blue), 800 MeV linac (red) and dedicated NF booster (orange) within the RAL site boundary.

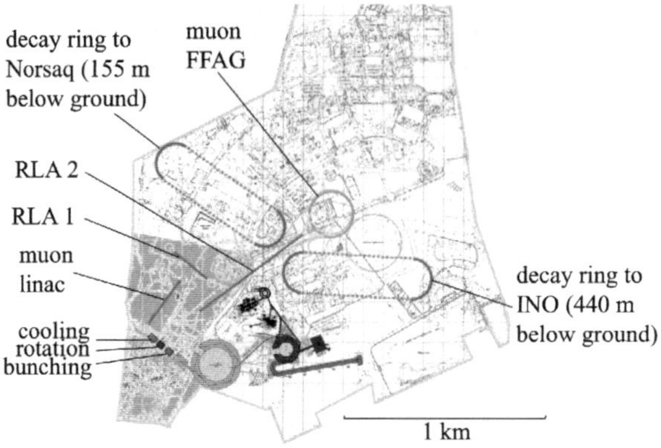

Figure 3. Schematic layout of the NF on the Harwell Oxford site. The proton driver is as shown in figure 2 and the Harwell Oxford site boundary is shown in light blue.

Table 1. Scenarios for bunch sharing between an upgraded ISIS and NF, assuming that bunches are transferred from the 3.2 GeV RCS and that 4 MW is required for the NF target.

≈3.2 GeV RCS design	Power at 3.2 GeV (MW)	Total number of bunches	Bunch spacing (ns)	Protons per bunch ($\times10^{13}$)	Number of bunches to ISIS	Power to ISIS (MW)	Number of bunches to NF	NF booster energy (GeV)
4SP	5	5	280	3.9	2	2	3	4.3
4SP	5	5	280	3.9	3	3	2	6.4
5SP	5	9	140	2.2	6	3.3	3	7.7
5SP	4	9	140	1.8	6	2.7	3	9.6

Booster RCS designs [6] have concentrated on achieving the necessary acceleration and bunch compression with present-day, cost-effective RCS technology. Although the preliminary lattice design has been produced, a great deal of work remains to be done to produce a full conceptual scenario, and no consideration has yet been given to beam transport to the pion-production target. FFAG options are yet to be explored, and would be based on technology that remains to be fully tested, but in principle would offer the advantage of allowing all the bunches to be extracted to the NF target with the same energy (unlike the RCS where the ~120 μs sequential extraction delay required by the NF baseline would give time for the main magnet field to vary between bunches).

Optimised longitudinal muon capture in the muon front end of the NF requires compression of the proton bunch length from the ≈100 ns for the neutron source to 1–3 ns at the NF target. Several methods have been proposed in order to reach this goal, based on either adiabatic compression during acceleration or fast phase rotation at the end of acceleration (or in an additional compressor ring).

3.3. *Summary*

The site-specific design at RAL is in a preliminary stage, and will require extensive effort on beam dynamics and accelerator engineering (and strategic research and development in a number of key areas such as high power front ends, RF systems, stripping foils, kickers and diagnostics) before it can be regarded as viable. The common proton driver could fit onto the RAL site, on land already set aside for large facilities and research expansion, but the complete NF would require the use of part of the Harwell Oxford Campus. A possible schematic layout of the NF on the Harwell Oxford site is shown in figure 3. It is hoped that a new (up to ≈180 MeV) linac will be installed on ISIS between 2014 and

2020, but further upgrade stages are unlikely to be realised in the foreseeable future unless a decision is made to site the NF at RAL, and funding for the common proton driver is forthcoming.

4. Plans at CERN

Following a rich history of neutrino physics using the CERN proton accelerators, new proposals are being made for experiments requiring higher beam power and either exploiting the existing machines or assuming the availability of new dedicated accelerators.

4.1. Neutrino experiments with existing accelerators

The CNGS experiment [7] is currently operating using a 500 kW proton beam from the SPS and sending neutrinos to the Gran Sasso underground laboratory 730 km away. The SPS performance is expected to improve by the end of the decade, as a result of the ongoing upgrade programme of the LHC injectors [8]. The recently started LAGUNA-LBNO Design Study [9] is aimed at making use of the increased SPS beam power, tentatively set at 750 kW, for generating a conventional ν_μ beam and sending it to a new underground experiment located in Pyhäsalmi (Finland), at a distance of 2300 km. Beyond the upgrade of the SPS and PS complex, which is foreseen in the context of the High Luminosity upgrade of the LHC, this proposal only assumes the construction of a new transfer line from the SPS to a new target area and decay tunnel oriented towards Finland.

4.2. Future high power proton drivers

New accelerators have to be built for reaching a higher beam power. Two possibilities are being considered with different proton beam energies and therefore with neutrinos of different energy spectra. The first one delivers 4 MW of beam power at 4–5 GeV, the second one 2 MW at 30–50 GeV. Both make use of a multi-GeV Superconducting Proton Linac (SPL). The SPL is the subject of active R&D [10] in collaboration with external partners (ESS, CEA-Saclay, CNRS-IN2P3, University of Rostock, Cockcroft Institute) and with the support of the European Commission in the frame of its 6th Framework Programme. Its low energy part up to 160 MeV is based on Linac4 which is presently in construction [11].

4.2.1. Low energy (5 GeV) proton driver

In this option, the entire beam acceleration is obtained in a ~500 m long high power linac providing 10^{14} p/p in beam pulses of 0.4-0.8 ms at a repetition rate of 50 Hz. The superconducting linac part (160 MeV to 5 GeV) uses two types of 5 cell superconducting elliptical cavities [10]. They operate at 704.4 MHz and their main characteristics are given in table 2.

Table 2. SPL cavities.

Section	Low energy	High energy
Energy range [MeV]	160–790	790–5000
$\beta_{geometrical}$	0.65	1
Number of cells/cavity	5	5
Accelerating gradient [MV/m]	19.3	25
Number of cavities/focusing period	3	8
Number of cavities	60	200

The linac beam is accumulated in a 200–300 m circumference fixed-energy ring, using charge-exchange injection. To generate a conventional low energy ν_μ beam from π decay, the beam is fast ejected from the accumulator onto the target. In that case, a fraction of the linac beam (~200 kW) could be diverted to a radioactive ion production system of ISOL-type to generate a beta-beam [12].

For a Neutrino Factory [13, 14], the accumulator is isochronous so that the chopped linac beam can be accumulated in bursts filling fractions of the circumference and separated by gaps, without using any RF system. Once accumulation is finished, bursts are transferred successively to the compressor ring where rotation takes place in the longitudinal phase plane with the energy stored in RF cavities. A few (3–6) bunches of ~2 ns rms length are finally ejected one by one onto the target. The circumference ratio between accumulator and compressor is such as to guarantee the arrival of the successive beam bursts at the correct location in the compressor without any resynchronization.

4.2.2. Medium energy (30–50 GeV) proton driver

In the context of the LAGUNA-LBNO design study (DS) mentioned in section 4.1, a High Power Proton Synchrotron (HP-PS), which would deliver 2 MW of beam power onto the target and decay tunnel first used by the SPS and aimed at the Pyhäsalmi (Finland) experiment, will be studied. The exact energy and cycling rate will be defined in consultation with the experimenters within the LAGUNA-LBNO DS. Thereafter the accelerator will be designed using the work done for PS2 [15, 16] and its layout on the CERN site will be drawn (a preliminary sketch is shown in figure 4). The injector will be a slower cycling, and hence lower power, version of the Superconducting Proton Linac. Contrary to PS2, the HP-PS will be dedicated to neutrino production and will not be connected to the SPS.

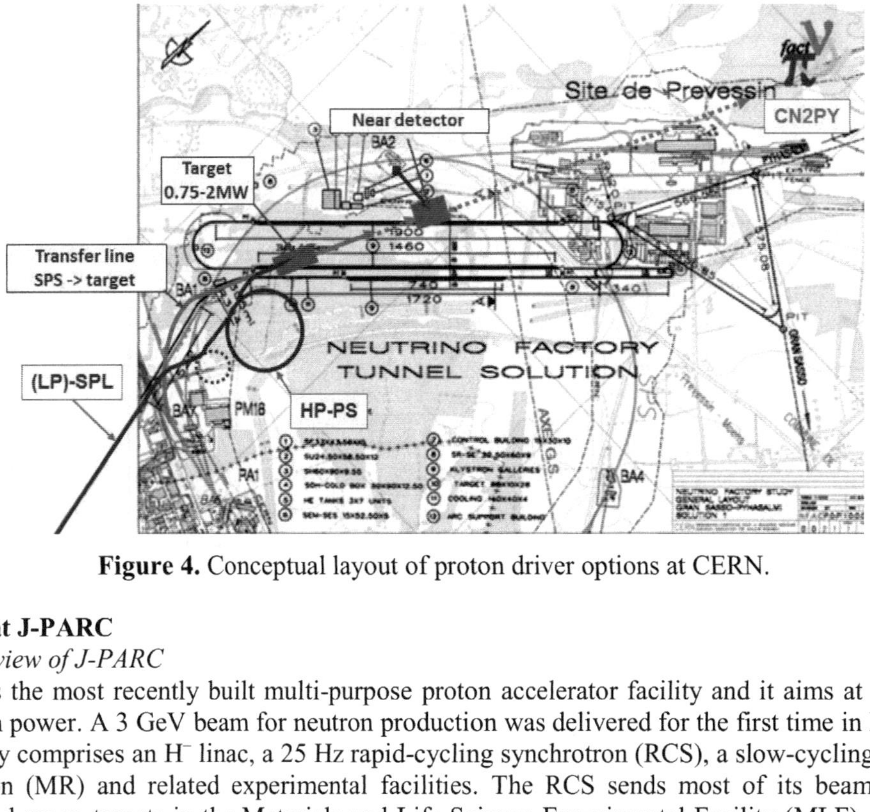

Figure 4. Conceptual layout of proton driver options at CERN.

5. Plans at J-PARC

5.1. Overview of J-PARC

J-PARC is the most recently built multi-purpose proton accelerator facility and it aims at megawatt-class beam power. A 3 GeV beam for neutron production was delivered for the first time in May 2008. The facility comprises an H^- linac, a 25 Hz rapid-cycling synchrotron (RCS), a slow-cycling main ring synchrotron (MR) and related experimental facilities. The RCS sends most of its beam pulses to neutron and muon targets in the Materials and Life Science Experimental Facility (MLF). A fraction of the RCS beam pulses is injected into the MR, which accelerates them up to 30 GeV and supplies particles to the hadron experimental hall using a slow extraction (SX) system, or to a neutrino production target using a fast extraction (FX) system. The neutrino beam is aimed at Super-Kamiokande, a large water Cherenkov detector located 295 km west from Tokai, for the long baseline neutrino oscillation experiment called T2K (Tokai to Kamioka). The layout of the MR is shown in figure 5. It has three-fold symmetry and a circumference of 1567.5 m. It is the first large proton accelerator using an imaginary gamma transition lattice. The three dispersion-free 116-m long straight sections, each of which consists of 3 FODO cells and matching sections to the arcs at the two ends, are dedicated to "injection and beam collimators", "slow extraction", and "rf cavities and fast extraction", respectively.

For cost-saving reasons, only the first part of the linac has presently been built, which limits its beam energy to 181 MeV (400 MeV nominal) and its peak current to 30 mA (50 mA nominal). These parameters must be brought up to their nominal values to reach the nominal beam powers of 1 MW at the RCS and 0.75 MW at the MR. An additional linac section using Annular-Coupled Structure (ACS) linac has to be installed in the downstream part of the existing linac to bring the beam energy up to 400 MeV. The present ion source and RFQ have to be replaced for the beam current to reach 50 mA. Design and manufacture of the accelerator components for both energy and intensity upgrade are now well in progress.

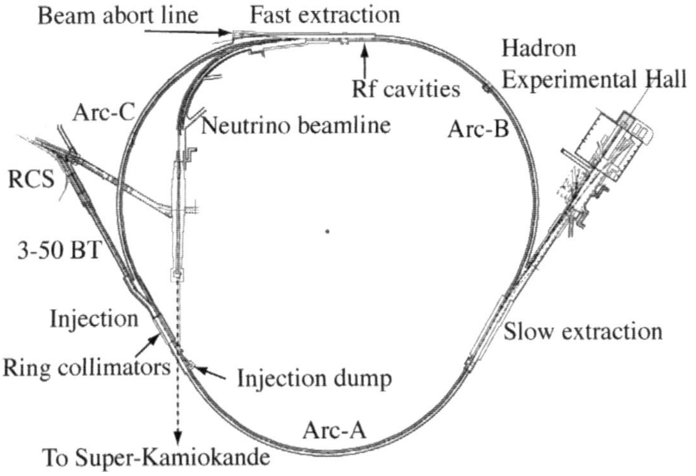

Figure 5. Layout of the J-PARC Main Ring and its experimental facilities.

5.2. Recent operating experience with the Main Ring

On March 7, 2011, the MR started operating with a cycle time of 3.04 s, delivering 145 kW of beam power to the T2K target until the morning of March 11, the day of the Great East Japan Earthquake. Since the start of physics data taking in January 2010, a total of 1.43×10^{20} protons has been delivered on the T2K target, 1.4 times the amount accumulated for K2K (KEK to Kamioka) during 4 years [17]. This has already led to an important physics result announced on June 15, 2011 [18]. Six electron neutrino candidate events have been identified, showing for the first time that muon neutrinos can transform into electron neutrinos through the quantum mechanical phenomena of neutrino flavour oscillation.

5.3. Consequences of the earthquake

According to the planned schedule, operation for physics was stopped in the morning of March 11. When the earthquake struck at 14:46 JST (Japan standard time), the linac was operating in an accelerator study mode while the RCS and MR were in maintenance. Lots of damage happened [19]. Subsidence occurred at many places. Underground water entered into the linac tunnels, but pumping with a diesel generator limited the water level to a maximum of 10 cm. The electrical and cooling infrastructures of the RCS were heavily damaged. In the MR tunnel, tens of cracks appeared and many of them let groundwater leak in. In contrast, most accelerator components, except linac beam monitors (BPMs and CTs), were not seriously hurt. Repairs are in progress and beam operation is planned to resume in December 2011 with a low current and a small duty cycle. The user program will then restart and 50 days of beam time are foreseen for the users until the end of March 2012.

5.4. Plans for the near future

Since the earthquake, the plan of the J-PARC intensity upgrade has been reconsidered. Both ACS and the new front-end part are now scheduled for installation during a 7-month shutdown corresponding with the 2013 summer maintenance period. Figure 6 shows the power upgrade plan for the RCS and the FX operation of the MR for the next three years, ultimately reaching the nominal specifications of 1 MW at 3 GeV and 750 kW at 30 GeV. Additional shields and new collimator units [20] will be installed for increasing the power loss capacity of the ring collimator section. More accelerating voltage will be obtained with the addition of 3 RF systems, bringing their total number up to nine.

For further increasing the MR beam power, a shorter cycle time of the order of 1 s is being considered. R&D has already started for the required magnet power supply and high-gradient RF cavity [21].

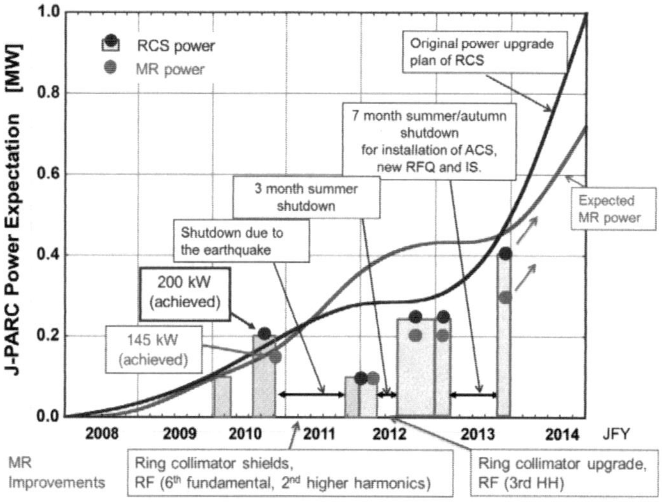

Figure 6. Power upgrade plan for the RCS and the FX operation of the MR for the next three years.

References

[1] Project X – Reference Design Report 2010 *FNAL Project X Document 776* (Nov 2010),
[2] Findlay D J S 2007 *Proc. PAC'07 (Albuquerque, USA, June 2007)* p 695.
[3] Warsop C M *et al.*2011 *Proc. IPAC'11 (San Sebastian, Spain, July 2011)* p 2760.
[4] Thomason J W G *et al.* 2008 *Proc. HB2008 (Nasville, USA, Augut 2008)* p 434.
[5] Plostinar D C *et al.* 2011 *Proc. IPAC'11 (San Sebastian, Spain, July 2011)* p 2100.
[6] Pasternak J and Jenner L J 2011 *Proc. IPAC'11 (San Sebastian, Spain, July 2011)* p 2751.
[7] Elsener K ed. 1998 The CNGS (Conceptual Technical Design), *CERN* 98-02.
[8] Garoby R *et al.* 2011 *Proc. IPAC'11 (San Sebastian, Spain, July 2011)* p 2517.
[9] LAGUNA-LBNO Design Study 2011 *European Commission 7th Framework Programme* - Grant agreement no: 284518.
[10] Gerigk F *et al.* 2010 *Proc. LINAC'10 (Tsukuba, Japan, Sept. 2010)* p 761.
[11] Vretenar M 2010 Linac4 *Proc. LINAC'10 (Tsukuba, Japan, Sept. 2010)* p 684.
[12] Final report of the EURISOL Design Study 2009 *EUROPEAN COMMISSION CONTRACT No. 515768 RIDS,* http://www.eurisol.org/doc_details.php?operation=download&docu=1075 .
[13] Aiba M 2008 *CERN-AB-2008-060 BI.*
[14] Appolonio M *et al.* 2009 Accelerator design concept for future neutrino facilities *J. Instrum.* **4** *P07001* pp 2-13.
[15] Benedikt M, Goddard B 2009 *Proc. PAC'09 (Vancouver, Canada, May 2009)* p 1828.
[16] Garoby R 2010 *CERN sLHC Project Report 0034.*
[17] Koseki T 2011 *Proc. IPAC'11 (San Sebastian, Spain, July 2011)* p 2499.
[18] http://www.kek.jp/ja/news/press/2011/J-PARC_T2Kneutrino.html
[19] Hasegawa K *et al.*2011 *Proc. IPAC'11 (San Sebastian, Spain, July 2011)* p 2727.
[20] Shirakata M *et al.*2011 *Proc. IPAC'11 (San Sebastian, Spain, July 2011)* p 3496.
[21] Ohmori C *et al.* 2011 *Proc. IPAC'11 (San Sebastian, Spain, July 2011)* p 2885.

Journal of Physics: Conference Series **408** (2013) 012017 doi:10.1088/1742-6596/408/1/012017

Prospects for CLFV experiments

S Mihara
High Energy Accelerator Research Organization (KEK)
1-1 Oho Tsukuba, 305-0801 Ibaraki, Japan

E-mail: satoshi.mihara@kek.jp

Abstract. Understanding lepton-flavor violation in the charged lepton sector (Charged Lepton-Flavor Violation, CLFV) is a complementary approach to neutrino physics experiments in order to understand elementary partcile behavir at high energy scale above TeV. A number of experiments are recetnly being conducted based on this perspective. They are expected to explore the physics beyond the Standard Model in a cooperative way with future neutrino experiments.

1. Introduction

As has been suggested by many theoretical models and proved by leading experiments, studying known particles with a great precision has a potential not only to test the Standard Model (SM) but also to find a clue to new physics beyond it. Neutrino experiments have actually played this role and will certainly keep their importance in future. Muon is also expected to take lead in this field thanks to its fruitful characteristics. The muon is in most cases produced as a decay product of a pion and decays to an electron and two neutrinos to conserve the lepton flavor. We can relatively easily produce muons in an accelerator and study precisely its properties by measuring the electron.

Neutrino oscillation, Lepton-Flavor Violation (LFV) in the neutral lepton sector, is now an established phenomenon by many experiments around the world; understanding details of the oscillation parameters is currently the most important subject as discussed in this workshop. In parallel to this, it is also important to understand the origin of neutrino mass, which is proved to be non-zero by finding oscillation of neutrinos between different flavors. One of the possibilities to explain the neutrino masses is the see-saw mechanism [1,2], in which the smallness of neutrino masses is naturally explained by large Majorana masses of their chiral partners (right-handed neutrinos). If these heavy neutrinos exist, they can contribute to enhance processes violating the lepton flavor in the charged lepton sector (Charged Lepton-Flavor Violation, CLFV) [2-4]. Because any kind of CLFV is strictly forbidden in the SM or extremely small even if we take into account the neutrino oscillation, observation of any CLFV processes would be a clear evidence of new physics, while improvements on existing limits would stringently constrain many of the new physics models beyond the SM.

2. Recent result from $\mu \rightarrow e\gamma$ search at PSI (MEG)

The MEG experiment at Paul Scherrer Institute is a precursor in CLFV search using muons. The experiment searches for the lepton-flavor-violating decay $\mu \rightarrow e\gamma$ which is forbidden within the SM of elementary particles. The experiment aims at exploring the decay mode with the world best sensitivity using the high intensity DC muon beam (3×10^7 positive muons per second) provided at PSI, and innovative detector system composed of a positron spectrometer and a photon detector in search of back-to-back, monoenergetic, time coincident photons and positrons.

Recently the MEG experiment updated the result by combining data samples taken in 2009 and 2010. The statistics corresponds to 1.8×10^{14} μ^+ decays. A likelihood analysis method was adopted with a blind procedure on examining the data. The probability density functions (PDFs) needed for the likelihood analysis were constructed using the real event data outside of the blind region. Details of detector performance and data analysis were presented in a separated presentation in this workshop [5]. They successfully obtained a 90% C.L. upper limit of 2.4×10^{-12} on the branching ratio of the $\mu^+\rightarrow e^+\gamma$ decay [6]. This gives the most stringent limit on the existence of this decay mode as present.

The result described above is not limited by the background yet, but by data statistics; they plan to increase statistics by continuing data acquisition in 2011 and 2012. Detector upgrade plan is also considered to further improve the sensitivity of the experiment.

3. Future experiments

There are a couple of experiments planned in the world with better sensitivities than the MEG experiment. The current best limit on the μ-e conversion is set to be 7×10^{-13} by the SINDRUM II experiment [7]. Two major experiments are in the planning phase to improve the physics sensitivity by more than an order of magnitude compared to the MEG experiment. They use the μ-e conversion process for this purpose. One is the Mu2e experiment at FNAL in the U.S. and the other is the COMET experiment at J-PARC in Japan. Both experiments set their goal of the branching ratio sensitivity at 10^{-16}, starting physics data acquisition around 2018.

3.1. μ-e conversion search experiments

An advantage of the μ-e conversion search experiment over the $\mu\rightarrow e\gamma$ search comes from the fact that the pulsed muon beam with high intensity can be used in the μ-e conversion search. This is simply because the μ-e conversion signal is characterized by only one energetic electron emerging from muonic atoms. The energy of the signal electron will be as large as the muon mass with a reduction corresponding to the binding energy of a muon in a muonic atom. This is 105 MeV for the case of aluminum used as a muon stopping target material. A possible background to the signal would come from decay-in-orbit electrons, which can extend to the signal region although its spectrum rapidly decreases in proportion to E_e^{-5}; the background can be suppressed sufficiently below the sensitivity with reasonable electron spectrometer resolution. Thus there is a lot of room for improvement in the physics sensitivity for CLFV in the μ-e conversion search. It should be also pointed out that the μ-e conversion can be induced not only by photon exchange but also by other particle exchanges. In the former case the μ-e conversion will generally have a branching ratio smaller by a factor of α compared to the branching ratio of the $\mu\rightarrow e\gamma$ decay, while in the latter case the μ-e conversion can be enhanced by an exchange of other particles than a photon. This means that measuring the μ-e conversion branching ratio with a similar sensitivity to that for the $\mu\rightarrow e\gamma$ decay is quite important to understand the physics behind them.

The muon beam used for the μ-e conversion search is required to have a width between two consecutive pulses as large as the life of muons in muonic atom (0.88 μsec for aluminum). The width of the pulse itself is required to be small enough compared to this scale in order to reject prompt background. This prompt background is dominated by pions that can enter and flash the detector. The detector electronics will be switched off during this flash and invoked after a few hundred nsec to measure electrons from muonic atoms. As long as the μ-e conversion search experiment employs this scheme, another requirement on the beam structure must be met; leakage of protons from pulses to between pulses needs to be minimized. The degree is call "beam extinction", which is defined as a ratio of the number of protons in between pulses to the number of protons in a pulse. In either of the Mu2e or COMET experiments, the extinction level needs to be suppressed well below 10^{-9} to 10^{-10}.

The Mu2e experiment uses proton beam provided at the FNAL accelerator facility. The proton beam accelerated to 8GeV is sent to a storage ring to shape the beam pulse structure to satisfy the requirement of the experiment. Then the beam is transferred to another ring before being extracted and transferred to the experiment. Figure 1 shows a set up of the Mu2e experiment. A pion production target is located in a solenoid magnet with gradient magnetic field to efficiently collect pions that decay to muons. These muons are transported through an S-shape curved solenoid; charge and momentum selections are performed during transportation by utilizing characteristics of beam optics in a curved solenoid. A muon stopping target composed of aluminum disks is located near the entrance of the spectrometer magnet. Muonic atoms are formed when muons are stopped in the target. Most of muons decay in the orbit or are captured by nucleus, emitting neutrinos in the final state to conserve the lepton flavor. However, if the μ-e conversion occurs, an energetic electron of 105 MeV will be emitted from the muonic atom. The detector located inside the spectrometer magnet will measure the electron momentum and energy to identify the signal. Design of the experiment is in progress with intensive R&D work. The Mu2e plans to start physics data acquisition in 2018 by using proton beam of 20kW. Target sensitivity of the experiment is 6×10^{-17} as 90% C.L. upper limit. Details of the experiment are presented in [8] in this workshop.

The COMET planned at J-PARC is a competing experiment to the Mu2e experiment. They also plan to achieve a sensitivity of 6×10^{-17} as 90% C.L. upper limit. The COMET experiment uses proton beam of 8 GeV provided at J-PARC. The anticipated beam power is 50kW accelerated in the main ring of the J-PARC. The beam is extracted directly from the ring using the slow extraction technique and transported to the experiment. A pion production target is located in a solenoid magnet with gradient magnetic field. Pions decay to muons during transportation through a C-shape curved solenoid; and muons are finally stopped in a muon stopping target as shown in Figure 2. The C-shape curved solenoid has larger separation power of momentum than the S-shape curved solenoid, but compensating vertical magnetic field is required simultaneously because the center of the muon beam trajectory will drift upward or downward depending on the field direction. This makes the magnet design complicated although this can be realized by tiling the coil winding in the curved section by a proper angle. Another C-shape magnet is connected to the target solenoid to perform momentum selection of electrons emitted from the muon stopping target. A spectrometer magnet containing an electron tracker and calorimeter to identify the signal electron ends the magnet chain. The COMET experiment also plans to start physics data acquisition in 2018 after an engineering run in 2017.

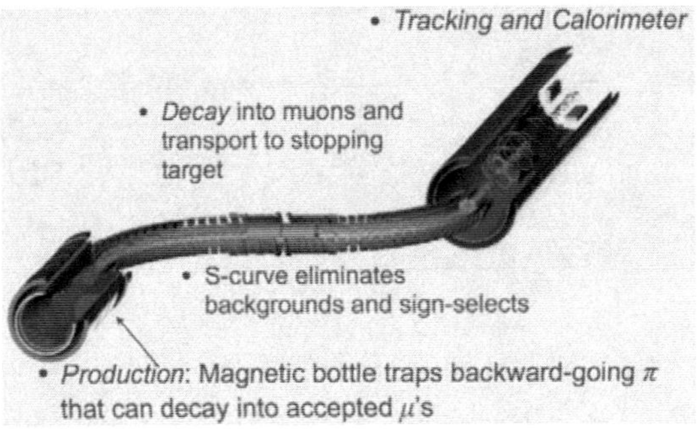

Figure 1. Set up of the Mu2e experiment

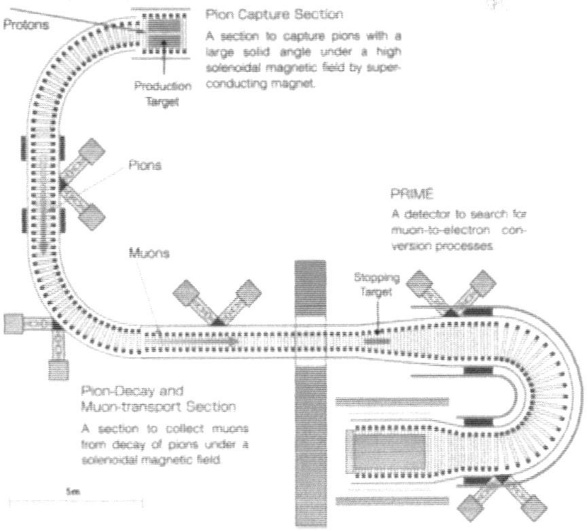

Figure 2. Set up of the COMET experiment

Intensive R&D work is in progress also in the COMET experiment. Development of superconducting wire tolerable in high radiation environment is conducted in collaboration with a wire manufacturer in Japan. Coil winding technique is also investigated by producing sample coils. These activities are summarized in presentations given in this workshop [10].

The COMET experiment successfully conducted the 1st beam extinction measurement using extracted bam in 2010 and obtained a preliminary result of $(5.6\pm0.7)\times10^{-7}$. In their careful analysis beam particle leakage corresponding to this extinction is attributed to pulse-forming inefficiency before acceleration and can be improved by modifying the injection scheme of protons from the booster ring to the main ring. A preliminary evaluation shows that improvement of 10^{-6} will be achieved with this new injection scheme that will enable them to achieve the required extinction level. This has to be confirmed after the J-PARC operation is restarted in 2012. The status of the experiment was presented and summarized concisely in [11].

3.2. Other future CLFV experiments

There are more CLFV experiments using muons are planned. One of them is the DeeMe experiment at J-PARC. The experiment plans to use the pion production target also as a muon stopping target. They have actually found by measuring electron energy spectrum with delayed timing that huge amount of muonic atoms are certainly formed in the pion production target. An electron spectrometer with kicker magnets will be constructed to sweep prompt background efficiently. A single event sensitivity of 2×10^{-14} is anticipated using a SiC target. Details of the experiment were presented in a separate presentation in this workshop [12].

Another effort in search of CLFV is being considered at PSI using a different decay mode of muons. The experiment intends to search for the decay $\mu^+\rightarrow e^+e^+e^-$ using the surface muon beam provided at PSI. Detector concept using silicon trackers with a solenoid magnet is presented in this workshop [13] in order to improve the sensitivity on the branching ratio beyond the current best limit of 1.0×10^{12} obtained by the SINDRUM experiment [14].

4. Summary

Prospects of CLFV experiments are summarized in this article. The MEG experiment at PSI is leading in this field and expected to provide fruitful result soon. Detector upgrade plan is considered to further improve the physics sensitivity of the experiment. There are a number of activities trying to search for CLFV. These experiments plan to utilize high-intensity proton machines to go beyond the MEG sensitivity by using large amount of muon samples. They will certainly play a key role to understand the physics of neutrino oscillation from a different point of view.

References

[1] Yanagida T 1979 *Proc. of the Workshop on Unified Theory and Baryon Number of the Universe* ed. Sawada O and Sugamoto A (Tsukuba: KEK) p 95
[2] Gell-Mann M, Ramond P, and Slansky R 1979 *Supergravity* ed. Van Nieuwenhuizen P and Freedman D (Amsterdam: North Holland)
[3] Barbieri R, Hall L and Strumia A 1995 *Nucl. Phys.* B **455** 219
[4] Hisano J, Nomura D and Yanagida T 1998 *Phys. Lett.* B **437** 351
[5] Raidal M *et al.* 2008 *Eur. Phys. J.* C **57** 13
[6] Nicolò D *these proceedings*
[7] Adam J *et al.* (MEG Collaboration) 2011 *Phys. Rev. Lett.* **107** 171801
[8] Burtl W *et al.* (SINDRUM II Collaboration) 2006 *Eur. Phys. J.* C **47** 337
[9] Bernstein B *these proceedings*
[10] Yoshida M *these proceedings*; Cook S *these proceedings*
[11] D'Arcy R *these proceedings*
[12] Mihara S *these proceedings*
[13] Berger N *these proceedings*
[14] Bellgardt U *et al.* (SINDRUM Collaboration) 1988 *Nucl. Phys.* B **299** 1

Constraints and tests of the TeV scale see-saw mechanism

Alejandro Ibarra

Physik-Department T30d, Technische Universität München,
James-Franck-Straße, 85748 Garching, Germany

E-mail: alejandro.ibarra@ph.tum.de

Abstract. The type I see-saw mechanism requires the existence of right-handed neutrinos with a Majorana mass much larger than the Dirac neutrino mass. Nevertheless, the right-handed neutrino mass scale is completely unknown and can range between 0 and the Planck mass. In this talk we discuss the constraints on the scenario where the right-handed neutrino mass lies at the TeV scale, motivated by the possibility of testing the mechanism of neutrino mass generation at the Large Hadron Collider.

1. Introduction

One of the simplest extensions of the Standard Model consists on introducing right-handed neutrinos [1]. Being singlets under the Standard Model gauge group, the most general Lagrangian contains two new terms: a Dirac mass term, coupling left- and right-handed neutrinos, and a Majorana mass term for the right-handed neutrinos:

$$-\mathcal{L} \supset \overline{\nu_L} M_D \nu_R + \frac{1}{2}\overline{\nu_R^C} M_M \nu_R + \text{h.c.} \tag{1}$$

In terms of the mass eigenstates ν_i and N_i this Lagrangian reads:

$$-\mathcal{L} \supset \frac{1}{2}\overline{\nu_i^C}\mathcal{M}_{\nu ij}\nu_j + \frac{1}{2}\overline{N_i^C}\mathcal{M}_{Nij}N_j + \text{h.c.} \tag{2}$$

If $M_M \gg M_D$, the light neutrino mass is related to the parameters of Eq. (1) through $\mathcal{M}_\nu \simeq -M_D M_M^{-1} M_D^T$, while the heavy neutrino mass through $\mathcal{M}_N \simeq M_M$.

Low energy experiments restrict the eigenvalues of $\mathcal{M}_\nu$ to be $\lesssim \mathcal{O}(0.1 \text{ eV})$, but not the size of the entries of M_D or M_M. In this talk, we will analyze the constraints on the scenario where the right-handed neutrino masses lie at the TeV scale, motivated by the exciting possibility of testing the mechanism of neutrino mass generation at collider experiments (see, *e.g.* [2]).

2. Scenarios with TeV mass right-handed neutrinos and large Yukawa couplings

From $\mathcal{M}_\nu \simeq -M_D M_M^{-1} M_D^T$, it follows that, naively, when $M_M = 1$ TeV then $m_D \approx 10^{-4}$ GeV in order to generate neutrino masses not larger than $\mathcal{O}(0.1 \text{ eV})$. This translates into a Yukawa coupling $\mathcal{O}(10^{-6})$ which induces low energy effects too suppressed to be observed, unless the right-handed neutrino has additional interactions, for instance when it is charged under $U(1)_{B-L}$. There is, however, a class of see-saw scenarios where the Yukawa couplings can be sizable while being the right-handed neutrinos at the TeV scale. To identify this class of scenarios we first note that there is a continuous family of Dirac neutrino masses which solves the equation $\mathcal{M}_\nu = -M_D M_M^{-1} M_D^T$, given by $M_D = iU^* \sqrt{D_m}\Omega\sqrt{D_M}V^\dagger$ [3], where U is the unitary matrix which diagonalizes $\mathcal{M}_\nu$, $\mathcal{M}_\nu = U^* D_m U^\dagger$, V the matrix which diagonalizes M_M, $M_M = V^* D_M V^\dagger$, and Ω is an arbitrary complex orthogonal matrix: $\Omega\Omega^T = 1$. In the case of two right-handed neutrinos, the matrix Ω reads [4]:

$$\Omega_{\rm NH} \equiv \begin{pmatrix} 0 & 0 \\ \cos\hat\theta & \pm\sin\hat\theta \\ -\sin\hat\theta & \pm\cos\hat\theta \end{pmatrix}, \qquad \Omega_{\rm IH} \equiv \begin{pmatrix} \cos\hat\theta & \pm\sin\hat\theta \\ -\sin\hat\theta & \pm\cos\hat\theta \\ 0 & 0 \end{pmatrix}, \tag{3}$$

for normal hierarchy (NH) and inverted hierarchy (IH), respectively. It is possible to exploit the freedom in the choice of Ω to generate a sizable m_D with entries in U of $\mathcal{O}(0.1)$, in D_m of $\mathcal{O}(10^{-10}$ eV) and in D_M of $\mathcal{O}(10^3$ GeV). Decomposing $\hat\theta = \omega - i\xi$ and taking $\xi \gg 1$ one finds:

$$\Omega_{\rm NH} \simeq \frac{e^{i\omega}e^{\xi}}{2}\begin{pmatrix} 0 & 0 \\ 1 & \mp i \\ i & \pm 1 \end{pmatrix}, \qquad \Omega_{\rm IH} \simeq \frac{e^{i\omega}e^{\xi}}{2}\begin{pmatrix} 1 & \mp i \\ i & \pm 1 \\ 0 & 0 \end{pmatrix}. \tag{4}$$

Thus, by adjusting ξ it is possible to construct a viable see-saw scenario with TeV mass right-handed neutrinos and arbitrarily large Dirac masses. It is also important to note that the matrix Ω introduces in general new sources of flavour structure in M_D which are unrelated to the flavour mixing observed in neutrino oscillation experiments. However, in the limit $\xi \gg 1$ the extra flavour structure is fixed, thus rendering a fairly predictive scenario.

A large Dirac neutrino mass can significantly modify the charged current (CC) and neutral current (NC) leptonic interactions, due to the large mixing between left- and right-handed neutrinos. The CC and NC interactions involving the light Majorana neutrinos have the form:

$$\mathcal{L}^\nu_{CC} = -\frac{g}{\sqrt{2}}\bar\ell\gamma_\rho\,[U_{\rm lep}]_{\ell i}\,\nu_i W^\rho + {\rm h.c.} \simeq -\frac{g}{\sqrt{2}}\bar\ell\gamma_\rho\left[\left(1-\frac{1}{2}(RV)(RV)^\dagger\right)U\right]_{\ell i}\nu_i W^\rho + {\rm h.c.}, \tag{5}$$

$$\mathcal{L}^\nu_{NC} = -\frac{g}{2c_w}\overline{\nu_i}\gamma_\rho\left[U^\dagger_{\rm lep}U_{\rm lep}\right]_{ij}\nu_j Z^\rho + {\rm h.c.} \simeq -\frac{g}{2c_w}\overline{\nu_i}\gamma_\rho\left[U^\dagger\left(1-(RV)(RV)^\dagger\right)U\right]_{ij}\nu_j Z^\rho + {\rm h.c.} \tag{6}$$

with $R^* \simeq M_D M_M^{-1}$. Furthermore, the left-right neutrino mixing gives rise to sizable CC and NC couplings of the heavy Majorana neutrinos N_j to the W and Z bosons:

$$\mathcal{L}^N_{CC} \simeq -\frac{g}{2\sqrt{2}}\bar\ell\gamma_\rho(RV)_{\ell i}(1-\gamma_5)N_i W^\rho + {\rm h.c.}, \tag{7}$$

$$\mathcal{L}^N_{NC} \simeq -\frac{g}{2c_w}\overline{\nu_{\ell L}}\gamma_\rho\,(RV)_{\ell i}N_i Z^\rho + {\rm h.c.}. \tag{8}$$

Thus, the combination RV parametrizes the effects of the heavy neutrinos in low energy phenomenology. Using the parametrization discussed above, this matrix can be cast as $RV \simeq -iU_{\rm lep}\sqrt{D_m}\Omega^*\sqrt{D_M^{-1}}$ [5], which allows to express it in terms of the measurable neutrino parameters, the neutrino Yukawa eigenvalue and the heavy neutrino masses. Explicitly [6]:

$$(RV)_{\alpha 1} \simeq -e^{i\omega} y v\sqrt{\frac{M_2}{(M_1+M_2)}}\sqrt{\frac{m_3}{m_2+m_3}}\left(U_{\alpha 3} + i\sqrt{\frac{m_2}{m_3}}U_{\alpha 2}\right), \tag{9}$$

while $(RV)_{\alpha 2} \simeq \pm i\sqrt{M_1/M_2}(RV)_{\alpha 1}$. The corresponding expressions for the inverted hierarchy can be easily obtained upon the substitution $m_{2,3} \to m_{1,2}$, $U_{\alpha 2,\alpha 3} \to U_{\alpha 1,\alpha 2}$ ($\alpha = e,\ \mu,\ \tau$).

A series of measurements restrict the couplings of the charged and neutral current interactions of the light and heavy neutrinos, namely the elements of the matrix RV. Alternatively, these constraints can be translated into constraints on the parameters of the full theory, which in this scenario are just the neutrino Yukawa eigenvalue and the right-handed neutrino masses.

3. Constraints on the parameters of the TeV scale see-saw mechanism

The most important constraint on this scenario comes from measurements of neutrinoless double beta decay ($(\beta\beta)_{0\nu}$). In the presence of additional CC interactions, the effective Majorana mass $|(m_\nu)_{ee}|$, which controls the $(\beta\beta)_{0\nu}$ rate, reads [7]:

$$|(m_\nu)_{ee}| \simeq \left| \sum_{i=1}^{3} U_{ei}^2 \, m_i - \sum_{k=1}^{2} \frac{M_a^2}{M_k} f(A) \, (RV)_{ek}^2 \right| , \tag{10}$$

where $M_a \approx 0.9$ GeV and $f(A) = 0.079$ for ^{76}Ge. Thus, assuming $M_2 \sim 1000$ GeV and $|(RV)_{e2}| \sim 10^{-2}$, the experimental constraint $|(m_\nu)_{ee}| < 0.35$ eV [8] requires $(M_2 - M_1)/M_1 \lesssim 10^{-2}$. Thus, in this scenario the constraints from $(\beta\beta)_{0\nu}$ require the right-handed neutrinos to form a pseudo-Dirac pair in order to suppress the otherwise large lepton number violating effects.

Furthermore, the heavy neutrinos have lepton flavour violating CC interactions, which can contribute via quantum effects to the rare lepton decays. In contrast to the standard contribution to the process $\mu \to e\gamma$ from the light neutrinos, which is very suppressed by the tiny factor $\Delta m^2/M_W^2$ due to the GIM mechanism, the contribution from the heavy neutrinos is not GIM suppressed and could lead to large rates, unless the couplings are small. More concretely, if the rare muon decay is dominated by the effects of the heavy neutrinos [9]:

$$\mathrm{B}(\mu \to e\gamma) \simeq \frac{3\alpha}{8\pi} |(RV)_{\mu 1}(RV)_{e1}|^2 \left| G\left(\frac{M_1^2}{M_W^2}\right) - G(0) \right|^2 , \tag{11}$$

where $G(x)$ is a monotonic function which takes values in the interval $[4/3, 10/3]$, with $G(x) \cong \frac{10}{3} - x$ for $x \ll 1$. Then, in order to satisfy the experimental upper bound from MEGA [10], $\mathrm{B}(\mu \to e\gamma) \leq 1.2 \times 10^{-11}$, the couplings must satisfy $|(RV)_{\mu 1}(RV)_{e1}| \lesssim 1.8 \times 10^{-4}$ (0.6×10^{-4}) for $M_1 = 100$ (1000) GeV [6].

Lastly, the sizable left-right neutrino mixing implies a 3×3 leptonic mixing matrix which is non-unitary. From present neutrino oscillation data and different measurements of electroweak processes (e.g., on $W^\pm$ decays, invisible Z decays, universality tests of EW interactions) it follows that $|(RV)_{e1}|^2 \lesssim 2 \times 10^{-3}$, $|(RV)_{\mu 1}|^2 \lesssim 0.8 \times 10^{-3}$, $|(RV)_{\tau 1}|^2 \lesssim 2.6 \times 10^{-3}$ for $M_1 \gtrsim 100$ GeV [11].

We show in Fig. 1 the summary of constraints on the parameter space of the TeV see-saw scenario when the light neutrino spectrum has NH (left panel) and IH (right panel) for $M_1 = 1000$ GeV and different Yukawa couplings [6]. From the plot it follows that the most stringent constraint on the parameter space comes from neutrino oscillation experiments and the non-observation of the process $\mu \to e\gamma$, which restricts the Yukawa coupling to be $\lesssim 0.1$. Furthermore, our analysis indicates that the present bound on the $\mu \to e\gamma$ decay rate will make very difficult the observation of heavy right-handed neutrinos at the LHC or the observation of deviations from the Standard Model predictions in the electroweak precision measurements.

4. Conclusions

We have studied the constraints on a see-saw scenario where the right-handed neutrino mass lies at the TeV scale. In this scenario the flavour structure of the neutrino Yukawa couplings is essentially determined by the low energy neutrino parameters, leading to fairly strong correlations among the new phenomena. We have shown that the constraints from neutrino oscillation experiments and the non-observation of the process $\mu \to e\gamma$ makes very difficult the observation of the heavy right-handed neutrinos at the LHC or the observation of deviations from the Standard Model predictions in the electroweak precision measurements. However, the experimental constraints on this scenario still allow i) for an enhancement of the rate of neutrinoless double beta decay, which thus can be in the range of sensitivity of the GERDA

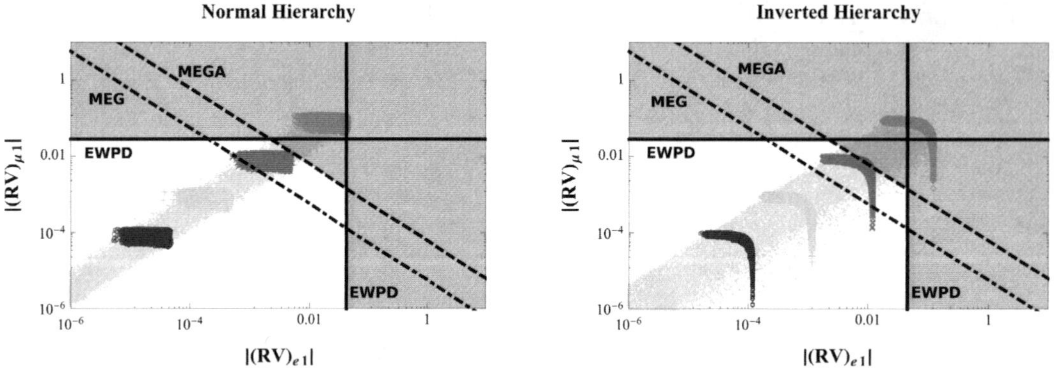

Figure 1. Constraints on the parameter space of the TeV see-saw scenario, spanned by $|(RV)_{e1}|$ and $|(RV)_{\mu 1}|$ when the light neutrino spectrum has normal hierarchy (left panel) and inverted hierarchy (right panel), for $M_1 = 1$ TeV and neutrino Yukawa eigenvalue $y = 0.001$ (blue $\circ$), $y = 0.01$ (green $+$), $y = 0.1$ (red $\times$) and $y = 1$ (orange $\diamondsuit$). The cyan points correspond to random values of $y \leq 1$. The grey areas are excluded by the present measurements of electroweak precision observables (solid lines) or by the constraint $B(\mu \to e\gamma) \leq 1.2 \times 10^{-11}$ from MEGA (dashed line). The dot-dashed line corresponds to $B(\mu \to e\gamma) = 10^{-13}$, which is the projected sensitivity of the MEG experiment.

experiment [12] even when the light Majorana neutrinos possess a normal hierarchical mass spectrum, and ii) for the predicted $\mu \to e\gamma$ decay rate to be within the sensitivity range of the MEG experiment [13].

Acknowledgments

The author would like to thank Emiliano Molinaro and Serguey Petcov for a very pleasant and fruitful collaboration. This work was partially supported by the DFG cluster of excellence "Origin and Structure of the Universe."

References

[1] For a review see, for example, Gonzalez-Garcia M C and Maltoni M 2008 *Phys. Rept.* **460** 1-129
[2] del Aguila F, Aguilar-Saavedra J A and Pittau R 2007 *JHEP* **0710** 047
[3] Casas J A and Ibarra A 2001 *Nucl. Phys.* B **618** 171-204
[4] Ibarra A and Ross G G 2004 *Phys. Lett.* B **591** 285; Ibarra A and Ross G G 2003 *Phys. Lett.* B **575** 279
[5] Ibarra A, Molinaro E and Petcov S T 2010 *JHEP* **1009** 108
[6] Ibarra A, Molinaro E and Petcov S T 2011 *Phys. Rev.* D **84** 013005
[7] Vergados J 1983 *Nucl. Phys.* B 218 109; Haxton W C and Stephenson J 1984 *Prog. Part. Nucl. Phys.* **12** 409; Blennow M *et al.* 2010 *JHEP* **1007** 096
[8] Klapdor-Kleingrothaus H V *et al.* 2001 *Eur. Phys. J.* A **12** 147-154
[9] Petcov S T 1977 *Sov. J. Nucl. Phys.* **25** 340 [Yad. Fiz. **25** (1977) 641]; Bilenky S M , Petcov S T and Pontecorvo B 1977 *Phys. Lett.* B **67** 309; Cheng T P and Li L F 1980 *Phys. Rev. Lett.* **45** 1908
[10] Brooks M L *et al.* [MEGA Collaboration] 1999 *Phys. Rev. Lett.* **83** 1521
[11] Antusch A, Baumann J P and Fernandez-Martinez E 2009 *Nucl. Phys.* B**810** 369
[12] Abt I *et al. Preprint* hep-ex/0404039; Smolnikov A A *Preprint* arXiv:0812.4194 [nucl-ex]
[13] Maki A 2008 *AIP Conf. Proc.* **981** 363-365

Neutrinoless double beta decay in the context of seesaw models

Jacobo López-Pavón

IPPP, Department of Physics, Durham University, South Road, Durham, DH1 3LE, UK

E-mail: jacobo.lopez-pavon@durham.ac.uk

Abstract. The phenomenology of neutrinoless double beta decay in the context of seesaw models is studied. We focus on the interplay between the contribution of the light mostly active neutrinos and the extra degrees of freedom of the models, clarifying why the light neutrino contribution should always be considered. We show how considering the relation between the mostly active neutrino and the New Physics contribution leads to strong constraints on the New Physics parameters or to a cancellation, which would result into unobservable process even if neutrinos are Majorana particles, depending on the mass regime of the extra states. Finally, we also discuss how non canonical versions of the seesaw could reconcile large rates of neutrinoless double beta decay with more stringent cosmological bounds on neutrino masses. In order to perform this analysis we compute the nuclear matrix elements as a function of the mass of the mediating fermion using the Interacting Shell Model (ISM) and estimating the associated uncertainty.

Neutrino oscillations, implying the massive nature of neutrinos, constitute an evidence for physics beyond the Standard Model (SM). Thus, models accommodating this neutrino masses become an important component in the search for New Physics. In this context, one of the most promising processes is the neutrinoless double beta decay ($0\nu\beta\beta$ decay), whose detection is the purpose of several ongoing and upcoming experiments. Since this process is lepton number violating, its observation would imply that neutrinos are Majorana fermions.

Among the more popular models for neutrino masses, we find the different types of seesaw mechanisms: type-I, where the SM field content is extended with fermion singlets, type-II, where the extension is done with heavy scalar triplets, and type-III, where the extra heavy particles included are fermion triplets. They are the extensions of the SM particle content that lead to the Weinberg $d = 5$ effective operator after the heavy fields have been integrated out.

All these extra degrees of freedom, required to induce the Majorana nature of the SM neutrinos, can also contribute to the $0\nu\beta\beta$ process. In the literature, it is common to study the effects of the SM neutrinos or the extra states introduced to account for their masses independently. However, these contributions are indeed related through the generation of neutrino masses. Taking into account this relation, a precious information about the origin of neutrino masses can be obtained.

In order to study the interplay between the contributions to the $0\nu\beta\beta$ decay rate of the SM neutrinos and the extra degrees of freedom, we have computed the nuclear matrix element

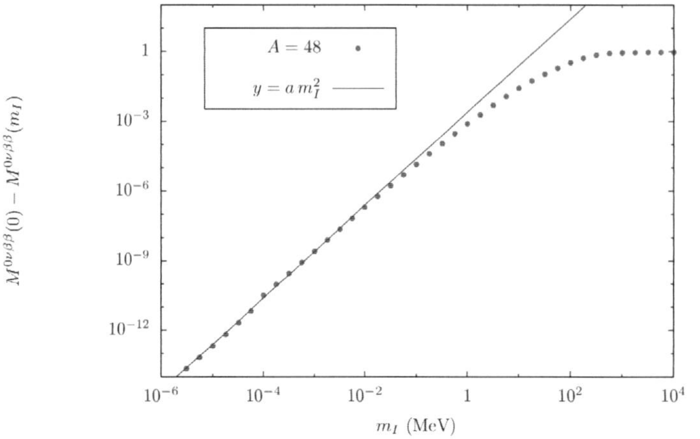

Figure 1. NME cancellation in the light mass regime. The example is taken for the ^{48}Ca decay, but the same quadratic dependence is seen for the remaining nuclei studied. The value of the parameter is $a = 2.51 \cdot 10^{-3}$ MeV^{-2}.

(NME) without any assumption on the mass of the neutrinos mediating the process:

$$\frac{\Gamma_{0\nu\beta\beta}}{\ln 2} = G_{01} \left| \sum_j U_{ej}^2 \frac{m_j}{m_e} M^{0\nu\beta\beta}(m_j) \right|^2 , \tag{1}$$

where G_{01} is a well-known kinematic factor, m_e the electron mass, m_j the mass of the j neutrino, $M^{0\nu\beta\beta}(m_j)$ the NME and U_{ej} are elements of the neutrino mixing matrix.

The calculation was performed using Interacting Shell Model (ISM) nuclear wavefunctions, one of the most popular methods employed to obtain $0\nu\beta\beta$ decay NMEs. The results [1],[2] show two distinct regions for the behavior of the NME as a function of the virtual neutrino mass: almost constant up to $m_i \simeq 100$ MeV and then decreasing quadratically as the neutrino mass increases beyond 100 MeV. The value for the mass for which the transition takes place, 100 MeV, corresponds to the typical momentum exchange of the decay, which is the momentum of the virtual neutrino $|\mathbf{p}|$. This scale is fixed by the typical distance between the two decaying nucleons, $r \simeq 1$ fm, and is usually referred as the nuclear scale. Notice that this behavior of the NME can easily be understood expanding the neutrino propagator $\frac{1}{p^2 - m_j^2}$.

Different approximations are required in order to perform the calculation, both obtaining the wavefunctions and in the treatment of the two-body transition operator, and consequently some uncertainties are induced into the NMEs. These can be estimated[1] in $\sim^{+25\ \%}_{-35\ \%}$ for light neutrino exchange ($m_j \leq 100$ MeV) and $\sim^{+35\ \%}_{-40\ \%}$ for heavy neutrinos ($m_j \geq 100$ MeV), more sensitive to the short range part of the transition operator.

With the NMEs we can now analyze the contributions to $0\nu\beta\beta$ decay of the different mechanisms that lead to Majorana neutrino masses. Within the type-I seesaw, depending on whether the extra mass eigenstates fall in the light or heavy neutrino mass regimes we can split their respective contributions to the amplitude:

$$A \propto \sum_i^{\text{light}} m_i U_{ei}^2 M^{0\nu\beta\beta}(0) + \sum_I^{\text{light}} m_I U_{eI}^2 M^{0\nu\beta\beta}(m_I) + \sum_I^{\text{heavy}} m_I U_{eI}^2 M^{0\nu\beta\beta}(m_I), \tag{2}$$

[1] See [1] for details.

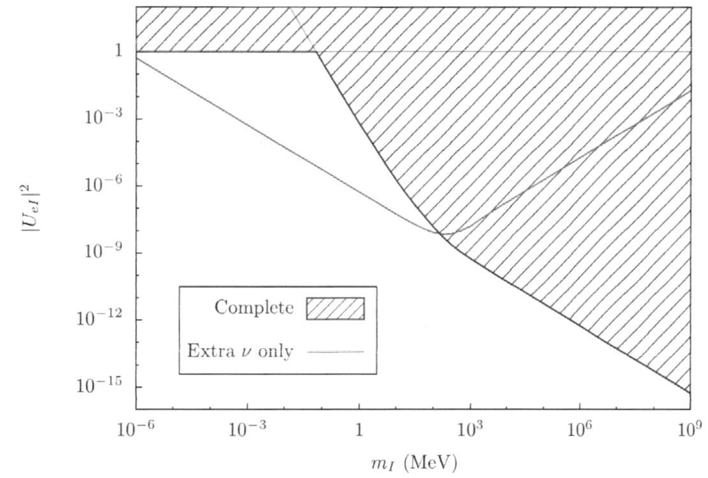

Figure 2. Bounds from CUORICINO on the extra neutrino mixing from $0\nu\beta\beta$ decay in ^{130}Te, with a 90 % CL half-life [7]. We have assumed the extra neutrinos are non hierarchical and show the bounds as a function of their common mass. We compare the case in which the contribution from the SM neutrinos is properly taken into account (striped area) to that in which only the extra contribution is considered (above the red line).

where capital letters denote the mass index of the mostly sterile states and lowercase letters the mostly active SM states. Notice that we can safely assumed $M^{0\nu\beta\beta}(m_i) = M^{0\nu\beta\beta}(0)$ for the NMEs associated to the SM neutrinos. Moreover, the diagonalization of the neutrino mass matrix provides the following useful relation:

$$U^{*}\text{diag}\left\{m_1, m_2, ..., m_n\right\} U^{\dagger} = \left(\begin{array}{cc} 0 & Y_N v/\sqrt{2} \\ Y_N^T v/\sqrt{2} & M_N \end{array} \right). \tag{3}$$

which relates the mostly active and extra degrees of freedom parameters. We can now distinguish three cases exhibiting very different phenomenologies depending on the mass regime of the extra mass eigenstates:

All extra mass states are light. In this case Eq. (3) implies

$$A \approx -\sum_I^{\text{light}} m_I U_{eI}^2 \left(M^{0\nu\beta\beta}(0) - M^{0\nu\beta\beta}(m_I)\right). \tag{4}$$

Since in this regime the NMEs are basically independent of the neutrino mass, $M^{0\nu\beta\beta}(m_I) \simeq M^{0\nu\beta\beta}(m_i) \simeq M^{0\nu\beta\beta}(0)$, the rate of $0\nu\beta\beta$ decay is very suppressed. Indeed, only the different neutrino masses in the NME prevent a full cancellation leading to a suppression of $\Delta m^2/p^2$ with $|p^2| \simeq (100\text{ MeV})^2$. The $\Delta m^2/p^2$ dependence of $M^{0\nu\beta\beta}(0) - M^{0\nu\beta\beta}(m_I)$ that drives the suppression to the $0\nu\beta\beta$ rate is depicted in Fig. 1. As it can be observed, the $0\nu\beta\beta$ decay becomes experimentally inaccessible in this regime, even if neutrinos are Majorana particles.

All extra mass states in the heavy regime. Now the NMEs for this extra states are very suppressed compared to the SM ones. Furthermore, Eq. (3) implies that

$$A \approx -\sum_I^{\text{heavy}} m_I U_{eI}^2 \left(M^{0\nu\beta\beta}(0) - M^{0\nu\beta\beta}(m_I)\right) \approx -\sum_I^{\text{heavy}} m_I U_{eI}^2 M^{0\nu\beta\beta}(0) \approx \sum_i^{\text{light}} m_i U_{ei}^2 M^{0\nu\beta\beta}(0). \tag{5}$$

Thus, the contribution from the light active neutrinos (first term) dominates the transition rate but can be used to set a strong bound on the mixing of the heavy neutrinos [3]. This leads to a much stronger constraint than the one usually considered in the literature, as it is shown in Fig. 2.

Extra mass states in the light and heavy regimes. In this scenario the leading terms stem from the light states:

$$A \propto \sum_i^{\text{light}} m_i U_{ei}^2 M^{0\nu\beta\beta}(m_i) + \sum_I^{\text{light}} m_I U_{eI}^2 M^{0\nu\beta\beta}(m_I). \tag{6}$$

Now it is possible to satisfy Eq. (3) even in a situation where $m_i U_{ei}^2 \ll m_I U_{eI}^2$, by canceling the contribution of the extra heavy states against that of the extra light ones while keeping the light neutrino masses small. This implies a certain level of fine-tuning. However, in such a situation, the contribution of the light extra states could dominate over that of the active. As an example, if we consider the controversial Heidelberg-Moscow claim for a positive $0\nu\beta\beta$ decay signal [4], the accommodation of this signal through only SM neutrinos would require, using our ISM NMEs, 0.24 eV $< m_{\beta\beta} < 0.89$ eV at 2σ. The interpretation of this claim as light active SM neutrinos is very disfavored by the constraints from cosmology and neutrino oscillation data [5, 6]. However, this signal could be accommodated in a model with heavier neutrinos (which are not bounded by cosmology) mediating the process. Indeed, we could reinterpret the result as $0.24 \text{ eV} < \left|\sum_I^{\text{heavy}} m_I U_{eI}^2\right| < 0.89 \text{ eV}$.

As for the type-II and type-III seesaws, current bounds from accelerator experiments place the extra degrees of freedom in the heavy regime. This effectively reduces the phenomenology to that which appears for the type-I seesaw with only heavy extra states[2]. In the same manner, combined type-I and type-II or type-III seesaw models resemble the situation of type-I seesaw with both light and heavy extra states. For instance, the Heidelberg-Moscow claim can be interpreted as $0.24 \text{ eV} < \left|m_{ee}^{\Delta,\Sigma}\right| < 0.89 \text{ eV}$ in this context, where m_{ee}^{Δ} and m_{ee}^{Σ} are the Majorana neutrino masses induced by the type-II and type-III seesaw models, respectively.

Acknowledgments

We would like to acknowledge financial support from the European projects EURONU (CE212372), EuCARD (European Coordination for Accelerator Research and Development, Grant Agreement number 227579) and LAGUNA (Project Number 212343).

References

[1] Blennow M, Fernandez-Martinez E, Lopez-Pavon J and Menendez J, *JHEP* **1007** (2010) 096 arXiv:1005.3240 [hep-ph].
[2] Blennow M, Fernandez-Martinez E, Lopez-Pavon J and Menendez J, http://wwwth.mppmu.mpg.de/members/blennow/nme mnu.dat, (2010).
[3] Xing Z z, *Phys. Lett. B* **679** (2009) 255 arXiv:0907.3014 [hep-ph].
[4] Klapdor-Kleingrothaus H V, Krivosheina I V, *Mod. Phys. Lett.* **A21** (2006) 1547-1566.
[5] Komatsu E *et al.* [WMAP Collaboration], *Astrophys. J. Suppl.* **192** (2011) 18. arXiv:1001.4538 [astro-ph.CO].
[6] Hannestad S, Mirizzi M, Raffelt G G, Wong Y Y Y, *JCAP* **1008** (2010) 001. arXiv:1004.0695 [astro-ph.CO].
[7] Arnaboldi C *et al.* [CUORICINO Collaboration], *Phys. Rev.* **C78** (2008) 035502. arXiv:0802.3439 [hep-ex].

[2] See [1] for details.

Journal of Physics: Conference Series **408** (2013) 012020 doi:10.1088/1742-6596/408/1/012020

Long-baseline sensitivity studies and comparison (discussion session)

Walter Winter

Institut für theoretische Physik und Astrophysik, Universität Würzburg, D-97074 Würzburg, Germany

E-mail: winter@physik.uni-wuerzburg.de

Abstract. In this discussion session, the sensitivity and optimization of future long-baseline experiments is addressed, with a special emphasis on *feasible* projects and the description in terms of the error on the parameters. In addition, a statement on the precision interesting for $\nu_e \to \nu_\tau$ and $\nu_\mu \to \nu_\tau$ oscillation measurements is obtained. A special topic is the impact of the recent T2K hint for non-zero θ_{13}.

1. Introduction and task assignment

The task assignment for this session has been the discussion of the sensitivity and optimization of future long-baseline experiments. One of the objectives has been the focus on *feasible* projects (*i.e.*, for beta beams), another the expression of the sensitivities in terms of the error on the parameters. In addition, a statement on the precision interesting for $\nu_e \to \nu_\tau$ and $\nu_\mu \to \nu_\tau$ oscillation measurements has been asked for, and a report on studies of such measurements for superbeam and neutrino factory. An important recent topic has been the 2.5σ hint from the T2K experiment for large θ_{13} [1]. This hint has strengthened the physics case for a new facility, especially since existing equipment will most likely not be sufficient to establish the mass hierarchy and CP violation at a high confidence level [2].

Now first of all, what does "feasible" mean? For the purpose of this session, only setups actively being studied, such as within the Euronu design study or the international design study for the Neutrino Factory (IDS-NF), have been selected. Another interpretation of "feasibility" is to consider setups only for which a cost estimate is in preparation. More specifically, three representatives have been chosen:

- LBNE as superbeam representative.[1]
- A CERN-SPS based beam as beta beam representative, possibly to be combined with the SPL superbeam.
- The IDS-NF baseline for the Neutrino Factory, possibly with a low energy alternative

The expression of the sensitivities in terms of the "error on the parameters" has been interpreted as a the expression in terms of the main impact parameters for the sensitivities, and systematics.

[1] Note that there are also different experimental proposals actively being studied for superbeams, such as J-PARC to Hyper-Kamiokande [3]. However, in terms of the parameters relevant for physics, such as baseline and energy, it is similar to the T2K experiment – although not in terms of performance, of course.

In order to discuss the comparison of facilities, an introductory talk to each kind has been asked for by three champions for the three classes of experiments: Jim Strait (superbeams) [4], Elena Wildner (beta beams) [5], and Ken Long (Neutrino Factory) [6]. The questions asked to the champions where:

- Optimization (L, E, *etc.*) of the setup: Is that the physics-wise optimal setup for that class?
- Under which boundary conditions was that obtained: physics-wise, *e.g.*, where in parameter space, and technology-wise, *e.g.*, constrained to some site?
- Does the optimization change for θ_{13}?
- What is the discovery reach for θ_{13}, mass hierarchy, and CP violation? What are the assumptions going into that? What is the performance for large θ_{13}?
- What are the critical systematics and other important impact factors for physics potential (for example, external knowledge on cross sections required, which cannot be obtained with near detectors)?

These talks where followed by a discussion session, the results of which are summarized below.

2. Discussion on comparison of facilities

The topics and results of the discussion session have especially lead to a number of re-formulated questions, which need to be addressed in the future.

One of the most relevant topics is the optimization of the setups for large θ_{13}. There are different ways to address this optimization, such as identifying the minimal requirements for an experiment to measure the main performance indicators, see, *e.g.*, Ref. [7] for beta beams and Ref. [8] for the Neutrino Factory, or the identification of the experiment parameters (L, E, off-axis angle, *etc.*) with the best performance in the δ_{CP}-direction, see, *e.g.*, Ref. [9] for the Neutrino Factory and Ref. [10, 11] for superbeams. In either case, the optimal experiment configuration for a specific performance indicator can be predicted as a function of θ_{13} and detector response. This prediction is quite straightforward for beta beams and the Neutrino Factory, because the beam spectrum can be analytically computed, whereas it depends on the target system and horn geometry for the superbeams. Therefore, the question for the optimal setup is not straightforward to answer for superbeams unless a particular beam configuration is used. In addition, note that if θ_{13} is known (within certain errors), the optimization direction in parameter space will change from θ_{13} into δ_{CP}. This has peculiar effects: For example, from Ref. [10] (Figs. 6 and 8), it can be read off that the "minimal" wide band superbeam for θ_{13} close to the Chooz bound has a baseline $L \simeq 700$ km similar to MINOS, which is just long enough to measure the mass hierarchy for any δ_{CP} and CP violation for 75% of all δ_{CP}. A longer baseline is mostly preferable for the mass hierarchy measurement if θ_{13} turns out to be smaller. Therefore, it is expected that a solid lower bound on θ_{13} will have a major impact on the optimization of future experiments. One of the questions to be addressed in the future was therefore: Does the optimization of the individual experiments change if the T2K hint is confirmed? Another one: What is the impact of prior θ_{13} (*e.g.*, from Daya Bay) and mass hierarchy (*e.g.*, from atmospheric neutrinos) measurements on sensitivities and optimization? From the conceptual point of view, a point no common agreement was reached on, was: Does a future experiment have to measure all parameters (such as δ_{CP} and mass hierarchy) in a self-consistent way, or is it better to rely on a combination of different strategies (such as a beam experiment for CP violation and atmospheric neutrinos for the mass hierarchy)?

Another important topic, which is especially relevant for large θ_{13}, is systematics. For example, the cross sections are an important ingredient to the normalization uncertainty, and may even affect the spectral shape. Let us consider a very simplified picture independent of nuclear models and other workarounds. Whereas the cross sections at the Neutrino Factory can be obtained in a self-consistent way at the near detectors (only the ν_μ cross sections are needed), the appearance channel cross sections for superbeams and beta beams have to be extracted from other experiments or, possibly, the ν_e contamination of the beam (superbeams), since a different flavor is present in the near detector. Here a combination of a superbeam and beta beam may be clearly synergistic, since a near detector in one experiment class can measure the cross sections needed for the appearance analysis of the other experiment class. It has been noted in the discussion that the assumptions for systematics are not transparent in experiment comparison plots, maybe even not comparable; these should be documented and made publically available. In addition, studies of the performance as a function of the exposure have been identified as interesting in that context, since these will illustrate when the systematics limitation becomes relevant and what the systematics-dominated limit will be.

As far as the performance indicators are concerned, the results of future experiments will be shown, among other sections, as fits in the θ_{13}-δ_{CP}-plane. However, even though recent global fits show already a slight (low confidence level) preference for some value of δ_{CP} [12], the value of δ_{CP} will not be well enough constrained to rely on in terms of the optimization of the experiment. Since the preferred value of δ_{CP} depends on the experiment class, it is therefore an important question how to quantify the precision on θ_{13} and δ_{CP}. As possible solutions, the performance of future facilities may either be shown as a function of the true δ_{CP}, see, *e.g.*, Refs. [13, 14] ("CP pattern"), or for certain benchmark sets of parameters in the θ_{13}-δ_{CP}-plane. The advantage of the first method is a more complete understanding of the parameter space, the advantage of the second method is that the performance indicator looks closer to the actually expected result. Other questions in this direction, to be addressed in the future, were: Can the θ_{13} precision from reactor experiments be easily exceeded by long-baseline experiments? What limits the θ_{13} precision at reactor experiments? Is θ_{13} or $\sin\theta_{13}$ (as it appears in the mixing matrix) the quantify of interest?

3. Precision interesting for oscillation searches with tau neutrinos

As far as the $\nu_e \to \nu_\tau$ and $\nu_\mu \to \nu_\tau$ *oscillation* measurements are concerned, the main focus has been on effects present in the neutrino propagation rather than neutrino production or detection. A qualitative statement can be already made from simple statistics considerations: given the relatively low detection efficiency of the ν_τ detectors if the hadronic decay channels of the τ cannot be considered, and the relatively small considered detector masses up to about 4 to 10 kt (such as for emulsion cloud chambers), the statistics will be much lower than in the leading $\nu_e \leftrightarrow \nu_\mu$ and $\nu_\mu \to \nu_\mu$ oscillation channels. For example, for the Neutrino Factory, magnetized emulsion cloud chambers with a fiducial mass of up to 5-10 kt have been proposed [15, 16], compared to 100 kt magnetized iron detectors for the leading channels. Given the difference in efficiency, about a factor of 50 to 100 more statistics can be accumulated in the leading channels compared to the ν_τ channels – not even to speak of the τ production threshold. Because of this huge difference, ν_τ detection sometimes helps, but is hardly competitive to other options, see, *e.g.*, Ref. [17] for standard oscillation physics and Ref. [18] for non-standard interactions. Therefore, a re-formulated version of the initial question concerning the precision interesting for ν_τ searches could be: what kind of new physics would be present in the ν_τ channels with a factor of 50-100 enhancement compared to the leading channels in spite of almost maximal θ_{23}? A possible answer to this question may be sterile neutrinos, especially if CP phases are present, see Ref. [19]. Another possible answer may be a chirally enhanced effect in $\epsilon^s_{\mu\tau}$ [20], which is, however, best observed in a near detector. Another potential issue in that context is the fact that there are relatively few studies for ν_τ oscillation searches for superbeams yet. The status of the $\nu_e \to \nu_\tau$ and $\nu_\mu \to \nu_\tau$ oscillation measurements has been reviewed by Toshihiko Ota [21]. The remaining question is, of course, if there are other new physics cases requiring

ν_τ oscillation searches, and if they affect the baseline optimization. For example, an interesting case may be phases from non-unitary, for which a ν_τ detector in a distance of 130 km has been discussed [22–24].

4. Summary and conclusions

In this session, the long-baseline sensitivity studies and comparisons have been discussed, and the precision interesting for $\nu_e \to \nu_\tau$ and $\nu_\mu \to \nu_\tau$ oscillation measurements. For the sensitivity studies, the main outcome has been that the optimization and performance for large θ_{13} has to be reviewed. In addition, systematical errors, which are important in that case, have to be identified, tested, and documented in a transparent way. The physics case for ν_τ searches relies on new physics potentially present in the ν_τ channels in spite of the large difference in statistics to the leading channels and the almost maximal value of θ_{23} (which typically leads to similar sized signals in the leading channels). So far only extremely few convincing cases have been identified.

References

[1] Abe K *et al.* (T2K Collaboration) 2011 *Phys. Rev. Lett.* **107** 041801 (*Preprint* `arXiv:1106.2822[hep-ex]`)
[2] Huber P, Lindner M, Schwetz T and Winter W 2009 *JHEP* **0911** 044 (*Preprint* `arXiv:0907.1896[hep-ex]`)
[3] Abe K *et al.* 2011 (*Preprint* `arXiv:1109.3262[hep-ex]`)
[4] Strait J *Superbeams* these proceedings
[5] Wildner E *Beta beams* these proceedings
[6] Long K *Neutrino Factory* these proceedings
[7] Winter W 2008 *Phys. Rev. D* **78** 037101 (*Preprint* `arXiv:0804.4000[hep-ph]`)
[8] Tang J and Winter W 2010 *Phys. Rev. D* **81** 033005 (*Preprint* `arXiv:0911.5052[hep-ph]`)
[9] Agarwalla S K, Huber P, Tang J and Winter W 2011 *JHEP* **1101** 120 (*Preprint* `arXiv:1012.1872[hep-ph]`)
[10] Barger V *et al.* 2006 *Phys.Rev. D* **74** 073004 (*Preprint* `hep-ph/0607177`)
[11] Barger V, Huber P, Marfatia D and Winter W 2007 *Phys. Rev. D* **76** 053005 (*Preprint* `hep-ph/0703029`)
[12] Schwetz T, Tortola M and Valle J 2011 (*Preprint* `arXiv:1108.1376[hep-ph]`)
[13] Winter W 2004 *Phys. Rev. D* **70** 033006 (*Preprint* `hep-ph/0310307`)
[14] Huber P, Lindner M and Winter W 2005 *JHEP* **0505** 020 (*Preprint* `hep-ph/0412199`)
[15] Donini A, Meloni D and Migliozzi P 2002 *Nucl. Phys. B* **646** 321–349 (*Preprint* `hep-ph/0206034`)
[16] Autiero D *et al.* 2004 *Eur. Phys. J. C* **33** 243–260 (*Preprint* `hep-ph/0305185`)
[17] Huber P, Lindner M, Rolinec M and Winter W 2006 *Phys. Rev. D* **74** 073003 (*Preprint* `hep-ph/0606119`)
[18] Kopp J, Ota T and Winter W 2008 *Phys. Rev. D* **78** 053007 (*Preprint* `arXiv:0804.2261[hep-ph]`)
[19] Donini A, Fuki K i, Lopez-Pavon J, Meloni D and Yasuda O 2009 *JHEP* **08** 041 (*Preprint* `arXiv:0812.3703[hep-ph]`)
[20] Antusch S, Blennow M, Fernandez-Martinez E and Ota T 2010 *JHEP* **1006** 068 (*Preprint* `arXiv:1005.0756[hep-ph]`)
[21] Ota T *Physics case for ν_τ detection* these proceedings
[22] Fernandez-Martinez E, Gavela M B, Lopez-Pavon J and Yasuda O 2007 *Phys. Lett. B* **649** 427–435 (*Preprint* `hep-ph/0703098`)
[23] Goswami S and Ota T 2008 *Phys. Rev. D* **78** 033012 (*Preprint* `arXiv:0802.1434[hep-ph]`)
[24] Meloni D, Ohlsson T, Winter W and Zhang H 2010 *JHEP* **1004** 041 (*Preprint* `arXiv:0912.2735[hep-ph]`)

Physics case for ν_τ detection

Toshihiko Ota
Max-Planck-Institut für Physik (Werner-Heisenberg-Institut)
Föhringer Ring 6, 80805 München, Germany
E-mail: toshi@mppmu.mpg.de

Abstract. In this talk, we discuss physics cases in which the detection of tau neutrino plays an important role. Some examples of the utility of ν_τ detection in new physics search are given.

1. Introduction

In the past decade, ν_τ-detection experiments achieved great success. The DONUT experiment observed neutrinos interacting with tau lepton through charged current [1], which was the first direct evidence of the existence of *tau neutrino*. The OPERA experiment reported the observation of a ν_τ event originating from $\nu_\mu \to \nu_\tau$ oscillation [2]. Experimental technique of ν_τ detection has made progress, which is nicely summarized in [3]. In this talk, we discuss the utility of ν_τ detection in future oscillation experiments. Of course, if an experiment is aimed at examining and confirming the consistency of the standard oscillation framework in the tau sector, ν_τ detection is essential on purpose. On the other hand, the general consensus holds that information of ν_τ-detection channel makes only a sub-dominant impact on determination of the standard oscillation parameters, because of the limited detector size. Here, we leave the detailed discussions on the standard oscillation to IDS-NF interim design report [4], and in the following, focus on the role of ν_τ detection in new physics search. At first sight, ν_τ detection seems to be helpful to reveal a new physics effect associated with tau flavour. However, since the maximal mixing between ν_μ and ν_τ transmits the effect from the tau sector to the muon sector, the ν_μ-detection channel also acquires a sensitivity to the new physics effect in the tau sector. Therefore, it is often the case that ν_τ detection is not an essential component to discover the tau-associated new physics signals. In this talk, we try to give some examples of physics cases in which information of ν_τ detection has a significant impact.

2. New physics search

We categorise new physics effects into two groups in terms of their origin:

(i) **High energy origin** — If the Standard Model (SM) is an effective theory of fundamental theories which are realized at the higher energy scales, the full Lagrangian at the electroweak scale is described with the SM Lagrangian and the series of the non-renormalizable terms which are the remnant of physics at the high energy scales. Here, we discuss the effect of Non-Standard neutrino Interactions (NSI). The most of the new physics effects categorised into this group (e.g., non-unitarity of the lepton mixing matrix induced by dimension six operators) can also be parameterized as NSIs. We discuss the current bounds and future prospect on the NSI search, especially focus on the usefulness of ν_τ detection.

	Beam (channel)	$2L2Q$	$4L$	NU
Conventional beam	$\pi\ (\mu \to \tau)$	$7.9 \cdot 10^{-5}$	n/a	$4.4 \cdot 10^{-6}$
Beta beam	$\beta\ (e \to \tau)$	$< 10^{-6}$	n/a	$1.0 \cdot 10^{-5}$
Neutrino factory	$\mu\ (\mu \to \tau)$	$< 10^{-6}$	$1.0 \cdot 10^{-3}\ (3.2 \cdot 10^{-5})$	$4.4 \cdot 10^{-6}$
	$\mu\ (e \to \tau)$	$< 10^{-6}$	$1.0 \cdot 10^{-3}\ (3.2 \cdot 10^{-5})$	$1.0 \cdot 10^{-5}$

Figure 1. Model independent bounds to the tau-associated source and detection NSIs. Note that the numbers correspond to the bound to $|\epsilon^s_{\alpha\tau}|^2$. Thanks to the chiral enhancement, the effect of $|\epsilon^s_{\mu\tau}|^2$ in conventional beam experiments is still allowed to be significantly large ($\sim 10^{-4}$). Table is taken from [5].

(ii) **Low energy origin** — New particles and interactions are introduced at the electroweak scale, which affect the nature of neutrino. Light sterile (SM singlet) states and a new long range force mediated by light gauge boson are categorized into this group. In this talk, we discuss the sensitivities to the parameters associated with sterile neutrinos.

2.1. High energy origin: NSI

If the NSIs — four-fermion interactions including a neutrino or two — are left in the low energy effective theory, the standard formula of neutrino oscillation must be modified. The initial and the final states are no more the pure flavour states but the flavour mixture states,

$$|\nu^s_\alpha\rangle = |\nu_\alpha\rangle + \sum_{\gamma\neq\alpha} |\nu_\gamma\rangle, \qquad \langle\nu^d_\beta| = \langle\nu_\beta| + \sum_{\gamma\neq\beta} \langle\nu_\gamma|, \tag{1}$$

and neutrinos receive the additional matter effect potential,

$$(V_{\rm NSI})_{\alpha\beta} = \sqrt{2} G_F N_e \epsilon^m_{\alpha\beta}, \tag{2}$$

in their propagation. These epsilon parameters are defined as the proportion of the non-standard effects to the standard ones. Currently, the effective interactions with neutrinos are not strongly constrained, because of the difficulty in their direct detection. The model independent bounds are investigated in [5, 6, 7]. The bounds to the propagation NSI are summarized as [6][1]

$$|\epsilon^m_{\alpha\beta}| < \begin{pmatrix} 4.2 & 0.33 & 3.0 \\ 0.33 & 0.068 & 0.33 \\ 3.0 & 0.33 & 21 \end{pmatrix}. \tag{3}$$

Figure 1 shows the constraints to the tau-associated NSIs at neutrino beam source and detection, which originate from the effective interactions with mass dimension six. Although these direct bounds shown in Eq. (3) still allow NSIs to take large values, it is hard to derive such large NSIs from high energy models. Since NSIs are related with the corresponding charged lepton processes in typical models, they are in general constrained from the observation of the charged sector. For example, in the minimal supersymmetric standard model (MSSM), NSIs are induced by the one-loop diagrams with the slepton mixings, which are similar to the corresponding charged lepton

[1] The NSI in the μ-τ sector is also constrained from the observation of atmospheric neutrinos [8, 9].

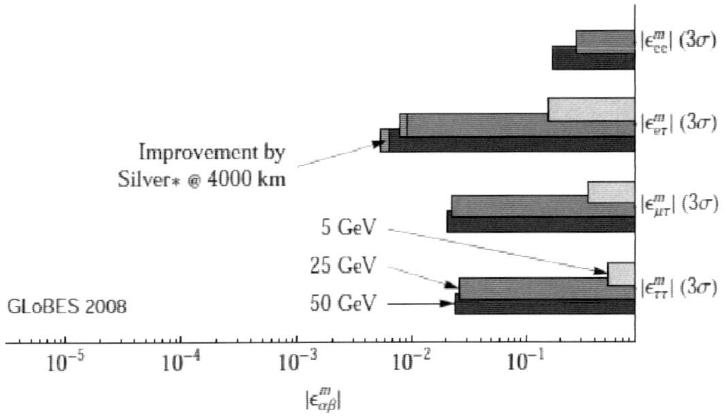

Figure 2. Sensitivities to NSIs in propagation at the IDS-NF baseline setup of neutrino factory. The silver channel $\nu_e \to \nu_\tau$ slightly improves the sensitivity to $\epsilon^m_{e\tau}$ but does not to the others. The figure is taken from [13]

flavour violating (LFV) processes. Therefore, NSI and charged LFV are strongly correlated with each other [10, 11]. The typical size of NSI in the MSSM is evaluated as [10]

$$|\epsilon^m_{\mu\tau}| \sim |\epsilon^s_{\mu\tau}| \text{ at neutrino factory} \simeq 10^{-3}\sqrt{\frac{\mathrm{Br}(\tau \to \mu\gamma)}{10^{-7}}}, \tag{4}$$

which is far below the upper bound of the model independent bound and also below the expected future sensitivity at neutrino factory (see Fig. 2 and [12, 13, 14]). Figure 2 also indicates that an additional ν_τ detector does not help to improve the sensitivity. In order to obtain large NSIs in a theoretically motivated model, it is necessary to invoke the mechanism to avoid the correlation with the charged lepton sector. Operators with Higgs doublets can provide only NSIs without the corresponding charged lepton processes by using the vacuum expectation value of Higgs doublet [15, 16]. However, the implementation of such higher dimensional operators to high energy models raises the non-trivial issue [17]. The use of the chiral enhancement mechanism might be another possibility to break the $SU(2)_L$ relation. The operators,

$$\mathcal{O}_{ED} = (\overline{L_\tau}\mu_R)(\overline{d_R}Q), \qquad \mathcal{O}_{EU} = (\overline{L_\tau}\mu_R)(\overline{Q}u_R), \tag{5}$$

contain both the NSI and the charged LFV. However, the NSI effect in pion decay gets enhanced and the LFV tau decays do not. After taking into account the constraints from the rare tau decays, the current bounds to $|\epsilon^s_{\mu\tau}|^2$ induced from these operators can become almost 10^{-4}, which is fairly in the range of the expected sensitivity of MINSIS proposal [18] — ν_τ near detector at NuMI. Finally, it should be repeated that the realization of the model independent upper bounds requires a lot to the models.

2.2. Low energy origin: sterile neutrino
The introduction of sterile neutrinos is motivated from both theoretically and experimentally aspects. A plausible motivation might be that all the three observations — appearance signal at LSND and MiniBooNE, recent re-calculation of reactor neutrino flux, and the measurement of neutrinos from gallium decays — suggest the same mass squared difference ~ 1 eV^2 [20, 21]. Here, following [19], we see the future sensitivity to the parameters associated with the sterile state and its improvement by the ν_τ-detection channel.

At present, the mixing angle between tau and sterile neutrinos is not strongly constrained, and neutrino factory is expected to improve the current bound significantly. Naively, an additional

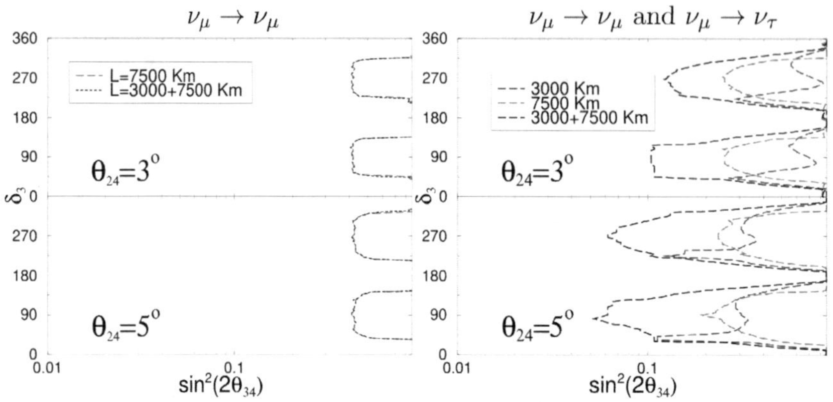

Figure 3. Discovery potential of the additional CP-violating phase δ_3 associated with a sterile neutrino. The synergy of two channels significantly improves the discovery reach. Plots are taken from [19].

ν_τ detector seems to assist in making the sensitivity better. However, the effect of ν_τ-detection information is quite limited [19, 22], because disappearance channel at magic baseline (L =7,500 km) is sensitive to the mixing angle and its large statistics dominate the sensitivity. On the other hand, Figure 3 shows that the discovery reach to the additional CP-violating phase associated with sterile state is improved by the synergy of the ν_μ and the ν_τ detection channels.

3. Summary

In this talk, we have tried to find possible physics cases in which ν_τ detection plays an important role. Here, we have seen two examples — i) Search for chiral enhanced source NSI at near detector of conventional beam experiments, and ii) Discovery of new CP-violating effects associated with sterile neutrinos at neutrino factories.

References

[1] Kodama K *et al.* (DONUT Collaboration) 2001 *Phys. Lett.* B **504** 218 (*Preprint* hep-ex/0012035)
[2] Agafonova N *et al.* (OPERA Collaboration) 2010 *Phys. Lett.* B **691** 138 (*Preprint* arXiv:1006.1623 [hep-ex])
[3] Migliozzi P and Terranova F 2011 *New J. Phys.* **13** 083016 (*Preprint* arXiv:1107.3018 [hep-ex])
[4] Choubey S *et al.* (IDS-NF Collaboration) 2011 *IDS-NF* **020** https://www.ids-nf.org
[5] Antusch S *et al.* 2010 *JHEP* **1006** 068 (*Preprint* arXiv:1005.0756 [hep-ph])
[6] Biggio C *et al.* 2009 *JHEP* **0908** 090 (*Preprint* arXiv:0907.0097 [hep-ph])
[7] Carpentier M and Davidson S 2010 *Eur.Phys.J.* C **70** 1071 (*Preprint* arXiv:1008.0280 [hep-ph])
[8] Gonzalez-Garcia M and Maltoni M 2008 *Phys.Rept.* **460** 1 (*Preprint* arXiv:0704.1800 [hep-ph])
[9] Mitsuka G *et al.* (Super-Kamiokande Collaboration) 2011 *Preprint* arXiv:1109.1889 [hep-ex]
[10] Bellazzini B *et al.* 2011 *JHEP* **1106** 104 (*Preprint* arXiv:1012.3759 [hep-ph])
[11] Ota T and Sato J 2005 *Phys.Rev.* D **71** 096004 (*Preprint* hep-ph/0502124)
[12] Ribeiro N *et al.* 2007 *JHEP* **0712** 002 (*Preprint* arXiv:0709.1980 [hep-ph])
[13] Kopp J *et al.* 2008 *Phys.Rev.* D **78** 053007 (*Preprint* arXiv:0804.2261 [hep-ph])
[14] Coloma P *et al.* 2011 *JHEP* **1108** 036 (*Preprint* arXiv:1105.5936 [hep-ph])
[15] Berezhiani Z and Rossi A 2002 *Phys.Lett.* B **535** 207 (*Preprint* hep-ph/0111137)
[16] Davidson S *et al.* 2003 *JHEP* **0303** 011 (*Preprint* hep-ph/0302093)
[17] Biggio C *et al.* 2009 *JHEP* **0903** 139 (*Preprint* arXiv:0902.0607 [hep-ph])
[18] Alonso R *et al.* 2010 *Preprint* arXiv:1009.0476 [hep-ph]
[19] Donini A *et al.* 2009 *JHEP* **0908** 041 (*Preprint* arXiv:0812.3703 [hep-ph])
[20] Giunti C and Laveder M 2011 *Preprint* arXiv:1107.1452 [hep-ph]
[21] Kopp J *et al.* 2011 *Phys.Rev.Lett.* **107** 091801 (*Preprint* arXiv:1103.4570 [hep-ph])
[22] Meloni D *et al.* 2010 *Phys.Rev.* D **82** 093008 (*Preprint* arXiv:1007.2419 [hep-ph])

Optimized Neutrino Factory for small and large θ_{13}

Sanjib Kumar Agarwalla

Instituto de Física Corpuscular, CSIC-Universitat de València,
Apartado de Correos 22085, E-46071 Valencia, Spain

E-mail: `Sanjib.Agarwalla@ific.uv.es`

Abstract. Recent results from long baseline neutrino oscillation experiments point towards a non-zero value of θ_{13} at around 3σ confidence level. In the coming years, further ratification of this result with high significance will have crucial impact on the planning of the future long baseline Neutrino Factory setup aimed to explore leptonic CP violation and the neutrino mass ordering. In this talk, we discuss the baseline and energy optimization of the Neutrino Factory including the latest simulation results on the magnetized iron neutrino detector (MIND) in the light of both small and large θ_{13}. We find that in case of small θ_{13}, baselines of about 2500 to 5000 km is the optimal choice for the CP violation measurement with E_μ as low as 12 GeV can be considered. However, for large θ_{13}, we show that the lower threshold and the backgrounds reconstructed at lower energies allow in fact for muon energies as low as 5 to 8 GeV at considerably shorter baselines, such as Fermilab to Homestake. This suggests that with the latest MIND simulation, low- and high-energy versions of the Neutrino Factory are just two different forms of the same experiment optimized for different regions of the parameter space.

1. Introduction and Motivation

The performance of future long baseline neutrino oscillation experiments to unravel leptonic CP violation (CPV) and the neutrino mass hierarchy (MH) is crucially dependent on the achievable event rates and hence strongly on the size of θ_{13} [1–4]. Recent T2K [5] and MINOS [6] data have provided new information on θ_{13}, indicating that $0.01 \lesssim \sin^2 2\theta_{13} \lesssim 0.1$ at 2σ confidence level with a best-fit value of 0.051 (0.063) for normal (inverted) mass ordering [7]. It is expected that in the next few years, this hint will be validated with high significance with more informations coming from the T2K, MINOS and upcoming reactor and NOνA experiments, opening up the possibility of observing CPV in the lepton sector if the Dirac CP phase, $\delta_{\rm CP}$, is not equal to 0° or 180°.

The prospects of the present generation long baseline experiments, T2K and NOνA in probing CPV and the mass ordering are very limited even if θ_{13} is large [8]. The experiment class, which can ultimately address these unsolved issues with unprecedented sensitivity for both small and large θ_{13}, is the Neutrino Factory [1,3]. In a Neutrino Factory, the neutrino beam is created from the decay of muons in flight in the straight sections of a storage ring. The International Neutrino Factory and Superbeam Scoping Study [1,9,10] has laid the foundations for the currently ongoing Design Study for the Neutrino Factory (IDS-NF) [11] which has come up with a first-version baseline setup of a high energy neutrino factory (HENF) with $E_\mu = 25\,\rm GeV$ and two baselines $L_1 \simeq 3\,000 - 5\,000\,\rm km$ and $L_2 \simeq 7\,500\,\rm km$ served by two racetrack-shaped storage rings [12].

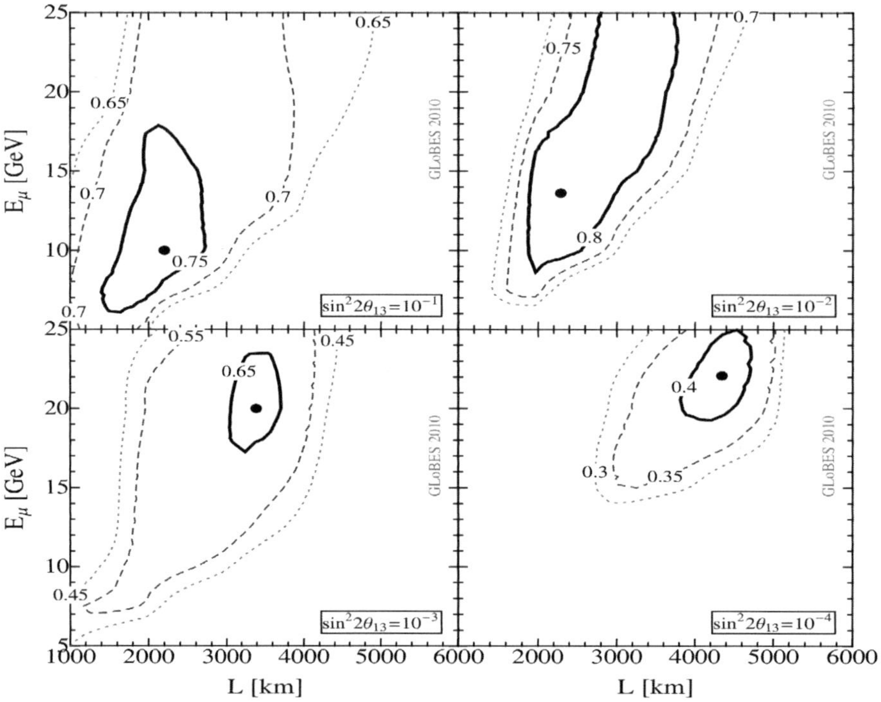

Figure 1. Fraction of $\delta_{\rm CP}$(true) for which CPV will be discovered at $3\,\sigma$ C.L. as a function of L and E_μ for the single baseline Neutrino Factory. The different panels correspond to different true values of $\sin^2 2\theta_{13}$, as given there. Here we consider 5×10^{21} useful muon decays in total with a 50 kt detector. The optimal performance is marked by a dot: (2200,10.00), (2288,13.62), (3390,20.00) and (4345,22.08) with regard to their best reaches of the fraction of $\delta_{\rm CP}$(true) at: 0.77, 0.84, 0.67 and 0.42.

The most promising avenue to explore CPV and $sgn(\Delta m^2_{31})$ at a Neutrino Factory is the sub-dominant $\nu_e \to \nu_\mu$ oscillation channel which produces muons of the opposite charge (wrong-sign muons) to those stored in the storage ring and these can be detected with the help of the charge identification capability of a magnetized iron neutrino detector (MIND) [13]. As the design of the Neutrino Factory matures, more realistic detector simulation and analysis technique have been developed which improves the efficiency of MIND at low energies and also lowers the threshold [14, 15]. These new simulations provide the full response matrices for all the signal and background channels. In the most recent analysis [15], low energy neutrino signal events down to 1 GeV were selected with an efficiency plateau of ~60% for ν_μ and ~80% for $\bar{\nu}_\mu$ events starting at ~5 GeV, while maintaining the background level at or below 10^{-4}. For the neutral current background, the impact of migration is non-negligible and it is peaked at lower energies. This feed-down is the strongest effect of migration and thus has potential impact on the energy optimization, since it penalizes neutrino flux at high energies, where there is little oscillation but a large increase in fed-down background. Now we discuss the simultaneous optimization of baseline and muon energy in a green-field Neutrino Factory scenario [3] using the new migration matrices for MIND detector [15].

2. Optimization of a green-field Neutrino Factory setup with new MIND detector

Here we are interested to optimize L and E_μ for the single baseline Neutrino Factory to have CPV discovery at $3\,\sigma$ C.L. for different fractions of true values of $\delta_{\rm CP}$ assuming that the true

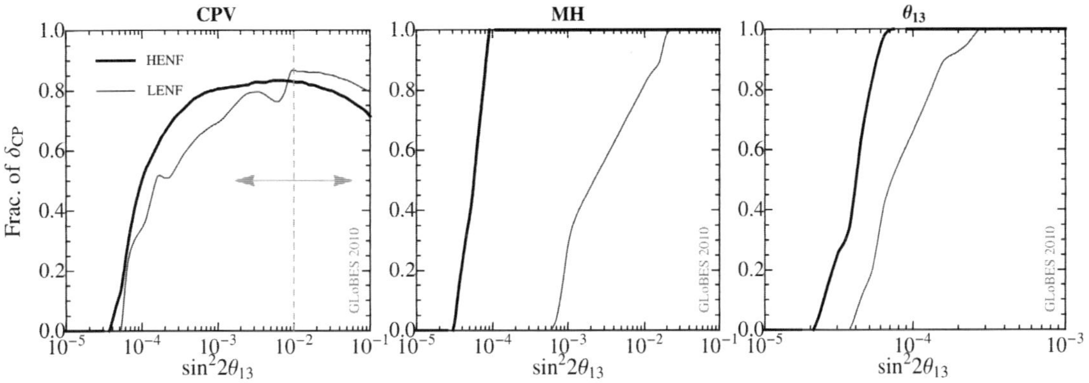

Figure 2. CP fractions for which a discovery at 3σ confidence level is possible as a function of true $\sin^2 2\theta_{13}$. From left to right for CPV, MH and θ_{13}. Here we compare the performances of a single baseline LENF with $E_\mu = 10$ GeV, $L = 2000$ km, and a 50 kt MIND detector, and a two baseline HENF with $E_\mu = 25$ GeV and a 50 kt MIND detector at each baseline, and $L_1 = 4000$ km, and $L_2 = 7500$ km.

value of $\sin^2 2\theta_{13}$ is known. The different panels of Fig. 1 depict the same for four different true values of $\sin^2 2\theta_{13}$. As expected, this figure clearly shows that the optimization strongly depends on the true value of $\sin^2 2\theta_{13}$ chosen. In case of large $\sin^2 2\theta_{13} \simeq 10^{-1}$, shorter baselines and lower energies are preferred. Even E_μ as low as 5 to 8 GeV at the Fermilab-Homestake baseline of about 1300 km is quite close to the optimal choice, which means that the MIND detector approaches the magnetized totally active scintillator detector performance of the low energy Neutrino Factory (LENF) [16]. Note that compared to earlier analyses without feed-down effect due to background migration, now too high E_μ are in fact disfavored in the large $\sin^2 2\theta_{13}$ case. For the other extreme, $\sin^2 2\theta_{13} \simeq 10^{-4}$, baselines between 4 000 and 5 000 km are preferred with $E_\mu \simeq 20 - 25$ GeV, which corresponds more to the choice of L_1 for HENF, such as the IDS-NF baseline. Including the other two panels, the optimal region within each panel moves from the lower left on the plots to the upper right as the true value of $\sin^2 2\theta_{13}$ decreases. This means that, depending on the choice of true $\sin^2 2\theta_{13}$, the optimization results in the LENF, the HENF, or an intermediate scenario, and that the low and high energy Neutrino Factories are just two versions of the same experiment optimized for different parts of the parameter space. Note that, all the figures presented in this talk have been taken from [3] and all the simulations have been performed using the GLoBES software [17, 18].

3. Summary and conclusions

In this talk, we have discussed the optimization of the Neutrino Factory based on the most up-to-date analysis of the MIND detector using response matrices for all the signal and background channels. The lower threshold and higher efficiencies compared to earlier simulations indicate that the MIND detector characteristics are getting more similar to the characteristics of the detectors proposed for the LENF (e.g., a magnetized TASD). For large $\sin^2 2\theta_{13}$, a single baseline Neutrino Factory with E_μ as low as 5 to 8 GeV and a baseline as short as Fermilab to Homestake (about 1300 km) might be sufficient. In the case of small $\sin^2 2\theta_{13} < 10^{-2}$, however, a two baseline Neutrino Factory with one baseline between about 2500 km and 5000 km, the other one at about the magic baseline 7500 km is ideal. The additional information coming from the second baseline which is at the magic distance (insensitive to δ_{CP}) helps to remove the degeneracies and it enhances the fraction of δ_{CP} for which CPV discovery is possible. Fig. 2 nicely depicts this message where we compare the performance of the optimal single baseline LENF with the optimal two baseline HENF for the same MIND detector. It is quite evident from this figure that for $\sin^2 2\theta_{13} \gtrsim 10^{-2}$ the low energy version can perform all of the required measurements, whereas for smaller values the HENF is clearly better. It suggests that this different optimization would be sufficient to compensate for the relative deterioration of performance at large $\sin^2 2\theta_{13}$ observed in the traditional HENF.

We conclude that the features of the new detector simulation that the backgrounds are typically reconstructed at lower energies and that the threshold is lower have significant impact on the optimization of E_μ. The recent simulation results for the MIND have reduced the performance margin between TASD and MIND considerably. Therefore, we suggest that the low- and high-energy Neutrino Factory should not be regarded as separate options and they merely correspond to two extreme corners of a common parameter space.

Acknowledgments
I am grateful to the conference organizers for the invitation. I would like to thank Patrick Huber, Jian Tang and Walter Winter for their collaboration. I would also like to thank Paul Soler for insightful discussions during NUFACT11. I acknowledge the support from the European Union under the European Commission Framework Programme 07 Design Study EUROnu, Project 212372 and the project Consolider-Ingenio CUP.

References

[1] Bandyopadhyay A *et al.* (ISS Physics Working Group) 2009 *Rept. Prog. Phys.* **72** 106201 (*Preprint* arXiv:0710.4947[hep-ph])
[2] Mezzetto M and Schwetz T 2010 *J. Phys. G* **37** 103001 (*Preprint* arXiv:1003.5800[hep-ph])
[3] Agarwalla S, Huber P, Tang J and Winter W 2011 *JHEP* **1101** 120 (*Preprint* arXiv:1012.1872[hep-ph])
[4] Agarwalla S, Li T and Rubbia A 2012 *JHEP* **1205** 154 (*Preprint* arXiv:1109.6526[hep-ph])
[5] Abe K *et al.* (T2K Collaboration) 2011 *Phys. Rev. Lett.* **107** 041801 (*Preprint* arXiv:1106.2822[hep-ex])
[6] Adamson P *et al.* (MINOS Collaboration) 2011 *Phys. Rev. Lett.* **107** 181802 (*Preprint* arXiv:1108.0015[hep-ex])
[7] Schwetz T, Tortola M and Valle J 2011 *New J. Phys.* **13** 109401 (*Preprint* arXiv:1108.1376[hep-ph])
[8] Huber P, Lindner M, Schwetz T and Winter W 2009 *JHEP* **0911** 044 (*Preprint* arXiv:0907.1896[hep-ph])
[9] Abe T *et al.* (ISS Detector Working Group) 2009 *JINST* **4** T05001 (*Preprint* arXiv:0712.4129[physics.ins-det])
[10] Apollonio M *et al.* (ISS Accelerator Working Group) 2009 *JINST* **4** P07001 (*Preprint* arXiv:0802.4023[physics.acc-ph])
[11] International design study of the neutrino factory, http://www.ids-nf.org
[12] Huber P, Lindner M, Rolinec M and Winter W 2006 *Phys. Rev.* D **74** 073003 (*Preprint* hep-ph/0606119)
[13] Cervera A, Donini A, Gavela M, Gomez Cadenas J, Hernandez P, Mena O and Rigolin S 2000 *Nucl. Phys.* B **579** 17-55 (*Preprint* hep-ph/0002108)
[14] Cervera A, Laing A, Martin-Albo J and Soler P 2010 *Nucl. Instrum. Meth.* A **624** 601-614 (*Preprint* arXiv:1004.0358[hep-ex])
[15] A. Laing 2010 *Optimization of Detectors for the Golden Channel at a Neutrino Factory,* Ph.D. Thesis, Glasgow University, Glasgow U.K.
[16] Geer S, Mena O and Pascoli S 2007 *Phys. Rev.* D **75** 093001 (*Preprint* hep-ph/0701258)
[17] Huber P, Lindner M and Winter W 2005 *Comput. Phys. Commun.* **167** 195 (*Preprint* hep-ph/0407333)
[18] Huber P, Kopp J, Lindner M, Rolinec M and Winter W 2007 *Comput. Phys. Commun.* **177** 432-438 (*Preprint* hep-ph/0701187)

Sterile neutrinos and IceCube

F Halzen[1] for the IceCube collaboration

[1] University of Wisconsin-Madison, IceCube Research Center, Madison, Wisconsin, USA

E-mail: halzen@icecube.wisc.edu

Abstract. Although the framework for oscillations of the three neutrino flavors in the Standard Model has been convincingly established, indications persist that it may be incomplete. Challenges are coming from the LSND and MiniBooNe short-baseline experiments, from the neutrino sources used in the Gallex and Sage solar neutrino experiments and, more recently, from an a-posteriori analysis of reactor neutrino experiments. One way to accommodate the reported "anomalies", if real, is to introduce one or more sterile neutrinos in the mass range $\delta m^2 \sim 1eV^2$. TeV atmospheric neutrinos propagating through the Earth undergo resonant oscillations in the presence of sterile neutrinos; a clear signature in a neutrino telescope like IceCube is the the change in shape of the zenith-energy distribution of the atmospheric neutrinos.

IceCube detects more than 100,000 atmospheric neutrinos per year. Statistics do not limit such a measurement, but the uncertainties in modeling the expectations of the conventional 3-flavor scenario, including the systematics of the detector, do. We review the status and future perspectives of understanding the zenith and energy response of IceCube in the TeV energy range.

1. Sterile neutrinos and IceCube

Neutrino experiments offer opportunities for new discoveries and, occasionally, total surprises. Examples of new physics include sterile neutrinos and additional degrees of freedom in the energy density of the Universe. Even though it is premature to motivate future facilities on the basis of present indications (which include some hints from short-baseline experiments [1] and reactor data [2]), recent developments underscore the possibility of unexpected discoveries, supporting the construction of neutrino facilities with the widest science reach. IceCube is such a facility [3]. IceCube measures the flux of atmospheric muons and neutrinos with high statistics in a high-energy range that has not been previously explored. The procedure is simply to compare data with expectations based on the extrapolation of lower-energy data. Any deviation observed is an opportunity for discovery in astrophysics or the physics of the neutrinos themselves.

The potential of IceCube to detect eV-mass sterile neutrinos has been recognized for some time [4, 5, 6]. The problem has been revisited [7, 8, 9] with the measurement of the atmospheric neutrino flux in the energy range 100 GeV–400 TeV with unprecedented statistics, taken when IceCube was half complete [10]. The signature for sterile neutrinos is the disappearance of $\nu_\mu(\bar{\nu}_\mu)$ resulting from mixing with a new species during propagation between production and detection. For $\delta m^2 \sim 1eV^2$, resonant oscillations are expected in IceCube for TeV atmospheric neutrinos at characteristic zenith angles. An additional signature is the appearance of an excess of $\nu_e(\bar{\nu}_e)$, with a particular energy and zenith dependence, due to a sterile neutrino acting as an intermediary between muon and electron neutrinos.

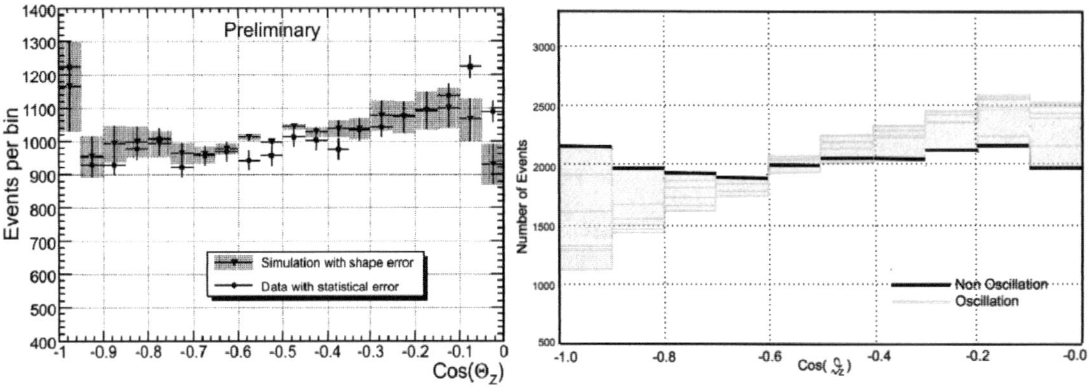

Figure 1. Zenith distribution $\cos(\theta_Z)$ of atmospheric neutrinos from [11]. *On the left*: Simulation has been normalized to the data. Error bars for data are statistical only. Error bars for simulation indicate the uncertainty in the shape of the predicted distribution (normalization uncertainty has been removed). These error bars do not include certain simulation errors discussed in the text. *On the right*: Illustration of the impact of adding a sterile neutrino into the conventional 3-flavor scheme on the $\cos(\theta_Z)$ distribution. For this particular model (green lines), $\Delta m^2_{41} = 0.4$ eV^2, $\sin^2\theta_{24} = 0.1$, and $\sin^2\theta_{34}$ is varied from 0 to 0.5 in increments of 0.1. Green and black lines include standard atmospheric neutrino oscillations and detector acceptance. (credit: A. Esmaili, Universidade Estadual de Campinas, Brazil).

However, before addressing the question of whether the data support evidence for a sterile neutrino, we are investigating the consistency between the data and expectation and the correctness of the model implemented in simulation. Given the high statistics, the focus is on the systematics of the experiment. The systematic issues are being studied with renewed emphasis; up to now, our priority has been the search for neutrino sources beyond the atmosphere, analyses where the simulation of the background is done using the data themselves. We discuss the status of this effort that will also result in improved sensitivity to, for instance, diffuse limits on cosmic neutrinos.

2. Atmospheric neutrinos

2.1. Zenith distribution

Figure 1 compares the observed to the predicted distribution of atmospheric ν_μ (plus $\bar{\nu}_\mu$) from data taken while IceCube operated in a partially completed 40-string configuration. Since the uncertainty in the normalization is large (see Ref. 11), it is more interesting to examine the shape of the zenith distribution. In Fig. 1, the predicted distribution has been normalized to the event rate in data. This plot is similar to Fig. 19 of Ref. 11, except shape uncertainties are included for the predicted flux. These shape uncertainties include contributions from uncertainty in the ratio of pions to kaons produced by the cosmic ray flux, uncertainties in the spectral slope of the cosmic ray flux, and uncertainties in the simulation of DOM sensitivity and ice properties. See [10, 11] for further discussion of systematic uncertainties associated with atmospheric neutrino measurements with the 40-string IceCube detector. Some of these uncertainties will be reduced as the detector and irreducible systematic uncertainties are better understood. It is important to note that uncertainties due to problems with simulation for the 40-string detector, recently discovered and discussed below, are not included in these error bars.

Several possibilities have been explored to resolve the apparent mismatch between data and

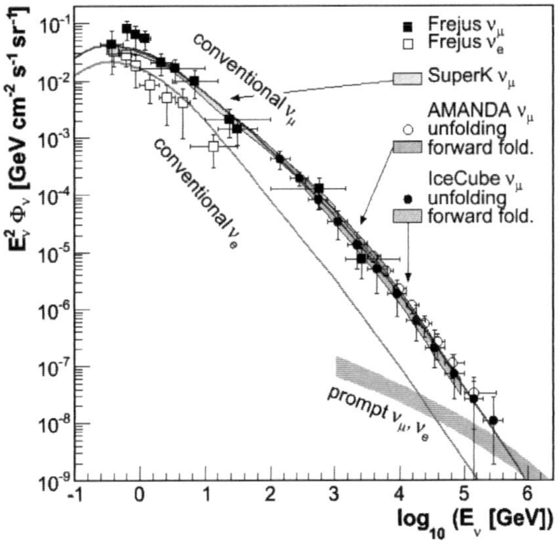

Figure 2. Measurements of the atmospheric neutrino energy spectrum; the Fréjus results [13], SuperK [14], AMANDA forward folding analysis [15] and unfolding analysis [16], IceCube (40 strings) forward folding analysis [10] and unfolding analysis [11]. All measurements include the sum of neutrinos and antineutrinos. The expectations for conventional ν_μ and ν_e flux are from [17]. The prompt flux is from [18].

simulation in the near horizontal region, including the candidates discussed in [11, 12]: regional and seasonal variations in atmospheric neutrino production, simulation of ice properties and photon propagation in the ice, and inadequate simulation of atmospheric muon backgrounds including various composition models of cosmic rays.

While these effects have some impact on the zenith distribution, we have found that an important change arose from improvements in the modeling of the rock layer below the detector and the ice/rock boundary, as well as the assignment of event weights according to model simulations. These event weights account for the probability of a simulated neutrino to survive propagation through the Earth and interact in or near the detector. Preliminary testing of improved simulation for the 59-string detector configuration indicates that these inadequacies in the modeling of the detector were a major contributor to the shape disagreement between the 40-string simulation and data. Atmospheric neutrino analyses and searches for a diffuse flux of astrophysical neutrinos, with data taken by the 59-string detector, are currently underway and are using this improved simulation.

2.2. Energy spectrum

Figure 2 compares the results of atmospheric neutrino measurements from about a GeV to 400 TeV. In IceCube's spectrum unfolding measurement of [11], events in the 90° to 97° zenith region were not used. However, in Fig. 2, this result has been scaled to the flux for the entire hemisphere, to allow a more direct comparison to the other results. IceCube's forward folding result from Ref. 10 is also shown. Both of these results are affected by the simulation issues discussed above. New analyses using improved simulation are underway using data taken while IceCube operated in the 59-string configuration.

3. Conclusion

IceCube has been stably operating with a completed detector since May 2011. Atmospheric neutrinos are collected at the rate of $O(10^5)$ per year. Our ability to constrain a variety of sterile neutrino models is only limited by systematic uncertainties. The calibration of the final detector will be superior to that of configurations operated during construction. A new generation of analysis tools are resulting in improved effective area, angular and energy resolution.

Acknowledgments

We acknowledge support from the following agencies: U.S. National Science Foundation, Office of Polar Programs; U.S. National Science Foundation, Physics Division; U. of Wisconsin Alumni Research Foundation; U.S. Department of Energy; U.S. National Energy Research Scientific Computing Center (NERSC); the LONI grid at UCLA; Swedish Research Council; K. & A. Wallenberg Foundation, Sweden; German Ministry for Education and Research; Deutsche Forschungsgemeinschaft; Fund for Scientific Research, Belgium; IWT-Flanders; Belgian Federal Science Policy Office (BELSPO); the Netherlands Organisation for Scientific Research; and the Swiss National Science Foundation (SNF).

References

[1] Aguilar-Arevalo A *et al.* (The MiniBooNE Collaboration) 2010 *Phys. Rev. Lett.* **105** 181801 (*Preprint* hep-ex/1007.1150) and references therein.
[2] Mueller T A *et al.* 2011 *Phys. Rev.* C **83** 054615 (*Preprint* hep-ex/1101.2663)
[3] Halzen F and Klein S R 2010 *Rev. Sci. Instrum.* **81** 081101 (*Preprint* astro-ph.HE/1007.1247)
[4] Smirnov A Y 2012 *these proceedings*
[5] Nunokawa H, Peres O L G and Zukanovich Funchal R 2003 *Phys. Lett.* B **562** 279 (*Preprint* hep-ph/0302039)
[6] Choubey S 2007 *J. High Energy Phys.* **0712** 014 (*Preprint* hep-ph/0709.1937)
[7] Kopp J, Maltoni M and Schwetz T 2011 *Phys. Rev. Lett.* **107** 091801 (*Preprint* hep-ph/1103.4570)
[8] Razzaque S and Smirnov A Y 2011 *J. High Energy Phys.* **1107** 084 (*Preprint* hep-ph/1104.1390)
[9] Barger V, Gao Y and Marfatia D 2011 *Phys. Rev.* D **85** 011302 (*Preprint* hep-ph/1109.5748)
[10] Abbasi R *et al.* (IceCube Collaboration) 2011 *Phys. Rev.* D **84** 082001
[11] Abbasi R *et al.* (IceCube Collaboration) 2011 *Phys. Rev.* D **83** 012001
[12] Abbasi R *et al.* (IceCube Collaboration) 2011 *Astrophys. J.* **732** 18
[13] Daum K *et al.* 1995 *Zeitschrift für Physik* C **66** 417
[14] Gonzalez-Garcia M C, Maltoni M and Rojo J 2006 *J. High Energy Phys.* **10** 75
[15] Abbasi R *et al.* (IceCube Collaboration) 2009 *Phys. Rev.* D **79** 102005
[16] Abbasi R *et al.* (IceCube Collaboration) 2010 *Astropart. Phys.* **34** 48
[17] Barr G D *et al.* 2004 *Phys. Rev.* D **70** 023006
[18] Enberg R, Reno M H and Sarcevic I 2008 *Phys. Rev.* D **78** 043005

Impact of nuclear effects in the measurement of neutrino oscillation parameters

Davide Meloni

Dipartimento di Fisica "E. Amaldi", Universitá degli Studi Roma Tre, Via della Vasca Navale 84, 00146 Roma

E-mail: meloni@fis.uniroma3.it

Abstract. In recent years the experimental study of neutrino oscillations has much contributed to our knowledge of particle physics by establishing non vanishing neutrino masses and by measuring or constraining the corresponding mixing angles. Within the domain of neutrino oscillations, the main goal of the next generation of facilities is the measurement of the mixing angle θ_{13} and the observation of leptonic CP violation, for which we have no hints at the moment. We discuss more in detail how various models of neutrino-nucleus cross section affect the forecasted precision measurement of θ_{13} and the CP violating phase δ [1].

1. Summary of the charged current neutrino-nucleus cross sections

In the quasi-elastic regime (QE), the doubly-differential cross section, in which a neutrino carrying initial four-momentum $k = (E_\nu, \mathbf{k})$ scatters off a nuclear target to a state of four-momentum $k^{'} = (E_\ell, \mathbf{k}^{'})$ can be written in Born approximation as follows:

$$\frac{d^2\sigma}{d\Omega dE_\ell} = \frac{G_F^2\, V_{ud}^2}{16\,\pi^2}\,\frac{|\mathbf{k}^{'}|}{|\mathbf{k}|}\, L_{\mu\nu}\, W_A^{\mu\nu}\ , \tag{1}$$

where G_F is the Fermi constant and V_{ud} is the CKM matrix element coupling u and d quarks. The leptonic tensor is completely determined by lepton kinematics, whereas the nuclear tensor $W_A^{\mu\nu}$, containing all the information on strong interactions dynamics, describes the response of the target nucleus, $W_A^{\mu\nu} = \sum_X \langle 0|J_A^{\mu\,\dagger}|X\rangle\, \langle X|J_A^\nu|0\rangle\ \delta^{(4)}(p_0+q-p_X)$, where $|0\rangle$ and $|X\rangle$ are the initial and final hadronic states carrying four momenta p_0 and p_X and J_A^μ is the nuclear electroweak current operator; the sum includes all hadronic final states. In the impulse approximation (IA) scheme, the nuclear current can be written as a sum of one-body currents, i.e. $J_A^\mu \to \sum_i J_i^\mu$, while the final state reduces to the direct product of the hadronic state produced at the weak vertex (with momentum $\mathbf{p}^{'}$) and that describing the $(A-1)$-nucleon residual system, with momentum $\mathbf{p}_{\mathcal{R}}$: $|X\rangle \to |i, \mathbf{p}^{'}\rangle \otimes |\mathcal{R}, \mathbf{p}_{\mathcal{R}}\rangle$.

1.1. *The Spectral Function approach*

Following Ref. [2], the final expression of the hadronic tensor can be cast in the following form:

$$W_A^{\mu\nu} \;=\; \frac{1}{2}\int d^3p\, dE\, P(\mathbf{p},E)\frac{1}{4\,E_{|\mathbf{p}|}\,E_{|\mathbf{p}+\mathbf{q}|}}\, W^{\mu\nu}(\tilde{p},\tilde{q})\ , \tag{2}$$

where $E_{\mathbf{p}} = \sqrt{|\mathbf{p}|^2 + m_N^2}$ and the function $P(\mathbf{p}, E)$ is the target *Spectral Function*, i.e. the probability distribution of finding a nucleon with momentum $\mathbf{p}$ and removal energy E in the target nucleus. The quantity $W^{\mu\nu}$ is the tensor describing the weak interactions of the i-th nucleon in free space; the effect of nuclear binding of the struck nucleon is accounted for by the replacement $q = (\nu, \mathbf{q}) \to \tilde{q} = (\tilde{\nu}, \mathbf{q})$ with $\tilde{\nu} = E_{|\mathbf{p}+\mathbf{q}|} - E_{|\mathbf{p}|}$. The second argument in the hadronic tensor is $\tilde{p} = (E_{|\mathbf{p}|}, \mathbf{p})$.

1.2. The Relativistic Fermi Gas

The RFG [3] model, widely used in MonteCarlo simulations, provides the simplest form of the Spectral Function, $P_{RFGM}(\mathbf{p}, E) = \left(\frac{6\pi^2 A}{p_F^3}\right) \theta(p_F - \mathbf{p})\, \delta(E_{\mathbf{p}} - E_B + E)$, where p_F is the Fermi momentum and E_B is the average binding energy, introduced to account for nuclear binding. Thus, in this model p_F and E_B are two parameters that are *adjusted* to reproduce the experimental data. For oxygen, the analysis of electron scattering data yields $p_F = 225$ MeV and $E_B = 25$ MeV.

1.3. The Relativistic Mean Field approach

Within the Relativistic Mean Field approximation we refer to the model described in [4], where, like in the previous cases, the nuclear current is written as a sum of single-nucleon currents. The wave functions for the target and the residual nuclei are described in terms of an independent-particle model. Then, the transition matrix elements can be cast in the following form:

$$J_N^{\mu}(\omega, \vec{q}) = \int d\vec{p}\bar{\psi}_F(\vec{p} + \vec{q})\hat{J}_N^{\mu}(\omega, \vec{q})\psi_B(\vec{p}) \;, \tag{3}$$

where ψ_B and ψ_F are the wave functions for initial bound and final outgoing nucleons, respectively, and $\hat{J}_N^{\mu}$ is the relativistic current operator. In particular, the relativistic bound-state wave functions (for both initial and outgoing nucleons) are obtained as a solution of the Dirac equation, in the presence of the same relativistic nuclear mean field potential, derived from a Lagrangian containing σ, ω and ρ mesons.

1.4. The Random Phase Approximation

The last model we want to take into account has been introduced in [5], where the hadronic tensor is expressed in terms of the nuclear response functions treated in the Random Phase Approximation (RPA). The response functions are related to the imaginary part of the corresponding full polarization propagators and the introduction of the RPA approximation means that the polarization propagators are the solutions of integral equations involving the bare propagators and the effective interaction between particle-hole excitations. Within this formalism, the authors of [5] were able to show that multinucleon terms sizably increase the genuine charged current QE cross section in such a way to reproduce the MiniBooNE results [6]. The mechanism responsible for the enhancement that brings the theoretical cross section into agreement with the data is multi nucleon knock out, leading to two particle-two hole (2p2h) nuclear final states. In the following, we will refer to this "generalized" QE cross section as RPA-2p2h whereas we adopt the short RPA for the genuine QE cross section.

1.5. Comparison of the cross sections

To summarize this section, we present in Fig.1 a comparison of the five total QE cross sections for the $\nu_\mu\,^{16}O \to \mu^- X$ process (left panel) and $\bar{\nu}_\mu\,^{16}O \to \mu^+ X$ (right panel), in the energy range $E_\nu \sim [0, 0.75]$ GeV. The curves have been computed using the dipole structure of the form factors and, in particular, a value of the axial mass close to $m_A \sim 1$ GeV. As it can be easily

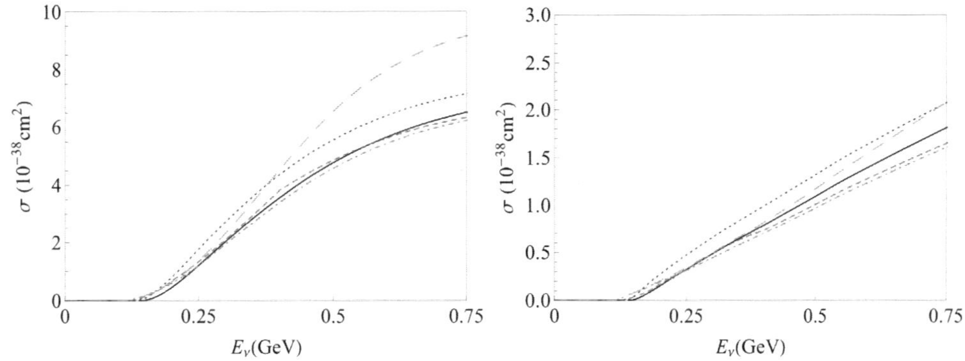

Figure 1. Total charged current QE cross sections for the $\nu_\mu^{16}O \to \mu^- X$ (left panel) and $\bar{\nu}_\mu^{16}O \to \mu^+ X$ (right panel) processes in the energy range $E_\nu \sim [0, 0.75]$ GeV. —— refer to the SF model, ······ to the RFG, - - - - lines to the RMF, —— to RPA and — — — to RPA-2p2h.

seen in the left panel, the RFG prediction sizably overestimates the SF, RMF and RPA results by roughly 15%, a fact that is well known to happen also for many other models with a more accurate description of the nuclear dynamics than the RFG approach. On the other hand, the inclusion of 2p2h contributions largely enhances the QE cross section in the RPA approximation for energies above $\sim$ 0.5 GeV, although it is smaller than the RFG at smaller energies. For antineutrinos (right panel) the observed pattern is almost the same.

2. The impact on the (θ_{13}-δ) measurement

To estimate the impact of different models of the cross section on the measurement of θ_{13} and leptonic CP violation we choose a $\gamma = 100$ β-Beam facility as a representative example because the neutrino flux from such a facility spans up to $\sim$ 0.7 GeV with the peak around 0.3 GeV and is thus mostly sensitive to the quasielastic region explored here. Details of the detector response can be found in Ref. [7]. It is important to notice that we are only using the quasielastic contribution to the neutrino cross section depicted in Fig.1. As an illustration we have focused on the dependence on the nuclear model adopted of two different observables, namely the *CP and* θ_{13} *discovery potentials*, defined as the values of the CP-violating phase δ_{CP} and θ_{13} for which respectively the hypothesis of CP conservation $\delta_{CP} = 0, \pm\pi$ or $\theta_{13} = 0$ can be excluded at 3σ after marginalizing over all other parameters. The CP discovery potential is shown in the left panel of Fig.2, where we superimposed the results obtained using the RFG cross section and the SF, RMF and RPA calculations. We clearly see that, for $\delta_{CP} \sim \pm 90^o$ (where the sensitivity is maximal) the RFG model gives a prediction which is around a factor 2 better than the SF, RMF and RPA models (which, as expected, behave almost in the same way) in $\sin^2 2\theta_{13}$ and around a 40% better than the RPA-2p2h. This is not surprising because the β-Beam facility used in our simulations mainly probes energies smaller than 0.5 GeV, where the RFG is still larger than any other model (see Fig.1). For the other points in the parameter space, the difference is less evident but still significant. The difference in the sensitivity to θ_{13} can be seen in the right panel of Fig. 2 where we show the results in the ($\sin^2 2\theta_{13}, \delta_{CP}$)-plane. In this case the predictions of the SF, RMF and RPA models differ by up to a factor of $\sim$ 60% compared to the RFG model for $\delta_{CP} \sim \pm 90^o$ while the difference is less pronounced for $\delta_{CP} \sim 0^o$. The RPA-2p2h results are more similar to the RFG for $\delta_{CP} \sim 0^o$ and to the other models for $\delta_{CP} \sim 180^o$. This is also evident in the right panel where we show the CP fraction.

Finally, it is interesting to comment on the effect of using different nuclear models also in the simultaneous determination of θ_{13} and δ_{CP}. We have performed such an analysis for the input

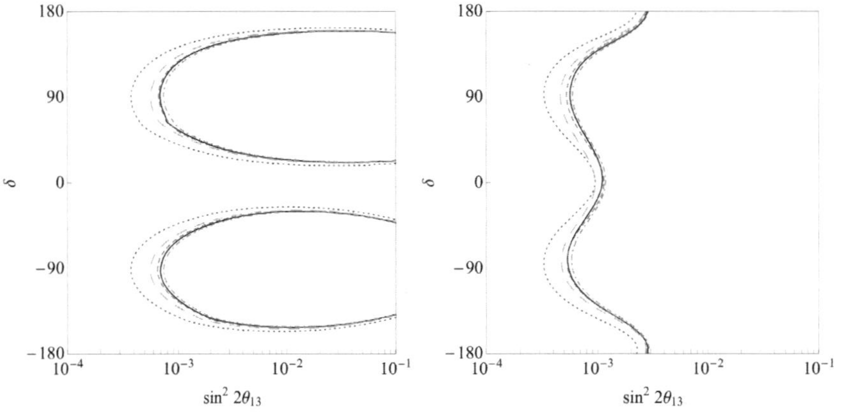

Figure 2. Left panel: CP discovery potential in the $(\theta_{13}, \delta_{CP})$ plane. —— refer to the SF model, $\cdots\cdots$ to the RFG, - - - - lines to the RMF, to RPA and to RPA-2p2h. Right panel: θ_{13} discovery potential.

value $(\theta_{13}, \delta_{CP}) = (0.9^o, 30^o)$ (not including degenerate solutions coming from our ignorance of the octant of θ_{23} [8] and in the neutrino mass ordering) finding that, using the RFG and RPA-2p2h models, we are able to reconstruct the true values of θ_{13} and δ_{CP} within reasonable uncertainties, whereas with the other models we can only measure two distinct disconnected regions (the fake one around the value of θ_{13} and $\delta_{CP} \sim 180^o$), which worsen the global sensitivity on those parameters. In conclusion, we have analyzed the impact of five different theoretical models for the neutrino-nucleus charged current QE cross sections on the forecasted sensitivity of the future neutrino facilities to the parameters θ_{13} and δ_{CP}. We have found that the sensitivities computed with the FG are better, by up to a factor of 2, with respect to the other models.

Acknowledgments

We acknowledge MIUR (Italy), for financial support under the contract PRIN08.

References

[1] E. Fernandez-Martinez, D. Meloni, *Phys. Lett. B* **697**, 477-481 (2011). (arXiv:1010.2329 [hep-ph]).
[2] O. Benhar, N. Farina, H. Nakamura, M. Sakuda and R. Seki, *Phys. Rev. D* **72**, 053005 (2005) (arXiv:hep-ph/0506116); O. Benhar and D. Meloni, *Nucl. Phys. A* **789**, 379 (2007) (arXiv:hep-ph/0610403); O. Benhar and D. Meloni, *Phys. Rev. D* **80**, 073003 (2009) (arXiv:0903.2329 [hep-ph]).
[3] R. A. Smith and E. J. Moniz, *Nucl. Phys. B* **43** (1972) 605 (Erratum-ibid. B **101** (1975) 547).
[4] C. Maieron, M. C. Martinez, J. A. Caballero and J. M. Udias, *Phys. Rev. C* **68**, 048501 (2003) (arXiv:nucl-th/0303075); M. C. Martinez, P. Lava, N. Jachowicz, J. Ryckebusch, K. Vantournhout and J. M. Udias, *Phys. Rev. C* **73**, 024607 (2006) (arXiv:nucl-th/0505008).
[5] M. Martini, M. Ericson, G. Chanfray and J. Marteau, *Phys. Rev. C* **80**, 065501 (2009) (arXiv:0910.2622 [nucl-th]); M. Martini, M. Ericson, G. Chanfray and J. Marteau, *Phys. Rev. C* **81** (2010) 045502 (arXiv:1002.4538 [hep-ph]).
[6] A. A. Aguilar-Arevalo *et al.* [MiniBooNE Collaboration], *Phys. Rev. D* **81** (2010) 092005 (arXiv:1002.2680 [hep-ex]).
[7] J. Burguet-Castell, D. Casper, E. Couce, J. J. Gomez-Cadenas and P. Hernandez, *Nucl. Phys. B* **725**, 306 (2005) (arXiv:hep-ph/0503021).
[8] D. Meloni, *Phys. Lett. B* **664**, 279-284 (2008). (arXiv:0802.0086 [hep-ph]).

Journal of Physics: Conference Series **408** (2013) 012025 doi:10.1088/1742-6596/408/1/012025

The reactor antineutrino anomaly

G Mention

CEA, Irfu, SPP, Centre de Saclay, F-91191 Gif-sur-Yvette, France

E-mail: guillaume.mention@cea.fr

Abstract. Recently, new reactor antineutrino spectra have been provided for ^{235}U, ^{239}Pu, ^{241}Pu, and ^{238}U, increasing the mean flux by about 3 percent. To a good approximation, this reevaluation applies to all reactor neutrino experiments. The synthesis of published experiments at reactor-detector distances below 100 m leads to a ratio of observed event rate to predicted rate of 0.976 ± 0.024. With our new flux evaluation, this ratio shifts to 0.943 ± 0.023, leading to a deviation from unity at 98.6% C.L. which we call the reactor antineutrino anomaly. The compatibility of our results with the existence of a fourth non-standard neutrino state driving neutrino oscillations at short distances is discussed. The combined analysis of reactor data, gallium solar neutrino calibration experiments, and MiniBooNE-ν data disfavors the no-oscillation hypothesis at 99.8% C.L. The oscillation parameters are such that $|\Delta m^2_{\rm new}| > 1.5$ eV^2 (95%) and $\sin^2(2\theta_{\rm new}) = 0.14 \pm 0.08$ (95%).

1. Introduction

Neutrino oscillation experiments over the last twenty years have established a picture of neutrino mixing and masses that explains the results of solar, atmospheric and reactor neutrino experiments [3]. These experiments are consistent with the mixing of ν_e, ν_μ and ν_τ with three mass eigenstates, ν_1, ν_2 and ν_3. In particular, the squared mass differences are required to be $|\Delta m^2_{31}| \simeq 2.4\,10^{-3}$ eV^2 and $\Delta m^2_{21}/|\Delta m^2_{31}| \simeq 0.032$.

Reactor experiments at distances below 100 m from the reactor core (ILL-Grenoble, Goesgen, Rovno, Krasnoyarsk, Savannah River and Bugey see [2] for references therein) have played an important role in the establishment of this pattern. The measured rate of $\bar{\nu}_e$ was found to be in reasonable agreement with that predicted from the reactor antineutrino spectra, though slightly lower than expected, with the measured/expected ratio at 0.976 ± 0.024, including recent revisions of the neutron mean lifetime [3] ($\tau_n = 885.7$ s).

2. New predicted cross section per fission

Fission reactors release about $10^{20}\,\bar{\nu}_e$ GW^{-1}s^{-1}, which mainly come from the beta decays of the fission products of ^{235}U, ^{238}U, ^{239}Pu, and ^{241}Pu. The emitted antineutrino spectrum is then given by: $S_{\rm tot}(E_\nu) = \sum_k f_k S_k(E_\nu)$ where f_k refers to the contribution of the main fissile nuclei to the total number of fissions of the k$^{\rm th}$ branch, and S_k to their corresponding neutrino spectrum per fission.

In preparation for the Double Chooz reactor experiment, we have re-evaluated the specific reactor antineutrino flux, S_k, (ν/fission), improving the electron to antineutrino data conversion [1]. The method relies on detailed knowledge of the decays of thousands of fission products, while the previous conversion procedure used a phenomenological model based on 30 effective beta branches. Both methods are constrained by the well-measured ILL spectrum of fission induced electrons that accompanies the antineutrinos [1]. This new approach provided a better handle on the systematic errors of the conversion. Although it did not reduce the final error budget, it led to a systematic shift of about 3% in the normalization of ^{235}U, ^{239}Pu, and ^{241}Pu antineutrino fluxes, respectively.

Accounting for new reactor antineutrino spectra [1] the normalization of predicted antineutrino rates, $\sigma_{f,k}^{\text{pred}}$, is shifted by +2.5%, +3.1%, +3.7%, +9.8% for k=^{235}U, ^{239}Pu, ^{241}Pu, and ^{238}U respectively. In the case of ^{238}U the completeness of nuclear databases over the years largely explains the +9.8% shift from the reference computations [1]. Our new computation takes into account the off-equilibrium correction [1] of the antineutrino fluxes.

3. Impact on past experimental results

In the eighties and nineties, experiments were performed at a few tens of meters from nuclear reactor cores at ILL, Goesgen, Rovno, Krasnoyarsk, Bugey (3 and 4) and Savannah River [2]. We only consider here experiments with baselines below 100 m to get rid of a possible (θ_{13}, Δm^2_{31}) driven oscillation effect at Palo Verde or CHOOZ. These experiments reported either the ratios (R) of the measured to predicted cross section per fission, or the observed event rate to the predicted rate. The prediction of the cross section per fission is defined as:

$$\sigma_f^{\text{pred}} = \int_0^\infty S_{\text{tot}}(E_\nu)\sigma_{\text{V-A}}(E_\nu)dE_\nu = \sum_k f_k \sigma_{f,k}^{\text{pred}}, \tag{1}$$

where the $\sigma_{f,k}^{\text{pred}}$ are the predicted cross sections for each fissile isotope, S_{tot} is the model dependent reactor neutrino spectrum for a given average fuel composition (f_k) and $\sigma_{\text{V}-\text{A}}$ is the theoretical cross section of reaction $\bar{\nu}_e + p \to e^+ + n$ (see [1, 2] for details). The new predicted cross section for any fuel composition can then be computed from Eq. (1).

The ratios of observed to predicted event rates (or cross section per fission), $R = N_{\text{obs}}/N_{\text{pred}}$, are summarized in Table 1. The observed event rates and their associated errors are unchanged with respect to the publications, the predicted rates are reevaluated separately in each experimental case. We observe a general systematic shift more or less significantly below unity. These reevaluations unveil a new *reactor antineutrino anomaly* [2], clearly illustrated in Fig. 1.

In order to quantify the statistical significance of the anomaly we can compute the weighted average of the ratios of expected over predicted rates, for all short baseline reactor neutrino experiments (including their possible correlations). We assume a 2.0% systematic uncertainty fully correlated among all 19 ratios in result of the common normalization uncertainty of the beta-spectra [2]. Potential experimental correlations coming from the use of similar or identical detector, detection technology and $\bar{\nu}_e$ source have also been taken into account as well as the slight non-gaussianity of the ratios R by enlarging to a small degree the error bars estimated through a Monte Carlo simulation. With the old antineutrino spectra the mean ratio is μ=0.976±0.024. With the new antineutrino spectra, we obtain μ=0.943±0.023, corresponding to a $-2.2\,\sigma$ effect. Clearly the new spectra induce a statistically significant deviation from the expectation.

Assuming the correctness of $\sigma_f^{\text{pred,new}}$ the anomaly could be explained by a common bias in all reactor neutrino experiments. The measurements used different detection techniques (scintillator counters and integral detectors). Neutrons were tagged either by their capture in metal-loaded scintillator, or in proportional counters, thus leading to two distinct systematics. As far as the neutron detection efficiency calibration is concerned, we note that different types of radioactive sources emitting MeV or sub-MeV neutrons were used (Am-Be, ^{252}Cf, Sb-Pu, Pu-Be). It should

Table 1. $N_{\rm obs}/N_{\rm pred}$ ratios based on old and new spectra. Off-equilibrium corrections have been applied when justified. The err column is the total error published by the collaborations including the error on $S_{\rm tot}$, the corr column is the part of the error correlated among experiments.

#	result	Det. type	τ_n (s)	^{235}U	^{239}Pu	^{238}U	^{241}Pu	old	new	err(%)	corr(%)	L(m)
1	Bugey-4	^{3}He+H_2O	888.7	0.538	0.328	0.078	0.056	0.987	0.942	3.0	3.0	15
2	ROVNO91	^{3}He+H_2O	888.6	0.614	0.274	0.074	0.038	0.985	0.940	3.9	3.0	18
3	Bugey-3-I	^{6}Li-LS	889	0.538	0.328	0.078	0.056	0.988	0.946	4.8	4.8	15
4	Bugey-3-II	^{6}Li-LS	889	0.538	0.328	0.078	0.056	0.994	0.952	4.9	4.8	40
5	Bugey-3-III	^{6}Li-LS	889	0.538	0.328	0.078	0.056	0.915	0.876	14.1	4.8	95
6	Goesgen-I	^{3}He+LS	897	0.620	0.274	0.074	0.042	1.018	0.966	6.5	6.0	38
7	Goesgen-II	^{3}He+LS	897	0.584	0.298	0.068	0.050	1.045	0.992	6.5	6.0	45
8	Goesgen-II	^{3}He+LS	897	0.543	0.329	0.070	0.058	0.975	0.925	7.6	6.0	65
9	ILL	^{3}He+LS	889	$\simeq 1$	—	—	—	0.832	0.802	9.5	6.0	9
10	Krasn. I	^{3}He+PE	899	$\simeq 1$	—	—	—	1.013	0.936	5.8	4.9	33
11	Krasn. II	^{3}He+PE	899	$\simeq 1$	—	—	—	1.031	0.953	20.3	4.9	92
12	Krasn. III	^{3}He+PE	899	$\simeq 1$	—	—	—	0.989	0.947	4.9	4.9	57
13	SRP I	Gd-LS	887	$\simeq 1$	—	—	—	0.987	0.952	3.7	3.7	18
14	SRP II	Gd-LS	887	$\simeq 1$	—	—	—	1.055	1.018	3.8	3.7	24
15	ROVNO88-1I	^{3}He+PE	898.8	0.607	0.277	0.074	0.042	0.969	0.917	6.9	6.9	18
16	ROVNO88-2I	^{3}He+PE	898.8	0.603	0.276	0.076	0.045	1.001	0.948	6.9	6.9	18
17	ROVNO88-1S	Gd-LS	898.8	0.606	0.277	0.074	0.043	1.026	0.972	7.8	7.2	18
18	ROVNO88-2S	Gd-LS	898.8	0.557	0.313	0.076	0.054	1.013	0.959	7.8	7.2	25
19	ROVNO88-3S	Gd-LS	898.8	0.606	0.274	0.074	0.046	0.990	0.938	7.2	7.2	18

be mentioned that the Krasnoyarsk, ILL, and SRP experiments operated with nuclear fuel such that the difference between the real antineutrino spectrum and that of pure ^{235}U was less than 1.5%. They reported similar deficits to those observed at other reactors operating with a mixed fuel. Hence the anomaly cannot be associated with a single fissile isotope neither with a single detection technique. All these elements argue against a trivial bias in the experiments, but a detailed analysis of the most sensitive of them, involving experts, would certainly improve the quantification of the anomaly. The other possible explanation of the anomaly is based on a real physical effect and is detailed in Section 5.

4. Other experimental results considered here

In addition to the rate information we included the shape on provided the Bugey-3 and ILL published data [2] in our combined analysis. We also considered the previously quoted anomalies affecting other short baseline electron neutrino experiments Gallex, Sage and MiniBooNE, reviewed in [4] to quantify the compatibility with these anomalies. We reanalyzed the Gallex and Sage calibration runs with ^{51}Cr and ^{37}Ar radioactive sources as in [5, 4] with an observed average deficit of $R_G = 0.86 \pm 0.06\,(1\sigma)$. However we included possible correlations between these four measurements leading to a slightly more conservative result, with the no-oscillation hypothesis disfavored at 97.7% C.L. We also reanalyzed the MiniBooNE electron neutrino excess assuming the very short baseline neutrino oscillation explanation of [4]. Further details of this analysis are provided in [2].

5. The fourth neutrino hypothesis

The reactor antineutrino anomaly could be explained through the existence of a fourth non-standard sterile neutrino. For simplicity we restrict our analysis to the 3+1 four-neutrino scheme. The three active neutrino masses are assumed to be separated from an isolated sterile neutrino mass by $|\Delta m^2_{\rm new}| \gg 10^{-2}$ eV^2, responsible for very short baseline reactor neutrino oscillations.

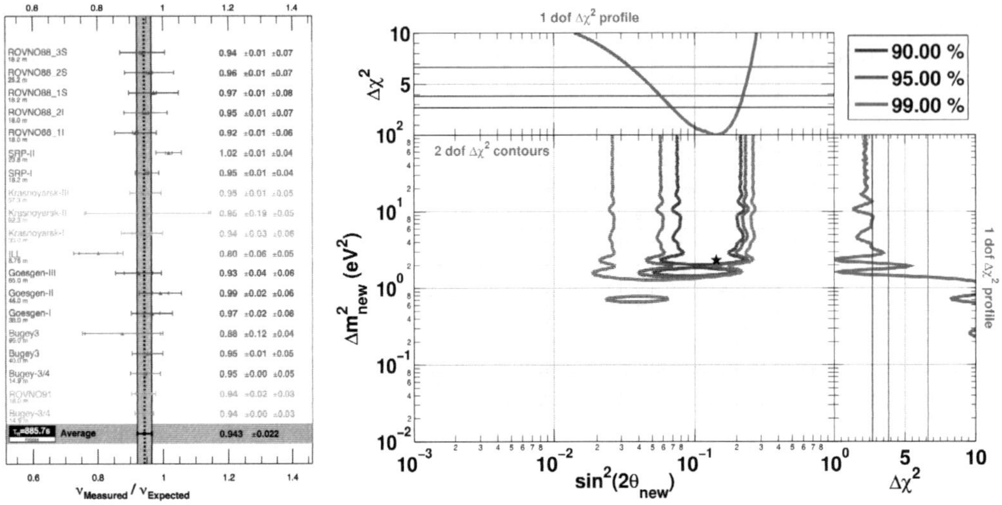

Figure 1. Left: weighted average (with correlations) of 19 measurements of reactor neutrino experiments operating at short baselines. A summary of experiment details is given in Table 1. Right: Allowed regions in the $\sin^2(2\theta_{\rm new}) - \Delta m^2_{\rm new}$ plane from the combination of reactor neutrino experiments, Gallex and Sage calibration sources experiments, MiniBooNE reanalysis, and the ILL and Bugey-3-energy spectra. The data are well fitted by the 3+1 neutrino hypothesis, while the no-oscillation hypothesis is disfavored at 99.8% C.L.

For energies above the inverse beta decay threshold and baselines below 100 m, we adopt the approximated oscillation formula, $P_{ee} = 1 - \sin^2(2\theta_{\rm new})\sin^2\left(\Delta m^2_{\rm new} L/4E_{\bar{\nu}_e}\right)$.

The ILL experiment may have seen a hint of oscillation in their measured positron energy spectrum [2], but Bugey-3's results do not point to any significant spectral distortion more than 15 m away from the antineutrino source (note that the large spatial extension of the Bugey nuclear core is sufficient to wash out most of the oscillation pattern). Hence, in a first approximation, hypothetical oscillations could be seen as an energy-independent suppression of the $\bar{\nu}_e$ rate by a factor of $\frac{1}{2}\sin^2(2\theta_{\rm new})$, thus leading to $\Delta m^2_{\rm new} \gtrsim 1$ eV^2. Considering the weighted average of all reactor experiments we get an estimate of the mixing angle, $\sin^2(2\theta_{\rm new}) \sim 0.115$. The ILL positron spectrum is thus in agreement with the oscillation parameters found independently in our re-analyses, mainly based on rate information. Because of the differences in the systematic effects in the rate and shape analyses, this coincidence is in favor of a true physical effect rather than an experimental anomaly.

The no-oscillation hypothesis is disfavored at 99.8% C.L. The significance is dominated by the gallium and reactor data. Allowed regions in the $\sin^2(2\theta_{\rm new}) - \Delta m^2_{\rm new}$ plane are displayed in Fig. 1, together with the marginal $\Delta\chi^2$ profiles for $|\Delta m^2_{\rm new}|$ and $\sin^2(2\theta_{\rm new})$. The combined fit leads to the following constraints on oscillation parameters: $|\Delta m^2_{\rm new}| > 1.5$ eV^2 (95% C.L.) and $\sin^2(2\theta_{\rm new}) = 0.14 \pm 0.08$ (95% C.L.).

References

[1] Mueller Th A *et al.* 2011 *Phys. Rev.* C **83** 054615 (*Preprint* arXiv:1101.2663 [hep-ex])
[2] Mention G *et al.* 2011 *Phys. Rev.* D **83** 073006 (*Preprint* arXiv:1101.2755 [hep-ex])
[3] Nakamura K *et al.* 2010 (Particle Data Group) *J. Phys.* G **37** 075021.
[4] Giunti C and Laveder M 2010 *Phys. Rev.* D **82** 053005 (*Preprint* arXiv:1005.4599 [hep-ph])
[5] Giunti C and Laveder M 2010 *Phys. Rev.* C **83** 065504 (*Preprint* arXiv:1006.3244v2 [hep-ph])

Journal of Physics: Conference Series **408** (2013) 012026 doi:10.1088/1742-6596/408/1/012026

MINOS search for sterile neutrinos

Alexandre B Sousa[1]

Department of Physics, Harvard University, Cambridge, Massachusetts 02138, USA

E-mail: asousa@physics.harvard.edu

Abstract. Using a NuMI beam exposure of 7.1×10^{20} protons-on-target, the MINOS long-baseline experiment has performed a search for active to sterile neutrino mixing over a distance of 735 km. Details of the analysis are provided, along with results from comparisons with standard three neutrino oscillations and fits to a 3+1 model including oscillations into one sterile neutrino. An outlook on the future sterile neutrino related contributions from MINOS and the proposed MINOS+ project is also presented.

1. Introduction

The possible existence of one or more light sterile neutrinos, in addition to the three active flavors, has been put forward as an explanation for the observation of an excess appearance of $\bar{\nu}_e$ in a $\bar{\nu}_\mu$ beam over a short baseline, as reported by the LSND experiment. Interest in sterile neutrinos has been rekindled with observations from antineutrino running in the MiniBooNE experiment that are consistent with active to sterile neutrino oscillations driven by at least one additional large mass-square difference $\mathcal{O}(1\,\text{eV}^2)$[1]. MINOS can probe active to sterile neutrino mixing by measuring the rate of neutral-current (NC) events at two locations, over a baseline of 735 km. Because NC cross-sections are identical among the three active flavors, NC event rates are unaffected by standard neutrino mixing. However, oscillations into a sterile non-interacting neutrino flavor would result in an energy-dependent depletion of NC events at the far site.

2. The NuMI beam and the MINOS detectors

MINOS measures neutrinos from the NuMI beam using two detectors: the 980 ton (27 ton fiducial) Near Detector (ND), located 1.04 km downstream of the beam target at Fermilab; and the 5.4 kton (4.0 kton fiducial) Far Detector (FD), placed 735 km downstream of the target in the Soudan Underground Laboratory, in Minnesota [2]. The energy resolution function for neutrino-induced hadronic showers is approximately $56\%/\sqrt{E}$. The results presented here were obtained using an exposure of 7.07×10^{20} protons on target taken exclusively with a beam configuration for which the peak neutrino event energy is 3.3 GeV. The NuMI beam includes a 1.3% $(\nu_e + \bar{\nu}_e)$ contamination arising from the decay of muons originating in kaon and pion decays.

[1] On behalf of the MINOS Collaboration.

3. Sterile neutrino analysis

In the MINOS detectors, NC interactions give rise to events with a short diffuse hadronic shower and either small or no tracks, whereas charged-current (CC) events typically display a long muon track accompanied by hadronic activity at the event vertex. The separation between NC and CC events proceeds through selection criteria based on topological variables: events crossing fewer than 47 planes for which no track is reconstructed are selected as NC; events crossing fewer than 47 planes that contain a track are classified as NC only if the track extends less than 6 planes beyond the shower. These selections result in an NC-selected sample with 89% efficiency and 61% purity. Highly inelastic ν_μ and $\bar{\nu}_\mu$ CC events, where the muon track is not distinguishable from the hadronic shower, are the main source of background for the NC-selected spectrum. Furthermore, the analysis classifies 97% of ν_e-induced CC events as NC, requiring the possibility of ν_e appearance to be considered when extracting results. The predicted NC energy spectrum in the FD is obtained using the ND data. An estimate of the ratio of events in the FD and ND as a function of reconstructed energy, E_{reco}, is calculated from Monte Carlo simulations. The ratio is multiplied by the observed ND energy spectrum to produce the predicted FD spectrum. To avoid biases, the analysis selections and procedures were determined prior to examining the FD data, following the precepts of a blind analysis. Figures 1 and 2 show the reconstructed energy spectra in each detector.

4. Results

4.1. Comparisons with three-flavor oscillation scenario

The selection procedures identify 802 NC interaction candidates in the FD, with $754 \pm 28(\text{stat}) \pm 37(\text{syst})$ events expected from standard three-flavor mixing (assuming $\theta_{13} = 0°$) [5]. The NC-like reconstructed energy spectrum is shown on Fig. 2. The agreement between the observed

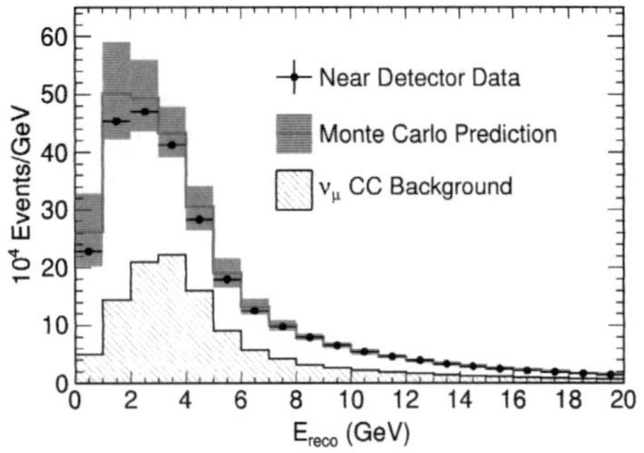

Figure 1. Reconstructed energy spectrum of NC-selected events in the ND compared to the Monte Carlo prediction shown with 1σ systematic errors (red band). Also displayed is the simulation of the background from misidentified CC events (hatched histogram).

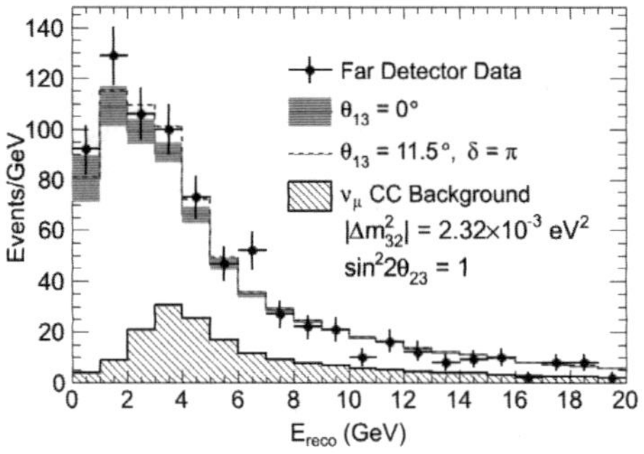

Figure 2. Reconstructed energy spectrum of NC-selected events in the FD compared with predictions for standard three-flavor mixing with and without ν_e appearance at the MINOS 90% C.L. limit [3].

and predicted NC spectra is quantified using the statistic $R \equiv \frac{N_{\text{data}} - B_{\text{CC}}}{S_{\text{NC}}}$, where N_{data} is the observed number of events, B_{CC} is the predicted CC background from all flavors, and S_{NC} is the expected number of NC interactions. The values of N_{data}, S_{NC} and contributions to B_{CC} for various reconstructed energy ranges are shown in Table 1. The values of R for each energy range show no evidence of a depletion in the NC rate at the FD, supporting the hypothesis that standard three-flavor oscillations explain the data.

Table 1. Values of the R statistic and its components for several reconstructed energy ranges. The numbers shown in parentheses include ν_e appearance with $\theta_{13} = 11.5°$ and $\delta_{CP} = \pi$. The displayed uncertainties are statistical, systematic, and the uncertainty associated with ν_e appearance.

E_{reco} (GeV)	N_{Data}	S_{NC}	$B^{\nu_\mu}_{\text{CC}}$	$B^{\nu_\tau}_{\text{CC}}$	$B^{\nu_e}_{\text{CC}}$
$0-3$	327	248.4	33.2	3.2	3.1 (21.5)
$3-120$	475	269.6	156.0	9.2	31.2 (53.8)
$0-3$	$R = 1.16 \pm 0.07 \pm 0.08 - 0.08(\nu_e)$				
$3-120$	$R = 1.02 \pm 0.08 \pm 0.06 - 0.08(\nu_e)$				
$0-120$	$R = 1.09 \pm 0.06 \pm 0.05 - 0.08(\nu_e)$				

4.2. Oscillation fits with a 3+1 sterile neutrino model

The data are compared with a neutrino oscillation model that allows admixture with one sterile neutrino. In this model, an additional mass scale Δm^2_{43} with magnitude $\mathcal{O}(1\,\text{eV}^2)$ is introduced, along with the assumption that no oscillation-induced change of the neutrino event rate is measurable at the ND site, but rapid oscillations are predicted at the FD location. Based on the magnitude of the systematic uncertainties at the ND, this approximation is assumed to be valid for $0.3 < \Delta m^2_{43} < 2.5\,\text{eV}^2$. A detailed description of this model is provided in Ref. [4] and references therein. Both the NC-selected energy spectrum shown in Fig. 2 and the CC-selected spectrum in the FD data are used in the fits to the oscillation models. Limits on the sterile mixing angles $\theta_{34} < 26°\,(37°)$ and $\theta_{24} < 7°\,(8°)$ are obtained at the 90% C.L.. The numbers in parentheses represent the limits extracted for maximal ν_e appearance. The latter result is presently the most stringent constraint on θ_{24} for $\Delta m^2_{43} \sim 1\,\text{eV}^2$. A comparison of the MINOS result with other disappearance results is shown in Fig. 3. The coupling between active and

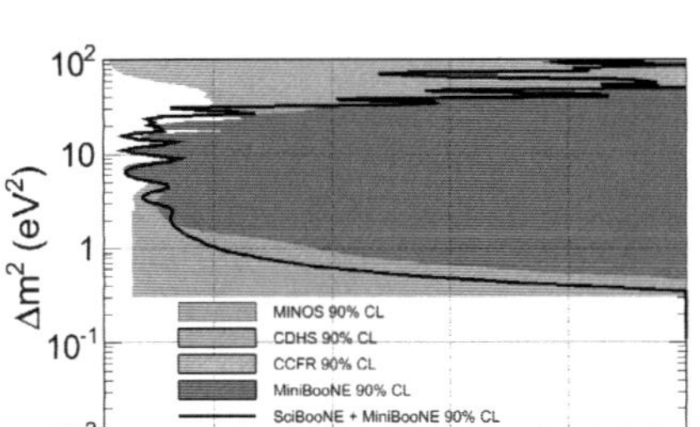

Figure 3. MINOS exclusion compared to MiniBooNE, CDHS, CCFR ν_μ disappearance results. The MINOS 90%CL excluded region is shown in green. Following the discussion in the text, the bound is assumed to be valid for $0.3 < \Delta m^2_{43} < 2.5\,\text{eV}^2$.

sterile neutrinos may also be quantified in terms of the fraction of disappearing ν_μ that oscillate into ν_s, $f_s \equiv P_{\nu_\mu \to \nu_s}/(1 - P_{\nu_\mu \to \nu_\mu})$, where the $P_{\nu_\mu \to \nu_x}$ refer to neutrino oscillation probabilities. MINOS places the most stringent constraint to date on this quantity, $f_s < 0.22\,(0.40)$ at the 90% C.L., where the number in parentheses denotes the limit assuming ν_e appearance.

A recent paper by Hernandez and Smirnov [6] analyzed the MINOS sterile neutrino search results in detail and points out that for large enough values of Δm_{43}, oscillations at the ND, even if not visible due to the size of systematic uncertainties on the ND energy spectrum, could affect the FD predicted spectrum and reduce the strength of the constraint on θ_{24} for $\Delta m^2_{43} > 1\,\text{eV}^2$. Moreover, in the same paper and elsewhere in these proceedings [7, 8], it is pointed out that the combination of the MINOS bound on θ_{24} with the Bugey bound on θ_{14} largely rules out the

LSND signal region for $0.3 < \Delta m^2_{43} < 1\,\text{eV}^2$. Motivated by these remarks, MINOS is working on the inclusion of ND oscillations into the sterile mixing models and will update the analysis with reviewed results in the near future.

5. Outlook

The valuable contributions that MINOS is making to the body of knowledge on sterile neutrino searches will be continued and improved upon by the proposed MINOS+ project. In MINOS+, the MINOS detectors will continue operation during the NOνA era (starting in 2013), using the NuMI beam upgraded to 700 kW of beam power. MINOS+ will run for three years in neutrino and three years in antineutrino mode. The expected sensitivities to θ_{24} are shown in Figs. 4 and 5 for neutrino and antineutrino running. MINOS+ can improve the MINOS limits by a a factor of two and place a stringent limit on antineutrino disappearance over short baselines for $\Delta m^2_{43} \sim 1\,\text{eV}^2$. Before the MINOS+ era, MINOS also plans to run the sterile neutrino

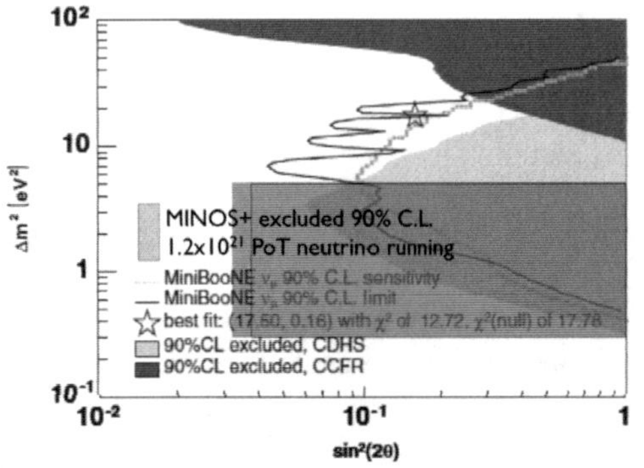

Figure 4. Expected MINOS+ 90% C.L. exclusion regions for the θ_{24} mixing angle compared to other disappearance measurements for neutrino running.

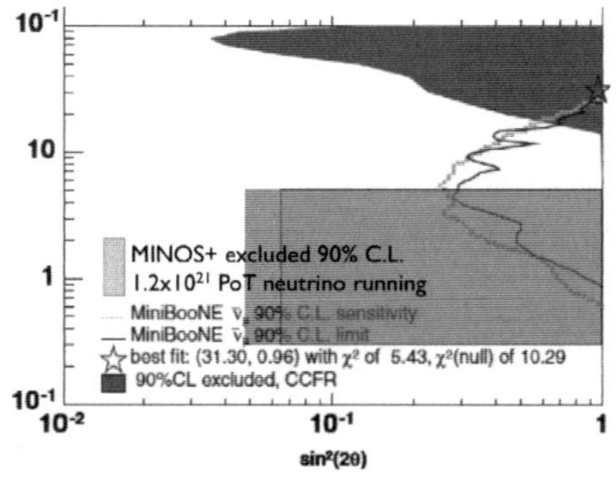

Figure 5. Expected MINOS+ 90% C.L. exclusion regions for the θ_{24} mixing angle compared to other disappearance measurements for antineutrino running.

analysis on the 3.3×10^{20} POT of antineutrino data already collected. The main challenges for this analysis arise from limited statistics and feed-down NC interactions from the higher energy neutrino component, which accounts for 58% of the beam composition when the NuMI horns focus π^- and K^-. Results of the analysis are expected during 2012.

6. References

[1] Durjcic Z, these Proceedings.
[2] Michael D G *et al.* (MINOS), 2008 *Nucl. Instrum. Meth.* A **596** 190.
[3] Adamson P *et al.* (MINOS), 2010 *Phys. Rev.* D **82** 051102(R).
[4] Adamson P *et al.* (MINOS), 2010 *Phys. Rev.* D **81** 052004.
[5] Adamson P *et al.* (MINOS), 2011 *Phys. Rev. Lett.* **107** 011802.
[6] Hernandez D and Smirnov A Y, 2012 *Phys. Lett.* B, **706** 360.
[7] Smirnov A Y, these Proceedings.
[8] Giunti C, these Proceedings.

XIII International Workshop on Neutrino Factories, Super beams and Beta beams (NUFACT11) IOP Publishing
Journal of Physics: Conference Series **408** (2013) 012027 doi:10.1088/1742-6596/408/1/012027

MiniBooNE oscillation results 2011

Zelimir Djurcic

Argonne National Laboratory, 9700 South Cass Avenue, Argonne, IL 60439, USA

E-mail: `zdjurcic@hep.anl.gov`

Abstract. The MiniBooNE neutrino oscillation search experiment at Fermilab has recently updated results from a search for $\bar{\nu}_\mu \to \bar{\nu}_e$ oscillations, using a data sample corresponding to 8.58×10^{20} protons on target in anti-neutrino mode. These high statistics results represent an increase in statistics of 52% compared to the results published in 2010. An excess of 57.7 ± 28.5 events is observed in the energy range 200 MeV $< E_\nu <$ 3000 MeV. The data favor LSND-like $\bar{\nu}_\mu \to \bar{\nu}_e$ oscillations over a background only hypothesis at 91.1% confidence level in the energy range $475 < E_\nu <$3000 MeV.

1. Introduction

Motivated by the LSND observation of an excess of observed $\bar{\nu}_e$ events above background prediction in a $\bar{\nu}_\mu$ beam [1], the MiniBooNE experiment was designed to test the neutrino oscillation interpretation of the LSND signal in both neutrino and anti-neutrino modes. The MiniBooNE collaboration has performed a search for $\nu_\mu \to \nu_e$ oscillations with 6.486×10^{20} protons on target (POT), the results of which showed no evidence of an excess of ν_e events for neutrino energies above 475 MeV [2]. Despite having observed no evidence for oscillations above 475 MeV, the MiniBooNE $\nu_\mu \to \nu_e$ search observed an excess of 128.8±43.4 events at low energy, between 200-475 MeV [3]. Although the excess is incompatible with LSND-type oscillations within the simple two neutrino oscillation framework, several hypotheses, including sterile neutrino oscillations with CP violation, anomaly-mediated neutrino-photon coupling, and many others, have been proposed that provide a possible explanation for the excess itself [4]. In some cases, these theories offer the possibility of reconciling the MiniBooNE ν_e excess with the LSND $\bar{\nu}_e$ excess. A search in antineutrino mode provides a more direct test of the LSND signal, which was observed with antineutrinos. The MiniBooNE collaboration has published a search for $\bar{\nu}_\mu \to \bar{\nu}_e$ oscillations with 5.66×10^{20} POT, the results of which showed an evidence of an excess of $\bar{\nu}_e$ events for neutrino energies above 475 MeV [6]. The allowed regions from the fit, shown in Fig. 1, are consistent with $\bar{\nu}_\mu \to \bar{\nu}_e$ oscillations in the 0.1 to 1 eV^2 Δm^2 range and consistent with the allowed region reported by the LSND experiment [1]. The data show an excess of 43.2 ± 22.5 events: 277 electron-like events have been observed in $200 < E_\nu < 3000$ MeV reconstructed energy range, compared to an expectation of $233.8 \pm 15.3(stat) \pm 16.5(syst)$ events [6]. In the energy range $475 < E_\nu < 1250$ MeV, the observed $\bar{\nu}_e$ events, when constrained by the $\bar{\nu}_\mu$ data events, have a $\chi^2/DF = 18.5/6$ and a probability of 0.5% for a background-only hypothesis.

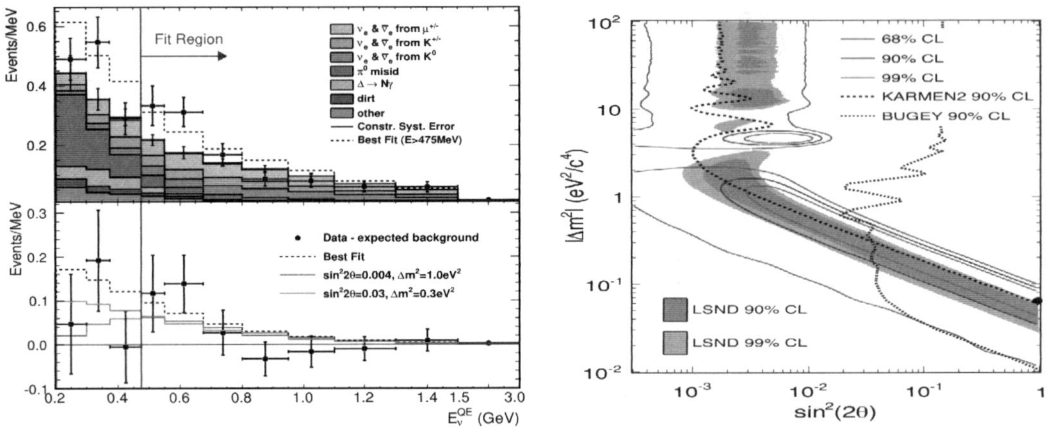

Figure 1. Results obtained with 5.66×10^{20} POT exposure. *Top Left*: Reconstructed E_ν distribution of $\bar{\nu}_e$ CCQE candidates in MiniBooNE anti-neutrino running. *Bottom Left*: The difference between the data and predicted backgrounds as a function of reconstructed neutrino energy. The error bars include both statistical and systematic components. Also shown in the figure are expectations from the best oscillation fit with $E_\nu > 475$ MeV, $(\Delta m^2, \sin^2 2\theta) = (0.064$ eV^2, 0.96) where the fit is extrapolated below 475 MeV and from other neutrino oscillation parameter sets in the LSND allowed region. *Right*: MiniBooNE 68%, 90%, and 99% C.L. allowed regions for events with $E_\nu > 475$ MeV within a two neutrino $\bar{\nu}_\mu \to \bar{\nu}_e$ oscillation model, obtained with 5.66×10^{20} POT exposure [6]. The shaded areas show the 90% and 99% C.L. LSND allowed regions. The black dot shows the best fit point.

2. New $\bar{\nu}_\mu \to \bar{\nu}_e$ results

In this report we describe the latest unpublished results that the MiniBooNE collaboration updated in the search for $\bar{\nu}_\mu \to \bar{\nu}_e$ oscillations, using a data sample corresponding to 8.58×10^{20} POT exposure. The analysis technique used here was already described [3, 5, 6] and assumes only $\bar{\nu}_\mu \to \bar{\nu}_e$ oscillations with no $\bar{\nu}_\mu$ disappearance and no ν_μ oscillations. The signature of $\bar{\nu}_\mu \to \bar{\nu}_e$ oscillations is an excess of $\bar{\nu}_e$-induced charged-current quasi-elastic (CCQE) events. Fig. 2 (top left) shows the reconstructed E_ν distribution of observed $\bar{\nu}_e$ CCQE candidates and background expectation. The data show an excess of 57.7 ± 28.5 events: 412 electron-like events have been

Table 1. The number of data, fitted background, and excess events in the $\bar{\nu}_e$ appearance analysis for different E_ν ranges. The uncertainties include both statistical and constrained systematic errors. All known systematic errors are included in the systematic error estimate.

E_ν range [MeV]	Data	Background	Event Excess
200 - 475	189	$150.4 \pm 12.3 \pm 13.9$	38.6 ± 18.5
475 - 675	80	$58.3 \pm 7.6 \pm 5.5$	21.7 ± 9.4
475 - 1250	168	$151.7 \pm 12.3 \pm 15.0$	16.3 ± 19.4
475 - 3000	223	$203.9 \pm 14.3 \pm 20.2$	19.1 ± 24.7
200 - 3000	412	$354.3 \pm 18.8 \pm 21.4$	57.7 ± 28.5

observed in $200 < E_\nu < 3000$ MeV reconstructed energy range, compared to an expectation of $354.3 \pm 18.8(stat) \pm 21.4(syst)$ events (see also Table 1). Fig. 2 (bottom left) shows the event excess as a function of E_ν. Using a likelihood-ratio technique, the best MiniBooNE oscillation

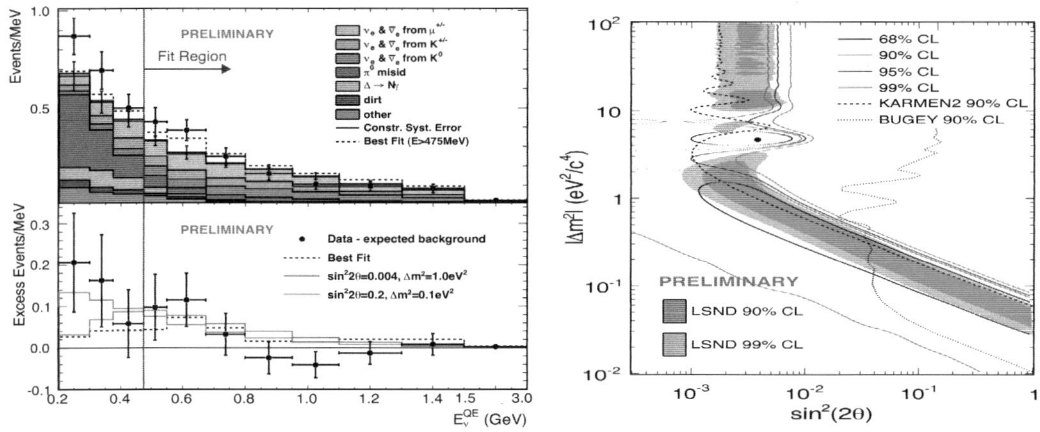

Figure 2. *Top Left*: Reconstructed E_ν distribution of $\bar{\nu}_e$ CCQE candidates in MiniBooNE antineutrino running. *Bottom Left*: The difference between the data and predicted backgrounds as a function of reconstructed neutrino energy. The error bars include both statistical and systematic components. Also shown in the figure are expectations from the best oscillation fit with $E_\nu > 475$ MeV, $(\Delta m^2, \sin^2 2\theta) = (4.64 \text{ eV}^2, 0.00337)$ where the fit is extrapolated below 475 MeV and from other neutrino oscillation parameter sets in the LSND allowed region. *Right*: MiniBooNE 68%, 90%, and 99% C.L. allowed regions for events with $E_\nu > 475$ MeV within a two neutrino $\bar{\nu}_\mu \to \bar{\nu}_e$ oscillation model. The black dot shows the best fit point.

fit for $475 < E_\nu < 3000$ MeV occurs at $(\Delta m^2, \sin^2 2\theta) = (4.64 \text{ eV}^2, 0.00337)$. The energy range $E_\nu > 475$ MeV has been chosen for the fit as this is the energy range MiniBooNE used for searching for oscillations in neutrino mode. Also, this energy range avoids the region of the unexplained low energy excess in neutrino mode [3]. The χ^2 for the best-fit point in the energy range of $475 < E_\nu < 1250$ MeV is 4.3 for 4 DF, corresponding to a χ^2-probability of 36%. The probability of the background-only fit relative to the best oscillation fit is 8.9%. Fig. 2 (right) shows the MiniBooNE 68%, 90%, and 99% C.L. closed contours for $\bar{\nu}_\mu \to \bar{\nu}_e$ oscillations in the $475 < E_\nu < 3000$ MeV energy range. Both the old and updated results are consistent with the LSND oscillation region although the updated result has a much reduced significance for LSND-like oscillation signal. The updated results shown an excess of 38.6 ± 18.5 events in the low energy region $200 < E_\nu < 475$ MeV, more prominent than in the previous result. Part of this excess may be attributed to the neutrino-mode low energy excess [3], given 22% neutrino contribution to the beam in antineutrino-mode. With the oscillation fit region extended down to 200 MeV, the MiniBooNE closed contours for $\bar{\nu}_\mu \to \bar{\nu}_e$ oscillations are similar as shown in Fig 3 (left). Fig 3 (right) shows the fit with the subtraction of 17 events expected assuming the low energy excess scales with neutrino component of the beam. The best oscillation fit point without and with this subtraction correspond to $(\Delta m^2, \sin^2 2\theta) = (4.64 \text{ eV}^2, 0.0045)$ and $(\Delta m^2, \sin^2 2\theta) = (4.64 \text{ eV}^2, 0.0037)$, respectively.

3. Conclusion and next steps

The MiniBooNE experiment observes an excess of 57.7 ± 28.5 events in the full energy range 200 MeV $< E_\nu <$ 3000 MeV for a data sample corresponding to 8.58×10^{20} POT. The allowed regions from the fit, shown in Fig. 2, are consistent with $\bar{\nu}_\mu \to \bar{\nu}_e$ oscillations in the 0.1 to 1 eV^2 Δm^2 range. The data favor LSND-like oscillations over a background only hypothesis at 91.1% confidence level. With new data update excess is indicated at low energy, as with neutrinos. The MiniBooNE antineutrino results are statistically limited and a larger data sample will be needed

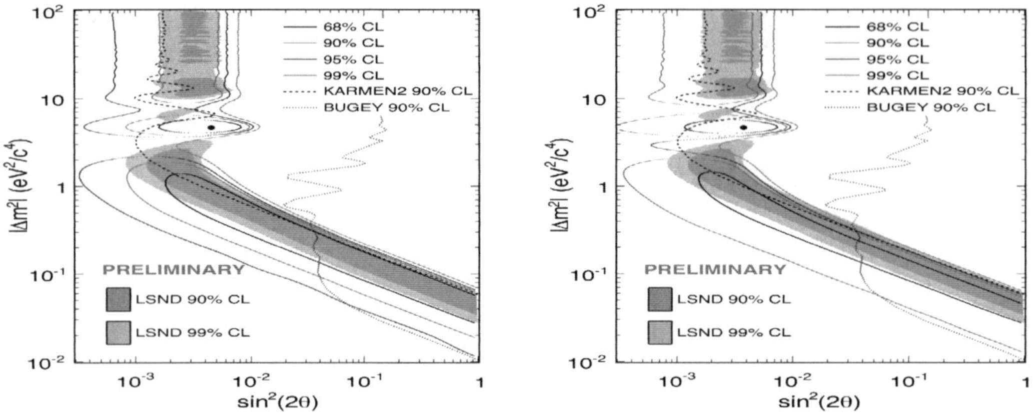

Figure 3. *Left*: MiniBooNE 90% and 99% C.L. allowed regions for events with $E_\nu > 200$ MeV within a two neutrino $\bar{\nu}_\mu \to \bar{\nu}_e$ oscillation model. The shaded areas show the 90% and 99% C.L. LSND allowed regions. *Right*: MiniBooNE 90% and 99% C.L. allowed regions for events with $E_\nu > 200$ MeV within a two neutrino $\bar{\nu}_\mu \to \bar{\nu}_e$ oscillation model, with the subtraction of the expected 17 event excess in the $200 < E_\nu < 475$ MeV low-energy region from the neutrino component of the beam.

to make more definitive statements. MiniBooNE's requested additional running to reach close to 15×10^{20} POT and significantly increase the current data statistics before anticipated accelerator shutdown at Fermilab expected in Spring 2012. In the next step a combined $\bar{\nu}_\mu \to \bar{\nu}_e$ and $\nu_\mu \to \nu_e$ analysis will be performed. In addition, several proposed short-baseline experiments [7] will be sensitive to $\bar{\nu}_\mu \to \bar{\nu}_e$ oscillations in the 0.1 to 1 eV^2 Δm^2 range.

Acknowledgments

We would like to acknowledge the support of Fermilab, the Department of Energy, and the National Science Foundation.

References

[1] Athanassopoulos C *et al.* 1995 *Phys. Rev. Lett.* **75** 2650; Athanassopoulos C *et al.* (LSND Collaboration) 1996 *Phys. Rev. Lett.* **77** 3082; Athanassopoulos C *et al.* (LSND Collaboration) 1998 *Phys. Rev. Lett.* **81** 1774; Athanassopoulos C *et al.* (LSND Collaboration) 1998 *Phys. Rev.* C **58**; Aguilar A *et al.* (LSND Collaboration) 2001 *Phys. Rev.* D **64** 112007

[2] Aguilar-Arevalo A A *et al.* (MiniBooNE Collaboration) 2007 *Phys. Rev. Lett.* **98** 231801

[3] Aguilar-Arevalo A A *et al.* (MiniBooNE Collaboration) 2009 *Phys. Rev. Lett.* **102** 101802; Djurcic Z 2009 *MiniBooNE Oscillation Results Proc. Rencontres de Moriond session Electroweak Interactions And Unified Theories Preprint* arXiv:0907.3747[hep-ex]

[4] Karagiorgi G *et al.* 2009 *Phys. Rev.* D **80** 073001; Harvey J A *et al.* 2007 *Phys. Rev. Lett.* **99** 261601; Harvey J A *et al.* 2008 *Phys. Rev.* D **77** 085017; Gninenko S N and Gorbunov D S 2010 *Phys. Rev.* D **81** 075013; Sorel M *et al.* 2004 *Phys. Rev.* D **70** 073004; Karagiorgi G *et al.* 2007 *Phys. Rev.* D **75** 013011; Melchiorri A *et al.* 2009 *JCAP 0901* 036; Maltoni M and Schwetz T 2007 *Phys. Rev.* D **76** 093005; Pas H *et al.* 2005 *Phys. Rev.* D **72** 095017; Goldman T *et al.* 2007 *Phys. Rev.* D **75** 091301; Barger V, Marfatia D and Whisnant K 2003 *Phys. Lett.* B **576** 303; Akhmedov E and Schwetz T *Preprint* arXiv:1007.4171[hep-ph]; Nelson A E and Walsh J 2008 *Phys. Rev.* D **77** 033001; Kostelecky V A and Mewes M 2004 *Phys. Rev.* D **69** 016005; Katori T *et al.* 2006 *Phys. Rev.* D **74** 105009

[5] Aguilar-Arevalo A A *et al.* 2009 *Phys. Rev. Lett.* **103** 111801

[6] Aguilar-Arevalo A A *et al.* 2010 *Phys. Rev. Lett.* **105** 181801; Djurcic Z 2011 *AIP Conf. Proc.* **1382** 91

[7] Baibussinov B *et al.* 2009 *Preprint* arXiv:0909.0355[hep-ex]; Stancu I *et al.* 2009 (*Preprint* arXiv:0910.2698[hep-ex]); Agarwalla S K and Huber P 2010 *Preprint* arXiv:1007.3228[hep-ph]

XIII International Workshop on Neutrino Factories, Super beams and Beta beams (NUFACT11) IOP Publishing
Journal of Physics: Conference Series **408** (2013) 012028 doi:10.1088/1742-6596/408/1/012028

The Status of the MicroBooNE Experiment

B J P Jones

Laboratory for Nuclear Science, Massachusetts Institute of Technology, 77 Massachusetts Ave., Cambridge, MA 02139, United States of America

E-mail: bjpjones@mit.edu

Abstract. The MicroBooNE experiment is an upcoming liquid argon TPC experiment which will pursue a program of short baseline neutrino physics in the Fermilab Booster Neutrino Beam, starting in early 2014. I give a description of the experiment and review the current status of various aspects of the MicroBooNE detector.

1. Introduction

The Booster Neutrino Beamline (BNB) is a conventional neutrino beam at Fermilab which supplies a family of experiments with neutrinos of peak energy around 1 GeV. The MiniBooNE experiment in the BNB has reported an excess of ν_e like events at low energy in both neutrino and antineutrino mode [1], and at intermediate energies in antineutrino mode [2]. Many explanations have been proposed, including new oscillation physics, photon production in neutrino scattering events and the presence of misunderstood instrumental effects. The MicroBooNE experiment aims to confirm or refute the existence of these excesses with a totally different detector technology. Unlike MiniBooNE, which cannot differentiate between photon and electron induced Cherenkov rings, the MicroBooNE liquid argon TPC will be able to distinguish between the two event types and hence clarify the nature of the excesses and the evidence they provide for new physics.

The MicroBooNE detector will sit 100 m upstream of MiniBooNE in the BNB and plans to accumulate 6.6×10^{20} protons on target in neutrino mode. This will provide enough neutrino interactions to identify the source of the low energy excess, if observed, as photons or electrons with a confidence of 4σ (photons) or 5σ (electrons). MicroBooNE will also accumulate similar statistics to MiniBooNE in order to search for sterile neutrino oscillations in neutrino mode. Further, MicroBooNE will perform world-leading measurements of neutrino cross sections on argon. As well as being of great importance to future liquid argon detectors, these cross section measurements, especially for low track energies, can provide insights into physics topics such as the strange spin of the proton, and the effects of short range nuclear correlations.

The detector is comprised of a stainless steel cryostat filled with 170 tons of liquid argon, of which 70 tons forms the active volume. This volume is instrumented with a three wireplane time projection chamber (TPC) of 8256 channels and an optical system comprising of 30 photomultiplier tubes assemblies.

In addition to pursuing a diverse program of short baseline neutrino physics, MicroBooNE is an important stepping stone towards a giant liquid argon detector for a future long baseline experiment such as LBNE [3].

2. Operating Principals of Liquid Argon TPCs

2.1. Charge collection

Neutrino interactions in MicroBooNE produce zero or more charged particles with a momentum in the beam direction. These particles traverse the active volume and liberate ionization electrons from nearby argon atoms. Since argon is inert, if impurity levels are kept below O(100 ppt), an electron drift of 2.5 m can be achieved using an electric field of 500 V/cm.

Along this side are arranged several wireplanes at different angles perpendicular to the drift direction. MicroBooNE has two induction planes and one collection plane. Voltages are induced upon the wires by charges passing an induction wire or being absorbed by a collection wire. By recording of the voltage across each wire, a 3D image of the original charge deposit can be constructed, with the wire numbers providing information about the charge distribution parallel to the wire planes, and the drift time providing the perpendicular coordinate.

2.2. Light collection

Charged particles moving through liquid argon also produce scintillation and Cherenkov light. MicroBooNE incorporates an optical system to measure this light, which gives several benefits.

The BNB is a pulsed beam, and has structure on a microsecond timescale. This is also the timescale associated with charge drift through the TPC volume, during which charge losses due to electron absorption, and dispersion via diffusion and space charge effects take place. In order to obtain the absolute drift distance and apply the corrections necessary for an accurate charge determination, a precise measurement of t_0 is required. The prompt scintillation light gives a t_0 measurement with a few ns precision.

An optical trigger also allows for a great simplification in trigger electronics due to the far smaller number of channels present in the optical system, provides stronger rejection of cosmic ray events falling outside the beam window and helps mitigate against TPC noise effects.

An interesting quality of the scintillation light in argon is that it is produced via two pathways, both involving molecular excimer states of argon, one mediated by a singlet excitation and one by a triplet state [4]. These excimers eventually decay into two argon atoms and one 128 nm photon, but have very different lifetimes. As such, there are two time constants associated with the scintillation light: a fast component with $\tau_F = 6ns$ and a slow component with $\tau_S = 1.6\mu s$. A competing excimer collision process results in scintillation quenching which is dependent upon the local excimer concentration, and hence the dE/dx of the ionizing particle. This process more strongly quenches the slow light, hence the ratio of prompt to delayed light is dependent upon the identity and energy of the ionizing particle. The exploitation of this dependence in augmenting the reconstruction of neutrino events is being investigated.

3. Current status of MicroBooNE subsystems

3.1. TPC

The MicroBooNE TPC system will consist of three wire planes, oriented vertically (3456 wires) and at ± 60 deg (2400 wires each). Each wire is tested for tensile strength and then wound onto a ferrule fixed to a carrier board at each end by an automated wire winding machine. Wire winding is set to begin this fall at Syracuse and Yale Universities, and boards will be affixed to the TPC frame at Fermilab in Spring 2012.

Simulations of electric field uniformity inside the field cage and the effects of field inhomogeneities upon charge trajectories have been performed. Image distortions due to field non-uniformity are expected to be no greater than 1 mm at any point in the active volume.

3.2. Light collection system

The MicroBooNE optical system is comprised of 30 8 inch cryogenic photomultiplier tubes (PMTs) mounted behind acrylic plates which are coated with a layer of tetraphenyl butadiene

Journal of Physics: Conference Series **408** (2013) 012028 doi:10.1088/1742-6596/408/1/012028

Figure 1. Left to Right: The MicroBooNE cryostat in situ with surrounding cryogenics, the BNL automated wire winding machine, a mechanical model of one optical system PMT assembly, cutaway of cryostat showing field cage and feedthrough arrangement

(TPB) enriched polystyrene. TPB is a wavelength shifter which absorbs light over a wide range of wavelengths and emits in the visible. One PMT, wavelength shifting plate, plastic support stand, cryogenic base and combined signal and high voltage cable comprises one self contained PMT assembly.

The PMT assemblies will be arranged on a rack which slides behind the TPC wireplanes, a position chosen to avoid exposing the PMTs to high field environments. Testing of each of the MicroBooNE PMTs in cryogenic conditions has begun at Fermilab in a liquid nitrogen test stand, and a prolonged vertical slice test of several full assemblies will be performed in liquid argon in the new year in order to test the stability and performance of the units.

3.3. Cryogenics
MicroBooNE employs an extensive system of cryogenics in order to maintain argon purity. Oxygen and water continuously outgas from the cryostat, and trace amounts can greatly reduce free electron lifetime. Concentration of these impurities must be held below 100 ppt. To achieve this, constant purification is performed, and the electron lifetime is measured in real time by lifetime monitors both inside the cryostat and in the recirculation system. Scintillation light is also quenched by nitrogen, and the dissolved nitrogen level must be kept below 100 ppb.

Temperature variations within the active volume can lead to an image distortion, swith the drift velocity having temperature dependence of the form $\delta \log(v_d)/\delta T = -1.9\%/K$. As such, MicroBooNE has the design specification that temperature variations must be kept below 1 K across the fiducial volume. Computational fluid dynamical simulations have been performed to model the convection currents and heat flow in the detector and confirm that uniformity to within 0.1 K can be achieved, with convection flow rates being negligible compared to the electron drift velocity.

3.4. Electronics
The noise specification for the MicroBooNE TPC readout electronics is that the equivalent noise charge must be kept below 660 e. MicroBooNE will employ both warm and cold electronics, in a scheme designed to minimize this TPC noise. Initial preamplification and signal shaping is performed by cold front end ASICs inside the cryostat. The signal is transported to several cryostat feedthroughs by a twisted pair, at which point an intermediate level of amplification is performed by warm interface electronics. Cables from the interface run 20-30 m to a DAQ system in the detector hall.

3.5. Computing and software

Simulation, reconstruction and data analysis jobs for MicroBooNE are performed using the LArSoft software package, which is a general purpose framework for liquid argon TPC experiments [5]. LArSoft is developed and maintained by collaborators from the participating experiments, currently ArgoNeut [6], MicroBooNE, LBNE LAr and some Fermilab test stands.

LArSoft interfaces with external event generators to simulate neutrino beam interactions [7] and cosmic ray events [8], and also includes internal event generators for validation studies. There is a full GEANT4 simulation, customized to provide fast handling of charge drift physics and voltage induction in the TPC wires. LArSoft also incorporates optical simulation capabilities, with both full photon tracking and fast volume sampling based simulation methods.

The vast amount of information recorded for a single neutrino interaction event in liquid argon makes the task of achieving an optimal event reconstruction formidable. The development of new algorithms by the LArSoft and MicroBooNE collaborations is ongoing, and whilst basic 3D tracking and shower finding methods are in place, work on reconstruction methods of increasing sophistication will continue into the analysis phase of the experiment.

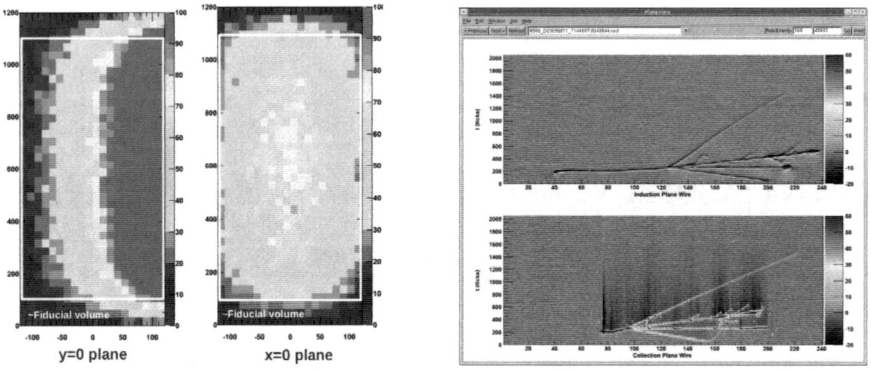

Figure 2. Left : Trigger sensitivity map for the x=0 and y=0 detector planes, produced using the LArSoft optical physics simulation. Right : Event display showing raw data for a typical neutrino event from the ArgoNeuT experiment

4. Summary

The MicroBooNE experiment is a 170 T liquid argon neutrino experiment which is set to begin running in the BNB at Fermilab in early 2014. Construction and testing of both the TPC and optical detector systems are underway, and engineering designs and prototypes for cryogenic systems and electronics are in place. The experiment will pursue a range of short baseline physics including cross section measurements and an exploration of the MiniBooNE anomalies with an emerging and powerful neutrino detection technology.

References

[1] Aguilar-Arevalo A A *et al.* (MiniBooNE Collaboration) 2009 *Phys. Rev. Lett.* **102** 101802
[2] Aguilar-Arevalo A A *et al.* (MiniBooNE Collaboration) 2010 *Phys. Rev. Lett.* **105** 181801
[3] Kisslinger L S 2011 (*Preprint* `arXiv:1108.4062 [hep-ph]`)
[4] Cennini P *et al.* 1999 *Nucl. Instrum. Meth. A* **432** 240–248
[5] Rebel B *et al.* (LArSoft Collaboration) URL `https://cdcvs.fnal.gov/redmine/projects/larsoftsvn`
[6] Spitz J (ArgoNeuT Collaboration) 2010 *J. Phys. Conf. Ser.* **203** 012108
[7] Andreopoulos C *et al.* 2010 *Nucl. Instrum. Meth. A* **614** 87–104
[8] Hagmann C *et al.* 2008 (CRY Collaboration) Tech. Rep. UCRL-TM-229453 LLNL

Journal of Physics: Conference Series **408** (2013) 012029 doi:10.1088/1742-6596/408/1/012029

Light sterile neutrinos from flavor symmetries

He Zhang

Max-Planck-Institut für Kernphysik, Postfach 103980, 69029 Heidelberg, Germany

E-mail: he.zhang@mpi-hd.mpg.de

Abstract. We study the realization of eV-scale sterile neutrinos in flavor symmetry theories. In particular, we focus on a popular flavor A_4 model, and point out that, eV-scale sterile neutrinos can easily be accommodated, while the exact tri-bimaximal mixing pattern is perturbed due to active-sterile mixing. We also illustrate the effects of light sterile neutrinos on the effective mass in neutrino-less double beta decay.

1. Introduction

Despite the successful achievements of neutrino oscillation experiments, there are experimental anomalies that cannot be explained within the standard three neutrino framework. In particular, the possible presence of sterile neutrinos points towards non-standard neutrino physics. The issue of the LSND and MiniBooNE results together with a recent re-evaluation of the anti-neutrino spectra of nuclear reactors [1–3] hint towards the presence of one or two sterile neutrino states with masses at the eV scale. Moreover, the light-element abundances from precision cosmology and Big Bang nucleosynthesis favor extra radiation in the Universe, which could be interpreted with the help of additional sterile neutrinos [4].

Although sterile neutrinos do not directly enter the weak interactions by definition, their admixture with active neutrinos would modify the neutrino flavor mixing and lead to rich experimental phenomena, e.g., the active-sterile oscillations [5], the contributions to the neutrino-less double beta decay amplitude, and the corrections to the beta decay spectra. From the theoretical side, a straightforward extension to the Standard Model to accommodate sterile neutrinos is the seesaw mechanism, in which right-handed neutrinos play the role of sterile neutrinos, whereas they are typically assumed to be many order of magnitude heavier than the Standard Model scale. Possible attempts yielding low-scale sterile neutrinos include the split seesaw mechanism in extra dimensions [6], the Froggatt-Nielsen (FN) mechanism [7, 8], flavor symmetries (e.g., Ref [9–11]), and the minimal extended type I seesaw [12]. Another kind of approaches frequently employed in explaining the neutrino mixing are flavor symmetries, e.g., models based on the tetrahedral group A_4. In this work, we illustrate the effects of light sterile neutrinos in $0\nu\beta\beta$, and then study a model combining two of the above-mentioned ideas, i.e., the FN mechanism and flavor symmetry, giving a non-trivial origin of eV-scale sterile neutrinos.

2. Sterile neutrinos in neutrino-less double beta decay

Assuming that there are $n_s = n - 3$ sterile neutrinos, the neutrino mass matrix then takes an $n \times n$ form, and can be diagonalized by using a unitary transformation as $m_\nu = U\text{diag}(m_1, m_2, ..., m_n)U^T$. The neutrino flavor eigenstates ν_f are therefore related to their mass

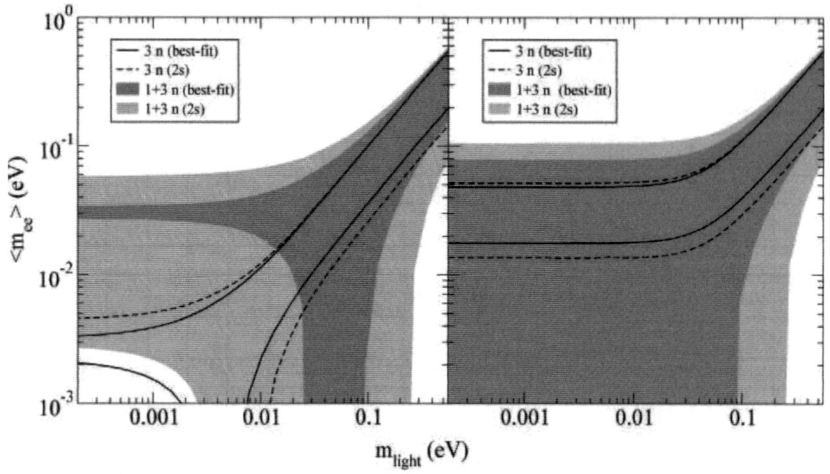

Figure 1. The allowed ranges in the $\langle m_{ee}\rangle - m_{\rm light}$ parameter space, both in the three-neutrino picture (unshaded regions) and with one sterile neutrino (shaded regions), for the normal mass ordering ($m_3 > m_2 > m_1$, left panel) or inverted mass ordering ($m_2 > m_1 > m_3$, right panel).

eigenstates ν_i via $\nu_f = \sum_i^n U_{fi}\nu_i$. In the 4×4 case (namely, $n_s = 1$), U is parameterized by $U = R_{34}\tilde{R}_{24}\tilde{R}_{14}R_{23}\tilde{R}_{13}R_{12}P$, where R_{ij} denote the rotations in ij space, and $P = {\rm diag}(1, e^{i\alpha/2}, e^{i(\beta/2+\delta_{13})}, e^{i(\gamma/2+\delta_{14})})$ contains the three Majorana phases α, β and γ. Note that the sterile neutrino masses could be either heavier or lighter than the active neutrinos. For the latter case, the active neutrino masses should be around the eV scale, which induces more tensions with cosmological observations on the total neutrino mass. We will thus in the following parts only consider the case $m_{\rm sterile} \gg m_{\rm active}$.

In the four neutrino mixing case, the effective neutrino mass in $0\nu\beta\beta$ is given by

$$\begin{aligned}\langle m_{ee}\rangle_{4\nu} &= \left|c_{12}^2 c_{13}^2 c_{14}^2 m_1 + s_{12}^2 c_{13}^2 c_{14}^2 m_2 e^{i\alpha} + s_{13}^2 c_{14}^2 m_3 e^{i\beta} + s_{14}^2 m_4 e^{i\gamma}\right| \\ &\simeq \left|c_{14}^2 \langle m_{ee}\rangle_{3\nu} + s_{14}^2\sqrt{\Delta m_{41}^2}e^{i\gamma}\right| , \qquad (1)\end{aligned}$$

where $\langle m_{ee}\rangle_{3\nu} = c_{12}^2 c_{13}^2 m_1 + s_{12}^2 c_{13}^2 m_2 e^{i\alpha} + s_{13}^2 m_3 e^{i\beta}$ is the standard expression for three active neutrinos. By using data from Ref. [2], the allowed range of $\langle m_{ee}\rangle_{4\nu}$ as a function of the lightest mass $m_{\rm light}$ are shown in Fig. 1. One observes from the plot that, in the presence of one sterile neutrino, the standard behavior can be completely mixed up. In particular, if the lightest neutrino mass is reasonably small, e.g. $m_{\rm light} < 0.01$ eV, the allowed range of $\langle m_{ee}\rangle_{4\nu}$ is dominated by the term $s_{14}^2\sqrt{\Delta m_{41}^2} \simeq 0.031$ eV, which means that $\langle m_{ee}\rangle_{4\nu}$ cannot vanish in the normal ordering case (the contribution of the two light active neutrinos cannot cancel that of the sterile neutrino). However, in the inverted ordering $\langle m_{ee}\rangle_{4\nu}$ can vanish even for very small active neutrino masses. The effective mass can also be zero in the regime where the active neutrinos are quasi-degenerate ($m_{\rm light} > 0.1$ eV). This modified behavior of $\langle m_{ee}\rangle$ is of particular interest in the inverted ordering case, since the usual lower bound on the effective mass (cf. the solid and dashed lines in the right panel of Fig. 1) is no longer valid (see, e.g., Ref. [13] for detailed discussions). A similar situation holds for two sterile neutrino case. If future $0\nu\beta\beta$ experiments measure a tiny effective mass and the neutrino mass hierarchy is confirmed to be inverted from long-baseline neutrino oscillations, the sterile neutrino hypothesis would be an attractive explanation for this inconsistency.

Table 1. Particle assignments of the A_4 model, in which an additional Z_3 symmetry is assumed in order to avoid mixing between the charged lepton and neutrino sectors. The $U(1)_{\rm FN}$ charge generates the hierarchy of charged lepton masses and regulates the scale of the sterile state.

Field	L	e^c	μ^c	τ^c	$h_{u,d}$	φ	φ'	ξ	ν_s
$SU(2)_L$	2	1	1	1	2	1	1	1	1
A_4	$\underline{3}$	$\underline{1}$	$\underline{1}''$	$\underline{1}'$	$\underline{1}$	$\underline{3}$	$\underline{3}$	$\underline{1}$	$\underline{1}$
Z_3	ω	ω^2	ω^2	ω^2	1	1	ω	ω	1
$U(1)_{FN}$	-	3	1	0	-	-	-	-	6

3. Flavor A_4 model of sterile neutrinos

We proceed to show the realization of light sterile neutrinos in a modified flavor A_4 symmetry model [14], which accommodates both the tri-bimaximal mixing and one or two light sterile neutrinos. The relevant particle assignments are shown in Table 1, where a flavon θ carrying a negative unit of the FN charge is also introduced, suppressing the corresponding mass term by powers of the small parameter $\langle\theta\rangle/\Lambda \equiv \lambda < 1$.

With the above particle assignments, along with the A_4 multiplication rules, we arrive at the invariant Lagrangian

$$\begin{aligned}\mathcal{L}_{\rm Y} &= \frac{y_e}{\Lambda}e^c(\varphi L)h_d + \frac{y_\mu}{\Lambda}\mu^c(\varphi L)'h_d + \frac{y_\tau}{\Lambda}\tau^c(\varphi L)''h_d + \frac{x_a}{\Lambda^2}\xi(Lh_uLh_u) + \frac{x_d}{\Lambda^2}(\varphi' Lh_uLh_u) \\ &+ \frac{x_e}{\Lambda^2}\xi(\varphi' Lh_u)\nu_s + \frac{x_f}{\Lambda^2}(\varphi'\varphi' Lh_u)\nu_s + \frac{x_s}{\Lambda}(\varphi\varphi) + \text{h.c.} + \ldots, \end{aligned} \tag{2}$$

where Λ is the cut-off scale, and the dots stand for higher dimensional operators. If one chooses the real basis for A_4, along with the flavon VEV alignments $\langle\xi\rangle = u$, $\langle\varphi\rangle = (v,0,0)$ and $\langle\varphi'\rangle = (v',v',v')$, then the charged-lepton mass matrix is diagonal, and the 4×4 neutrino mass matrix reads

$$M_\nu^{4\times4} = \begin{pmatrix} a+\frac{2d}{3} & -\frac{d}{3} & -\frac{d}{3} & e \\ \cdot & \frac{2d}{3} & a-\frac{d}{3} & e \\ \cdot & \cdot & \frac{2d}{3} & e \\ \cdot & \cdot & \cdot & m_s \end{pmatrix}, \tag{3}$$

where $a = 2x_a\frac{uv_u^2}{\Lambda^2}$, $d = 2x_d\frac{v'v_u^2}{\Lambda^2}$, $e = \sqrt{2}x_e\frac{uv'v_u}{\Lambda^2}\lambda^6$, and $m_s = x_s\frac{v^2}{\Lambda}\lambda^{12}$ have dimensions of mass. Note that the active-sterile neutrino couplings and the sterile neutrino mass are both suppressed by powers of λ.

Assuming $v'/\Lambda \approx v/\Lambda \approx u/\Lambda \approx \langle\theta\rangle/\Lambda \simeq 10^{-1.5}$ together with the cut-off scale $\Lambda \simeq 10^{12.5}$ GeV, one can estimate that $a \sim d \sim e \simeq 0.1$ eV, and $m_s \simeq 10^{0.5}$ eV with the assumption that the Yukawa couplings $x_{a,d,e}$ are of order 1. As a result, the choice of the mass scales allows the hierarchies $a \sim d \sim e < m_s$. We can then approximately diagonalize $M_\nu^{4\times4}$ (i.e., $M_\nu^{4\times4} = U{\rm diag}(m_1,m_2,m_3,m_4)U^T$), and obtain

$$U \simeq \begin{pmatrix} \frac{2}{\sqrt{6}} & \frac{1}{\sqrt{3}} & 0 & 0 \\ -\frac{1}{\sqrt{6}} & \frac{1}{\sqrt{3}} & -\frac{1}{\sqrt{2}} & 0 \\ -\frac{1}{\sqrt{6}} & \frac{1}{\sqrt{3}} & \frac{1}{\sqrt{2}} & 0 \\ 0 & 0 & 0 & 1 \end{pmatrix} + \begin{pmatrix} 0 & 0 & 0 & \frac{e}{m_s} \\ 0 & 0 & 0 & \frac{e}{m_s} \\ 0 & 0 & 0 & \frac{e}{m_s} \\ 0 & -\frac{\sqrt{3}e}{m_s} & 0 & 0 \end{pmatrix} + \begin{pmatrix} 0 & -\frac{\sqrt{3}e^2}{2m_s^2} & 0 & 0 \\ 0 & -\frac{\sqrt{3}e^2}{2m_s^2} & 0 & 0 \\ 0 & -\frac{\sqrt{3}e^2}{2m_s^2} & 0 & 0 \\ 0 & 0 & 0 & -\frac{3e^2}{2m_s^2} \end{pmatrix}, \tag{4}$$

where the first term reproduces the tri-bimaximal mixing, while the higher order corrections second and third terms are suppressed by the ratio e/m_s. Furthermore, the eigenvalues are

$$m_1 = a + d\,, \quad m_2 = a - \frac{3e^2}{m_s}\,, \quad m_3 = -a + d\,, \quad m_4 = m_s + \frac{3e^2}{m_s}\,. \tag{5}$$

One immediately observes that the active neutrino mass spectrum is slightly modified by the sterile neutrino, while $m_4 \simeq m_s$ is stablized at the eV scale. Compared to the parametrization of U, we obtain

$$\sin^2\theta_{12} \simeq \frac{1}{3} - \frac{2}{3}\left(\frac{e}{m_s}\right)^2\,, \quad \sin^2\theta_{23} \simeq \frac{1}{2} + \frac{1}{2}\left(\frac{e}{m_s}\right)^2\,, \tag{6}$$

and a vanishing θ_{13}. In addition, the active-sterile mixing angles are of similar magnitude and given by

$$\sin^2\theta_{14} \simeq \sin^2\theta_{24} \simeq \sin^2\theta_{34} \simeq \left(\frac{e}{m_s}\right)^2\,. \tag{7}$$

Therefore, the desired active-sterile mixing $\sin^2\theta_{14} \sim 0.01$ could be obtained by taking, e.g., $m_s \sim 1$ eV and $e \sim 0.1$ eV.

Note that, the flavor model under discussion can be easily generalized to a seesaw framework. A detailed survey could be useful and will be elaborated on in future.

4. Conclusion

We have pointed out that the presence of light sterile neutrinos would significantly change the effective mass measured in neutrino-less double beta decay experiments. In particular, the sterile neutrino contributions to the effective mass could lead to a vanishing $\langle m_{ee}\rangle$, even if three active neutrinos feature an inverted mass ordering, which is clearly different from the predictions of the standard three neutrino picture. We also presented a flavor A_4 model, which accommodates both a light sterile neutrino and deviations from the exact tri-bimaximal mixing. We further stress that a keV sterile neutrino playing the role of warm dark matter could also be included in the flavor A_4 model by choosing smaller FN charges.

Acknowledgments

This work was supported by the ERC under the Starting Grant MANITOP and by the Deutsche Forschungsgemeinschaft in the Transregio 27 "Neutrinos and beyond – weakly interacting particles in physics, astrophysics and cosmology".

References

[1] Mention G *et al* 2011 *Phys. Rev.* D **83** 073006 (*Preprint* arXiv:1101.2755 [hep-ex])
[2] Kopp J, Maltoni M and Schwetz T 2011 *Phys. Rev. Lett.* **107** (*Preprint* arXiv:1103.4570 [hep-ph])
[3] Giunti C and Laveder M 2011 *Preprint* arXiv:1109.4033 [hep-ph]
[4] Hamann J, Hannestad S, Raffelt G G, Wong Y Y Y 2011 *Preprint* arXiv:1108.4136 [astro-ph.CO]
[5] Zhang H 2007 *Mod. Phys. Lett. A* **22** 1341-1348 (*Preprint* hep-ph/0606040)
[6] Kusenko A, Takahashi F and Yanagida T T 2010 *Phys. Lett. B* **693** 144-148 (*Preprint* arXiv:1006.1731 [hep-ph])
[7] Froggatt C D and Nielsen H B 1979 *Nucl. Phys. B* **147** 277
[8] Merle A and Niro V 2011 *JCAP* **1107** 023 (*Preprint* arXiv:1105.5136 [hep-ph])
[9] Mohapatra R N 2001 *Phys. Rev.* D **64** 091301 (*Preprint* hep-ph/0107264)
[10] Shaposhnikov M 2007 *Nucl. Phys.* B **763** 49-59 (*Preprint* hep-ph/0605047)
[11] Lindner M, Merle A and Niro V 2011 *JCAP* **1101** 034 (*Preprint* arXiv:1011.4950 [hep-ph])
[12] Barry J, Rodejohann W and Zhang H 2011 *JHEP* **1107** 091 (*Preprint* arXiv:1105.3911 [hep-ph])
[13] Rodejohann W 2011 *Preprint* arXiv:1106.1334 [hep-ph]
[14] Altarelli G and Feruglio F 2005 *Nucl. Phys.* B **720** 64-88 (*Preprint* hep-ph/0504165)

Implications of new generations on neutrino masses

A Aparici, J Herrero-Garcia, N Rius and A Santamaria

Depto.de Fisica Teorica, and IFIC, Universidad de Valencia-CSIC
Edificio de Institutos de Paterna, Apt. 22085, 46071 Valencia, Spain

E-mail: juan.a.herrero@uv.es

Abstract. We explore the possible implications that new families, that are being searched for at the LHC, would have on neutrino masses. In particular, we have explored the possibility that the smallness of the observed neutrino masses is naturally understood in a modified version of the Standard Model (SM) with complete extra generations of fermions, i.e., that have right-handed neutrinos, in which neutrino masses are generated at two loops. With one extra family it is not possible to fit the observed spectrum of masses and mixings. However, the radiative mass generated provides an important constraint in these kind of models, so the neutrino masses do not exceed their cosmological bound. Within the context of two extra families, we analyse the allowed parameter space and the possible phenomenological signals.

1. Introduction

One of the most natural extensions of the Standard Model (SM) is the addition of extra sequential generations [1]. We analyse the relation between new generations and neutrino masses, one of the strongest evidence for physics beyond the SM, in view of the possibility that the former are discovered at the LHC.

Global fits of models with additional generations to the electroweak data have been performed and they favour no more than five generations. It should be kept in mind that most of the fits make some simplifying assumptions on the mass spectrum of the new generations and do not consider Majorana neutrino masses for the new generations or the possibility of breaking dynamically the gauge symmetry via the condensation of the new generations' fermions; all these would give additional contributions.

LEP II limits on new generation leptons are: $m_{\ell'} > 100.8$ GeV and $m_{\nu'} > 80.5\,(90.3)$ GeV for pure Majorana (Dirac) particles. When neutrinos have both Dirac and Majorana masses, the bound on the lightest is 63 GeV. For stable neutrinos LEP I measurement of the invisible Z width, $\Gamma_{\rm inv}$, implies $m_{\nu'} > 39.5\,(45)$ GeV for pure Majorana (Dirac) particles. Recent LHC bounds on the quarks are $m_{t'} > 450$ GeV, assuming $t' \to bW$, which in principle is plausible as EW fits bound the splitting within the doublets to be $|m_{t'} - m_{b'}| < 80$ GeV, and $m_{b'} > 385$ GeV. For the leptons, the splitting can be larger, $|m_{\ell'} - m_{\nu'}| < 140$ GeV.

The most striking effect of new generations is the enhancement of the Higgs-gluon-gluon vertex which arises from a triangle diagram with all quarks running in the loop, by a factor of 9 (25) in the presence of a fourth (fifth) generation. Thanks to that, LHC has been able to exclude, at 95% C.L., a fourth-generation SM-like Higgs in the range $120 < m_H < 600$ GeV. Stability of the Higgs potential, and triviality, together with these severe bounds, imply that if new generations exist, they will presumably come together with a new non-standard scalar

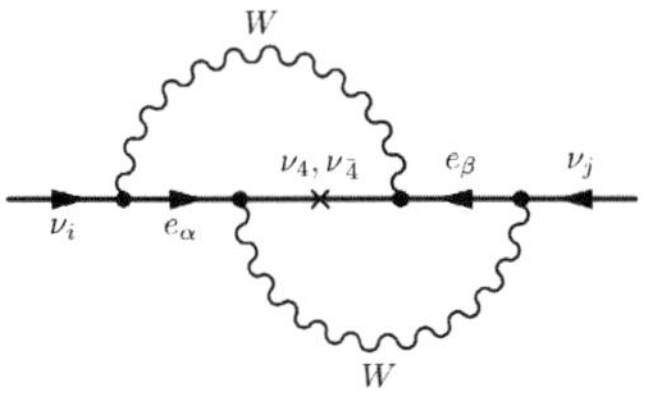

Figure 1. Two-loop diagram contributing to neutrino masses in the four-generation model.

sector, for instance, a strongly coupled composite Higgs, an extended Higgs sector or a Higgs with invisible decays, for example, to dark matter. Of course, if both a new family and the Higgs with a 3 generations' Standard Model cross section are detected, new physics would be needed to reconcile these facts.

In this talk (see [2] for further details and a complete list of references) we focus on how neutrino masses can be naturally generated at two loops by adding extra families and singlets. Recall that right-handed neutrinos do not have gauge charges and are not needed to cancel anomalies, therefore their number is not linked to the number of generations.

2. Four generations

We extend the SM by adding a complete fourth generation, i.e., with one right-handed neutrino ν_R, with a Majorana mass term [3]. We denote the new charged lepton E and the new neutrino ν_E. y_i with $i = e, \mu, \tau, E$ are the Yukawas of the doublets with ν_R. After spontaneous symmetry breaking (SSB) the mass matrix for the neutral leptons at tree level is a 5×5 Majorana symmetric matrix which has only one right-handed neutrino Majorana mass term. Therefore, it leads to three massless Weyl neutrinos and two massive Majorana ν_4 and $\nu_{\bar{4}}$, of masses $m_{4,\bar{4}} = \frac{1}{2}\left(\sqrt{m_R{}^2 + 4m_D^2} \mp m_R\right)$, where $m_D/v = \sqrt{\sum_i y_i^2} \approx y_E$, with $v = \langle \phi^{(0)} \rangle$, and $\tan^2\theta = m_4/m_{\bar{4}}$. Phenomenologically $m_R < 20$ TeV, so $m_{4,\bar{4}} > 63$ GeV. If $m_R \ll m_D$, we have $m_4 \approx m_{\bar{4}}$ and $\tan\theta \approx 1$ (pseudo-Dirac limit) while when $m_R \gg m_D$, $m_4 \approx m_D^2/m_R$, $m_{\bar{4}} \approx m_R$ and $\tan\theta \approx m_D/m_R$ (see-saw limit).

Since lepton number is broken by the ν_R Majorana mass, there is no symmetry which prevents the tree-level massless neutrinos from gaining Majorana masses; they are generated at two loops by the diagram of figure 1. The eigenvalues of the light neutrino mass matrix are proportional to m_μ^4, m_τ^4, m_E^4 which gives a huge hierarchy between neutrino masses:

$$\frac{m_2}{m_3} \leq \frac{1}{4y_E^2}\left(\frac{m_\tau}{m_E}\right)^2\left(\frac{m_\tau}{m_{\bar{4}}}\right)^2 \leq \frac{10^{-7}}{y_E^2}, \qquad (1)$$

where we have used that $m_E, m_{\bar{4}} \geq 100\,\text{GeV}$. To overcome this huge hierarchy very small values of y_E are needed, which would imply that the heavy neutrinos are not mainly ν_E but some combination of the three known ones ν_e, ν_μ, ν_τ; however this is not possible since it would yield observable effects in a variety of processes, like $\pi \to \mu\nu$, $\pi \to e\nu$, $\tau \to e\nu\nu$... These require that $y_{e,\mu,\tau} < 10^{-2} y_E$, so $y_E \approx 1$. Therefore, the simplest version of the model is unable to accommodate the observed spectrum of neutrino masses and mixings.

However, notice that whenever a new generation and a right-handed neutrino with Majorana mass at (or below) the TeV scale are added to the SM, the two-loop contribution to neutrino masses is always present and provides an important constraint for this kind of SM extensions. In figure 2 it is shown the upper bound in the Yukawa (y_i, with $i = e, \mu, \tau$) versus Majorana mass plane, so $m_\nu < 0.3$ eV, as implied by cosmology. It is clearly seen that, for most of the parameter space ($m_R > 0.01$ GeV), it is more restrictive than universality or LFV constraints.

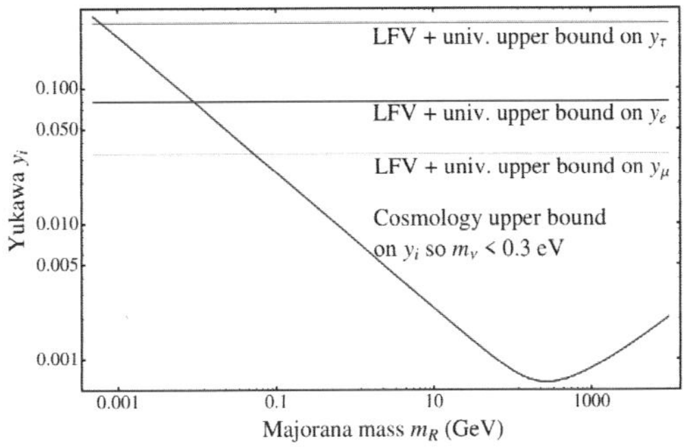

Figure 2. General constraint on light neutrino masses in four generations' models with heavy Majorana neutrinos. We show the bounds that LFV and universality from pion and tau decays set on the Yukawas at 90% C.L.: $y_\tau < 0.3$, $y_e < 0.08$ and $y_\mu < 0.033$.

It could be argued that new generation neutrinos may be Dirac particles, and the bound would not apply. However, if light neutrinos turn out to be Majorana, it is natural that the heavy ones are Majorana as well, since lepton number is not conserved anymore. For instance, in a four generations context with light neutrino masses obtained via the see-saw mechanism, as long as it is not completely decoupled from the light neutrinos, the fourth right-handed neutrino will always acquire a Majorana mass at the two-loop level, even if it is set to zero by hand at the Lagrangian level (in particular at the TeV scale for typical GUT scale Majorana masses of the first three right-handed neutrinos), so the previous constraint will in general apply to the light neutrinos' Yukawas.

3. The five-generation model

We add two families to the SM and two right-handed neutrinos, ν_{4R} and ν_{5R}. We denote the two charged leptons by E and F. The model, contrary to the four-generation case, has additional sources of CP violation in the leptonic sector, however, for simplicity we take all neutrino Yukawas y_α and y'_α real. For simplicity, we also choose $\sum_\alpha y_\alpha y'_\alpha = 0$.

The tree-level spectrum contains three massless neutrinos and four heavy Majorana neutrinos. The model should be compatible with the observed universality of fermion couplings and have small rates of lepton flavour violation (LFV) in the charged sector, which requires $y_e, y_\mu, y_\tau, y'_e, y'_\mu, y'_\tau \ll y_E, y_F, y'_E, y'_F$. In addition, it should fit the observed pattern of masses and mixings, for instance, reproducing the tribimaximal (TBM) mixing structure. A successful choice of the Yukawas to obtain normal hierarchy (NH) (see [2] for an analysis of inverted hierarchy (IH) and more details), i.e, $m_3 \approx \sqrt{|\Delta m^2_{31}|} \approx 0.05$ eV, $m_2 \approx \sqrt{\Delta m^2_{21}} \approx 0.01$ eV, will be $y_\alpha = y_E(\epsilon, \epsilon, -\epsilon, 1, 0)$ and $y'_\alpha = y'_F(0, \epsilon', \epsilon', 0, 1)$. Assuming that $m_{E,F} \gg m_{4,\bar{4},5,\bar{5}} \gg m_W$, we find:

$$m_2 \approx \frac{3g^4}{2(4\pi)^4 m_W^4} \epsilon^2 m_{4D}^2 m_{4R} m_E^2 \ln \frac{m_E}{m_{\bar{4}}} \quad (2)$$

$$m_3 \approx \frac{g^4}{(4\pi)^4 m_W^4} \epsilon'^2 m_{5D}^2 m_{5R} m_F^2 \ln \frac{m_F}{m_{\bar{5}}} , \quad (3)$$

and the required ratio $m_3/m_2 \approx 5$ can be easily accommodated.

The new upper bound on LFV $B(\mu \to e\gamma) < 2.4 \times 10^{-12}$ is translated into $\epsilon < 0.02$. We display in figure 3 (a) $B(\mu \to e\gamma)$ versus the mass of the heavy neutrinos m_4 in the NH case.

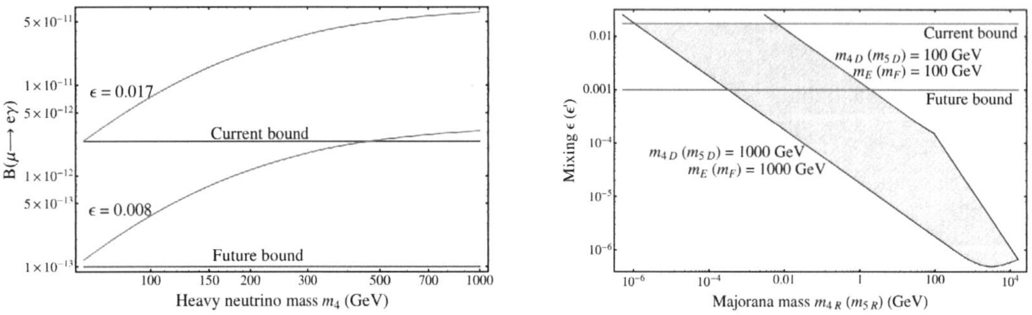

Figure 3. a) Left: $B(\mu \to e\gamma)$ against m_4 for different values of ϵ. We also display present and future limits. b) Right: Parameter space (region between the curves) that predicts correct heavy and light neutrino masses. We also show the current $\mu \to e\gamma$ bound and the expected μ–e conversion limit.

We also display present and near future limits. Violations of universality constrain the model in both hierarchies. For example, from pion decay, we obtain $\epsilon' < 0.04$.

Regarding lepton number violation signals at the LHC, for instance, if the lightest fourth generation lepton is the charged one, there is a striking signal: lepton number violating like-sign fourth generation lepton pair-production, through $q\bar{q}' \to W^\pm \to E^\pm \nu_{4,\bar{4}} \to E^\pm E^\pm W^\mp$ or via W fusion, $q\bar{q} \to W^\pm W^\pm q' q' \to E^\pm E^\pm jj$. Also the Higgs could be detected in this way: $gg \to H \to \nu_4 \nu_4 \to E^\pm E^\pm W^\mp W^\mp$, giving a rather unusual signal, with hardly any background.

With respect to neutrinoless double beta decay, there are contributions of the new heavy neutrinos, however, we obtain the same combination of parameters as in the light neutrino masses expressions, $m_{4R}\epsilon^2$, leading to unobservable effects in $0\nu\beta\beta$ when the former are fitted (of course the light neutrino contribution could be observed in the future, if the IH spectrum is realized).

To summarize the phenomenology of the model we present in figure 3 (b) the allowed regions in the $\epsilon - m_{4R}$ plane which lead to $m_3 \sim 0.05\,\mathrm{eV}$ varying the charged lepton masses $m_E\,(m_F)$ and the Dirac neutrino masses $m_{4D}\,(m_{5D})$ between 100 GeV–1 TeV, and imposing the bound on the neutrino mass, $m_4 > 63\,\mathrm{GeV}$. We also plot the present bounds on the mixings $\epsilon\,(\epsilon')$ from $\mu \to e\gamma$ and future much stronger limits from μ–e conversion if expectations are attained. So with five generations and two singlets all current data can be accomodated in the shaded region of the parameter space between the curves of figure 3 (b), which will be probed in the near future.

Acknowledgments

This work has been partially supported by FPA-2007-60323, FPA-2008-03373, CSD2007-00060, CSD2007-00042, PROMETEO/2009/116, PROMETEO/2009/128 and MRTN-CT-2006-035482 (FLAVIAnet). A.A. and J.H.-G. are supported by the MICINN under the FPU program.

References

[1] Holdom B *et. al.*, *PMC Phys.* A **3** (2009) 4.
Novikov V A *et. al.*, *Phys. Atom. Nucl.* **73** (2010) 636–642.
Carpenter L M *et. al.*, *Phys. Rev.* D **82** (2010) 114019.
Kribs G D *et. al.*, *Phys. Rev.* D **76** (2007) 075016. **PDG** Coll., *J. Phys.* G **37** (2010) 075021.
[2] Aparici A *et. al.*, JHEP **1107** (2011) 122.
[3] Petcov S T *et. al.*, *Phys. Lett.* B **143** (1984) 175.
Babu K S *et. al.*, *Phys. Rev. Lett.* **61** (1988) 674, *Phys. Lett.* B **218** (1989) 233–237.

Are there consistent models giving observable NSI?

Enrique Fernandez Martinez
CERN Physics Department, Theory Division, CH-1211 Geneva 23, Switzerland
E-mail: enfmarti@cern.ch

Abstract. While the existing direct bounds on neutrino NSI are rather weak, order 10^{-1} for propagation and 10^{-2} for production and detection, the close connection between these interactions and new NSI affecting the better-constrained charged letpon sector through gauge invariance make these bounds hard to saturate in realistic models. Indeed, Standard Model extensions leading to neutrino NSI typically imply constraints at the 10^{-3} level. The question of whether consistent models leading to observable neutrino NSI naturally arises and was discussed in a dedicated session at NUFACT 11. Here we summarize that discussion.

1. Direct bounds on neutrino NSI

The formalism of non-standard neutrino interactions (NSI) is a very widespread and convenient way of parameterizing the effects of new physics in neutrino oscillations [1, 2]. Even though present data constrain NSI to be a subleading effect in neutrino oscillation experiments, the possibility of their eventual detection and interference with standard neutrino oscillation measurements has triggered a considerable interest in the community. Present bounds on NSI are reviewed here and the possibility of saturating them in particular Standard Model (SM) extensions is discussed.

1.1. CC-like NSI

NSI for source and detector charged-current processes, referred as charged-current-like (CC-like) NSI are discussed below. Leptonic NSI are given by the effective Lagrangian density:

$$\mathcal{L}^{\ell}_{\rm NSI} = -2\sqrt{2}G_F\varepsilon^{\alpha\beta P}_{\gamma\delta}[\bar{\ell}_\alpha\gamma^\mu P\ell_\beta][\bar{\nu}_\gamma\gamma_\mu P_L\nu_\delta] \tag{1}$$

where G_F is the Fermi constant and P is either P_L or P_R and, due to Hermiticity, $\varepsilon^{\alpha\beta P}_{\gamma\delta} = \varepsilon^{\beta\alpha P*}_{\delta\gamma}$. For NSI at neutrino production via muon decay $\alpha = \mu$ and $\beta = e$. Notice that $\alpha = \beta = e$ would instead correspond to NSI with electrons in matter. In a similar fashion, the CC-like NSI with quarks are given by:

$$\mathcal{L}^{q}_{\rm NSI} = -2\sqrt{2}G_F\varepsilon^{qq'P}_{\alpha\beta}V_{qq'}[\bar{q}\gamma^\mu Pq'][\bar{\ell}_\alpha\gamma_\mu P_L\nu_\beta] + \text{h.c.} \tag{2}$$

where q is an up-type and q' is a down-type quark. Only $q = u$ and $q' = d$ are of practical interest for neutrino oscillations, due to their contributions to charged-current interactions with pions and nuclei. In [3] bounds on these production and detection NSI were derived from constraints on the comparison of different measurements of G_F: through μ decay (affected by leptonic NSI of Eq. (1)), β decays (affected by quark NSI of Eq. (2)) and the kinematic measurements of the masses of the gauge bosons M_Z and M_W (not affected by neutrino NSI). Universality tests of G_F from π and τ decays as well as short baseline neutrino oscillation experiments such as KARMEN or NOMAD, were also considered. Notice that in order to derive those bounds only one NSI parameter was switched on at a time in order to avoid the relaxation of the bounds through cancellations among different parameters. Here are the most stringent derived bounds:

$$|\varepsilon^{\mu e}_{\alpha\beta}| < \begin{pmatrix} 0.025 & 0.030 & 0.030 \\ 0.025 & 0.030 & 0.030 \\ 0.025 & 0.030 & 0.030 \end{pmatrix}, |\varepsilon^{ud}_{\alpha\beta}| < \begin{pmatrix} 0.041 & 0.025 & 0.041 \\ \substack{1.8\cdot 10^{-6} \\ 0.026} & 0.078 & 0.013 \\ \substack{0.087 \\ 0.12} & \substack{0.013 \\ 0.018} & 0.13 \end{pmatrix} \tag{3}$$

Whenever two values are quoted, the upper value refers to left-handed NSI and the lower to right-handed NSI. These bounds stem from the direct effect of the corresponding effective operator and are rather weak. However, Eqs. (1) and (2) are not gauge invariant and can be related to flavour changing processes involving charged leptons when promoting the neutrino fields to full lepton doublets which would lead to much tighter constraints. Possible ways to evade these stronger constraints are discussed in Sect. 2.

1.2. NC-like NSI

The neutral-current-like (NC-like) NSI effective Lagrangian density is defined as:

$$\mathcal{L}^M_{\text{NSI}} = -2\sqrt{2}G_F\varepsilon^{fP}_{\alpha\beta}[\bar{f}\gamma^\mu Pf][\bar{\nu}_\alpha\gamma_\mu P_L\nu_\beta], \text{f} = \text{e, u, d} \tag{4}$$

This type of NSI is the most extensively studied in the literature, since the constraints on the CC-like NSI are generally stronger. Indeed, the bounds from [4, 5, 6, 7], but discarding the loop constraints on $\varepsilon^{fP}_{e\mu}$ given the discussion in [8], result in the following $\mathcal{O}(10^{-1})$ constraints:

$$|\varepsilon^{e}_{\alpha\beta}| < \begin{pmatrix} \substack{0.06 \\ 0.14} & 0.10 & \substack{0.4 \\ 0.27} \\ 0.10 & 0.03 & 0.10 \\ \substack{0.4 \\ 0.27} & 0.10 & \substack{0.16 \\ 0.4} \end{pmatrix}, |\varepsilon^{u}_{\alpha\beta}| < \begin{pmatrix} \substack{1.0 \\ 0.7} & 0.05 & 0.5 \\ 0.05 & \substack{0.003 \\ 0.008} & 0.05 \\ 0.5 & 0.05 & \substack{1.4 \\ 3} \end{pmatrix}, |\varepsilon^{d}_{\alpha\beta}| < \begin{pmatrix} \substack{0.3 \\ 0.6} & 0.05 & 0.5 \\ 0.05 & \substack{0.003 \\ 0.015} & 0.05 \\ 0.5 & 0.05 & \substack{1.1 \\ 6} \end{pmatrix} \tag{5}$$

Again, the operator in Eq. (4) is not gauge invariant. The simplest gauge invariant extension of Eq. (4) promoting the left-handed fields to full SU(2) doublets would lead to flavour changing operators with charged leptons, allowing to derive much stronger bounds than the ones above.

2. Models to evade charged lepton NSI

In [9, 10] extensions of the SM leading to neutrino NSI but avoiding their charged current counterparts were studied both through $d = 6$ and $d = 8$ operators. Only two examples of SM extensions giving rise to matter NSI but avoiding their charged lepton counterpart exist at $d = 6$. The most direct one involves an antisymmetric 4-lepton operator, generated from the exchange of virtual singly charged scalar fields. In the second possibility NSI are induced through the $d = 6$ operator modifying the neutrino kinetic terms, generated by the exchange of virtual fermionic singlets. The latter operator generates the NSI in an indirect way, i.e. after canonical normalisation of the neutrino kinetic terms. In the first case, NSI with electrons in matter appear through the operator [11, 12, 13]:

$$\mathcal{L}^{d=6,as}_{NSI} = c^{d=6,as}_{\alpha\beta\gamma\delta}(\overline{L^c}_\alpha i\sigma_2 L_\beta)(\bar{L}_\gamma i\sigma_2 L^c_\delta) \tag{6}$$

which is generated after integrating out a charged scalar SU(2) singlet coupling to the SM lepton doublets:

$$\mathcal{L}^S_{int} = -\lambda^i_{\alpha\beta}\overline{L}^c_\alpha i\sigma_2 L_\beta S_i = \lambda^i_{\alpha\beta}S_i(\overline{\ell}^c_\alpha P_L\nu_\beta - \overline{\ell}^c_\beta P_L\nu_\alpha) \tag{7}$$

Integrating out the heavy scalars S_i generates the $d = 6$ operator of Eq. (6) at tree level. We find that, given the antisymmetric flavour coupling of Eq. (7), the only NSI induced are those between ν_μ and ν_τ with electrons in matter. Bounds on these NSI can be derived from μ and τ decays:

$$|\varepsilon^{m,e_L}_{\mu\mu}| < 8.2 \cdot 10^{-4}, |\varepsilon^{m,e_L}_{\tau\tau}| < 8.4 \cdot 10^{-3}, |\varepsilon^{m,e_L}_{\mu\tau}| < 1.9 \cdot 10^{-3} \tag{8}$$

NSI in production and detection are also induced with similar strengths. The second realization of NSI at $d = 6$ is via the operator:

$$\mathcal{L}^{d=6}_{kin} = -c^{d=6,kin}_{\alpha\beta} (\bar{L}_\alpha \cdot H^\dagger)\, i\partial\!\!\!/ \,(H \cdot L_\beta) \tag{9}$$

which induces non-canonical neutrino kinetic terms [14, 15, 16] though the vev of the Higgs field. After diagonalising and normalising the neutrino kinetic terms, a non-unitary lepton mixing matrix is produced from this operator and hence non-standard matter interactions as well as related NSI at the source and detector are induced. The tree level generation of this operator, avoiding a similar contribution to charged leptons that would lead to flavour changing neutral currents, requires the introduction of SM-singlet fermions which couple to the Higgs and lepton doublets via the Yukawa couplings (see e.g. [17]). Electroweak decays set the following bounds on the operator of Eq. (9) due to the effects a deviation from unitarity of the mixing matrix would imply [18, 16, 9]:

$$\frac{v^2}{2}|c^{d=6,kin}_{\alpha\beta}| < \begin{pmatrix} 4.0 \cdot 10^{-3} & 1.2 \cdot 10^{-4} & 3.2 \cdot 10^{-3} \\ 1.2 \cdot 10^{-4} & 1.6 \cdot 10^{-3} & 2.1 \cdot 10^{-3} \\ 3.2 \cdot 10^{-3} & 2.1 \cdot 10^{-3} & 5.3 \cdot 10^{-3} \end{pmatrix} \tag{10}$$

These bounds can be translated to constraints in the matter NSI that would be induced by a non-unitary mixing:

$$|\varepsilon_{\alpha\beta}| < \frac{v^2}{4}|\left(\frac{n_n}{n_e} - \delta_{\alpha e} - \delta_{e\beta}\right) c^{d=6,kin}_{\alpha\beta}| \tag{11}$$

Since the ratio of the neutron to electron density $\frac{n_n}{n_e}$ is in general close to 1, this implies that the bounds on $|\varepsilon_{e\mu}|$ and $|\varepsilon_{e\tau}|$ are significantly stronger than the bounds on the individual $\varepsilon_{\alpha\beta}$. For the main constituents of the Earth's crust and mantle the factor $\frac{n_n}{n_e} - 1$ means an additional suppression of two orders of magnitude of the NSI coefficient [9]. Thus, even if two $d = 6$ possibilities exist to induce neutrino NSI while avoiding the corresponding charged lepton NSI, they cannot saturate the mild direct bounds from Eqs. (3) and (5) and stronger $\mathcal{O}(10^{-3})$ bounds apply. In [9, 10] the generation of matter NSI avoiding similar operators involving charged leptons was also studied through $d = 8$ operators. In [9] the analysis was restricted to new physics realizations that did not involve cancellations between diagrams involving different messenger particles and, under those conditions, the bounds derived for the $d = 6$ realizations can be translated also to the $d = 8$ operators and, even if charged fermion NSI are avoided, no large neutrino NSI are obtained. In [10] it was shown that, allowing cancellation between diagrams involving different messengers, large neutrino NSI could be obtained by tunning away all other dangerous contributions to the charged lepton sector. However, this cancellation is only exact at zero momentum transfer, therefore constraints can be derived when probing larger energies at colliders, as studied in [19], and strong constraints $\varepsilon < 10^{-2} - 10^{-3}$ are again recovered through LEP2 data. Finally, the analyses of [9, 10] were limited to tree level effective operators. Therefore, they do not cover the possibility of loop-induced NSI or lighter mediators that cannot be integrated out of the theory. These can thus be ways out of the stringent constraints derived and potential sources of large NSI. Unfortunately, neither the loop generation nor lighter mediators seem to provide ways in which to evade gauge invariance and avoid the more stringent bounds leaking from the charged lepton sector. The generation of NSI via loop processes was studied in detail in [20] and applied in particular to the MSSM contributions. It was found that charged lepton flavour-violating processes constrain the induced NSI to the $\sim 10^{-3}$ level, as with the other alternatives explored. Concerning lighter mediators, this is a way out of the effective theory treatment presented in [9, 10], but gauge invariance still poses strong constraints and, indeed, flavour-conserving interactions, much more weakly constrained in the charged lepton sector than the flavour-violating counterparts, are usually considered instead.

3. Conclusions

We have reviewed the present model independent bounds on NSI. We find that production and detection NSI are bounded to be $< \mathcal{O}(10^{-2})$. Conversely the bounds on matter NSI are around one order of magnitude weaker. Saturating these mild direct bounds could lead to large observable signals at neutrino experiments. Matter NSI are however related to production and detection NSI and to flavour changing operators for charged leptons through gauge invariance. Exploring gauge invariant realizations we find that gauge invariant NSI are constraint to be $\mathcal{O}(10^{-3})$ making them very challenging, but not impossible, to probe at present [21, 22] facilities with dedicated new detectors or future neutrino oscillation facilities such as the Neutrino Factory [23, 24, 25, 26, 27, 28]. Furthermore, the stringent constraints from gauge invariance mainly apply to flavour-violating processes, thus, fully $SU(2)_L$ invariant operators can be considered with $\mathcal{O}(10^{-2})$ entries in some of the diagonal elements, see e.g. [29].

References

[1] Wolfenstein L 1978 *Phys. Rev.* D **17** 2369
[2] Mikheyev S P and Smirnov A Y 1985 *Sov. J. Nucl. Phys.* **42** 913–917
[3] Biggio C, Blennow M and Fernandez-Martinez E 2009 *JHEP* **08** 090
[4] Davidson S, Peña-Garay C, Rius N and Santamaria A 2003 *JHEP* **03** 011
[5] Barranco J, Miranda O G, Moura C A and Valle J W F 2006 *Phys. Rev.* D **73** 113001
[6] Barranco J, Miranda O G, Moura C A and Valle J W F 2008 *Phys. Rev.* D **77** 093014
[7] Bolanos A *et al.* 2009 *Phys. Rev.* D **79** 113012
[8] Biggio C, Blennow M and Fernandez-Martinez E 2009 *JHEP* **03** 139
[9] Antusch S, Baumann J P and Fernandez-Martinez E 2009 *Nucl. Phys.* B **810** 369–388
[10] Gavela M B, Hernandez D, Ota T and Winter W 2009 *Phys. Rev.* D **79** 013007
[11] Bilenky M S and Santamaria A 1994 *Nucl. Phys.* B **420** 47–93
[12] Bergmann S, Grossman Y and Pierce D M 2000 *Phys. Rev.* D **61** 053005
[13] Cuypers F and Davidson S 1998 *Eur. Phys. J.* C **2** 503–528
[14] De Gouvea A, Giudice G F, Strumia A and Tobe K 2002 *Nucl. Phys.* B **623** 395–420
[15] Broncano A, Gavela M B and Jenkins E E 2003 *Phys. Lett.* B **552** 177–184
[16] Antusch S, Biggio C, Fernandez-Martinez E, Gavela M B and Lopez-Pavon J 2006 *JHEP* **10** 084
[17] Abada A, Biggio C, Bonnet F, Gavela M B and Hambye T 2007 *JHEP* **12** 061
[18] Tommasini D, Barenboim G, Bernabeu J and Jarlskog C 1995 *Nucl. Phys.* B **444** 451–467
[19] Davidson S and Sanz V 2011 (*Preprint* arXiv: 1108.5320 [hep-ph])
[20] Bellazzini B, Grossman Y, Nachshon I and Paradisi P 2011 *JHEP* **1106** 104
[21] Alonso R *et al.* 2010 (*Preprint* arXiv: 1009.0476 [hep-ph])
[22] Antusch S, Blennow M, Fernandez-Martinez E and Ota T 2010 *JHEP* **06** 068
[23] Fernandez-Martinez E, Gavela M B, Lopez-Pavon J and Yasuda O 2007 *Phys. Lett.* B **649** 427–435
[24] Goswami S and Ota T 2008 *Phys. Rev.* D **78** 033012
[25] Kopp J, Ota T and Winter W 2008 *Phys.Rev.* D **78** 053007
[26] Antusch S, Blennow M, Fernandez-Martinez E and Lopez-Pavon J 2009 *Phys. Rev.* D **80** 033002
[27] Gago A, Minakata H, Nunokawa H, Uchinami S and Zukanovich Funchal R 2010 *JHEP* **1001** 049
[28] Coloma P, Donini A, Lopez-Pavon J and Minakata H 2011 *JHEP* **1108** 036
[29] Carpentier M and Davidson S 2010 *Eur.Phys.J.* C **70** 1071–1090

XIII International Workshop on Neutrino Factories, Super beams and Beta beams (NUFACT11) IOP Publishing
Journal of Physics: Conference Series **408** (2013) 012032 doi:10.1088/1742-6596/408/1/012032

Long-range interaction effects on neutrino oscillation

Hye-Sung Lee

Department of Physics, Brookhaven National Laboratory, Upton, NY 11973, USA

E-mail: `hlee@bnl.gov`

Abstract. Motivated by the recent anomaly in the muon neutrino and anti-muon neutrino disappearance experiments, we consider a long-range interaction with an extremely light gauge boson and extraordinarily weak coupling. A long-range interaction, consistent with current bounds, could have very pronounced effects on atmospheric neutrino disappearance that will be studied with the IceCube DeepCore array, currently in operation, and can have a significant effect on future high-precision long-baseline oscillation experiments. *This invited talk was based on the paper with Hooman Davoudiasl and William Marciano [1].*

1. Introduction

MINOS is a long baseline neutrino oscillation experiment from the Fermilab to Soudan Mine in Minnesota. It can study the muon neutrino and anti-muon neutrino oscillations over a baseline of $L = 735$ km with $E_\nu \sim$ GeV scale. The recent results of the MINOS ν_μ and $\bar{\nu}_\mu$ disappearance experiments show disagreement [2]. This cannot be explained by the standard oscillation picture since $P(\nu_\mu \to \nu_\mu) = P(\bar{\nu}_\mu \to \bar{\nu}_\mu)$ in true vacuum by the CPT invariance, and the matter effect in the ν_μ and $\bar{\nu}_\mu$ disappearance experiments at the MINOS are negligible.

It is likely due to the poor statistics especially in the $\bar{\nu}_\mu$ sector ($N_\nu \sim 2000$, $N_{\bar{\nu}} \sim 100$). Nevertheless, it is still possible that this is a hint of new physics that can affect neutrino oscillations differently from the anti-neutrino oscillation. In this talk, we go over a possible explanation and briefly discuss its implications for other neutrino experiments. A detailed discussion can be found in Ref. [1].

2. Long-range interaction and neutrino oscillation

What type of new physics would be able to explain the MINOS anomaly? If we look at the standard model interactions, the W boson gives a potential only to the electron neutrinos, and it can affect neutrino oscillations involving the electron neutrino. But the Z boson gives a flavor-universal potential to the neutrinos making no effect in neutrino flavor oscillations:

$$H_{\rm SM} = V_W \ (1,0,0) + V_Z \ (1,1,1)$$

with $V_W = \sqrt{2} G_F n_e$, $V_Z = -G_F n_n / \sqrt{2}$ where $n_e(n_n)$ is the electron(neutron) number density.

We would like to consider a lepton flavor-dependent long-range interaction (LRI). We assume a lepton flavor-dependent $U(1)$ gauge symmetry with an almost massless gauge boson Z'. (Although this is our set up, it may not be necessary to consider an Abelian gauge symmetry or even a LRI to address the MINOS anomaly.)

$$H_{\rm LRI} = V_{Z'} \begin{pmatrix} Q_e & & \\ & Q_\mu & \\ & & Q_\tau \end{pmatrix}$$

There are related works prior to our study about the LRI effects on neutrino oscillations and the MINOS anomaly explanation with a new interaction. (For a limited list, see Refs. [3, 4, 5, 6, 7, 8, 9, 10, 11].)

Under the new potential, the effective muon neutrino survival probability in the 2-flavor oscillation limit can be described by the effective mass splitting and mixing angle [3]: $\Delta\tilde{m}^2_{23} = \Delta m^2_{23}([\xi - \cos(2\theta_{23})]^2 + \sin^2(2\theta_{23}))^{1/2}$ and $\sin^2(2\tilde{\theta}_{23}) = \sin^2(2\theta_{23})/([\xi - \cos(2\theta_{23})]^2 + \sin^2(2\theta_{23}))$ with $\xi \equiv -2W_\tau E_\nu/\Delta m^2_{23}$ and $W_\tau = Q_\tau V_{Z'}$ (potential energy difference in $\nu_\mu - \nu_\tau$). The ξ flips sign for $\bar{\nu}$ causing different effects on ν and $\bar{\nu}$ unless $\sin^2(2\theta_{23}) = 1$.

In 1955, T.D. Lee and C.N. Yang discussed that LRIs get constraints from the Eötvös-type experiments [12]. The coupling constants (fine-structure constants) are constrained to be very small [13]: $\alpha' < 10^{-47}$ (for Q = baryon number), $\alpha' < 10^{-49}$ (for Q = lepton number). So it is clear that we need astronomical sources to have any meaningful effects, which is possible since we are dealing with a LRI. We assume $m_{Z'} < 1/AU \sim 10^{-18}$ eV to include the Sun (but not much smaller to avoid galaxy and beyond).

For an anomaly-free $U(1)'$ charges, we choose $Q = (B-L) + (L_\mu - L_\tau) = B - L_e - 2L_\tau$ or $H_{\rm LRI} = V_{Z'}\ (-1, 0, -2)$. Neutrons in the Sun and the Earth are the source of the new potential (due to $B-L$), and the flavor-dependent charges $(L_\mu - L_\tau)$ affect ν flavor oscillation.

The Sun and the Earth have neutrons of $N_n^\odot = 1.7 \times 10^{56}$ and $N_n^\oplus = 1.8 \times 10^{51}$, respectively, and the new potential by the neutrons is given by $V_{Z'} \sim (\alpha'/10^{-50}) \times \mathcal{O}(10^{-12}\ \text{eV})$. Since the MINOS ν oscillation is relevant to $\Delta m^2_{23}/E_\nu \sim \mathcal{O}(10^{-12}\ \text{eV})$, a LRI with $\alpha' \sim \mathcal{O}(10^{-50})$ level can affect MINOS experiments significantly. In other words, the ν oscillation experiments are a good probe of an extremely weak LRI.

We performed a rough fit to the MINOS data (for both the neutrinos and anti-neutrinos) with the LRI, and found the best-fit point at $\Delta m^2_{23} = 2.4\times10^{-3}\ \text{eV}^2$, $\sin^2(2\theta_{23}) = 0.9$, $\alpha' = 1.0\times10^{-52}$ (or $W_\tau = 5.6 \times 10^{-14}$ eV), using the simplified MINOS data from Ref. [9]. (The MINOS data constrains $\alpha' < 5 \times 10^{-52}$ at roughly 3σ level.)

This best-fit point is not positively ruled out by solar+KamLAND ν and atmospheric ν data, although there is a tension between the potential for MINOS data and SK atmospheric ν data (See Appendix A). This best-fit with LRI does not really improve the goodness-of-fit of the MINOS data over the SM fitting. We rather take it as a motivated benchmark point to explore other experiments to test the LRI idea. For this purpose, we want to consider two setups, the IceCube DeepCore experiment, and the future long baseline neutrino experiment (LBNE).

3. Implications for IceCube DeepCore and future LBNE experiments

DeepCore is composed of 6 additional densely instrumented strings plus 7 nearby IceCube strings, and it is located in the bottom center of the IceCube experiment at the South Pole. It was recently commissioned and the analysis of the first year data is in progress (see F. Halzen's talk). DeepCore is a high statistics detector that can study the atmospheric neutrino with precision [14]. It can trigger $\mathcal{O}(10^5)$ atmospheric neutrinos per year in $E_\nu \approx 1-100$ GeV, which is complementary to the original IceCube (optimized for $E_\nu > 10$ TeV).

We would like to see the LRI effects on the atmospheric muon neutrino disappearance at DeepCore. Figure 1 shows the numerical results without a full detector simulation. The black dashed curve is for the unoscillated case, the black solid curve is for the standard oscillation, and the red solid curve is for the MINOS best-fit case. It shows that LRI effects on the

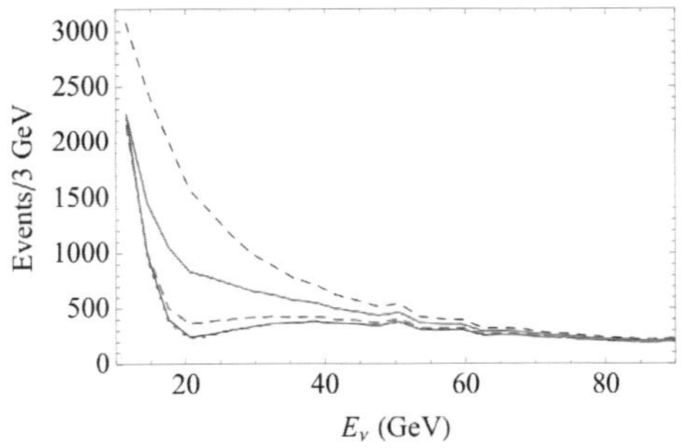

Figure 1. Atmospheric muon neutrino disappearance at DeepCore. The curves are the unoscillated oscillation (black dashed), standard oscillation (black solid), and the MINOS best-fit (red solid).

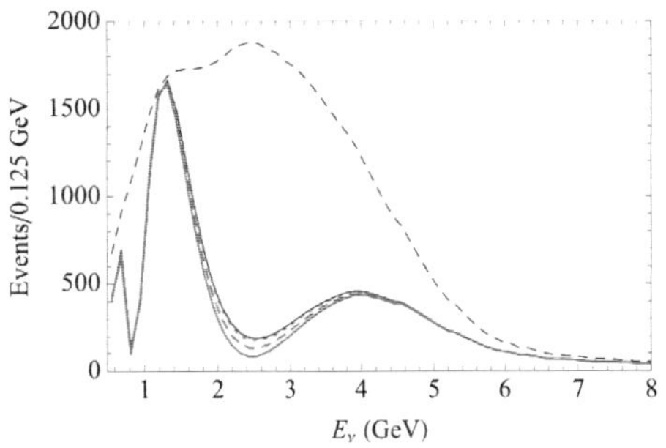

Figure 2. The ν_μ disappearance at the future LBNE (L = 1300 km, 200 kton water Cherenkov detector, 5-years run). The shapes and colors of the curves are the same as Fig. 1.

DeepCore atmospheric muon neutrino disappearance experiment are quite distinct from the standard oscillation.

Another interesting aspect that we may be able to observe at the DeepCore experiment is the annual modulation in the new potential. Since the distance between the Sun and the Earth changes over the year (about 3% level), the new potential would have an annual modulation, which means that the effective oscillation parameters might change with seasons. With an assumption of uniform flux (before oscillation) over the year, our calculation shows it can give a percent level annual modulation in the muon neutrino disappearance experiment, with a total number of a few thousands per year.

Now, we consider the implication for the future LBNE. Figure 2 shows the muon neutrino disappearance experiments for the envisioned LBNE at DUSEL. It assumes the baseline of $L = 1300$ km and 200 kton water Cherenkov detector [15]. As it shows, for the first minimum, the muon neutrino with the new potential has fewer events than the standard oscillation, which means larger $\sin^2(2\theta_{23})$. The corresponding result for the muon antineutrino has more events (or smaller $\sin^2(2\theta_{23})$) at the first minimum. The LBNE can tell the different LRI effects on ν and $\bar{\nu}$. The observation of the annual modulation effect, however, may need enhanced capabilities compared to what we considered.

4. Implications for the charged lepton sector

Let us make a brief comment on the implications of the LRI for the charged lepton sector. We have a lepton-flavor-dependent $U(1)$ gauge symmetry with a nearly massless gauge boson Z'. When we have a flavor-dependent U(1) gauge symmetry, Z'-mediated flavor changing neutral currents (FCNCs) at tree-level are present, in general. For instance, μ-e-Z' vertex may exist in the mass eigenstate.

The question is whether we can neglect this Z'-mediated FCNC because it has a tiny coupling ($\alpha' < 10^{-50}$) as constrained by the Eötvös-type experiments. We cannot neglect this, in general, because of the Goldstone boson equivalence principle. For instance, the $\mu \to eZ'$ decay is enhanced since $m_{Z'} \ll m_\mu$. Then, $\Gamma(\mu \to eZ') \sim \alpha' m_\mu (m_\mu^2/m_{Z'}^2)$ can lead to a too fast muon decay because of the extremely small $m_{Z'}$. Thus, explicit model buildings with Higgs sector should address this issue.

5. Summary

MINOS data may be a hint for the new physics which acts differently on neutrinos and antineutrinos. The lepton-flavor-dependent LRI with $\alpha' \sim 10^{-52}$ is such a possibility since neutrino oscillations are sensitive to it. The IceCube DeepCore experiment, which is an ongoing experiment, can test this possibility. Especially, if the annual modulation is observed, it can point to the solar origin of the new potential. A future LBNE can test if the LRI gives different effects on the neutrino and antineutrino oscillations. LRI effects on the charged lepton sector are not negligible, in general, and depend on the explicit Higgs sector model building.

Acknowledgments

This work was supported in part by the U.S. Department of Energy under Grant Contract No. DE-AC02-98CH10886. It is a great pleasure to thank my collaborators H. Davoudiasl and W. Marciano.

Appendix A. Constraints on the new potential

Our MINOS best-fit is at $\alpha' = 1.0 \times 10^{-52}$ or at the potential energy $W_\tau = 5.6 \times 10^{-14}$ eV.

The constraint on the LRI from the solar neutrino and KamLAND reactor neutrino was studied in Refs. [5, 6], which is, for our model, $\alpha' < 5 \times 10^{-52}$ at 3σ level.

The constraints from the atmospheric neutrino data measured at the Super-Kamiokande is given in terms of the $\epsilon_{\tau\tau}$ with $\epsilon_{\tau\tau} < 0.2$ at 95% CL [16]. Its corresponding value for the potential energy ($W_\tau = \epsilon_{\tau\tau}\sqrt{2}G_F n_e$) depends on n_e (electron number density). The Earth core ($R < 3400$ km) has $n_e \approx 12$ g/cm^3 ($W_\tau < 9 \times 10^{-14}$ eV) and the mantle ($R > 3400$ km) has $n_e \approx 5$ g/cm^3 ($W_\tau < 4 \times 10^{-14}$ eV). Our MINOS best-fit potential energy is somewhere in the middle, which shows some tension though it calls for a more detailed analysis in order to pinpoint a precise constraint on the LRI.

References

[1] Davoudiasl H, Lee H S and Marciano W J 2011 *Phys. Rev.* D **84** 013009
[2] Adamson P *et al* (MINOS collaboration) 2011 *Phys. Rev. Lett.* **107** 021801
[3] Joshipura A S and Mohanty S 2004 *Phys. Lett.* B **584** 103
[4] Grifols J A and Masso E 2004 *Phys. Lett.* B **579** 123
[5] Gonzalez-Garcia M C, de Holanda P C, Masso E and Zukanovich Funchal R 2007 *JCAP* **0701** 005
[6] Bandyopadhyay A, Dighe A and Joshipura A S 2007 *Phys. Rev.* D **75** 093005
[7] Engelhardt N, Nelson A E and Walsh J R 2010 *Phys. Rev.* D **81** 113001
[8] Mann W A, Cherdack D, Musial W and Kafka T 2010 *Phys. Rev.* D **82** 113010
[9] Kopp J, Machado P A N and Parke S J 2010 *Phys. Rev.* D **82** 113002
[10] Heeck J and Rodejohann W 2011 *J. Phys.* G **38** 085005
[11] Samanta A 2011 *JCAP* **1109** 010
[12] Lee T D and Yang C N 1955 *Phys. Rev.* **98** 1501
[13] Dolgov A D 1999 *Phys. Rept.* **320** 1
[14] Fernandez-Martinez E, Giordano G, Mena O and Mocioiu T 2010 *Phys. Rev.* D **82** 093011
[15] Diwan M 2010 *Talk at DURA Annual Meeting and DUSEL PDR Rollout, FermiLab, September 2-3, 2010* (`http://www.dusel.org/workshops/fallworkshop10/milind_dura_ralk.pdf`)
[16] Friedland A, Lunardini C and Maltoni M 2004 *Phys. Rev.* D **70** 111301

Non standard neutrino interactions at LEP2 and the LHC

Sacha Davidson[1] **and Veronica Sanz**[2]
[1] IPNL, Université de Lyon, Université Lyon 1, CNRS/IN2P3, 4 rue E. Fermi 69622 Villeurbanne cedex, France
[2] Department of Physics and Astronomy, York University, Toronto, ON, Canada

E-mail: s.davidson@ipnl.in2p3.fr, vsanz@yorku.ca

Abstract. We consider Non-Standard neutrino Interactions (NSI) connecting two neutrinos with two first-generation fermions (e, u or d), which we assume to arise at dimension eight due to New Physics. The coefficient is normalised as $4\varepsilon G_F/\sqrt{2}$. We explore signatures of NSI-on-electrons at LEP2, and of NSI-on-quarks at the LHC, treating the NSI as contact interactions at both energies. In models where the coefficients of dangerous dimension six operators are suppressed by cancellations, LEP2 provides interesting bounds on NSI operators ($\varepsilon \lesssim 10^{-2} - 10^{-3}$), which arise because $\sqrt{s} \sim 200$ GeV, and the cancellation applied at zero momentum transfer. At the LHC, we use the Equivalence Theorem, which relates the longitudinal W to the Higgs, to estimate the rate for $\overline{q}q \to W^+W^-e^+_\alpha e^-_\beta$ induced by NSI. We find that the cross-section is small, but that the outgoing particles have very high $p_T > 400$ GeV, which reduces the issue of backgrounds. In a conservative scenario, we find that the LHC at 14 TeV and with 100 fb^{-1} of data would have a sensitivity to $\varepsilon \gtrsim 3 \times 10^{-3}$.

1. Introduction

Many extensions of the Standard Model, such as Supersymmetry or leptoquarks, naturally induce low energy contact interactions of the form:

$$\varepsilon^{fX}_{\alpha\beta} \frac{4G_F}{\sqrt{2}} (\overline{f}\gamma^\rho P_X f)(\overline{\nu}_\alpha \gamma_\rho \nu_\beta) \tag{1}$$

where $f \in \{u, d, e\}$ is a first generation charged Standard Model fermion, and $\alpha, \beta \in \{e, \mu, \tau\}$. These are referred to as Non Standard neutrino Interactions (NSI) [1, 2], and can be generated by $SU(3) \times SU(2) \times U(1)$ gauge invariant operators at dimension six or higher. Future neutrino facilities, such as a Neutrino Factory [1], could be sensitive to such interactions with $\varepsilon \gtrsim 10^{-4}$. Formalism and current bounds [3] on NSI are reviewed in the paper on which this proceedings is based [4].

This proceedings reports on a preliminary exploration [4] of the complementarity of current collider experiments and future neutrino facilities to these neutral current operators. We focus on neutral current NSI induced at dimension eight, from operators such as

$$\frac{1}{\Lambda_8^4} (\overline{q}\gamma^\rho P_L q)(\overline{H\ell}_\alpha \gamma_\rho H\ell_\beta) \tag{2}$$

where q and ℓ are SM doublets, and H is the Higgs. Comparing with eqn (1) gives $v^2/\Lambda_8^4 = \varepsilon/v^2$, where $v = \langle H \rangle = 174$ GeV. So $\Lambda_8 \lesssim 2$ TeV, suggesting that new mediating particles, if weakly coupled or in loops, are kinematically accessible to the LHC. In this case, their discovery prospects are model-dependent, and have been widely studied[5]. Here, to retain some degree of model independence, we consider effective operators at colliders (LEP2, LHC). So we are assuming heavy New Physics with $\gtrsim 1$ couplings. We also assume that the New Physics does not generate "dangerous" dimension six operators involving two charged leptons instead of two neutrinos, because these are strictly constrained [2, 6].

In section 2, we argue that in some models, such as those considered by Gavela *et.al.* [6], the dimension eight NSI interaction of the form (1) is accompanied by dimension eight contact interactions involving charged leptons rather than neutrinos, with coefficients $\sim s/\Lambda_8^4$, $(t-u)/\Lambda_8^4$, where s, t and u are the Mandelstam variables.

In section 3, we use the Equivalence Theorem to replace $\langle H \rangle \nu_\alpha \to W^+ e_\alpha^-$, and study the prospects for detecting $q\bar{q} \to W^+W^-\ell_\alpha^+\ell_\beta^-$ at the LHC. Rough estimates suggest that couplings $\gtrsim 1$ are required for NSI to function as dimension eight contact interactions at LHC energies (assuming $\varepsilon \gtrsim 10^{-4}$). Such contact interactions would induce few events ($\sigma(pp \to W^=W^-\ell^+\ell^-) \sim 10^{-3}$ fb $\times(10^{-4}/\varepsilon)^2$), but at very high p_T where Standard Model backgrounds are negligeable. We estimate that the LHC could be sensitive to $\varepsilon \gtrsim 3 \times 10^{-3}$.

2. LEP2 bounds on dimension 8 derivative operators

Some NSI models, which suppress dangerous dimension six operators via a cancellation [6], could give rise to four fermion contact interactions with coefficients $\propto \{s,t,u\}/\Lambda_8^4$. Such contact interactions are subject to LEP2 bounds. We present estimates of the translation of the LEP bounds to the ε coefficient of NSI.

As an example of a cancellation in the coefficient of a dangerous dimension six operator, consider a model containing an SU(2) doublet vector and scalar with large masses m_V,m_S, and couplings

$$h(\bar{e}\ell)\tilde{S}_2^\dagger + g(\overline{e^c}\gamma_\mu\ell)V_2^\mu \tag{3}$$

The exchange of these two bosons gives the operators

$$-\frac{h^2}{m_S^2 - t}(\bar{e}\ell)(\bar{\ell}e) + \frac{g^2}{m_V^2 - u}(\overline{e^c}\gamma^\mu\ell)(\bar{\ell}\gamma_\mu e^c) = \left(-\frac{g^2}{m_V^2 - u} + \frac{h^2}{2(m_S^2 - t)}\right)(\bar{\ell}\gamma^\mu\ell)(\bar{e}\gamma_\mu e) \tag{4}$$

where t, u are Mandelstam variables. At zero momentum transfer, the coefficient can be cancelled by choosing $g^2/m_V^2 = h^2/(2m_S^2)$. However, at dimension 8, operators arise such as

$$\frac{g^2(t/m_S^2 - u/m_V^2)}{m_V^2}(\bar{\ell}\gamma^\mu\ell)(\bar{e}\gamma_\mu e) \tag{5}$$

We consider two independent dimension eight operators, with coefficients $\propto s/\Lambda_8^4$ and $\propto (t-u)/\Lambda_8^4$, and that have four visible fermion legs.

The LEP2 experiments searched for dimension six four fermion contact interactions in the channels $e^+e^- \to e^+e^-, \mu^+\mu^-, \tau^+\tau^-, \bar{q}q$ (where $q \in \{u,d,s,c,b\}$ is a light quark. The published bounds [7, 8, 9] can be translated (to within factors of 2, see [4]) to bounds on dimension eight operators, $\propto s/\Lambda_8^4$ or $(t-u)/\Lambda_8^4$, with the same external legs. If the NP inducing the derivative operators has $\mathcal{O}(1)$ couplings to the Higgs, then bounds on the ε coefficient of the dimension eight NSI operator, can be obtained by taking $s \sim v^2$. See [4] for details.

The resulting limits are listed in figure 1. Although the bounds are quoted with two significant figures, they are merely order of magnitude estimates, as various constants could appear in the passage between four-charged-lepton-derivative operators and NSI operators.

The OPAL experiment saw one $e^+e^- \to e^\pm\mu^\mp$ event at $\sqrt{s} = 189 - 209$ GeV, and published limits [10] on $\sigma(e^+e^- \to e^\pm\mu^\mp, e^\pm\tau^\mp, \tau^\pm\mu^\mp)$ at LEP2 energies. This analysis allows to set the bounds on ε for flavour-changing NSI given in figure 1.

$(\bar{e}\gamma^\mu P_X e)(\bar{\ell}\gamma_\mu P_Y \ell)$	ε
$e^+e^- \to e^+e^-$	
XY=LL	$\lesssim 3.7 \times 10^{-3}$
LL	$\lesssim 5.6 \times 10^{-3}$
RL	$\lesssim 4.7 \times 10^{-3}$
RL	$\lesssim 2.4 \times 10^{-3}$
$e^+e^- \to \mu^+\mu^-$	
XY=LL	$\lesssim 5.9 \times 10^{-3}$
LL	$\lesssim 4.3 \times 10^{-3}$
RL	$\lesssim 9.1 \times 10^{-3}$
$e^+e^- \to \tau^+\tau^-$	
XY=LL	$\lesssim 6.2 \times 10^{-3}$
LL	$\lesssim 1.1 \times 10^{-2}$
RL	$\lesssim 9.1 \times 10^{-3}$
RL	$\lesssim 1.8 \times 10^{-2}$

$(\bar{e}\gamma^\mu P_X e)(\bar{\ell}\gamma_\mu P_Y \ell)$	ε
$e^+e^- \to e^\pm\mu^\mp$	
$\forall\ XY$	$\lesssim 8.7 \times 10^{-3}$
$e^+e^- \to e^\pm\tau^\mp$	
$\forall\ XY$	$\lesssim 1.6 \times 10^{-2}$
$e^+e^- \to \tau^\pm\mu^\mp$	
$\forall\ XY$	$\lesssim 1.5 \times 10^{-2}$

$(\bar{e}\gamma^\mu P_L e)(\bar{q}\gamma_\mu P_Y q)$	$\varepsilon \lesssim$
Y=L	$\lesssim 4.8 \times 10^{-3}$
Y=R	$\lesssim 1.8 \times 10^{-2}$

Figure 1. Estimated bounds on ε, from LEP2, obtained using the assumptions outlined in the text. These could apply to NSI models where the coefficient of dimension six operators is suppressed by a cancellation.

3. LHC discovery reach

As mentioned before, the bound on $\Lambda_8 \lesssim 2TeV$ suggests that the New Physics particles responsible of NSIs would show up as resonances at the LHC, provided the interactions are perturbative. We want to take a more pessimistic approach, and study the prospects at the LHC for NSI involving quarks, in the improbable, but compartively model-independent, scenario that NSIs appear as contact interactions.

In a gauge invariant dimension eight NSI operator, the $H_0^* H_0 \bar{\nu}_\alpha \nu_\beta$ interaction is accompagnied by $H^+H^-\bar{e}_\alpha e_\beta$, which could be expected to reincarnate, after electroweak symmetry breaking, as a vertex involving $W^+W^-\bar{e}_\alpha e_\beta$. This expectation can be formalised, at energies $\gg m_W$, via the Equivalence Theorem[11], which identifies the Goldstone $H^\pm$ with the longitudinal componentof the $W^\pm$.

In the Equivalence Theorem limit, the partonic cross-section for $\bar{q}q \to e_\alpha^+ e_\beta^- W^+W^-$ is calculable, and small due to four body phase space. We obtain $\sigma(pp \to H^+H^-e_\alpha^+e_\beta^-)$ plotted in figure 2.

To determine the reach of the LHC in $\epsilon_{\alpha\beta}$, we need to consider backgrounds, such as the $t\bar{t}$ + jets, and signal selection. We simulated NSI events using FeynCalc [12] and MadGraph v5 [13], to obtain the p_T distribution of the signal leptons Fig. 3. We also plot the p_T of objects in a $t\bar{t}$ sample. The signal and background distributions are well separated, so asking for a p_T cut of order 400 GeV would reduce the $t\bar{t}$ backgrounds to less than 10^{-5} fb and keeps 70% of our signal. In particular, this suggests sensitivity to NSI interactions producing taus. However, if such high-p_T events were seen, boosted taus would be hard to tag, although they may show up as fat jets.

Summarizing, the cross-section is small, but the signal is characterized by well-separated, highly boosted objects in a pretty spherical event. Using these characteristics, especially a cut

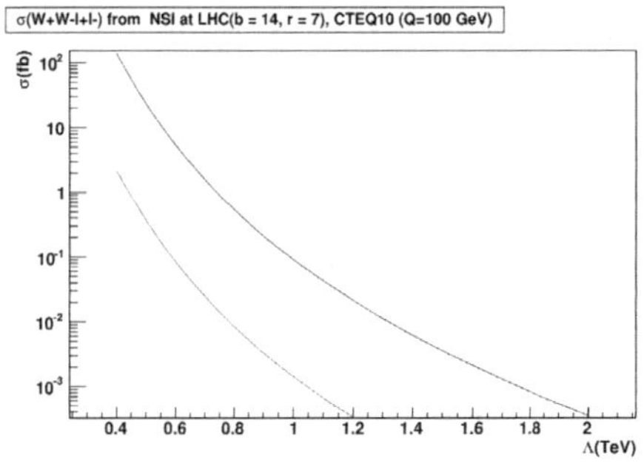

Figure 2. $\sigma(pp \to W^+W^-e^+_\alpha e^-_\beta)$ in fb, at the LHC with 14 TeV or 7 TeV, due to a contact interaction with coefficient $1/\Lambda_8^4$.

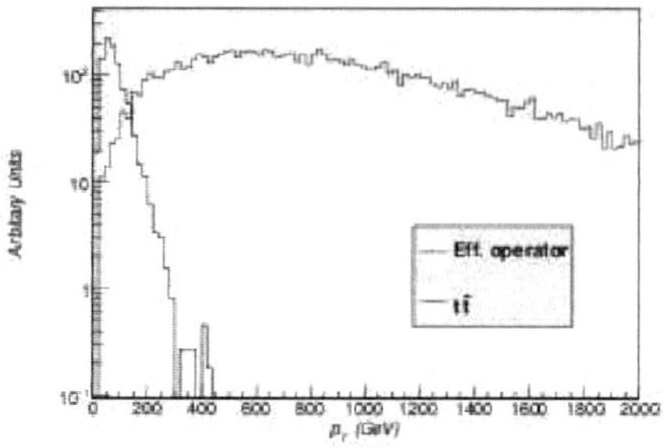

Figure 3. p_T of leptons at the 14 TeV LHC from NSI (flat) and $t\bar{t}$ (peaked), normalised to fit both lines in the same scale.

on p_T, we showed that the largest background, from $t\bar{t}$ can be reduced below the signal. The LHC reach depends on the value of ε, but if we assume the NSI signal is background-free, and ask for 100 events at a luminosity $\mathcal{L}$ (in fb^{-1}), then the reach in ε is

$$\varepsilon \sim 3 \times 10^{-2}/\sqrt{\mathcal{L}} \tag{6}$$

For example, for a luminosity of 100 fb^{-1}, the 14 TeV LHC could be sensitive to NSI-induced contact interactions corresponding to

$$\varepsilon \gtrsim 3 \times 10^{-3} \quad . \tag{7}$$

References

[1] see *e.g.* Bandyopadhyay A *et al.* (ISS Physics Working Group Collaboration) 2009 *Rept. Prog. Phys.* **72** 106201
[2] Antusch S, Baumann J P and Fernandez-Martinez E 2009 *Nucl. Phys.* B **810** pp 369-388
[3] Biggio C, Blennow M and Fernandez-Martinez E 2009 *JHEP* **0908** 090 (*Preprint* arXiv:0907.0097 [hep-ph])
[4] Davidson S and Sanz V *Non-Standard Neutrino Interactions at Colliders Preprint* arXiv:1108.5320 [hep-ph]
[5] See for example the ATLAS and CMS Technical Design Reports and references therein;
Aad G *et al.* (ATLAS Collaboration) *Preprint* arXiv:0901.0512 [hep-ex];
Bayatian G L *et al.* (CMS Collaboration) 2007 *J. Phys. G* **34** 995
[6] Gavela M B, Hernandez D, Ota T and Winter W 2009 *Phys. Rev.* D **79** 013007
[7] Bourilkov D 2001 *Phys. Rev.* D **64** 071701 (*Preprint* hep-ph/0104165)
[8] Schael S *et al.* (ALEPH Collaboration) 2007 *Eur. Phys. J.* C **49** 411 (*Preprint* hep-ex/0609051)
[9] Abbiendi G *et al.* (OPAL Collaboration) 2004 *Eur. Phys. J.* C **33** (2004) 173 (*Preprint* hep-ex/0309053)
[10] Abbiendi G *et al.* (OPAL Collaboration) 2001 *Phys. Lett.* B **519** pp 23-32 (*Preprint* hep-ex/0109011)
[11] Cornwall J M, Levin D N and Tiktopoulos G 1974 *Phys. Rev.* D **10** 1145;
Lee B W, Quigg C and Thacker H B 1977 *Phys. Rev.* D **16** 1519;
Chanowitz M S and Gaillard M K 1985 *Nucl. Phys.* B **261** 379;
Veltman H G J 1990 *Phys. Rev.* D **41** 2294
[12] See http://www.feyncalc.org/ for documentation.
[13] Alwall J, Herquet M, Maltoni F, Mattelaer O and Stelzer T 2011 *MadGraph 5: Going Beyond JHEP* **1106** 128 (*Preprint* arXiv:1106.0522 [hep-ph])

Journal of Physics: Conference Series **408** (2013) 012034 doi:10.1088/1742-6596/408/1/012034

If θ_{13} is large, then what?

Hisakazu Minakata

Department of Physics, Tokyo Metropolitan University, Hachioji, Tokyo 192-0397, Japan

E-mail: hisakazu.minakata@gmail.com

Abstract. As indicated by the recent data obtained by the T2K and the MINOS experiments θ_{13} can be large, even be comparable to the Chooz limit. Assuming that it will be confirmed by ongoing reactor and accelerator experiments I discuss its possible implications in the context of how to explore the remaining unknowns in the lepton flavor mixing. While it opens wide range of possibilities to explore CP and the mass hierarchy, I emphasize that the large θ_{13} allows us to take "all in one" (everything in a lunch box) approach.

1. It was a speculative title when it was proposed

You might have suspected that my title is too timely, but as a matter of fact, it was a speculative one when it was proposed to one of the conveners in early May, much earlier than the announcement of ν_e appearance events at T2K [1]. The resulted six clean events which fulfill the criteria for electron-like events in SK strongly suggest that θ_{13} has a nonzero value. Their best fit value is around $\theta_{13} \sim 10°$ close to the Chooz limit [2]. It was soon followed by the MINOS collaboration which also obtained a relatively large value of θ_{13} [3]. In fact, by taking account of these new data from T2K and MINOS into the global analysis the Bari group concluded that it implies evidence for nonzero θ_{13} at confidence level (CL) higher than 3σ [4]. The best fit value is $\theta_{13} \simeq 8°$. See also [5]. I believe it an urgent matter to discuss what are the consequences of suggested large values of θ_{13} though it must be confirmed by the further running of these experiments as well as the ongoing reactor θ_{13} experiments [6, 7, 8].

2. Why θ_{13} so large?

I stated in a talk in Neutrino 2008 in Christchurch that "my personal prejudice is that θ_{13} is relatively large. If it is true it further encourages us to proceed to search for lepton CP violation and determination of the mass hierarchy" [9]. Here is a brief recapitulation of the very simple reasoning which led me to the above statement. The lepton flavor mixing matrix, the MNS matrix [10], U_{MNS}, is a "relative matrix" $U_\ell^\dagger U_\nu$, where U_ℓ and U_ν denote the unitary matrices which diagonalize the charged lepton and the neutrino mass matrices, respectively. The MNS matrix is composed of the three angles θ_{ij} $(i, j = 1-3)$ and a CP violating phase δ_ℓ, the lepton analogue of the quark Kobayashi-Maskawa phase δ_q [11]. Two of the angles are known to be large, $\theta_{12} \simeq 34°$ and $\theta_{23} \simeq 45°$. Then, it is highly unlikely that only the remaining angle θ_{13} is extremely small, unless there is a good reason for tuning only θ_{13} into a small value. That is, if a symmetry exists such that θ_{13} vanishes at the symmetry limit, there is a natural reason why only θ_{13} is small among the three mixing angles. There exists many examples for such symmetry including $\mu - \tau$ symmetry and discrete symmetries. For a review see [12]. It is my prejudice that the large θ_{13} suggested by T2K and MINOS measurement, if confirmed, would give a tough time for these symmetries which aims at explaining small θ_{13}, unless a large correction to the symmetry limit is shown to be naturally induced.

3. Wide range of possibilities are now open for measuring CP and mass hierarchy

Without having compelling reasons for large θ_{13}, one had to take into account the possibility that θ_{13} is small in thinking about ways for measurement of CP phase and determination of the neutrino mass hierarchy. That is why many of the existing proposals for such apparatus naturally include those that have sensitivities to a tiny value of θ_{13} with varying limiting values. They include, for example (sorry for my arbitrary choice), neutrino factory [13, 14], beta beam [15], superbeam enriched with two detectors [16], and their combination [17]. For their sensitivities see [18]. I wonder, however, that racing for θ_{13} sensitivity could have produced an atmosphere that the better facility is the one which has sensitivities to smaller values of θ_{13}. In seeing the T2K and MINOS indication for large θ_{13}, I believe that it is the time to start thinking about changing the gear to more relevant things.

What is good if θ_{13} is as large as $\sim 10°$? My first answer is that a wide range of possibilities becomes open for strategies and setups aiming at exploring lepton CP violation through neutrino oscillation. I believe that in the context of conventional superbeam [19] unambiguous evidence for lepton CP violation requires a megaton scale water Cherenkov detector, or a $\sim$ 100 kton liquid Ar detector [20, 21, 22]. Yet, the suggested large value of θ_{13}, still need to be confirmed and assumed here, stimulates to examine other possibilities such as reactor-accelerator combined method [23], combining different accelerator measurement [24, 25, 26], or the one adding one more new facility which may or may not require an extensive cost [17, 27]. Some more exotic options are discussed in [9]. It is notable that the CERN-MEMPHYS project, despite an almost vacuum setting, has a reasonable sensitivity to the mass hierarchy to $\sin^2 2\theta_{13} \simeq 0.03$ before combining with atmospheric neutrino data [28].

Sometime ago the CP sensitivity achievable by the reactor-accelerator combined method was examined in detail [29]. The results obtained in this analysis indicate that combining data taken and to be taken by all planned experiments is not enough to guarantee discovery of CP violation in δ_ℓ coverage of more than 30% in most of the allowed region of θ_{13} even though one assumes that the mass hierarchy is known. It should be noticed, as observed and emphasized in [23], that the CP sensitivity is severely damaged by the parameter degeneracy due to unknown neutrino mass hierarchies [30]. For a recent overview of the parameter degeneracy see [31].

4. "All in one" approach becomes feasible

I believe that the one of the most important implications of the assumed large value of θ_{13} is that one can think of "all in one" (*obento* in Japanese, everything in a lunch box) approach. As we saw in the above, a traditional approaches for CP and the mass hierarchy entailed in aggressive proposals [13, 15, 16, 17] to include a very small value of θ_{13} within the scope. Though powerful and the redundancy is of course welcome, they may not be the optimal cost-effective apparatus for large θ_{13}. To my opinion, what is good in living in the world with large θ_{13} is that one can achieve the same goal by less expensive "all in one" setting. More concretely, I mean by it a single detector (such as a megaton water Cherenkov detector) assuming prior existence of intense neutrino beam. Timely enough the LOI of Hyper-Kamiokande (Hyper-K) project has just appeared [32]. Therefore, let us briefly discuss the spirit of the "all in one" approach by taking this concrete setting.

With intense neutrino beam from 1.66 MW proton driver at J-PARC it is demonstrated that Hyper-K with 0.56 megaton fiducial mass has a superb performance for discovery of CP violation, which covers most of the region of θ_{13} allowed at 3σ CL [4]. See Fig. 24 of [32] which assumes the known mass hierarchy and for 1.5 (3.5) years of neutrino (antineutrino) running.

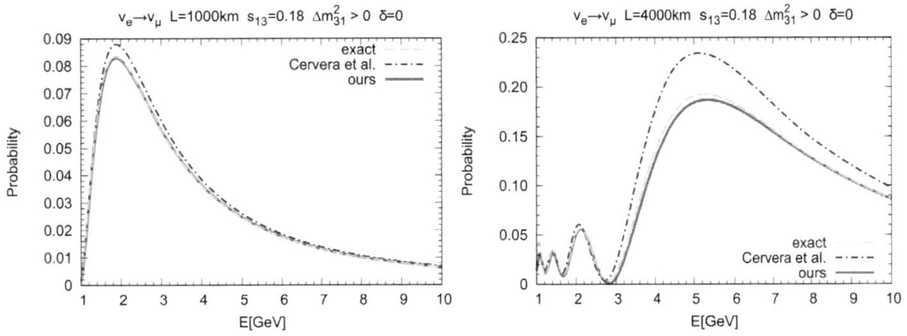

Figure 1. Comparison between the exact oscillation probability $P(\nu_e \rightarrow \nu_\mu)$ computed numerically as a function of energy (- - -), calculated by using the second-order formula in [14] (— · —), and with our formula with large θ_{13} corrections (——). The left and right panels are for baselines $L = 1000$ km and $L = 4000$ km for which the matter density is taken as 2.8 g/cm^3 and 3.6 g/cm^3, respectively.

It shows a remarkable accuracy of determination of δ_ℓ with 1σ error smaller than 20 degree for $\sin^2 2\theta_{13} \gtrsim 0.03$. However, if the mass hierarchy is unknown the analysis suffers from the sign-Δm^2 degeneracy [30].

While allowing clean measurement of CP violation, a drawback of setting with such short baseline is that it is difficult to determine the mass hierarchy, another important goal of future neutrino experiments. That is why many proposals for determining all the unknowns in neutrino parameters exploit baselines of several thousand km. If θ_{13} is large, however, the mass hierarchy can be determined by high-statistics observation of atmospheric neutrinos, the idea discussed by many authors. See [33, 34] for early references. The analysis presented in [32] indicates that the sensitivity to the mass hierarchy depend very much on which octant θ_{23} lives but it can be carried out by $\sim 3-10$ years running of Hyper-K for large θ_{13}. It should also be mentioned that having other capabilities such as proton decay discovery is of crucial importance. I emphasize that the realization of "all in one" approach is, of course, not limited to this particular version discussed above. But, it is certainly encouraging to see the concrete example appeared which is armed with the realistic simulations based on their long-term experiences at Super-K.

5. Large-θ_{13} perturbation theory of neutrino oscillation

Is theoretical treatment of neutrino oscillation well oiled when a large value of θ_{13} close to the Chooz limit is established? For example, does the well known treatment of the degeneracy [31] need modification? If yes, to what extent?

To answer these questions a framework dubbed as "$\sqrt{\epsilon}$ perturbation theory" ($\epsilon \equiv \Delta m^2_{21}/\Delta m^2_{31}$) has been formulated [35] by assuming $s_{13} \sim \sqrt{\epsilon} \simeq 0.18$, which roughly corresponds to the Chooz limit. By doing so one can systematically compute the large-θ_{13} corrections to the formula in [14]. While large corrections arise in certain limited region of energy and baseline as indicated in Fig. 1, one can show on general ground that the correction terms are of order $\sim \epsilon^2 \simeq 10^{-3}$. By general argument one can show that they are the terms proportional to either s^4_{13} or ϵs^2_{13} [35]. Therefore, the large-θ_{13} correction to the degenerate solutions obtained with the formula in [14], generally speaking, is not sizable. See [35] for more detail.

6. Conclusion

Assuming that the T2K and MINOS hints for large θ_{13} will be confirmed, I argued that we are at the turning point of thinking of our strategy of how to explore the remaining unknowns in the lepton flavor mixing. In particular, it allows us to rely on "all in one" approach. I hope that the personal view can be strengthened by further exploration toward Nufact 2012.

Acknowledgments

This work was supported in part by KAKENHI, Grant-in-Aid for Scientific Research No. 23540315, Japan Society for the Promotion of Science.

References

[1] Abe K *et al.* (T2K Collaboration) 2011 *Phys. Rev. Lett.* **107** 041801 (*Preprint* arXiv:1106.2822[hep-ex])
[2] Apollonio M *et al.* (CHOOZ Collaboration) 2003 *Eur. Phys. J.* C **27** 331 (*Preprint* hep-ex/0301017);
See also Boehm F *et al.* (Palo Verde Collaboration) 2001 *Phys. Rev.* D **64** 112001 (*Preprint* hep-ex/0107009);
For a bound from accelerator experiment, see Ahn M H *et al.* (K2K Collaboration) 2004 *Phys. Rev. Lett.* **93** 051801 (*Preprint* hep-ex/0402017);
Adamson P *et al.* (MINOS Collaboration) 2010 *Phys. Rev.* D **82** 051102 (*Preprint* arXiv:1006.0996[hep-ex])
[3] Adamson P *et al.* (MINOS Collaboration) 2011 *Phys. Rev. Lett.* **107** 181802
[4] Fogli G L, Lisi E, Marrone A, Palazzo A and Rotunno A M 2011 *Phys. Rev.* D **84** 053007
[5] Schwetz T, Tortola M and Valle J W F 2011 *New J. Phys.* **13** 109401 (*Preprint* arXiv:1108.1376[hep-ph])
[6] Ardellier F *et al.* (Double Chooz Collaboration) *Preprint* hep-ex/0606025
[7] Ahn J K *et al.* (RENO Collaboration) *Preprint* arXiv:1003.1391[hep-ex]
[8] Guo X *et al.* (Daya Bay Collaboration) *Preprint* hep-ex/0701029
[9] Minakata H 2008 *J. Phys. Conf. Ser.* **136** 022017
[10] Maki Z, Nakagawa M and Sakata S 1962 *Prog. Theor. Phys.* **28** 870
[11] Kobayashi M and Maskawa T 1963 *Prog. Theor. Phys.* **49** 652
[12] Mohapatra R N and Smirnov A Y 2006 *Ann. Rev. Nucl. Part. Sci.* **56** 569 (*Preprint* hep-ph/0603118)
[13] Geer S 1998 *Phys. Rev.* D **57** 6989 (1999 Erratum-ibid. D **59** 039903) (*Preprint* hep-ph/9712290);
De Rujula A, Gavela M B and Hernandez P 1999 *Nucl. Phys.* B **547** 21 (*Preprint* hep-ph/9811390)
[14] Cervera A *et al.* 2000 *Nucl. Phys.* B **579** 17 (2001 Erratum-ibid. B **593** 731) (*Preprint* hep-ph/0002108)
[15] Zucchelli P 2002 *Phys. Lett.* B **532** 166;
Bouchez J, Lindroos M and Mezzetto M 2004 *AIP Conf. Proc.* **721** 37 (*Preprint* hep-ex/0310059);
Burguet-Castell J, Casper D, Gomez-Cadenas J J, Hernandez P and Sanchez F 2004 *Nucl. Phys.* B **695** pp 217-240 (*Preprint* hep-ph/0312068)
[16] Ishitsuka M *et al.* 2005 *Phys. Rev.* D **72** 033003 (*Preprint* hep-ph/0504026);
Kajita T *et al.* 2007 *Phys. Rev.* D **75** 013006 (*Preprint* hep-ph/0609286);
Dufour F *et al.* 2010 *Phys. Rev.* D **81** 093001 (*Preprint* arXiv:1001.5165[hep-ph])
[17] De Bellefon A *et al.* *Preprint* hep-ex/0607026
[18] Bandyopadhyay A *et al.* (ISS Physics Working Group) 2009 *Rept. Prog. Phys.* **72** 106201
[19] Minakata H and Nunokawa H 2000 *Phys. Lett.* B **495** 369 (*Preprint* hep-ph/0004114);
Sato J 2001 *Nucl. Instrum. Meth.* A **472** 434 (*Preprint* hep-ph/0008056);
Richter B *Preprint* hep-ph/0008222
[20] Itow Y *et al.* *Preprint* hep-ex/0106019; For update: http://neutrino.kek.jp/jhfnu/loi/loi.v2.030528.pdf
[21] Beavis D *et al.*, *Preprint* hep-ex/0205040;
Diwan M V *et al.* 2003 *Phys. Rev.* D **68** 012002 (*Preprint* hep-ph/0303081);
Barger V *et al.* (*Preprint* arXiv:0705.4396[hep-ph])
[22] See e.g. Meregaglia A and Rubbia A 2006 *JHEP* **0611** 032 (*Preprint* hep-ph/0609106)
[23] Minakata H and Sugiyama H 2004 *Phys. Lett.* B **580** 216 (*Preprint* hep-ph/0309323)
[24] Minakata H, Nunokawa H and Parke S J 2003 *Phys. Rev.* D **68** 013010 (*Preprint* hep-ph/0301210)
[25] Barger V, Marfatia D and Whisnant K 2003 *Phys. Lett.* B **560** pp 75-86 (*Preprint* hep-ph/0210428)
[26] See e.g. Huber P, Lindner M, Rolinec M, Schwetz T and Winter W 2004 *Phys. Rev.* D **70** 073014;
Mena O, Nunokawa H and Parke S J 2007 *Phys. Rev.* D **75** 033002 (*Preprint* hep-ph/0609011)
[27] Jansson A, Mena O, Parke J S and Saoulidou N 2008 *Phys. Rev.* D **78** 053002
[28] Campagne J E, Maltoni M, Mezzetto M and Schwetz T 2007 *JHEP* **0704** 003 (*Preprint* hep-ph/0603172)
[29] Huber P, Lindner M, Schwetz T and Winter W 2009 *JHEP* **0911** 044 (*Preprint* arXiv:0907.1896[hep-ph])
[30] Minakata H and Nunokawa H 2001 *JHEP* **0110** 001 (*Preprint* hep-ph/0108085)
[31] Minakata H and Uchinami S 2010 *JHEP* **1004** 111 (*Preprint* arXiv:1001.4219[hep-ph])
[32] Abe K *et al.*, *Letter of Intent: The Hyper-Kamiokande Experiment Preprint* arXiv:1109.3262[hep-ex]
[33] Peres O L G and Smirnov A Y 2004 *Nucl. Phys.* B **680** pp 479-509 (*Preprint* hep-ph/0309312)
[34] Bernabeu J, Palomares Ruiz S and Petcov S T 2003 *Nucl. Phys.* B **669** pp 255-276 (*Preprint* hep-ph/0305152)
[35] Asano K and Minakata H 2011 *JHEP* **1106** 022 (*Preprint* arXiv:1103.4387[hep-ph])

Mass hierarchy determination using current and upcoming neutrino experiments

Suprabh Prakash[a], **Sushant Raut**[a] **and S Uma Sankar**[a,b]

[a] Department of Physics, Indian Institute of Technology Bombay, Mumbai 400076, India
[b] Department of Theoretical Physics, Tata Institute of Fundamental Research, Mumbai 400005, India

E-mail: suprabh@iitb.ac.in

Abstract. Determination of mass hierarchy is a crucial open ended question in neutrino physics today. In the wake of recent T2K [1], MINOS [2] and DChooz [3] results which support a high best-fit to θ_{13}, it is worthwhile to explore the potential of the upcoming and already running neutrino experiments for determining mass hierarchy. We consider the long baseline experiments NOνA and T2K and the reactor neutrino experiments Daya Bay, DChooz and RENO. In our analysis we find out that NOνA by itself is insufficient and would require another long baseline experiment such as T2K. This work assumes that by the time reactor neutrino experiments are over, we will have a precise θ_{13} information. Final results are presented as hierarchy exclusion plots in $\sin^2 2\theta_{13} - \delta_{CP}$ plane.
Supported by EUCARD, Neu2012

1. Introduction

The electron appearance probability is the best bet to find out the unknowns in neutrino physics as it is sensitive to all the neutrino parameters.

$$\begin{aligned}\mathcal{P}\left(\nu_\mu \to \nu_e\right) &= \sin^2 2\theta_{13} \sin^2 \theta_{23} \frac{\sin^2 \hat{\Delta}(1-\hat{A})}{(1-\hat{A})^2} + \alpha^2 \sin^2 2\theta_{12} \cos^2 \theta_{13} \cos^2 \theta_{23} \frac{\sin^2 \hat{\Delta}\hat{A}}{\hat{A}^2} \\ &+ \alpha \cos\theta_{13} \sin 2\theta_{12} \sin 2\theta_{13} \sin 2\theta_{23} \cos(\hat{\Delta} + \delta_{CP}) \frac{\sin \hat{\Delta}\hat{A}}{\hat{A}} \frac{\sin \hat{\Delta}(1-\hat{A})}{1-\hat{A}}\end{aligned}$$

where $\hat{\Delta} = \Delta_{31} L/4E$, $\hat{A} = A/\Delta_{31}$, $\alpha = \Delta_{21}/\Delta_{31}$. Here $A = 2EV_{CC}$; where V_{CC} is the matter dependent potential. In this work, we address the question of finding out the sign of Δ_{31}. This is termed as neutrino mass hierarchy: Normal ($\Delta_{31} > 0$) or Inverted ($\Delta_{31} < 0$).

2. Problem of degeneracy

The problem with the electron appearance channel is that it is plagued with degeneracies. That is one can have a situation where

- P(NH, ${\theta_{13}}^1$)=P(IH, ${\theta_{13}}^2$) for the same δ_{CP}: hierarchy-θ_{13} degeneracy
- P(NH, ${\delta_{CP}}^1$)=P(IH, ${\delta_{CP}}^2$) for the same θ_{13}: hierarchy-δ_{CP} degeneracy

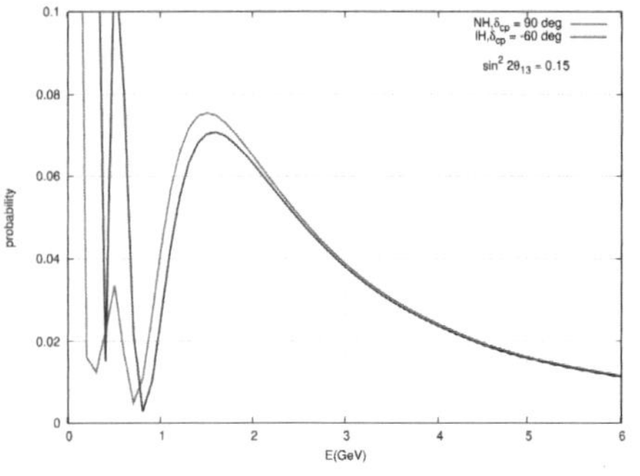

Figure 1. Plot showing the hierarchy-δ_{CP} degeneracy for NOνA.

Figure 1 shows hierarchy-δ_{CP} degeneracy. Here it is shown that P(NH, ${\delta_{CP}}^1 = 90°$) $\sim$ P(IH, ${\delta_{CP}}^2 = -60°$) for $\sin^2 2\theta_{13} = 0.15$. One can generate a similar plot for hierarchy-θ_{13} degeneracy. Since all three parameters (θ_{13}, δ_{CP} and hierarchy) are unknown, measurement becomes very difficult. However there are ways in which this problem can be handled.

3. Reactor neutrino experiments: Daya Bay, DChooz and RENO

The reactor neutrino experiments intend to measure a non-zero value of θ_{13}. They will use the electron disappearance channel which is independent of CP violating phase δ_{CP}. These experiments are built for very short baselines (very small matter effects) and therefore are insensitive to hierarchy-δ_{CP} degeneracy. This makes an independent θ_{13} measurement possible. The proposed reactor neutrino experiments are Daya Bay, DChooz and RENO.

4. Long baseline experiments: NOνA and T2K

The long baseline experiments considered are the ones which are sure to be installed as of now viz. NOνA and T2K. A brief description of the two experiments follow. Greater details can be found from their websites [4][5]. NOνA will use the NuMI beam from Fermilab which peaks around 1.6 GeV. The detector for NOνA is a Totally Active Scintillator Detector (TASD) of 15 kton fiducial volume, placed 810 km away from the beam source. The beam consists primarily of muon neutrinos and NOνA will look for electron events in the detector. The NuMI beam power has been assumed to be 0.7 MW. For T2K, a beam peaking at 0.6 GeV will come from the J-PARC accelerator at Tokai. A 22.5 kton fiducial volume Water Cerenkov detector is placed 295 km away at Kamioka. T2K will also look for electron neutrinos in a beam which was initially muon neutrinos. The beam power for T2K has been assumed to be 750 MW.

5. Simulation details

We use the software GLoBES [6][7] for our analysis. The runtime for NOνA is $(6\nu + 3\bar{\nu})$ for all the plots shown. The runtime for T2K is $(3\nu + 4\bar{\nu})$ for all the plots shown. Only electron appearance events have been considered as signal events. The backgrounds for this channel is formed by misidentified muons, NC events and intrinsic beam ν_e. We have done marginalization wherever required. Priors were added for θ_{13}, θ_{23} and Δ_{31} with $\sigma\left(\sin^2 2\theta_{13}\right) = 0.01$, $\sigma\left(\sin^2 2\theta_{23}\right) = 0.02$ and $\sigma\left(\Delta_{31}\right) = 0.03 \times \left(\Delta_{31}\right)$. A prior of $\sigma\left(\sin^2 2\theta_{13}\right) = 0.01$ effectively takes into account the data due to reactor neutrinos experiments. We have taken care in defining Δ_{31}^{NH} and Δ_{31}^{IH} in terms of the measured quantity Δ_{atmos}.[8]

$$\Delta_{31} = \Delta_{atmos} + \left(\cos^2\theta_{12} - \cos\delta\sin\theta_{13}\sin 2\theta_{12}\tan\theta_{23}\right)\Delta_{21}$$

where $\Delta_{atmos} = \pm 2.4 \times 10^{-3}\text{eV}^2$; + for NH, - for IH. We have assumed a 5% systematics uncertainty on both signal and background.

Journal of Physics: Conference Series **408** (2013) 012035 doi:10.1088/1742-6596/408/1/012035

6. Hierarchy exclusion plots

The following plots show the hierarchy excluding capabilities.

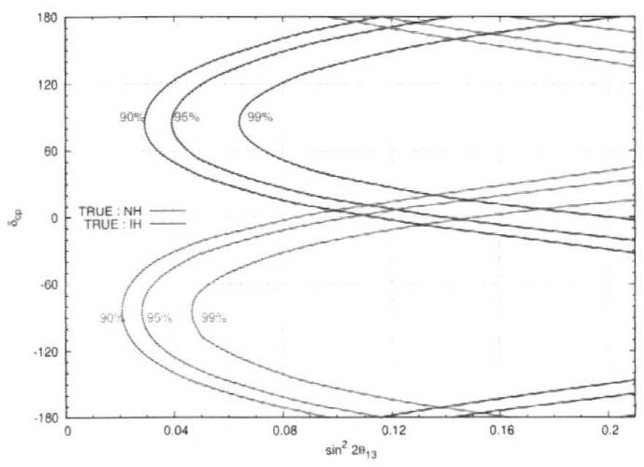

Figure 2. Hierarchy exclusion plots (without systematics) for NOνA only.

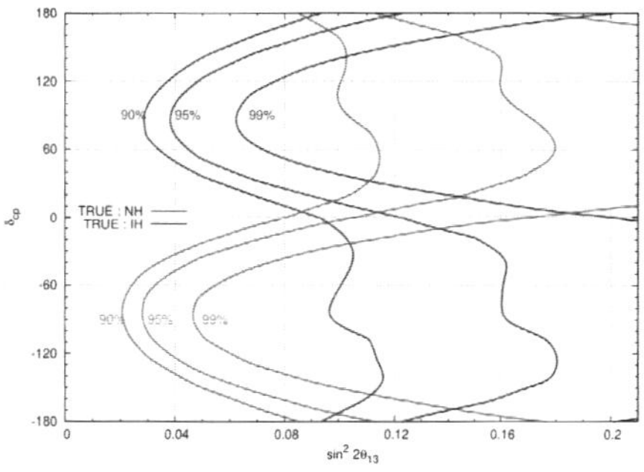

Figure 3. Hierarchy exclusion plots (with systematics) for the combined setup of NOνA and T2K.

The plots show different confidence level regions and for all ($\delta_{CP} - \sin^2 2\theta_{13}$) points on the right of a line, the wrong hierarchy can be excluded from the right hierarchy with that confidence level. It can be seen from this plot that NOνA performs very well only for a particular range of true δ_{CP} depending on the true mass hierarchy. For NH, this range is $-120°$ to $-60°$ and for IH, this range is $60°$ to $120°$. In δ_{CP} ranges other than these, NOνA does not perform well because of the hierarchy-δ_{CP} degeneracy. Similar studies were also done in [9].

7. How and why does T2K data help?

It can be seen from figure 3 that at 90% C.L., T2K data boosts significantly NOνA's ability for determining mass hierarchy for NH as the true hierarchy in the upper half plane (UHP) of [0,180°] and for IH in the lower half plane (LHP) of [-180°,0]. This can be understood by analysing the error plots (figures 4 and 5) due to individual runs of NOνA and T2K with NH as the true hierarchy and IH as the test hierarchy for the following true points: $\sin^2 2\theta_{13} = 0.05$, $\delta_{CP} = -90°$ and $\sin^2 2\theta_{13} = 0.15$, $\delta_{CP} = 90°$. For NH as the true hierarchy, the favorable plane for NOνA is LHP.

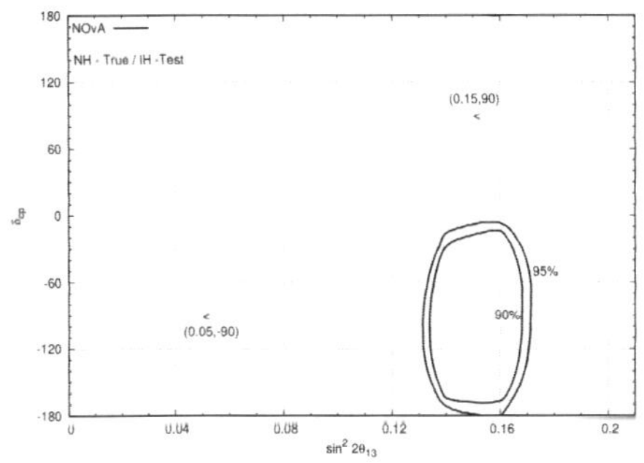

Figure 4. Error plots: NOνA.

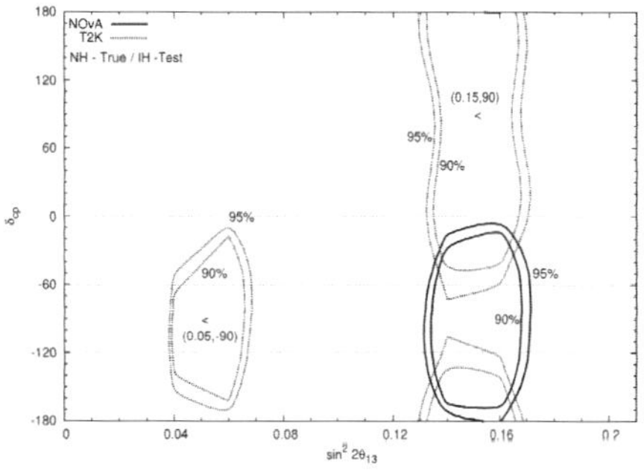

Figure 5. Error plots: NOνA and T2K; individually.

In LHP, NOνA by itself performs very well and T2K data does not have significant effects.

This is evident from the error plots. For the true point ($\sin^2 2\theta_{13} = 0.05$, $\delta_{CP} = -90°$) (δ_{CP} in the LHP), there is no error in the NOνA measurements (no contours at 90% and 95% C.L.). However when the true point is $\sin^2 2\theta_{13} = 0.15$, $\delta_{CP} = 90°$ (δ_{CP} in the UHP), NOνA gives totally wrong results. True point lies out of 90% and 95% C.L. contours. A wrong hierarchy and wrong δ_{CP}-plane fakes the true point. This is hierarchy-δ_{CP} degeneracy. However the error plots for T2K correctly predicts the true point. This is because it experiences less hierarchy-δ_{CP} degeneracy as it is a shorter baseline and appreciable matter effects do not develop. Thus T2K data provide the additional $\Delta\chi^2$ at the wrong hierarchy-wrong δ_{CP} point thereby increasing the sensitivity to hierarchy exclusion. This point can be elaborated more by figures 6 and 7 (for an updated analysis, see [10]).

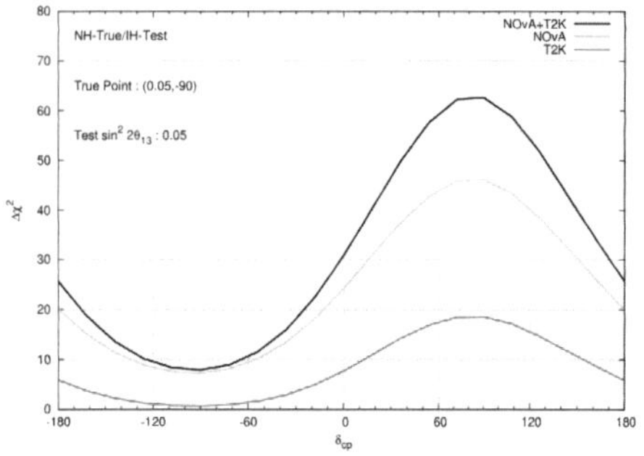

Figure 6. $\Delta\chi^2$ vs. δ_{CP} for $\sin^2 2\theta_{13} = 0.05$, $\delta_{CP} = -90°$: Moving along $\sin^2 2\theta_{13} = 0.05$ in the error plot plane.

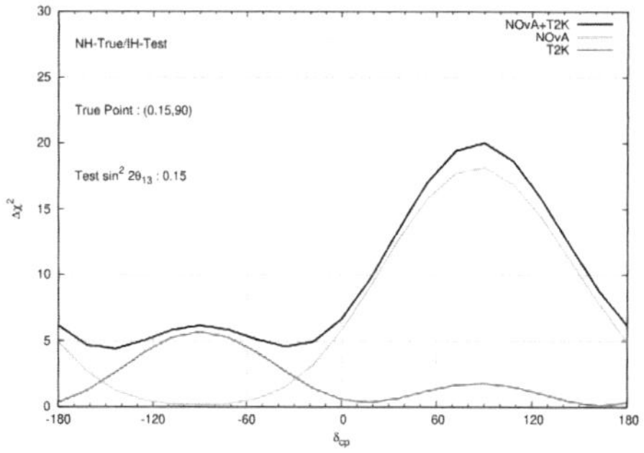

Figure 7. $\Delta\chi^2$ vs. δ_{CP} for $\sin^2 2\theta_{13} = 0.15$, $\delta_{CP} = 90°$: Moving along $\sin^2 2\theta_{13} = 0.15$ in the error plot plane.

8. Results

The exclusion plots show that the combined setup of the LBL experiments - NOνA and T2K and the reactor neutrino experiments will be able to exclude the wrong hierarchy from the right one @ 90%C.L. for all δ_{CP} provided $\sin^2 2\theta_{13} > 0.115$. This is the best that we can achieve with the present statistics.

References

[1] Abe K *et al.* (T2K Collaboration) 2011 *Phys. Rev. Lett.* **107** 041801 (*Preprint* `arXiv:1106.2822[hep-ex]`)
[2] Adamson P *et al.* (MINOS Collaboration) 2011 *Phys. Rev. Lett.* (*Preprint* `arXiv:1108.0015[hep-ex]`)
[3] Abe Y *et al.* (Double Chooz Collaboration) 2012 (*Preprint* `arXiv:1207.6632[hep-ex]`)
[4] `http://www-nova.fnal.gov/`
[5] `http://jnusrv01.kek.jp/public/t2k/`
[6] Huber P, Lindner M and Winter W 2005 *Comput. Phys. Commun.* **167** 195 (*Preprint* `hep-ph/0407333`)
[7] Huber P, Kopp J, Lindner M, Rolinec M and Winter W 2007 *Comput. Phys. Commun.* **177** 432–438 (*Preprint* `hep-ph/0701187`)
[8] Nunokawa H, Parke S J and Zukanovich Funchal R 2005 *Phys. Rev. D* **72** 013009 (*Preprint* `hep-ph/0503283`)
[9] Huber P, Lindner M, Schwetz T and Winter W 2009 *JHEP* **11** 044 (*Preprint* `arXiv:0907.1896[hep-ph]`)
[10] Prakash S, Raut S K and Sankar S U 2012 (*Preprint* `arXiv:1201.6485[hep-ph]`)

Journal of Physics: Conference Series **408** (2013) 012036 doi:10.1088/1742-6596/408/1/012036

Long-baseline super-beam experiments in Europe within LAGUNA

Pilar Coloma[a], **Tracey Li**[b], **Silvia Pascoli**[c]

[a]Departamento de Física Teórica & Instituto de Física Teórica, Universidad Autónoma de Madrid. 28049 Cantoblanco, Madrid (Spain)
[b]Instituto de Física Corpuscular, Edificio Institutos de Investigación, Paterna, Valencia E-46071 (Spain)
[c]IPPP, Department of Physics, Durham University, Durham DH1 3LE (United Kingdom)

E-mail: [a]`p.coloma@uam.es`, [b]`tracey.li@ific.uv.es`, [c]`silvia.pascoli@durham.ac.uk`

Abstract. We explore the physics reach of several possible configurations for a Super-Beam experiment in Europe, focusing on the possibilities of discovering θ_{13}, CP violation in the leptonic sector and the ordering of neutrino mass eigenstates. We consider the three different detector technologies: Water Čerenkov, Liquid Argon and Liquid Scintillator, and seven possible sites in Europe which would be able to host such a detector underground. The distances to these sites from CERN, where the beam would be originated, go from 130 km to 2300 km. The neutrino flux is optimized in each case as to match the first oscillation peak for each of the baselines under consideration. We also study the impact of several experimental factors in the performance of each detector technology.

1. Introduction and simulation details

Future long-baseline neutrino oscillation experiments are being designed to measure the neutrino oscillation parameters θ_{13}, δ and the mass hierarchy. We evaluate the performance of such an experiment within LAGUNA [1], optimizing and comparing the performances of seven different baselines and three detector technologies. In particular, we focus on a Super-Beam (SB) experiment based at CERN. SB fluxes have been optimized for each baseline according to the first oscillation peak [2]. In the lower energy configuration for the beam (for $L = 130$ km) the fluxes have been obtained assuming 0.56×10^{23} Protons on Target (PoT) per year, with an energy of 4.5 GeV. We assume 2 years of ν running and 8 years of $\bar{\nu}$ running in this case, as in [3]. Higher energy fluxes (for setups with $L > 130$ km) correspond to the CERN high-power PS2 configuration [4]: 3×10^{21} PoT per year, with an energy of 50 GeV. We assume $5+5$ years in this case, though, since we have checked that an asymmetric configuration is only marginally beneficial for the θ_{13} discovery potential of the experiment.

We use the GLoBES software [5,6] in all our simulations. Marginalization has been performed over solar and atmospheric parameters, according to their present best-fits and 1σ errors [7]. Whenever we marginalize over θ_{13} and δ we leave them completely free. A conservative 5% uncertainty has been assumed over the matter density, taken from the Preliminary Reference Earth Model (PREM) profile [8]. Full details used for the simulation of the detectors response can be found in [9]. Information about the Liquid Argon (LAr) detector has been obtained from [10,11] with migration matrices provided by L. Esposito and A. Rubbia. Information about the Liquid Scintillator (LS) detector has been obtained from [12–14]. The rejection capability for the Neutral-Current (NC) events at the LS detector is currently under study. Therefore, we have considered that a certain percentage of the total amount of unoscillated NC events will be misidentified as Charged-Current (CC) events and we have varied it within 0.5% and 5%. Finally, for the WC detector we have considered two different configurations: for the lower energy beam configuration we have used the same details as in [3]; however, for the higher energy setups we have followed [10]. The migration matrices in this case have been kindly provided by the LBNE collaboration [15], as well as the rejection efficiencies for the NC background events. For all detectors we assume the level of the intrinsic electron background to be the same as the signal efficiency.

2. Optimization studies

Full results of the optimization studies can be found in [9]. We studied the effects of the NC background, the intrinsic beam background, the values of the systematic errors, the ratio of $\nu : \bar{\nu}$ running, and the possibility of tau detection. The intrinsic beam background is only relevant in case of small θ_{13} ($\sin^2 2\theta_{13} < 10^{-2}$). For high energy configurations of the beam, we find that equal ν and $\bar{\nu}$ running times give better results than asymmetric configurations. Finally, tau detection is not only very challenging, but it does not significantly improve the sensitivity to oscillation parameters for any of our setups.

In Fig. 1, we show our results for the impact of systematic errors (left panel) and the NC background (right panel) on the CP discovery potential for the LAr and LS detectors, respectively. Results are shown for the Pyhäsalmi baseline, but a similar dependence is expected for the rest of locations. In the left panel, the first (second) value in the legend corresponds to the systematic uncertainty over the signal (background), which are taken to be completely uncorrelated. It can be seen that, while the signal systematics is mostly relevant in the region where $\sin^2 2\theta_{13} \gtrsim 10^{-2}$, background systematics are only relevant below $\sin^2 2\theta_{13} \sim 10^{-2}$. In the right panel, the value in the legend indicates the percentage of the total number of unoscillated NC events that are misidentified as CC events. It can be seen that the results in this case are worsened around a factor of 2 when the NC background is increased from a 0.5% to a 5%.

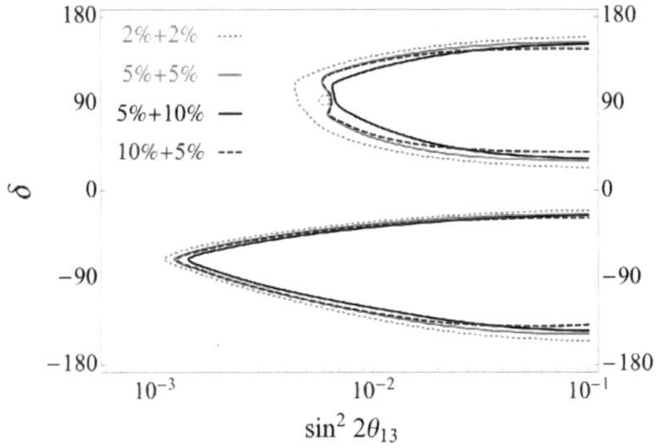

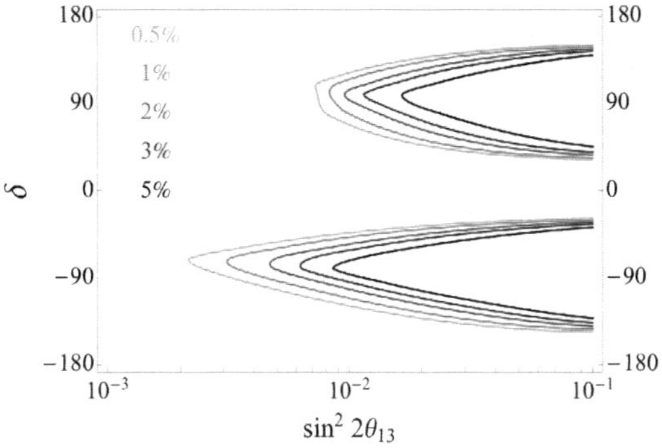

Figure 1. Effect of systematic uncertainties (left) and the NC backgrounds (right) on the CP discovery potential of a LAr and LS detector, respectively. Results are shown for the detector placed at Pyhäsalmi. In the region to the right of each line, the CP conservation hypothesis ($\delta = 0, \pi$) can be excluded at 3σ (1 d.o.f.).

3. Comparison of baselines and detectors

In the following, we compare the performance for the different baselines and detectors under consideration. We do not display the results for both the 630 km and 665 km baselines since the results obtained are virtually identical; therefore the results labelled '665' should be understood to also apply to the 630 km baseline. It should be noted that, in the absence of a proper simulation of the WC detector when exposed to a SB with mean neutrino energy around 1.5 GeV, we have used the same migration matrices and rejection efficiencies as for the higher energy setups. These matrices were obtained for a multi-GeV SB and therefore may be pessimistic to describe the performance of the WC for the setups with $L = 630$ and 665 km. This should be taken into account when comparing the results obtained for different baselines. We have also considered a range of values for the NC background in the LS detector. The results for the LS detector in the optimistic scenario (with 0.5% of NC events as background) are very similar to those obtained for the LAr detector and therefore will not be shown here. As it was shown in the previous section, the results for the LS in the pessimistic scenario where a 5% of the NC events are misidentified as CC events are generally around a factor of 2 worse than the ones for the LS with 0.5% NC background, and can be found in [9].

Results for the θ_{13} discovery potential: We find that the most relevant factor for this observable is the energy of the beam. Higher neutrino energies imply larger cross sections and, therefore, a larger number of events at the detector. Besides, the θ_{13} discovery potential of a given facility is not so wildly affected by the presence of degeneracies as the CP or the mass hierarchy discovery potentials. Therefore, in this case we find that the results are usually better for longer baseline setups, for which the spectrum peaks at higher energies. The only exception to this rule is the case of the WC placed at $L = 130$ km. In this case, the huge fiducial mass of the detector and its excellent performance in the sub-GeV regime overcome the statistical limitations, giving excellent results for this observable. We find that for the LAr, the WC and the LS with 0.5% NC background a non-vanishing θ_{13} can be discovered at the 3σ CL if $\sin^2 2\theta_{13} \gtrsim 6 \times 10^{-3}$, for any value of δ, with the exception of the setups with LAr or LS placed at $L = 130$ km, for which this value is shifted to $\sin^2 2\theta_{13} \gtrsim 2 \times 10^{-2}$.

Results for the CP discovery potential: In Fig. 2 we show the results for the 3σ CP discovery potential, for the LAr and the WC detectors. The baseline dependence is similar to that of the θ_{13} discovery due to the effect of larger cross sections at higher neutrino energies. However, the CP discovery potential is more affected by degeneracies: this can be clearly observed in the region where $\delta \sim +90°$, where matter effects move the sign degeneracies to CP-conserving values of δ and therefore the CP discovery potential is severely worsened. Finally, as it can be seen from the figure, the best results for this observable are obtained for the WC detector placed at 130 km from the source. This is again due to its excellent performance in the low energy regime, together with its much larger fiducial mass. In addition, the absence of matter effects reduces the effect of degeneracies and the CP discovery potential is practically unaffected. It should be noticed that both the LAr and the LS detectors give null results for this observable when they are placed at 130 km, as it can be seen from the left panel in the figure.

Results for the mass hierarchy discovery potential: We find that the length of the baseline is crucial in order to determine the mass hierarchy, regardless of the detector technology. The best results are obtained for the LAr, WC and LS with 0.5% NC background placed at 2300 km: the sensitivity to the mass hierarchy in this case reaches $\sin^2 2\theta_{13} \sim 8 \times 10^{-4}$ if $\delta \sim -140°$ and $\sin^2 2\theta_{13} \sim 5 \times 10^{-3}$ if $\delta = +90°$. The results for the second baseline in length, 1570 km, are very close to these and reach $\sin^2 2\theta_{13} \sim 1.5 \times 10^{-3}$ if $\delta \sim -140°$ and $\sin^2 2\theta_{13} \gtrsim 6 \times 10^{-3}$ if $\delta = +90°$. The results obtained at 665 and 650 km are around an order of magnitude worse than these, and the rest of baselines lie in between. Finally, the hierarchy discovery potential is zero for any of the detectors placed at $L = 130$ km due to the absence of matter effects.

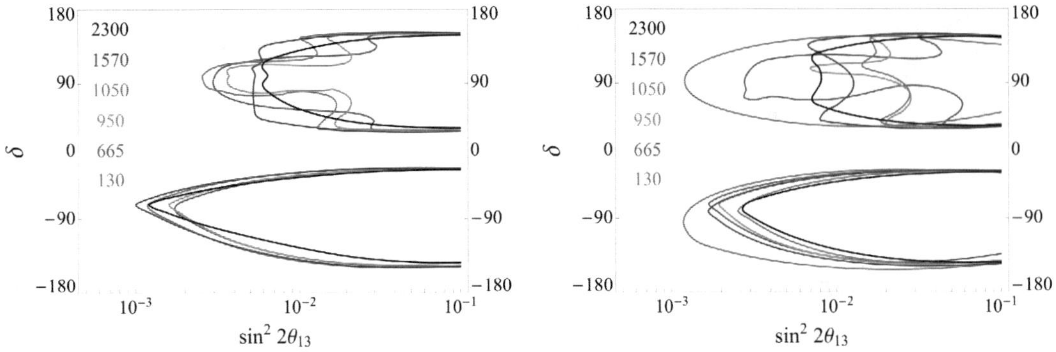

Figure 2. CP discovery potential for the LAr (left) and the WC (right) detectors, for each of the baselines indicated in the legend. In the region to the right of each curve the CP conservation hypothesis ($\delta = 0$ or π) can be excluded at 3σ (1 d.o.f.).

Acknowledgements

This work was funded by the European Community under the European Commission Framework Programme 7 Design Study LAGUNA (Project Number 212343). PC acknowledges financial support from Comunidad Autónoma de Madrid, and from projects HEPHACOS S2009/ESP-1473 (CAM, Spain) and FPA2009-09017 (DGI del MCyT, Spain).

References

[1] Angus D *et al.* 2010 The LAGUNA design study- towards giant liquid based underground detectors for neutrino physics and astrophysics and proton decay searches *Preprint* physics.ins-det/1001.0077 .

[2] Longhin A 2010 Optimization of neutrino fluxes for European super-beams *PoS* ICHEP2010 325 `http://irfu.cea.fr/en/Phocea/Pisp/index.php?id=72`.

[3] Campagne J E, Maltoni M, Mezzetto M, and Schwetz T 2007 Physics potential of the CERN-MEMPHYS neutrino oscillation project *JHEP* 0704:003.

[4] Rubbia A 2010 A CERN-based high-intensity high-energy proton source for long baseline neutrino oscillation experiments with next- generation large underground detectors for proton decay searches and neutrino physics and astrophysics *Preprint* hep-ph/1003.1921.

[5] Huber P, Lindner M, and Winter W 2005 Simulation of long-baseline neutrino oscillation experiments with GLoBES *Comput. Phys. Commun.* **167** 195.

[6] Huber P, Kopp J, Lindner M, Rolinec M, and Winter W 2007 New features in the simulation of neutrino oscillation experiments with GLoBES 3.0 *Comput. Phys. Commun.* **177** 432–438.

[7] Schwetz T, Tortola M, and Valle J W F 2011 Global neutrino data and recent reactor fluxes: status of three-flavour oscillation parameters *New J. Phys.* **13** 063004.

[8] Dziewonski A M and Anderson D L 1981 Preliminary reference earth model *Phys. Earth Planet Interiors* **25** 297–356.

[9] Coloma P, Li T and Pascoli S 2012 A comparative study of long-baseline superbeams within LAGUNA for large θ_{13} *Preprint* hep-ph/1206.4038.

[10] Akiri T *et al.* [LBNE Collaboration] 2010 The 2010 Interim Report of the Long-Baseline Neutrino Experiment Collaboration Physics Working Groups *Preprint* hep-ex/1110.6249.

[11] Barger V, Huber P, Marfatia D, and Winter W 2007 Which long-baseline neutrino experiments are preferable? *Phys. Rev.* D **76** 053005.

[12] Mollenberg R *et al.* 2011 Reconstruction of GeV neutrino events in LENA AIP Conf. Proc. **1382** (2011) 141.

[13] Mollenberg R 2010 *Private communication.*

[14] Wurm M *et al.* 2011 The next-generation liquid-scintillator neutrino observatory LENA *Preprint* astro-ph.IM/1104.5620.

[15] Whitehead L 2011 *Private communication.*

MINERνA status and event reconstruction

Gabriel N Perdue, on behalf of the MINERνA Collaboration
University of Rochester, Rochester, New York 14610 USA
E-mail: perdue@fnal.gov

Abstract. MINERνA (Main INjector ExpeRiment ν-A) is a few-GeV neutrino scattering experiment that began taking data in the NuMI beam at Fermilab (FNAL) in the Fall of 2009. MINERνA employs a fine-grained detector, with an eight ton active target region composed of plastic scintillator. It also uses nuclear targets composed of carbon, iron, and lead placed upstream of the active region to measure ν-A dependence. The experiment will provide important inputs for neutrino oscillation experiments and a pure weak probe of nuclear structure. We offer a set of initial kinematic distributions of interest and provide a summary of current operations and event reconstruction status.

1. Introduction

MINERνA is an on-axis neutrino-nucleus scattering experiment at Fermilab's NuMI (Neutrinos at the Main Injector) beamline that will measure interaction cross-sections and event kinematics in exclusive and inclusive states to high precision. The experiment will also examine nuclear effects and parton distribution functions (PDF's) using a variety of targets materials.

1.1. Precision Neutrino Oscillation Experiments and Nuclear Effects

Interaction details are crucial for neutrino energy estimation and in background separation in oscillation experiments. For example, in a ν_μ disappearance experiment, the two-flavor disappearance relation is show in Eq. 1:

$$P\left(\nu_\mu \to \nu_\mu\right) = 1 - \sin^2\left(2\theta_{23}\right)\sin^2\left(\frac{1.27\Delta m^2_{23}(eV^2)L(km)}{E_\nu(GeV)}\right) \quad (1)$$

Experiments cannot directly measure the neutrino energy that appears in this relation, and visible energy is a function of flux, cross-section, and detector response. Interactions occur in dense nuclear matter, making final state interactions (FSI) significant in producing the observed particles. Near-to-Far Detector ratios cannot entirely eliminate the associated uncertainties because the energy spectra at the two detectors are different due to beam, oscillation, matter, and even nuclear effects if the detector materials differ.

Understanding these effects is especially important in light of recent "hints" of a large θ_{13} [1]. As Figure 1 illustrates, for a fixed value of the CP violating phase, δ, the ν-$\bar{\nu}$ oscillation asymmetries are smaller at larger values of θ_{13}. Therefore, for larger values of θ_{13}, experiments are measuring a smaller difference between two larger numbers and systematic errors increase in relative importance. In this case, MINERνA cross-section and kinematic measurements increase in value.

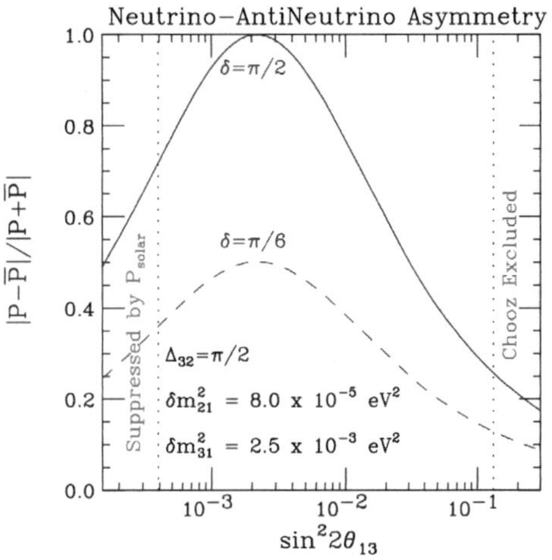

Figure 1. Oscillation asymmetry between neutrinos and anti-neutrinos as a function of $\sin^2 2\theta_{13}$ for different CP violating phases. Figure adapted from [2].

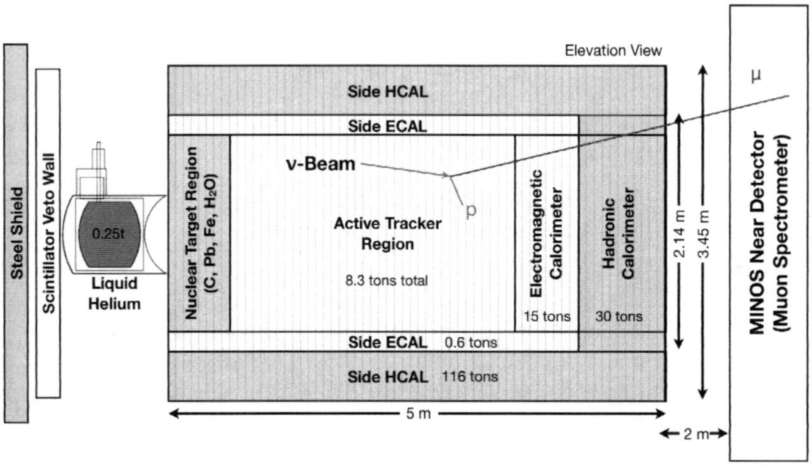

Figure 2. The MINERνA detector as of Summer 2011. The water target was not installed by that date.

Additionally, MINERνA supports a nuclear effects program complementary to charged lepton scattering measurements. Many quantities of interest carry large uncertainties: axial form factors as a function of A and momentum transfer (Q^2), quark-hadron duality, x-dependent nuclear effects, etc.

2. The MINERνA Detector and Operations

MINERνA is a horizontal stack of 120 similar modules weighing approximately two tons each. Each module has an inner detector (ID) composed of triangular plastic scintillator strips and an outer detector (OD) steel frame instrumented with plastic scintillator bars. Most modules feature an ID composed of two stereoscopic planes of scintillator, but some in the nuclear target and calorimetric regions of the detector give up one or both scintillator planes for target or absorber materials. See Figure 2 for a schematic of the detector.

Figure 3 illustrates integrated data collection and livetime. MINERνA began its physics run in March, 2010 in the NuMI beam forward horn current (FHC) mode, focusing π^+ mesons ("neutrino mode"). Reverse horn current (RHC) mode data, focusing π^- mesons ("antineutrinos mode"), was taken prior to March, 2010 and again in Winter 2010-2011.

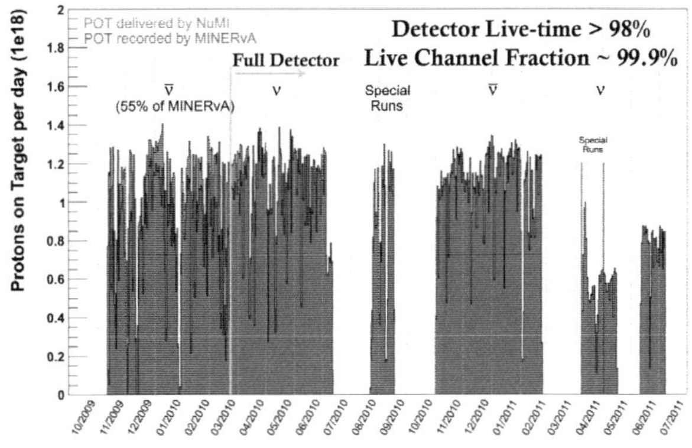

Figure 3. Accumulated protons on target (P.O.T.) and live-time. Prior to March 2010, we integrated data using a partially constructed version of the detector.

3. Particle ID

There are a number of exciting analyses underway in MINERνA, but early results will focus on charged-current reactions. As such, muon identification is important. Figure 4 illustrates muon classification and reconstruction efficiency in neutrino Monte Carlo (MC). MINERνA uses GENIE for event generation [4]. Current reconstruction efforts are focused on muon tracks matched into the MINOS near detector [3], and future efforts will begin to emphasize particles stopping in MINERνA.

The first step in a stopped muon analysis is Michel electron identification. Figure 5 provides an illustration of an event candidate from data and Figure 6 is a data to MC comparison of Michel energies and lifetimes in the MINERνA tracker. Table 1 gives the results of fits to the muon lifetime for the plot on the right in Figure 6.

4. Conclusion

MINERνA is actively collecting data and building software infrastructure for physics analysis. Operations are stable and we are progressing in particle identification and event reconstruction.

Acknowledgments

This work was supported by DOE Grant No. DE-FG02-91ER40685.

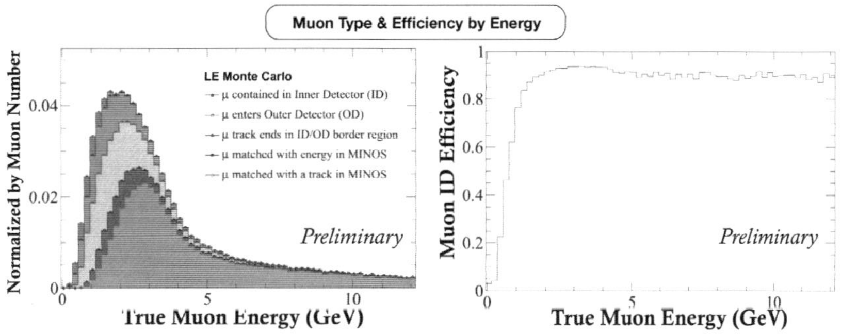

Figure 4. Muon reconstruction topologies (left) and efficiency (right) in MINERνA. Reconstruction efficiencies for particles matched into MINOS include the effect of reconstruction efficiency in that detector.

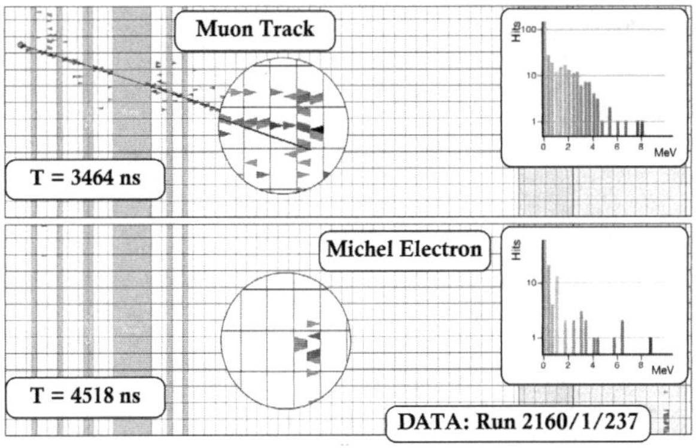

Figure 5. A Michel electron candidate in MINERνA data.

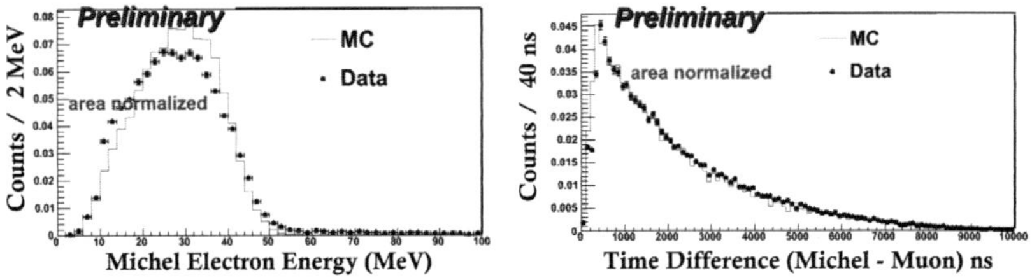

Figure 6. A Michel electron energies and lifetimes in MINERνA.

Table 1. Michel electron lifetime fits. MC is background-free μ^-, while data contains a small μ^+ contamination. The nominal μ^- lifetime in carbon is 2026 ns.

	Muon Lifetime in Plastic (ns)
MC	2100 ± 10
Data	2120 ± 20

References

[1] Abe K *et al.* (The T2K Collaboration.) 2011 *Preprint* arXiv:1106.2822[hep-ex]
[2] Nunokawa H *et al.* 2008 *Progress in Particle and Nuclear Physics* **60** 2
[3] Michael D *et al.* 2008 *Nucl. Inst. and Meth. A* **596**
[4] http://www.genie-mc.org/

XIII International Workshop on Neutrino Factories, Super beams and Beta beams (NUFACT11) IOP Publishing
Journal of Physics: Conference Series **408** (2013) 012038 doi:10.1088/1742-6596/408/1/012038

Neutrino cross section measurements @ SciBooNE

C Mariani
Columbia University, New York, NY 10027, USA
E-mail: mariani@nevis.columbia.edu

Abstract. We report measurements of cross sections of neutrinos of 0.7 GeV average energy scattering off a carbon target cross sections with by the SciBooNE experiment at Fermilab. These measurements are important inputs for current and future accelerator-based neutrino oscillation experiments in the interpretation of neutrino oscillation signals.

1. Introduction
The measurement of neutrino mixing angle θ_{13} is one of the most important goals in current neutrino experiments. For the current and next generation of long baseline neutrino oscillation experiments, T2K, NOvA and LBNE, the precise measurement of neutrino-nucleus cross sections in the few GeV energy range is an essential ingredient in the interpretation of neutrino oscillation signals.

2. SciBooNE experiment
The SciBooNE experiment is designed to measure neutrino cross sections on carbon near one GeV region. The experiment collected data in 2007 and 2008 with neutrino (0.99×10^{20} POT) and antineutrino (1.53×10^{20} POT) beams in the FNAL Booster Neutrino Beam line (BNB). The BNB uses a primary proton beam with kinetic energy 8 GeV and a beryllium target placed inside an aluminum horn as a neutrino production target. The SciBooNE detector is located 100 m downstream from the neutrino production target. The neutrino flux in the SciBooNE detector is dominated by muon neutrinos. The flux-averaged mean neutrino energy is 0.7 GeV in neutrino running mode and 0.6 GeV in antineutrino running mode

The SciBooNE detector consists of three detector components; SciBar, Electromagnetic Calorimeter (EC) and Muon Range Detector (MRD). SciBar is a fully active and fine grained scintillator detector that consists of 14,336 bars arranged in vertical and horizontal planes. SciBar is capable of detecting all charged particles and performing dE/dx-based particle identification. The EC is located downstream of SciBar. The detector is a spaghetti calorimeter with thickness of $11X_0$ and is used to measure π^0 and the intrinsic ν_e component of the neutrino beam. The MRD is located downstream of the EC in order to measure the momentum of muons up to 1.2 GeV/c using their range. It consists of 2-inch thick iron plates sandwiched between layers of plastic scintillator planes.

3. Charged current neutrino interactions
The SciBooNE collaboration reported a charged current quasi-elastic (CCQE) cross section measurement in [1]. We developed two distinct CCQE data sets, one with tracks matched

between SciBar and the MRD and the other with tracks contained within SciBar. To simulate neutrino-nuclear scattering, SciBooNE uses the NEUT [2] generator Monte Carlo (MC) simulation. In the SciBar-contained sample, the muons from charged-current neutrino candidates are tagged with particle identification based on energy deposition along the track and by searching for their decay electrons using SciBar's multi-hit TDCs. Events are further classified based on the number and type of tracks coming from the neutrino vertex. Removing the MRD track-matching cut allows the reconstruction of backwards-going tracks, thus expanding the Q^2, range open to the analysis. The neutrino vertex is defined using the timing of hits within the muon track, and the location of the tagged decay electron. Figure 1 shows the SciBooNE's CCQE cross section measurement and compares it with MiniBooNE and NOMAD CCQE measurements. We note that the SciBooNE and MiniBooNE results are obtained directly from the measured event yields and proton-on-target-normalized neutrino flux predictions, not by extracting via a cross-section ratio with a different (assumed) cross-section. These are the world's first such POT-normalized neutrino CCQE cross-section measurements.

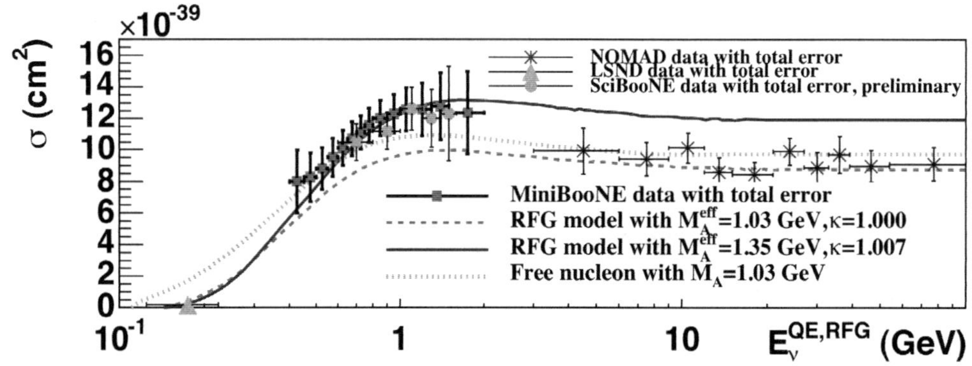

Figure 1. CCQE cross-section versus neutrino energy for the SciBooNE, MiniBooNE [3] and NOMAD [4] experiments.

In addition to the CCQE cross section the SciBooNE collaboration has reported in [5] a measurement of inclusive charged current interactions of muon neutrinos on carbon with an average energy of 0.7 GeV. Measuring p_μ and θ_μ, we have extracted the CC interaction rates and cross sections using flux and neutrino interaction simulations. We extract the ν_μ CC interaction rates by fitting muon kinematics, with precision of 6-15% for the energy dependent and 3% for the energy integrated analyses. We confirmed that the distributions after fitting well reproduce the predicted distributions with both NEUT and NUANCE [6] based simulations. We have also extracted CC inclusive interaction cross sections from the observed rates, with a precision of 10-30% for the energy dependent and 8% for the energy integrated analyses. This is the first measurement of the CC inclusive cross section on carbon around 1 GeV.

4. Neutrino charged current - pion production

4.1. Charged current coherent production in neutrino and antineutrino mode

The experimental signature of charged current coherent pion production is: 1) the existence of two and only two tracks originating from a common vertex, both consistent with minimum ionizing particles, a muon and a charged pion, 2) no nucleon recoil in the final state, 3) small momentum transfer. We searched for charged current coherent pion production with two distinct data samples; averaged energy 1.1 GeV and 2.3 GeV. Figure 2 shows the Q^2 distribution in the coherent pion sample for the neutrino energy of 1.1 GeV (left plot) and 2.3 GeV (right plot).

There is a data deficit in the low Q^2 region. No evidence for coherent pion production was observed. We evaluated the ratio of the charged current coherent pion production cross section to the charged current inclusive cross section to be: $(0.16 \pm 0.17(stat)^{+0.30}_{-0.27}(sys)) \times 10^{-2}$ at 1.1 GeV, and $(0.68 \pm 0.32(stat)^{+0.39}_{-0.25}(sys)) \times 10^{-2}$ at 2.2 GeV [7]. We confirmed the K2K [8] results and we set 90% confidence level upper limits on the cross section ratio of charged current coherent pion production to the total charged current cross section at 0.67×10^{-2} at a mean neutrino energy of 1.1 GeV and 1.36×10^{-2} at a mean neutrino energy of 2.2 GeV.

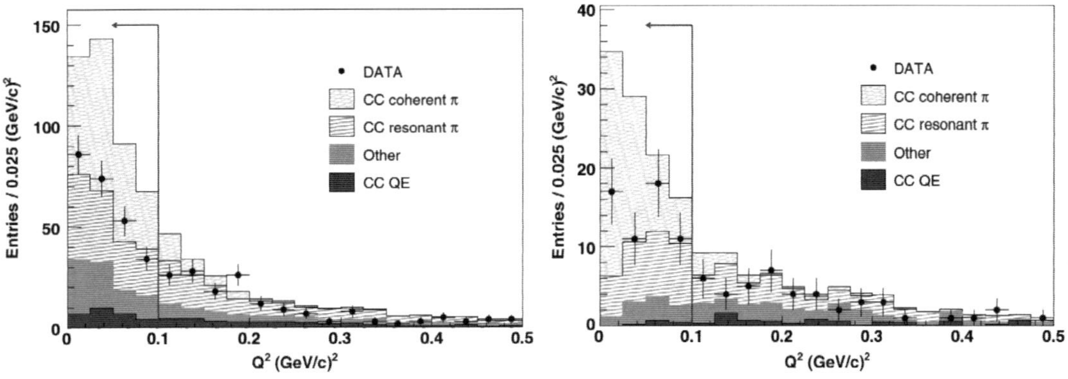

Figure 2. Reconstructed Q^2 distributions for two distinct subsamples; mean neutrino energy of 1.1 GeV (right) and 2.3 GeV (left). Figures are from [7].

SciBooNE has also performed a search for charged current coherent pion production in anti-neutrino mode running following the analysis performed on neutrino data. The only difference from the neutrino mode analysis is in the treatment of the neutrino (wrong-sign) contribution since SciBooNEs $\bar{\nu}$ beam contains a relatively large fraction of neutrinos. The preliminary results of charged current coherent pion production cross section ratio in SciBooNEs anti-neutrino mode running data is obtained to be $\frac{\sigma(\bar{\nu}CCcoh-\pi)+r\cdot\sigma(\nu CCcoh-\pi)}{\sigma(\bar{\nu}CC)+r\cdot\sigma(\nu CC)} = (1.131 \pm 0.34(stat)^{+0.31}_{-0.36}(sys)) \times 10^{-2}$ at the mean anti-neutrino energy of 1.0 GeV. $r = \frac{\int \Phi(\nu)dE}{\int \Phi(\bar{\nu})dE} = 0.19$ is a ratio of anti-neutrino and neutrino fluxes in the Booster Neutrino Beam based on SciBooNEs flux prediction [5].

4.2. Neutral pion production

The experimental signature of charged current neutral pion production is constituted by two electromagnetic cascades initiated by the conversion of the π^0 decay gamma rays, with an additional muon in the final state from charged current processes. The obtained preliminary results of charged current π^0 cross section is $5.6 \pm 1.9(stat) \times 10^{-40} cm^2$ per nucleon at the mean neutrino energy of 0.9 GeV. The systematic error is currently under evaluation. The measured cross section result agrees with the NEUT prediction within statistical uncertainty.

5. Neutral current neutrino interactions

The characteristic topology of neutral current π^0 production events is two gamma rays from π^0 decay in the final state. We reconstruct gamma rays converting in SciBar and select events with two reconstructed gamma rays and no muons. A neutral current π^0 interaction here is defined as neutral current neutrino interaction in which at least one π^0 is emitted in the final state from the target nucleus. We obtained $(7.7 \pm 0.5(stat) \pm 0.5(sys)) \times 10^{-2}$ as the ratio of the neutral current neutral pion production to total charged current cross section as in [9]. We measured the ratio of cross sections in order to minimize systematic uncertainty due to the neutrino flux prediction. The result agrees with the Rein- Sehgal model implemented in our neutrino interaction simulation program with nuclear effects. The spectrum shape of the π^0 momentum and angle also agree with the model.

The SciBooNE collaboration performed a search for neutral current coherent pion production using the neutral current π^0 sample described before. In neutral current coherent pion production, there is no recoil nucleon in the final state since the π^0 is produced by the neutrino interacting with the whole nucleus. To separate the neutral current coherent π^0 events from the neutral current resonant π^0 events, recoil protons in the final state are used. We measured the ratio of the neutral current coherent π^0 production to total charged current cross sections to be $(1.16 \pm 0.24) \times 10^{-2}$. The error includes both the statistical and systematical errors. The ratio of charged current coherent π^+ to neutral current coherent π^0 production is calculated to be $0.14^{+0.30}_{-0.28}$, using our published charged current coherent pion measurement.

6. K^+ production at BNB

SciBooNE reported a K^+ production cross section and rate measurements in interactions of protons on the MiniBooNE beryllium target in the BNB in [10]: this was done using high energy muons produced by neutrinos interacting in the SciBar detector. Using observed neutrino and antineutrino events in SciBooNE, we measure $\frac{d^2\sigma}{dpd\Omega} = (5.34 \pm 0.76)$ mb/(GeV/c $\times$ sr) for $p + Be \rightarrow K^+ + X$ at mean K^+ energy of 3.9 GeV and angle (with respect to the proton beam direction) of 3.7 degrees, corresponding to the selected K^+ sample.

Acknowledgements

We acknowledge the support of Fermilab, MEXT, JSPS (Japan), the INFN (Italy), the Ministry of Science and Innovation and CSIC (Spain), the STFC (UK), and the DOE and NSF (USA).

References

[1] Alcaraz-Aunion J L and Walding J (SciBooNE Collaboration) 2009 *AIP Conf. Proc.* **1189** 145–150 (*Preprint* arXiv:0909.5647 [hep-ex])
[2] Mitsuka G 2008 *AIP Conf. Proc.* **981** 262–264
[3] Aguilar-Arevalo A A *et al.* (MiniBooNE Collaboration) 2010 *Phys. Rev. D* **81** 092005 (*Preprint* arXiv:1002.2680 [hep-ex])
[4] Lyubushkin V *et al.* (NOMAD Collaboration) 2009 *Eur. Phys. J. C* **63** 355–381 (*Preprint* arXiv:0812.4543 [hep-ex])
[5] Nakajima Y *et al.* (SciBooNE Collaboration) 2011 *Phys. Rev. D* **83** 012005 (*Preprint* arXiv:1011.2131 [hep-ex])
[6] Casper D 2002 *Nucl. Phys. Proc. Suppl.* **112** 161–170 (*Preprint* arXiv:0208030 [hep-ph])
[7] Hiraide K *et al.* (SciBooNE Collaboration) 2008 *Phys. Rev. D* **78** 112004 (*Preprint* arXiv:0811.0369 [hep-ex])
[8] Hasegawa M *et al.* (K2K) 2005 *Phys. Rev. Lett.* **95** 252301 (*Preprint* arXiv:0506008 [hep-ex])
[9] Kurimoto Y *et al.* (SciBooNE Collaboration) 2010 *Phys. Rev. D* **81** 033004 (*Preprint* arXiv:0910.5768 [hep-ex])
[10] Cheng G *et al.* (SciBooNE Collaboration) 2011 *Phys. Rev. D* **84** 012009 (*Preprint* arXiv:1105.2871 [hep-ex])

Journal of Physics: Conference Series **408** (2013) 012039 doi:10.1088/1742-6596/408/1/012039

Neutrino detection in the ArgoNeuT LAr TPC

Ornella Palamara

INFN - Laboratori Nazionali del Gran Sasso, S.S. 17bis Km 18+910, Assergi (L'Aquila), Italy

E-mail: ornella.palamara@lngs.infn.it

Abstract. ArgoNeuT, a Liquid Argon Time Projection Chamber (LAr-TPC), has collected thousands of neutrino and anti-neutrino events between 0.1 and 10 GeV in the NuMI beamline at Fermilab (FNAL). Among other issues, the experiment will measure the cross section of the neutrino and anti-neutrino Charged Current Quasi-Elastic (CC QE) interaction on Ar target and analyze the vertex activity associated with such events. ArgoNeuT data analysis and FSI studies in progress are described and first measurements of inclusive muon neutrino Charged Current differential cross sections on Argon are reported.

1. Introduction

In the recent years, due to the increasing interest on LAr-TPC technology in the US, a dedicated experiment (ArgoNeuT) has been included as a first step in a graded program towards the realization of a kiloton-scale precision LArTPC-based detector to be used for understanding accelerator and atmospheric-based neutrino oscillations, proton decay, and supernova burst/diffuse neutrinos. The ArgoNeuT detector has been exposed to ν_μ and $\bar{\nu}_\mu$ in the NuMI beam (LE beam option) at Fermilab from September 2009 to February 2010. The ArgoNeuT physics goals are:

(1) Measure ν charged-current cross-sections in the few GeV range. In particular, the CC QE channel and CC Resonant channel, with unprecedented sensitivity to products of FSI (vertex activity characterization). The ArgoNeuT expected precision for CC inclusive data is around 15%. The CC QE interaction is the golden channel for long baseline neutrino oscillation experiments. However, the CC QE cross section is known with only 20-30% precision over most of energies. Furthermore, only a single (preliminary) measurement of the neutrino cross section on Argon target has ever been taken (at much higher energy, 28 GeV)[1];

(2) Study and optimize e/γ separation: excellent signal (CC ν_e) efficiency and superior background (NC π^0) rejection is important for future ν_e appearance searches;

(3) Develop reconstruction techniques useful for future LAr-TPCs experiments.

2. Detector design

The ArgoNeuT TPC is a 47×40×90 cm^3 (Drift×Vertical×Beam coordinate) rectangular volume containing 175 l of LAr active volume. The TPC is positioned in a vacuum jacketed cryostat. The TPC consists of two wire planes with 240 wires each, separated by 4mm and oriented at $\pm 60^o$. A 500 V/cm electric field in the drift region, between the cathode and first plane of wires, is established.

During the physics run, the detector was located just upstream of the MINOS Near Detector in the NuMI beam line. Since ArgoNeuT's TPC is too small to contain the majority of the muons produced in neutrino interactions from the NuMI beam, the information from the MINOS near detector are used in the ArgoNeuT data analysis. This is an enormous advantage for ArgoNeuT since MINOS, in addition to provide information on the momentum of the escaping muon, can also determine the sign of the muon from its magnetized detector.

ArgoNeuT collected about 1.35×10^{20} protons on target (POT) with the low energy NuMI configuration (0.1×10^{20} POT in neutrino mode and 1.25×10^{20} POT in antineutrino mode). The total collected statistics amounts to ~6600 ν_μ CC and ~4900 $\bar{\nu}_\mu$ CC events. This represents the first sample of thousands of (anti-)neutrino LAr-TPC events collected in a low-energy (<E>$\simeq$4 GeV) neutrino beam ever.

3. ν event reconstruction

The core features of the LAr-TPC technique can be summarized in its high resolution 3D imaging accompanied by a good calorimetric reconstruction of the ionizing event. The combination of these aspects certifies the excellent particle identification capability of this detector. The development of the ArgoNeuT analysis tools for reconstructing particle tracks in LAr-TPCs is performed in collaboration with MicroBooNE [2] and LBNE [3], through the common LArSoft [4] software toolkit.

Figure 1 and 2 depict examples of neutrino events collected in the ArgoNeuT TPC. Signals from the two wire planes combined with timing information provide a three dimensional picture of the neutrino event with complete calorimetric information. The reconstruction chains starts with a hit construction and identification from the individual wire raw data signal information. Then the hits are clustered with other nearby hits and two dimensional tracks are reconstructed. Then, a three dimensional track reconstruction is performed and finally particle identification (from energy deposition dE/dx along the track) is accomplished for stopping tracks. As mentioned previously, ArgoNeuT uses the data from the MINOS detector to obtain the information on muons that escape the detector. At this step, identified muon tracks in ArgoNeuT are matched to muon tracks in MINOS. The matching criteria are based on the angle between the ArgoNeuT and MINOS tracks as well as the radial difference between the projected-to-MINOS ArgoNeuT track and the candidate MINOS track (details on the matching requirements can be found in [5]). Once the matching is successful, the charge and momentum of the muons are extracted from MINOS data.

In particular, in the case of a CC QE interaction the neutrino energy can be reconstructed from the lepton and proton energy and angle with respect to beam axis. The muon and proton are identified in a CC QE event with a combination of dE/dx and range measurements in ArgoNeuT. The downstream magnetized MINOS near detector is used to sign-select and fully reconstruct the muon momentum, while the proton(s) (contained >50% of the times) is reconstructed with ArgoNeuT alone.

Outcomes from the reconstruction of events can be found in [6]. The complete reconstruction of the available sample of (anti)-neutrino events is in progress.

4. FSI studies

In ν interactions, when the target nucleon is bound in the parent nucleus, the Relativistic Fermi Gas Model (RFG) is usually adopted to describe the nuclear initial state. Final state particles (hadrons) produced at the primary neutrino collision undergo effects of strong interactions inside the target nucleus (Final State Interactions, FSI). In this case the absence of well defined models makes the treatment of these Nuclear Effects the main potential source of systematic uncertainty. FSI can lead to the emission or absorption of protons and neutrons and the production of nuclear fragments, α particles and low energy γs; less frequently pions. Although FSI can mask the identity of the primary vertex with consequent misleading event classification, they are usually neglected because their products are not detectable, unless a high quality detector is in use (like a LAr-TPC, that provides unique imaging and identification capabilities). ArgoNeuT events clearly show the presence of FSI products (e.g. multi-p accompanying the leading muon in CC QE events and γ activity in the volume around the vertex). A detailed reconstruction (and MC simulation with different ν events generators, NUANCE, GENIE and FLUKA) of the activity associated to the interaction vertex is in progress. Preliminary results have been obtained from

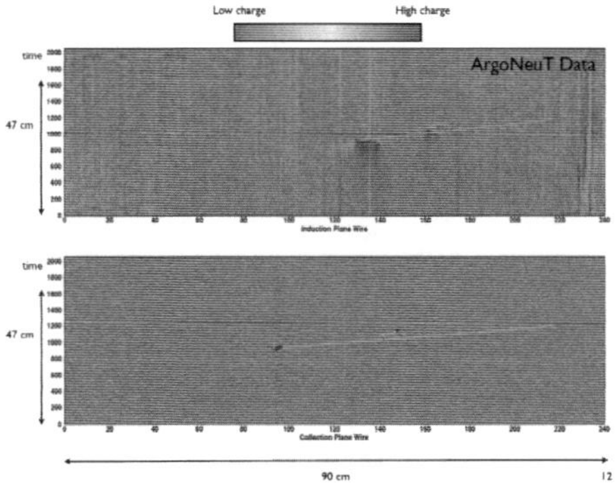

Figure 1. A neutrino event as seen in ArgoNeuT's induction and collection plane views. An escaping μ track and a contained p track at the vertex characterize this CC QE event.

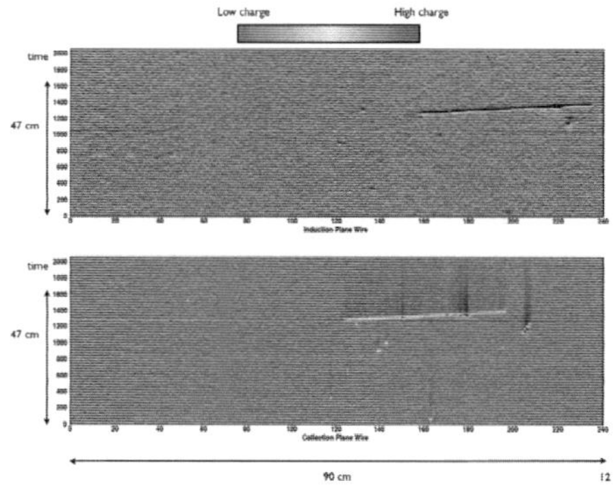

Figure 2. A single muon event. The muon track escapes ArgoNeuT and is matched with a negative charged muon track in MINoS. No proton is present at the vertex and γ-activity around the vertex is clearly visible.

the analysis of the easiest topology, one track events, i.e. single muon events, like the one shown in Figure 2. Single μ^- (μ^+) events can be produced by ν_μ (anti-ν_μ) CC interactions (mainly QE or RES with the proton/pion absorbed due to FSI). Few tens of single muon events (40 % μ^- and 60% μ^+) are expected in the ArgoNeuT ν-mode run data sample, using NUANCE MC generator and FLUKA MC generator (work done in collaboration with G. Battistoni [7]), which includes a full simulation of FSI. The number of events with this topology found in real data is consistent with expectations. Many of these events show e.m. activity (possibly due to γs from nuclear de-excitation converting in the LAr volume) around the vertex, as shown in Figure 2. Complete analysis of single tracks events, including muon sign determination from MINOS reconstruction and evaluation of background event from NC interactions (single π events faking single μ topology) is in progress. The substantially larger statistics collected in the anti-ν mode run will allow to perform a more precise measurement.

The interesting output of this analysis will be the measurement of $\frac{N_{\mu^-}}{N_{\mu^+}}$, useful for FSI studies, being an indication of nucleon charge exchange in Ar nuclei.

5. Inclusive ν_μ CC differential cross section

An important goal of the ArgoNeuT experiment is to characterize the features of the neutrino cross sections on Ar target. The first natural measurement to be done is an all-inclusive charge-current measurement. Indeed, such a measurement is totally independent on channel definitions and is minimally sensitive to FSI (see previous Section). All the different types of CC interactions have in common an outgoing muon. In addition, further cross-section measurements can be compared to the CC-inclusive one to help disentangling the effects of FSIs and nuclear modeling. In order to provide kinematic information, the differential cross section, as a function of the outgoing muon angle and momentum is a very useful measurement for theoretical predictions. Normally, the double differential cross-section ($\frac{d\sigma}{d\theta_\mu dp_\mu}$) would be calculated, but in order to do

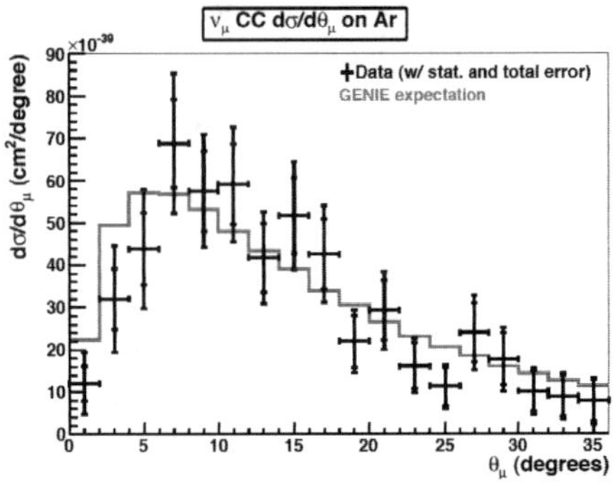

Figure 3. The ν_μ CC differential cross section (per argon nucleus) in muon angle.

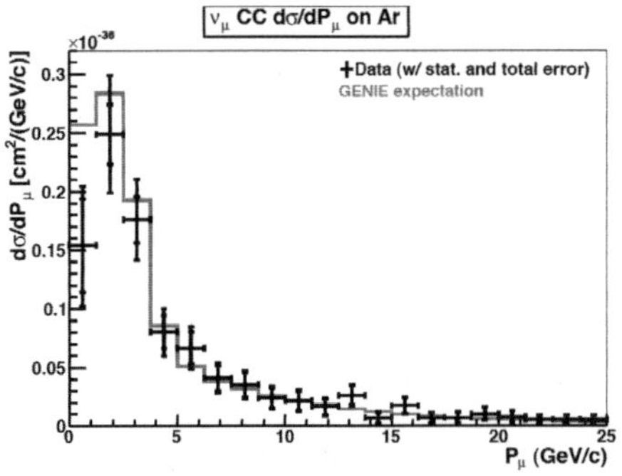

Figure 4. The ν_μ CC differential cross section (per argon nucleus) in muon momentum.

so, high statistics are required. Since ArgoNeuT collected only two weeks of neutrino data, one dimensional differential cross sections ($\frac{d\sigma}{d\theta_\mu}$, $\frac{d\sigma}{dp_\mu}$) are reported. All the details of the analysis and results of the CC-inclusive differential cross-section measurements can be found in [5] and will be available in a dedicated paper in preparation. Figure 3 and 4 shows the CC-inclusive differential cross-section as a function of the muon angle and of the muon momentum respectively. The data are consistent with GENIE Monte Carlo expectation across the full range of kinematics sampled, $0^o<\theta_\mu<36^o$ and $0<p_\mu<25$ GeV/c. An additional 1.25×10^{20} POT taken in anti-ν mode run is currently being analyzed.

6. Conclusion

ArgoNeuT is a fully operational LAr-TPC: during the ν-run, large samples of neutrino/antineutrino events have been collected for the first time ever in a low-Energy neutrino beam. ArgoNeuT data analysis is in progress: first neutrino CC-inclusive analysis just completed. Extensive real data analysis is invaluable in improving the LAr-TPC technique. Highly sophisticated/detailed MC codes are needed for FSI studies and are currently under test/optimization.

Acknowledgments

Acknowledgments The ArgoNeuT Collaboration gratefully acknowledge the cooperation of the MINOS Collaboration in providing their data for use in the analysis.

References

[1] de la Ossa M 2007 *Study of accelerator neutrino interactions in a liquid argon TPC* Ph.D. Thesis University of Granada
[2] Chen H *et al.* (MicroBooNE Collaboration) 2011 FERMILAB-PROPOSAL-0974
[3] http://lbne.fnal.gov
[4] https://cdcvs.fnal.gov/redmine/projects/larsoftsvn/wiki
[5] Spitz J 2011 *Measuring Muon-Neutrino Charged-Current Differential Cross Sections with a Liquid Argon Time Projection Chamber* PhD Dissertation Yale University
[6] Palamara O 2011 *The ArgoNeuT LAr-TPC: A dedicated experiment for neutrino cross section measurement at FNAL Nucl. Phys. Proc. Suppl.* **217** 189-192
[7] Battistoni G private communication

Journal of Physics: Conference Series **408** (2013) 012040 doi:10.1088/1742-6596/408/1/012040

The nucleon axial mass and the MiniBooNE CCQE neutrino-nucleus data

J Nieves[1], I Ruiz Simo[2] and M J Vicente Vacas[3]

[1] Instituto de Física Corpuscular (IFIC), Centro Mixto Universidad de Valencia-CSIC, Institutos de Investigación de Paterna, E-46071 Valencia, Spain

[2] Departamento de Física Atómica, Molecular y Nuclear, Universidad de Granada, E-18071 Granada, Spain.

[3] Departamento de Física Teórica and IFIC, Centro Mixto Universidad de Valencia-CSIC, Institutos de Investigación de Paterna, E-46071 Valencia, Spain

E-mail: jmnieves@ific.uv.es

Abstract. We analyze the MiniBooNE CCQE $d\sigma/dT_\mu d\cos\theta_\mu$ data using a theoretical model that has proved to be quite successful in the analysis of nuclear reactions with electron, photon and pion probes. We find that RPA and multinucleon knockout turn out to be essential for the description of the MiniBooNE data. We show these measurements are fully compatible with former determinations of nucleon axial mass M_A, in contrast with several previous analyses, which have suggested an anomalously large value. We find, $M_A = 1.08 \pm 0.03$ GeV. We also argue that the procedure, commonly used to reconstruct the neutrino energy for QE events from the muon angle and energy, could be unreliable for a wide region of the phase space, due to the large importance of multinucleon events.

1. Introduction

The interaction of neutrinos with nuclei at intermediate energies provides relevant information on the axial hadronic currents. The predicted cross sections for charged current (CC) quasielastic (QE) scattering are very similar for most theoretical models, however, they are clearly below the recently published MiniBooNE data [1]. Actually, the cross section per nucleon on ^{12}C is clearly larger than for free nucleons. The discrepancy is large enough to provoke much debate and theoretical attention. Some works try to understand these new data in terms of a larger value of M_A (nucleon axial mass) around 1.3–1.4 GeV [1, 2, 3, 4]. These large values are not only difficult to accommodate theoretically, but are also in conflict with the value for $M_A = 1.03 \pm 0.02$ GeV that is usually quoted as the world average [5, 6].

In most theoretical works QE is used for processes where the gauge boson W is absorbed by just one nucleon, which together with a lepton is emitted (see Fig. 1a). However, in the recent MiniBooNE measurements, QE is related to processes in which only a muon is detected. This latter definition could make sense because ejected nucleons are not detected in that experiment, but includes multinucleon processes (see Fig. 1b) and others like pion production followed by absorption[1]. However, it discards pions coming off the nucleus, since they will give rise to additional leptons after their decay (see Fig. 1c). In any case, their experimental results cannot

[1] Note, MiniBooNE analysis Monte Carlo corrects for those events.

be directly compared to most previous calculations, as it was first pointed out by M. Martini et al. [9, 10], in which only the one-body QE contribution is considered. In this talk, we present

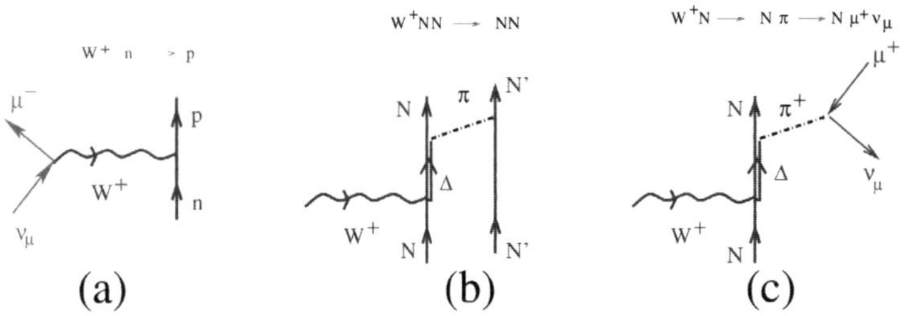

Figure 1. Mechanisms for W absorption inside of a nucleus.

a microscopic calculation of the CCQE-like double differential cross section $\frac{d\sigma}{dT_\mu d\cos\theta_\mu}$ measured by MiniBooNE and we will use these data to extract M_A. We will estimate the CCQE-like cross section by the sum of the theoretical QE (Fig. 1a) cross section and that induced by multinucleon mechanisms, as the one depicted in Fig. 1b, where the gauge boson is being absorbed by two or more nucleons without producing pions.

2. MiniBooNE CCQE-like cross sections and multinucleon mechanisms

First we pay attention to the total cross section and compare our results with the unfolded data of Ref. [1]. Our microscopic model, derived in Refs. [7] and [8], starts from a relativistic local Fermi gas (LFG) picture of the nucleus, which accounts for Pauli blocking and Fermi motion. The QE contribution was studied in [8] incorporating several nuclear effects. The main one is the medium polarization (RPA), including Δ-hole degrees of freedom and explicit π and ρ meson exchanges in the vector-isovector channel of the effective nucleon-nucleon interaction. The model for multinucleon mechanisms (not properly QE but included in the MiniBooNE data [1]) is fully discussed in Ref. [7]. The model includes one, two, and even three-nucleon mechanisms, as well as the excitation of Δ isobars. There are no free parameters in the description of nuclear effects, since they were fixed in previous studies of photon, electron, and pion interactions with nuclei [11, 12, 13, 14]. This theoretical model has proved to be quite successful in the study of nuclear reactions with photon [11], pion [12, 13] and electron [14] probes.

Up to neutrino energies around 1 GeV, the predictions of our model compare rather well, taking into account experimental and theoretical uncertainties, with the recent data published by the SciBooNE collaboration for total neutrino inclusive cross sections [15]. Results are displayed in Fig. 2 taken from Ref. [7]. There, it can also be appreciated how at larger energies, we underestimate the cross section. Indeed, we could observe that some $WNN \rightarrow NN\pi$ contributions neglected in our model, become relatively important at these higher energies.

Our predictions [7] for the flux-unfolded muon neutrino's CCQE-like cross section on ^{12}C measured in [1] are depicted in Fig. 3. The first observation is that our QE curve misses the data-points, being our predicted QE cross sections significantly smaller than those reported by the MiniBooNE collaboration. However, when multinucleon knock out contributions are added to the QE prediction of [8], we obtain the solid green line in a better agreement with the MiniBooNE data. In these calculations, M_A is fixed to 1.05 GeV. Thus we confirm the findings of [9, 10] on the crucial role played by the multinucleon mechanisms in the CCQE-like MiniBooNE data, and that when these latter processes are considered, high values of M_A in the 1.3-1.4 GeV range are not needed to describe the data. Our evaluation of these pionless multinucleon emission contributions to the cross section is fully microscopical and it contains terms, which were either not considered or only approximately taken into account in [9].

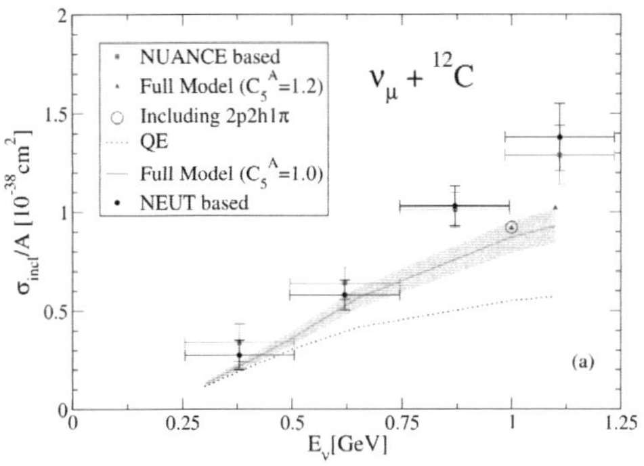

Figure 2. SciBooNE neutrino CC inclusive interaction cross section per nucleon [15], together with the QE and full model (including the theoretical uncertainty band) results of Ref. [7]. See this latter reference for further details.

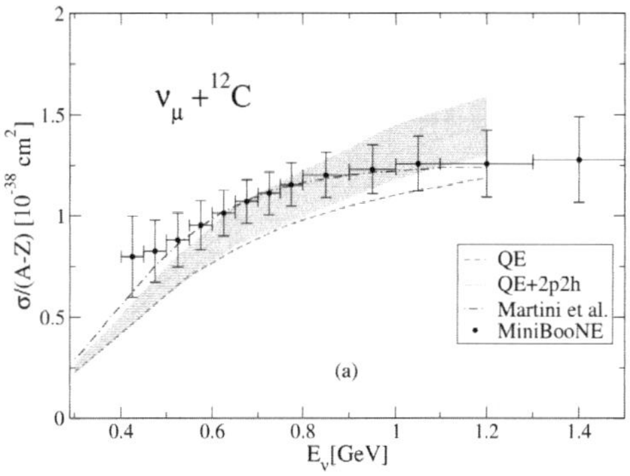

Figure 3. Flux-unfolded MiniBooNE ν_μ CCQE cross section per neutron, together with different theoretical predictions from Ref. [7]. Data points are taken from Ref. [1]. We also show the results (blue dash-dotted line) obtained in Ref. [10].

3. Extracting M_A from MiniBooNE data

The MiniBooNe data include energy and angle distributions as well, and therefore provide a much richer information. Furthermore, the unfolded cross section is not a very clean observable after noticing the importance of multinucleon mechanisms, because the unfolding itself is model dependent and assumes that the events are purely QE. In Fig. 4, we show our results for the MiniBooNE neutrino flux folded CCQE-like $d\sigma/dT_\mu d\cos\theta_\mu$ distribution with $M_A = 1.049$ GeV (value used in our previous works). The full (QE+multinucleon mechanisms) model agrees remarkably well with these data, despite the fact that no parameters have been fitted to data, beyond a global scale, λ.

The inclusion of this scale λ takes into account the global normalization uncertainty of around 10% acknowledged in [1]. Details can be found in [16]. Though the consistency of MiniBooNE data with standard values of M_A has been established now, one could still go further and use our full model to fit the data letting M_A to be a free parameter. We get $M_A = 1.08 \pm 0.03$ GeV and $\lambda = 0.92 \pm 0.03$ with a strong correlation between both parameters. The inclusion of multinucleon mechanisms and RPA is essential to obtain axial masses consistent with the world average. This can be appreciated in Fig. 5, where one can see that RPA strongly decreases the cross section at low energies, while multinucleon mechanisms accumulate their contribution at low muon energies and compensate that depletion. Therefore, the final picture is that of a delicate balance between a dominant single nucleon scattering, corrected by collective effects, and other mechanisms that involve directly two or more nucleons. Both effects can be mimicked by using a large M_A value (LFG entry in Table 1). We also see that the proportion of multinucleon events contributing to the QE-like signal is quite large for low muon energies and thus, the algorithm used to reconstruct the neutrino energy is badly suited for this region. This could

Table 1. Fit results for various models.

LFG	$\lambda = 0.96 \pm 0.03$	$M_A = 1.32 \pm 0.03$ GeV	$\chi^2/\#$ bins $= 33/137$
Full	$\lambda = 0.92 \pm 0.03$	$M_A = 1.08 \pm 0.03$ GeV	$\chi^2/\#$ bins $= 50/137$

have consequences in the determination of the oscillation parameters.

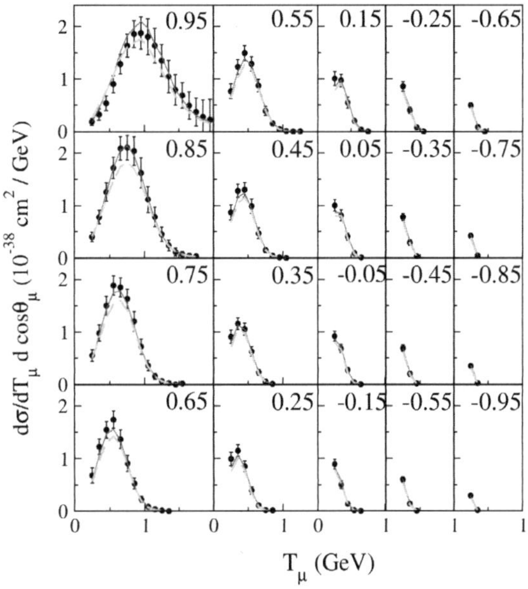

Figure 4. $d\sigma/dT_\mu d(\cos\theta_\mu)$ for different angular bins (labeled by its cosinus central value). Experimental points are taken from Ref. [1]. Green-dashed line (no fit) is the full model prediction (including multinucleon mechanisms and RPA) of Ref. [7] and calculated with $M_A = 1.049$ GeV. Red-solid line is best fit ($M_A = 1.32$ GeV) for the model without RPA and without multinucleon mechanisms.

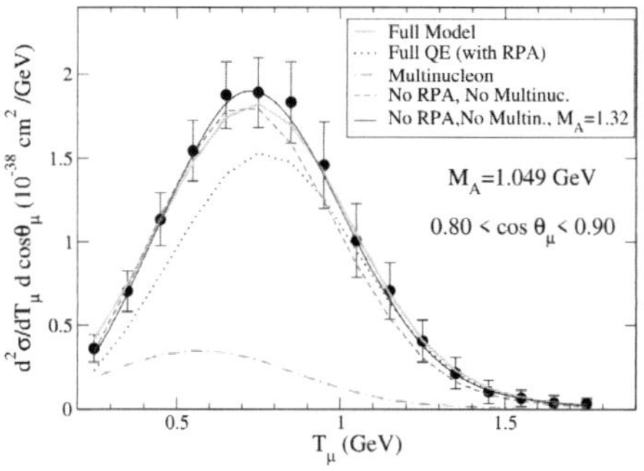

Figure 5. Muon angle and energy distribution for the $0.80 < \cos\theta_\mu < 0.90$ bin. Data from Ref. [1] and calculation with $M_A = 1.32$ GeV are multiplied by 0.9. In the other curves a value of $M_A = 1.049$ GeV was used.

Acknowledgments

This research was supported by DGI and FEDER funds, under contracts FIS2008-01143/FIS, FIS2006-03438 and CSD2007-00042, by Generalitat Valenciana contract PROMETEO/2009/0090 and by the EU HadronPhysics2 project, grant agreement n. 227431.

References

[1] Aguilar-Areval A A *et al.* [MiniBooNE Collaboration] 2010 *Phys. Rev.* D **81** 092005.
[2] Butkevich A V 2010 *Phys. Rev.* C **82** 055501.
[3] Benhar O, Coletti P and Meloni D 2010, *Phys. Rev. Lett.* **105** 132301.
[4] Juszczak C, Sobczyk J T and Zmuda J 2010 *Phys. Rev. C* **82** 045502.
[5] Bernard V, Elouadrhiri L and Meissner U G 2002 *J. Phys.* G **28**.
[6] Lyubushkin V *et al.* [NOMAD Collaboration] 2009 *Eur. Phys. J. C* **63** 355.
[7] Nieves J, Ruiz Simo I and and Vicente Vacas M J 2011 *Phys. Rev.* C **83** 045501.
[8] Nieves J, Amaro J E and Valverde M 2004 *Phys. Rev.* C **70** 055503.
[9] Martini M, Ericson M, Chanfray G and Marteau J 2009 *Phys. Rev.* C **80** 065501.
[10] Martini M, Ericson M, Chanfray G and Marteau J 2010 *Phys. Rev.* C **81** 045502.
[11] Carrasco R C and Oset E 1992 *Nucl. Phys.* A **536** 445.
[12] Nieves J, Oset E and Garcia Recio C 1993 *Nucl. Phys. A* **554** 509.
[13] Nieves J, Oset E and Garcia Recio C 1993 *Nucl. Phys. A* **554** 554.
[14] Gil A, Nieves J and Oset E 1997 *Nucl. Phys.* A **627** 543.
[15] Nakajima Y *et al.* [SciBooNE Collaboration] 2011 *Phys. Rev.* D **83** 012.
[16] Nieves J, Ruiz Simo I and and Vicente Vacas M J 2011 *Phys. Lett.* B **707** 72.

Two particle-two hole excitations in charged current quasielastic neutrino-nucleus interactions

Marco Martini
CEA/DAM/DIF, F-91297 Arpajon, France
E-mail: marco.martini@cea.fr

Abstract. We review the theoretical status of the models including the multinucleon emission channel in the calculation of quasielastic neutrino cross sections at MiniBooNE kinematics.

The MiniBooNE collaboration has recently measured the neutrino charged current (CC) quasielastic (QE) cross section on ^{12}C for a neutrino beam with average energy of 788 MeV [1]. The comparison of these results with a prediction based on the relativistic Fermi gas model using the standard value of the axial cut-off mass $M_A = 1.03$ GeV/c^2 reveals a substantial discrepancy. In the same model the larger value of $M_A = 1.35$ GeV/c^2 is needed to account for data. For nuclear physicists accustomed to the complexity of the nuclear system, this anomaly is likely to reflect the many-body aspect of the problem. Indeed Martini *et al.* have pointed out [2] that, depending on the detection method, certain types of inelastic events can simulate quasielastic ones. As the MiniBooNE experimental set-up defines a "quasielastic" event as one in which only a muon is detected, the ejection of a single nucleon (a genuine quasielastic event) is only one possibility, and one must in addition consider events involving a correlated nucleon pair from which the partner nucleon is also ejected. This leads to the excitation of 2 particle-2 hole (2p-2h) states; 3p-3h excitations are also possible. The inclusion in the quasielastic cross section of events in which several nucleons are ejected, leads to an excess over the genuine quasielastic value. Martini *et al.* have argued that this is the likely explanation of the anomaly showing that their evaluation can account for the excess in the cross section without any modification of the axial mass. After this suggestion, a number of articles have discussed the problem of multinucleon emission and whether it could account for the anomaly. The aim of this work is to review the actual theoretical status on this subject.

The theoretical studies of 2p-2h in connection with CCQE at MiniBooNE kinematics are actually performed essentially by three groups. There are the works of Martini *et al.* [2, 3, 4], the ones of Amaro *et al.* [5, 6] and the ones of Nieves *et al.* [7, 8]. Recently it appears also a paper of Bodek *et al.* [9] considering a more phenomenological approach related to electron scattering. Two-nucleon emission via two-body currents is not a new idea related to neutrino-nucleus interaction, it was studied since the eighties in connection to the electron scattering, photon and pion absorption. It is for example well known that the absorption of pions by nuclei is a two-nucleon mechanism. Concerning electron scattering, several theoretical calculations have shown that 2p-2h are needed to reproduce the inclusive cross section. In particular they contribute to fill the so called "dip" region between the quasielastic and the delta peaks. They affect the isospin spin-transverse (i.e. the magnetic) response. An exhaustive bibliography on these studies is beyond the purposes of this work. In the following we just mention the sources and the references of the calculations of the three groups mentioned above. The np-nh contributions in the papers of Martini *et al.* are obtained starting from the microscopic calculations of pion absorption at threshold [10], of pion and photon absorption [11] and of the transverse response in electron scattering [12]. They give two different parametrizations of this np-nh contribution (which leads to practically identical neutrino total "quasielastic" cross section), one more related to [10], the other to [12]. In the following we present the results just obtained with this second parametrization which is characterized by a more satisfactory energy and momentum transfer dependence. We refer to [2] for details. The 2p-2h contributions considered by Amaro *et al.* are taken from the full relativistic model of [13] related to the electromagnetic responses. The work of Nieves *et al.* can be considered as a generalization of [14], developed for the electron scattering, to the neutrino scattering. The contributions related to the Δ-MEC excitations are taken, as in the case of Martini *et al.* from [11]. Considering these three different models, it exists some differences already of level of genuine QE which can be particularly important when one compares the double differential cross sections. Amaro *et al.* considered the relativistic superscaling approach based on the superscaling behavior exhibited by electron scattering data. The models of Martini *et al.* and Nieves *et al.* are more similar: they start from a local Fermi gas picture of the nucleus. They consider medium polarization and collective effects through the random phase approximation (RPA) including Δ-hole degrees of freedom, π and ρ meson exchange and g' Landau-Migdal parameters in the effective $p-h$ interaction. The treatment of Nieves *et al.* is relativistic while the original one of Martini *et al.* was non relativistic. Recently it has been improved through relativistic corrections [4]. Concerning the 2p-2h sector we remind that there exist several contributions to two-body currents, see for example [7, 12, 13]. There are the so called pion-in-flight term, the contact term and the Δ-intermediate state or Δ-MEC term. At level of terminology, some authors refer just to the first two terms as Meson Exchange Currents contributions, like in [2], but actually the most current convention consists into include the Δ-term into MEC. In order to preserve the gauge invariance of the theory also the nucleon-nucleon (NN) correlation contributions must be taken into account. In an infinite system they give rise to divergences. The different prescriptions to regularize them lead to a model dependence of the results. In this connection Amaro *et al.* consider only the MEC contributions and not the correlations and the correlations-MEC interference terms. Correlations and interference are present both in Martini *et al.* and Nieves *et al.* even if Martini *et al.* consider only the Δ-MEC. On the other hand the treatment of Amaro *et al.* is fully relativistic as well as the one of Nieves *et al.* while the results of Martini *et al.* are related to a non-relativistic reduction of the two-body currents. Amaro *et al.* consider the contribution only in the vector sector while Martini *et al.* and Nieves *et al.* also in the axial one. The virtual meson exchanged between the two nucleons is just the pion for Amaro *et al.* while Martini *et al.* and Nieves *et al.* includes also g' Landau-Migdal parameters. Nieves *et al.* consider also the ρ meson exchange contribution.

Before to review the results obtained in neutrino scattering, let's analyze the transverse response in electron scattering. In Fig.1 we present the transverse (e, e') response on carbon calculated in the model of Martini *et al.* for some values of the momentum transfer q as a function of the energy transfer ω together with the experimental results [15, 16]. Note that there are already some differences between the two data sets. Concerning the results at $q = 300$ MeV/c one can compare the Martini *et al.* 2p-2h contribution with the one of Fig. 41 of [14]. One can observe that they are of the same order of magnitude. At higher q np-nh seems to be needed in order to reproduce data in particular in the dip region. Bodek *et al.* [9] parametrized this enhancement in terms of a correction to the magnetic form factor.

Turning to neutrino scattering, Fig.2 shows the "quasielastic" and the inclusive CC cross section as measured by MiniBooNE [1] and SciBooNE [17] respectively. The theoretical results

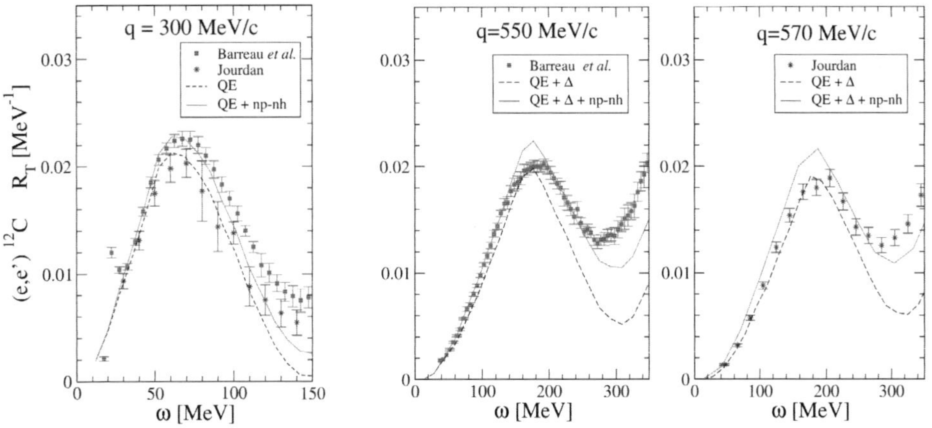

Figure 1. (e,e') transverse responses of ^{12}C. Experimental data are taken from [15] and [16]. The theoretical calculations are performed in the model of Martini *et al.*

of Martini *et al.* are also reported. As already mentioned the inclusion of the multinucleon component leads to an agreement with the "quasielastic" MiniBooNE data. Also the inclusive cross section, which is less sensitive to the background subtraction with respect to exclusive channels, can be reproduced, at least up to $E_\nu \simeq 1$ GeV once the np-nh channel is added. Similar conclusions both for QE and inclusive cross section have been obtained by Nieves *et al.* [7]. Concerning the MiniBooNE "QE" total cross section, it can actually be reproduced without any change in the axial mass also by Bodek *et al.* which on the other hand agrees also with NOMAD data. Also for Amaro *et al.* the agreement with the total "QE" cross section improves, at least at high neutrino energy, adding to the genuine QE in the SuSA approach the MEC contribution. Nevertheless the shape of this cross section is somewhat different with respect to the other calculations, maybe owing to the absence of NN correlations contributions. The total cross section has been analyzed also in the antineutrino mode by Martini *et al.* [3], Nieves *et al.* [7] and Bodek *et al.* [9]. The effect of np-nh as compared to one nucleon ejection should be somewhat less important for antineutrino because the spin isospin response which is affected by the 2p-2h piece has less weight for antineutrino owing to the negative axial-vector interference. This effect is evident in Martini *et al.* and Bodek *et al.*, less in Nieves *et al.* Let's turn to the MiniBooNE flux averaged "QE" double differential cross section. This is a directly measured quantity, hence free from the uncertainty of neutrino energy reconstruction which is a model dependent procedure. Theoretical calculations on 2p-2h contributions to this cross section have been performed by Amaro *et al.* [5, 6], Nieves *et al.* [8] and Martini *et al.* [4]. Two examples of results of Martini *et al.* are shown in Fig. 3. For a complete comparison in all the cosθ and T_μ range measured by MiniBooNE see [4]. The agreement with data is good in all this range once the multinucleon component is incorporated. It confirm that it is possible to account for the MiniBooNE results without any modification of the axial mass. Similar results have already been obtained by Nieves *et al.* [8]. Although slightly lower with respect to the Martini *et al.* calculations they are also fully compatible with MiniBooNE using the standard value of the axial mass. In some kinematical regions the agreement with MiniBooNE arises in Martini *et al.*, as well as in Nieves *et al.*, from the opposite actions of the RPA quenching and the enhancement from the np-nh contribution. For Amaro *et al.* [5, 6] the inclusion of 2p-2h MEC tends to improve the agreement with data at low angles, but it is not sufficient to account for discrepancy at higher angles. The absence of NN correlation contributions might be responsible for residual disagreement. A treatment of these contributions performed by the same group in the electron scattering sector [18] seems to suggest that they are of the same order

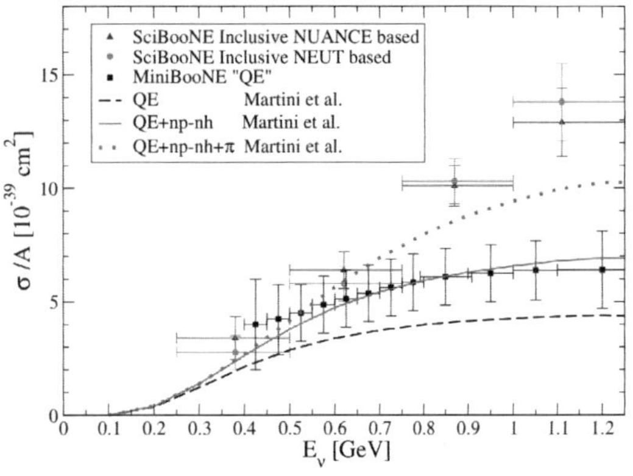

Figure 2. Inclusive and "quasielastic" cross sections measured by SciBooNE [17] and MiniBooNE [1] compared to Martini *et al.* calculations [2].

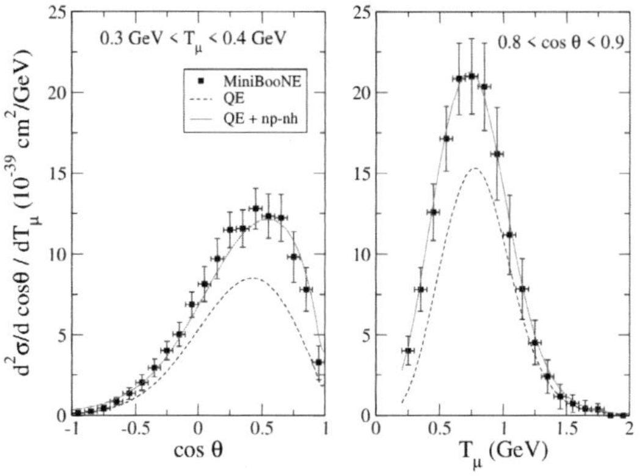

Figure 3. MiniBooNE [1] flux-averaged double differential QE cross section for some values of T_μ and $\cos\theta$. Theoretical calculations are the ones of Martini *et al.* [4].

of magnitude as the one of 2p-2h MEC and that they are supposed to have a greater impact for large neutrino-muon angles. A further enhancement of the cross section expected from these contributions would lead to an agreement with MiniBooNE.

In conclusion all the actual theoretical models treating the quasielastic with the inclusion of the multinucleon emission channel suggest that the MiniBooNE results can be explained without any modification of the axial mass. The complexity of the nuclear dynamics needed to reproduce the data should be taken into account in the algorithm to reconstruct the neutrino energy which is crucial in neutrino oscillation experiments.

Acknowledgments

I thank Magda Ericson for useful discussions, the NUFACT11 organizers and the WG2 conveners for the opportunity to present this review.

References

[1] Aguilar-Arevalo A A *et al.* [MiniBooNE Collaboration] 2010 *Phys. Rev.* D **81** 092005
[2] Martini M, Ericson M, Chanfray G and Marteau J 2009 *Phys. Rev.* C **80** 065501
[3] Martini M, Ericson M, Chanfray G and Marteau J 2010 *Phys. Rev.* C **81** 045502
[4] Martini M, Ericson M and Chanfray G 2011 *Phys. Rev.* C **84** 055502
[5] Amaro J E, Barbaro M B, Caballero J A, Donnelly T W and Williamson C F 2011 *Phys. Lett.* B **696** 151
[6] Amaro J E, Barbaro M B, Caballero J A, Donnelly T W and Udias J M 2011 *Phys. Rev.* D **84** 033004
[7] Nieves J, Ruiz Simo I and Vicente Vacas M J 2011 *Phys. Rev.* C **83** 045501
[8] Nieves J, Ruiz Simo I and Vicente Vacas M J 2012 *Phys. Lett.* B **707** 72
[9] Bodek A, Budd H and Christy M E 2011 *Eur. Phys. J.* C **71**, 1726
[10] Shimizu K and Faessler A 1980 *Nucl. Phys.* A **333** 495
[11] Oset E and Salcedo L L 1987 *Nucl. Phys.* A **468** 631
[12] Alberico W M, Ericson M and Molinari A 1984 *Annals Phys.* **154** 356
[13] De Pace A, Nardi M, Alberico W M, Donnelly T W and Molinari A 2003 *Nucl. Phys.* A **726** 303
[14] Gil A, Nieves J and Oset E 1997 *Nucl. Phys.* A **627** 543
[15] Barreau P *et al.* 1983 *Nucl. Phys.* A **402**, 515
[16] Jourdan J 1996 *Nucl. Phys.* A **603**, 117
[17] Nakajima Y*et al.* [SciBooNE Collaboration] 2011 *Phys. Rev.* D **83** 012005
[18] Amaro J E, Maieron C, Barbaro M B, Caballero J A and Donnelly T W 2010 *Phys. Rev.* C **82** 044601

XIII International Workshop on Neutrino Factories, Super beams and Beta beams (NUFACT11) IOP Publishing
Journal of Physics: Conference Series **408** (2013) 012042 doi:10.1088/1742-6596/408/1/012042

Confronting electron- and neutrino-nucleus scattering

Omar Benhar

INFN and Department of Physics
"Sapienza" Università di Roma. I-00185 Roma, Italy

E-mail: omar.benhar@roma1.infn.it

Abstract. The analysis of the sample of charged current quasi elastic events collected by the MiniBooNE Collaboration suggests that the scheme successfully employed to describe electron-nucleus scattering fails to explain neutrino-nucleus cross sections. I argue that, due to flux average, the double differential neutrino-nucleus cross section does not allow for a clearcut determination of the dominant reaction mechanism. A systematic study of the large body of electron scattering data may help to identify the processes, other than single nucleon knockout, contributing to the observed neutrino cross section.

Electron-nucleus scattering cross sections are usually analyzed at fixed beam energy, E_e, and electron scattering angle, θ_e, as a function of the electron energy loss ω. As an example, Fig. 1 shows the typical behavior of the inclusive cross sections at beam energy around 1 GeV. It is apparent that the different reaction mechanisms, yielding the dominant contributions to the cross section at different values of ω (corresponding to different values of the Bjorken scaling variable $x = Q^2/2m\omega$, where m is the nucleon mass and $Q^2 = 4E_e(E_e - \omega)\sin^2\theta_e/2$) can be easily identified.

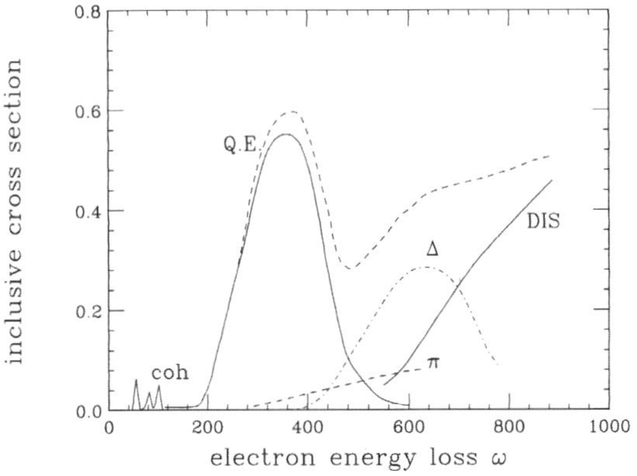

Figure 1. Typical behavior of the inclusive electron-nucleus scattering cross section at beam energy around 1 GeV, as a function of the electron energy loss ω.

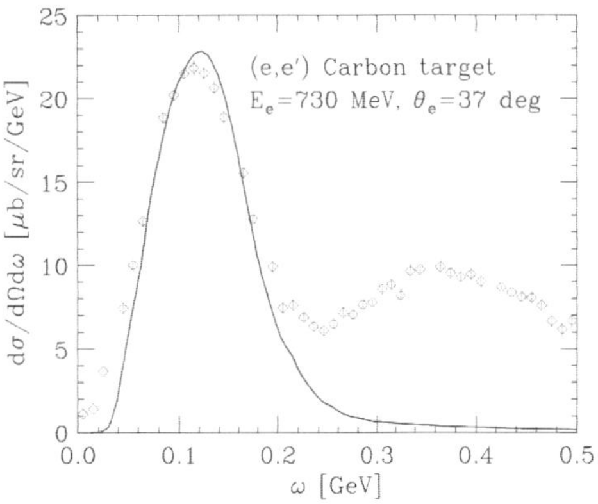

Figure 2. Quasi elastic electron-carbon cross section at beam energy $E_e = 730$ MeV and electron scattering angle $\theta_e = 37$ deg, plotted as a function of the energy loss ω [2]. The data points are from [3].

The bump centered at $\omega \sim Q^2/2m$, or $x \sim 1$, the position and width of which are determined by the momentum and removal energy distribution of the struck particle, corresponds to single nucleon knockout, while the structure visible at larger ω, or lower x, reflects the onset of coupling to two-nucleon currents, arising from meson exchange processes, excitation of nucleon resonances and deep inelastic scattering.

The quasi elastic electron-nucleus cross section can be described at quantitative level within the Impulse Approximation (IA), applicable when the magnitude of the three-momentum transfer, $|\mathbf{q}|$, is small compared to the average distance between nucleons in the target nucleus [1]. As an example, Fig. 2 shows the inclusive electron-carbon cross section at beam energy $E_e = 730$ MeV and electron scattering angle $\theta_e = 37$ deg, plotted as a function of the energy loss ω [2]. The data points are taken from [3], while the theoretical results have been obtained within the approach described in [1, 4], using a state-of-the-art parametrization of the vector nucleon form factors, extracted from the measured electron-proton electron-deuteron cross sections (for a review, see, e.g., [5]).

Applying the same scheme employed to obtain the solid line of Fig. 2 to neutrino scattering one gets the results shown in Fig. 3. The data points represent the double differential CCQE cross section of [6] averaged over the MiniBooNE neutrino flux, the mean energy of which is $\langle E_\nu \rangle = 788$ MeV, plotted as a function of the kinetic energy of the outgoing muon at different values of the muon scattering angle. The solid lines show the results of calculations performed using the same nuclear spectral functions and vector form factors employed in the calculation of the electron scattering cross section of Fig. 2 and the dipole parametrization of the axial form factor, with the axial mass $M_A = 1.03$ MeV [2] obtained from the world average of low statistics deuterium data [7, 8, 9].

Comparison between Figs. 2 and 3 indicates that the electron and neutrino cross sections corresponding to the same target and *seemingly* comparable kinematical conditions (the position of the QE peak in Fig. 2 corresponds to kinetic energy of the scattered electron $\sim$ 610 MeV) cannot be explained using the same theoretical approach and the value of the axial mass resulting from deuterium measurements.

The authors of [6, 10, 11] suggested that the measured CCQE cross sections may be explained by advocating a larger value of M_A, that should be regarded as an *effective* axial mass, modified by nuclear effects not included in the Fermi gas model employed in data analysis. However, most theoretical models of the CCQE cross section (for recent reviews see [12]) fail to support this explanation. In addition, the analysis of the large body of electron scattering data provide overwhelming evidence that the vector form factors are *not* modified in the nuclear medium.

In [2, 13] it is argued that the interpretation of the data sample of [6] as CCQE events may be hampered by the fact that, as the energy of the incoming beam is not fixed, the observed energy of the outgoing charged lepton *does not* uniquely determine the energy transfer to the target.

The implications of flux average can be easily understood considering the neutrino cross section at muon energy corresponding to the maximum of the spectrum shown in the upper panel of Fig. 3, i.e. $T_\mu = 0.55$ GeV and $\cos\theta = 0.75$. In this kinematics, $x = 1$ and 0.5 correspond to neutrino energy $E_\nu = 0.788$ and 0.975 GeV, respectively. As the values of the MiniBooNE flux corresponding to these energies are within less than 20% of one another, flux integration leads to collect contributions from different regimes, i.e. different reaction mechanisms, with about the same probability.

This feature can be best illustrated using the measured electron-carbon scattering cross sections. Figure 4 shows the data from [3, 14], taken at electron scattering angle $\theta_e = 37$ deg and beam energies ranging between 0.730 and 1.501 GeV, plotted as a function of the energy of the outgoing electron. It clearly appears that the energy bin corresponding to the top of the quasi elastic peak at $E_e = 0.730$ GeV, shown by the shaded area, receives significant contributions

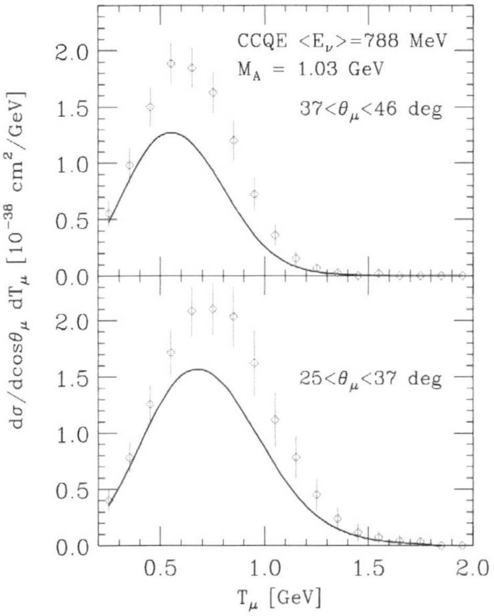

Figure 3. Flux averaged double differential CCQE cross section measured by the MiniBooNE collaboration [6], shown as a function of the kinetic energy of the outgoing muon. The upper and lower panels correspond to different values of the muon scattering angle. The theoretical results have been obtained in [2], using the formalism described in [1, 4], the same vector form factors employed in the calculation of the electron scattering cross section of Fig. 2, and a dipole parametrizaition of the axial form factor with $M_A = 1.03$ MeV.

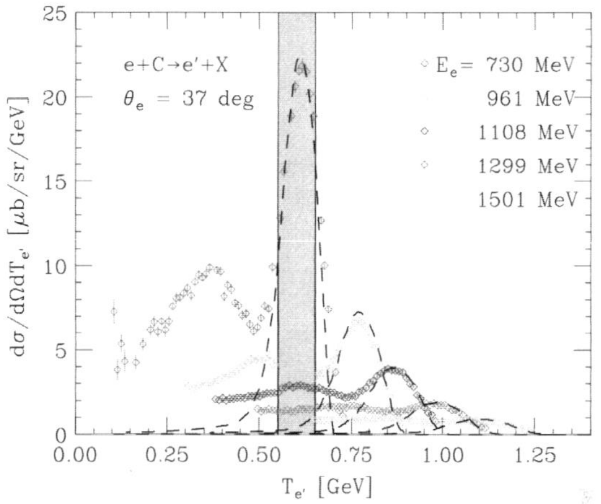

Figure 4. Inclusive electron-carbon cross sections at $\theta_e = 37$ deg and beam energies ranging between 0.730 and 1.501 GeV [3, 14]. The dashed lines represent the single nucleon knock out contribution, computed within the approach described in [1, 4]. The shaded area shows the energy bin corresponding to the peak of the cross section of Fig. 2.

from cross sections corresponding to different beam energies and different values of x.

The data displayed in Fig. 4 strongly suggest that the description of the flux-integrated CCQE neutrino nucleus cross section requires the inclusion of reaction mechanisms other than single nucleon knockout. According to the authors of [15, 16], the most important competing mechanism is multinucleon knockout, leading to two particle-two hole (2p2h) final states. Note that in neutrino experiments this final states cannot be distinguished from the one particle-one hole final states associated with single nucleon knockout.

Multinucleon knockout is known to occur due to i) initial state nucleon-nucleon correlations, ii) final state interactions between the struck nucleon and the spectator particles and iii) coupling to the two-body nuclear electroweak current.

Correlations between nucleons in the target ground state give rise to the tail extending to large ω, clearly visible in Fig. 2. However, their contribution, strongly constrained by semi-inclusive $(e, e'p)$ data [17] turns out to be quite small (less than 10% of the integrated spectrum). In principle, this reaction mechanism might be clearly identified detecting two nucleons moving in opposite directions with momenta much larger than the Fermi momentum (~ 250 MeV).

Final state interactions are not expected to play a relevant relevant in this context, as their main effects, which amounts to a shift and a redistribution of the inclusive strength, mostly affects the region of low energy loss, corresponding to $x > 1$.

The most important correction is likely to arise from processes involving the nuclear two-body current, as advocated by the authors of [15, 16]. It is long known that inclusion of these processes is needed to explain the nuclear electromagnetic response in the transverse channel [18]. In addition, their contribution turns out to decrease as the momentum transfer increases. This behavior may explain why the CCQE data at high neutrino energies collected by the NOMAD collaboration [9] can be described without any modification of the value of the nucleon axial mass.

In conclusion, comparison between electron and neutrino scattering data suggest that the analysis of the flux-averaged neutrino cross sections requires the development of models including a variety of relevant reaction mechanisms. While the approaches developed in [15, 16] certainly represent an important step towards the achievement of this goal, more theoretical work is still needed, e.g. to extend the applicability of the models to the regions of pion production and deep inelastic scattering.

The critical requirement to be met in developing new theoretical approaches will be consistency, i.e. the ability to describe different kinematical regimes using the same dynamical model. Within *ab initio* nuclear many-body theory this amounts to requiring that the target initial and final states be obtained from the same hamiltonian, fitted to the properties of exactly solvable few-nucleon systems, which also largely determines the structure of the vector two-body current through the continuity equation.

Acknowledgments
The results discussed in this paper have been obtained in collaboration with Pietro Coletti and Davide Meloni. An illuminating discussion with Georgia Karagiorgi and Camillo Mariani is also gratefully acknowledged. Finally, I would like to express to the organizers of NUFACT11 my deepest appreciation for setting up a most enjoyable and fruitful meeting.

References

[1] Benhar O, Day D and Sick I 2008 *Rev. Mod. Phys.* **80** 189
[2] Benhar O, Coletti P and Meloni D 2010 *Phys. Rev. Lett.* **105** 132301
[3] O'Connell J S *et al.* 1987 *Phys. Rev. C* **35** 1063
[4] Benhar O, Farina N, Nakamura H, Sakuda M and Seki R 2005 *Phys. Rev. D* **72** 053005
[5] Perdrisat C, Punjabi V and Vanderhaeghen M 2007 *Prog. Part. Nucl. Phys.* **59** 694
[6] Aguilar-Arevalo A A *et al.* (MiniBooNE Collaboration) 2010 *Phys. Rev. D* **81** 092005
[7] Bernard V *et al.* 2002 *J. Phys. G* **28** R1
[8] Bodek A, Avvakumov S , Bradford R and Budd H, 2008 *Eur. Phys. J. C* **53** 349
[9] Lyubushkin V *et al.* (NOMAD Collaboration) (2009) *Eur. Phys. J. C* **63** 355
[10] Gran R *et al.* (K2K Collaboration) 2006 *Phys. Rev. D* **74** 052002
[11] Aguilar-Arevalo A A *et al.* (MiniBooNE Collaboration) 2008 *Phys. Rev. D* **100** 032301
[12] Sanchez F, Sorel M and Alvarez-Ruso M 2010 *Proc. 6th Int. Work. on Neutrino-Nucleus Interactions in the Few-GeV Region (NUINT-09)*, AIP Conf. Proc. **1189**
[13] Benhar O 2010 *P*reprint nucl-th/1012.2032v1
[14] Sealock R M *et al.* 1989 *Phys. Rev. Lett.* **62** 1350
[15] Martini M, Ericson M, Chanfray G and Marteau J 2010 *Phys. Rev. C* **81** 045502
[16] Nieves J Ruiz-Simo I R and Vicente-Vacas M J 2012 *P*hys. Lett. B **707** 72
[17] Rohe D *et al.* (JLAB E97-006 Collaboration) 2004 *Phys. Rev. Lett.* **93** 182501
[18] Carlson J and Schiavilla R 1998 *Rev. Mod. Phys.* **70** 743

Charged-current and neutral-current coherent pion productions — Theoretical status

Satoshi X Nakamura

Excited Baryon Analysis Center, Jefferson Lab, Newport News, VA, 23606, USA

E-mail: satoshi@jlab.org

Abstract. The CC and NC coherent pion productions have been studied with PCAC-based models and microscopic models. Current status of the theoretical studies is reviewed.

1. Introduction

Recent neutrino oscillation experiments have driven quite a few nuclear theorists to study neutrino-nucleus interaction. Theoretical study of coherent pion productions (cohπ) has been also on the trend. The charged-current (CC) cohπ^+ is $\nu_l + A_{g.s.} \to l^- + A_{g.s.} + \pi^+$, while the neutral-current (NC) cohπ^0 is $\nu_l + A_{g.s.} \to \nu_l + A_{g.s.} + \pi^0$, where ν_l is the neutrino of $l = e, \mu, \tau$, and $A_{g.s.}$ is a nucleus in its ground state. The process takes place under small momentum transfer from the lepton. The NC cohπ^0 needs to be understood for the ν_e appearance event, because when one of two photons from $\pi^0 \to 2\gamma$ is not detected, the NC cohπ^0 can fake the ν_e appearance event. Theoretically, there are two main approaches to cohπ. One of them is based on the partially conserved axial current (PCAC) [1, 2, 3, 4, 5, 6], and the other is based on a microscopic dynamical model [7, 8, 9, 10, 11, 12, 13, 14]. In the following, I discuss both of the approaches, and then comparison among them and with recent experimental data.

2. PCAC-based models

The prominent PCAC-based model was proposed by Rein and Sehgal [1] (RS model). The model is based on the fact that a cohπ amplitude is proportional to the divergence of the axial-current in the limit of $q^2 \to 0$ where q is the momentum transfer from the lepton. The PCAC dictates that the divergence of the axial-current is proportional to the pion field. Thus, the cohπ amplitude and the elastic π-nucleus scattering amplitude is related. Continuation of the cross section to $q^2 \neq 0$ is parametrized solely by a dipole-type function. The elastic pion-nucleus cross section is simply given by the product of the pion-nucleon elastic cross section, the square of the nuclear form factor, and the pion absorption factor. The RS model works well for high energy neutrino ($E_\nu \gtrsim$ 2 GeV), from medium to heavy nuclei [15]. The RS model has been used in many analyses of neutrino experiments because of its success and simplicity.

However, recent neutrino experiments use neutrino beam of $E_\nu \lesssim$ 2 GeV and relatively light nuclei (*e.g.*, ^{12}C) as a target, for which the validity of the RS model is questionable. Indeed, recent experiments [16, 17] found that CC cohπ^+ cross section of the RS model is significantly larger than their data. Rein and Sehgal corrected the RS model by taking account of the finite muon mass whose effect can be significant for low E_ν [2]. They found 25% reduction of the CC cohπ^+ cross sections for $E_\nu = 1.3$ GeV. Further improvements were attempted by Hernández *et al.* [3]. They pointed out that the elastic pion-nucleus scattering model in the RS model is rather off data. They improved the model by considering previously ignored t-dependence of the pion-nucleon cross sections, and by using a more realistic pion absorption factor. It was found that, although the pion-nucleus elastic scattering is significantly improved, the cohπ cross sections from the improved RS model are still rather different from those of microscopic calculations.

Meanwhile, another avenue for the PCAC-based model was explored [4, 5, 6]. There, the PCAC relation was used to relate the axial current-nucleus interaction to the pion-nucleus interaction (nuclear PCAC), rather than the axial current-nucleon interaction to the pion-nucleon interaction (nucleon PCAC); the latter was employed in the RS model. Thus, they can use the pion-nucleus cross section data directly to calculate the cohπ cross sections. They found that this change yields a drastic reduction ($\sim 1/2$) of the cohπ cross sections, now the size is comparable to those of the microscopic calculations. Strictly speaking, however, the nuclear PCAC relates a cohπ amplitude to the half off-shell pion-nucleus scattering amplitude [18]. Therefore, the use of pion-nucleus cross section may call for some justification.

3. Dynamical microscopic models

A description of cohπ with a dynamical microscopic model consists of three basic ingredients: elementary amplitudes (neutrino-induced pion production off a single nucleon); nuclear medium effect on the Δ-propagation; pion-nucleus final state interaction. In the following, I will discuss each of them used in the previous works.

3.1 Elementary amplitude : The most important mechanism for the pion production is the Δ-excitation. Thus [7, 8, 9] used only the s-channel Δ diagram for the elementary amplitude. Among the $AN\Delta$ (A:axial-current; N:nucleon) form factors, the so-called $C_5^A(q^2)$ gives a dominant contribution so that the cohπ cross section is roughly $\sigma_{\text{coh}\pi} \propto |C_5^A|^2$. In [7, 8, 9], the Goldberger-Treiman (GT) value for $C_5^A(0)$ was taken: $C_5^A(0) = g_{\pi N\Delta} f_\pi/\sqrt{6}m_N = 1.2$. Martini *et al.* [12] used s- and u-channel nucleon and Δ mechanisms as the elementary amplitudes; the value of $C_5^A(0)$ is the GT value. Meanwhile, elementary amplitudes used in [10, 11, 14] consists of s- and u-channel nucleon and Δ mechanisms, t-channel π-exchange mechanism, a contact and a pion pole terms [19]. Because of the background terms, the value of $C_5^A(0)$ has to be smaller than the GT value by 20% in order to fit neutrino data. So far, all elementary amplitudes are based on tree-level calculations. On the other hand, Sato *et al.* proposed a model (SL model [20]) in which tree-level weak pion production mechanisms, which are similar to those of [19], are dressed by multiple πN rescattering to yield a unitary elementary amplitude. The SL amplitude contains significant pion cloud effect, which is unique in this model. Bare $AN\Delta$ couplings were fixed using the constituent quark model. Our group used the elementary amplitudes from the SL model [13]. All elementary amplitudes used in various groups fairly reproduce available data. Unfortunately, available data are rather limited, and not enough to test each elementary amplitude model. More good quality data would help.

3.2 Medium effect on Δ : In the vacuum, the Δ propagates by repeating the decay into πN state and formation of Δ (self energy). In a nuclear medium, this process becomes more complicated. Some πN states are forbidden by the Pauli blocking so that the Δ decay is suppressed. A pion emitted from the Δ can be absorbed by another nucleon, forming multiparticle-multihole configurations or RPA(random phase approximation)-type processes. The Δ mass and width can shift due to these medium effects. Oset *et al.* calculated the medium effects on the Δ propagation using a many-body approach [21], and parametrized the shift of the Δ mass and width as a function of the density. Their result has been employed by many people for calculating the cohπ cross sections [7, 8, 9, 10, 11, 12, 14]. However, non-locality due to the recoil of the Δ was not explicitly taken into account in these calculations of cohπ. Leitner *et al.* [22] pointed out this, and showed that the non-local effect can reduce the total cohπ cross sections by half. Hernández *et al.* argued that the non-local effect is effectively (partially) taken into account in the course of fitting the medium shift of the Δ to observables [14]. Our group [13] considered the non-locality of the Δ-propagation as well as several medium effects explicitly. Multiparticle-multihole effects are simulated by a phenomenological spreading potential, parameters of which were fitted to pion-nucleus elastic and total (elastic+inelastic) cross sections. It is important to fit the total cross sections because the spreading potential describes pion absorptions.

3.3 Final state interaction (FSI) : In [7, 9, 10, 11, 14], pion-nucleus FSI is considered in distorted pion wave function obtained by solving the Klein-Gordon equation with an optical potential. The optical potential is based on the Δ-hole model and the Lorentz-Lorentz correction, and reproduces spectra of pionic atoms and low-energy pion-nucleus elastic cross sections [23]. In Martini *et al.*'s approach [12], the elastic pion-nucleus cross section is obtained by solving a RPA equation with a bare polarization propagator as the driving term. The bare polarization propagator contains various nuclear excitation such as multiparticle-multihole configurations. The model well describes the elastic total cross sections in the Δ region. Our group constructed a pion-nucleus optical potential from πN amplitudes of the SL model, combined with the medium effects on the Δ. A pion-nucleus scattering amplitude is obtained by solving the Lippmann-Schwinger equation with the optical potential. The model well describes the elastic differential cross sections and the total cross sections in the Δ region.

3.4 Coherent pion production : Combining the above-discussed three ingredients, one can calculate cohπ cross sections. First let us discuss a photo-cohπ^0. This is a great testing ground for microscopic models of cohπ because data of good quality are available. Also, some dynamical models like ours [13] can predict cohπ cross sections without any adjustable parameters; all parameters have been fitted to pion-nucleus scattering data. Thus we have tested our prediction for photo-cohπ with data, and found a good agreement [13]. It is noteworthy that the medium effects on Δ and FSI reduce the cross sections by about half to reach the good agreement. This success serves as a good basis to apply our model to the neutrino-cohπ. The pion momentum distribution in CC cohπ^+ at $E_\nu = 1$ GeV is shown in Fig. 1. Significant medium effects are observed here. The dashed curve contain neither the Δ spreading potential nor the FSI. Including these nuclear effects lowers the dashed curve into the solid curve. The dash-dotted curve contains only resonant amplitude, showing significant contribution from non-resonant unitary amplitude. This is in contrast with calculations with tree-level elementary amplitude [19], because the non-resonant amplitude hardly contributes.

4. Comparison among models and data

At NUINT09, CC and NC cohπ cross sections from many theoretical models and Monte Carlo (MC) codes were compared [24]. Although details are different, recent theoretical models [4, 5, 6, 9, 10, 11, 12, 13, 14] give fairly similar CC and NC cohπ cross sections. However, the MC code outputs, which are based on the RS model with different modifications, are very different from those of the recent theoretical models.

There have been large discrepancies between experimental data and theoretical calculations on the cohπ cross sections. The latest data from SciBooNE [25] gave the ratio between CC cohπ^+ and NC cohπ^0: $\sigma_{CC}/\sigma_{NC} = 0.14^{+0.30}_{-0.28}$. On the other hand, all recent theoretical calculations gave $\sigma_{CC}/\sigma_{NC} = 1.5 \sim 2$, which is expected from the isospin factor.

One may suspect that this unsatisfactory situation is due to using the RS model in the analyses of the NC data. In fact, several authors [11, 13, 14] have shown that η-distribution ($\eta \equiv E_\pi(1 - \cos\theta_\pi)$) of the RS model rather deviates from those of their microscopic models (Fig. 2). The η-distribution has been used in the analyses of the NC data in order to decompose π^0 events into each production mechanism. The use of the RS model in the analyses could have

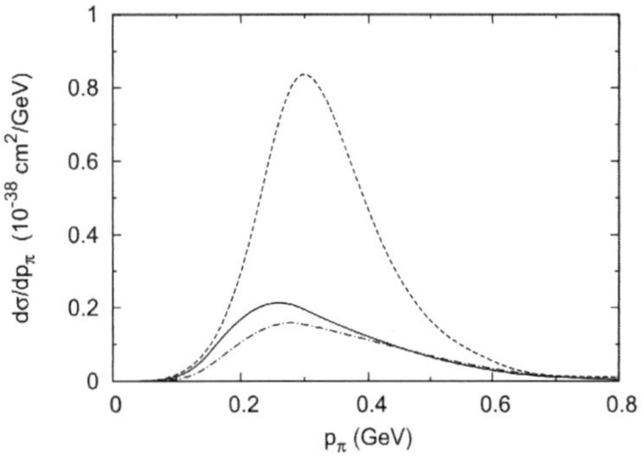

Figure 1. Pion momentum distribution in CC cohπ at $E_\nu = 1$ GeV from [13]. See the text for more.

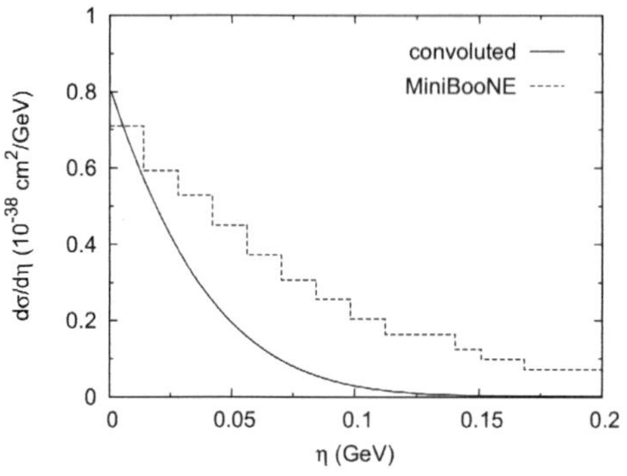

Figure 2. η-distribution for NC cohπ^0 on ^{12}C. The model calculation [13] convoluted with MiniBooNE flux (——); MC output (- - - -).

overestimated the NC cohπ^0 cross sections. It is very interesting to re-analyze the data with a more realistic cohπ model.

Acknowledgments
I thank the NUFACT11 organizers for the invitation and the financial support for my participation in the workshop. This work is supported by the U.S. Department of Energy, Office of Nuclear Physics Division, under Contract No. DE-AC05-06OR23177 under which Jefferson Science Associates operates Jefferson Lab.

References
[1] Rein D and Sehgal L M 1983 *Nucl. Phys.* B **223** 29
[2] Rein D and Sehgal L M 2007 *Phys. Lett.* B **657** 207
[3] Hernández E, Nieves J and Vicente-Vacas M J 2009 *Phys. Rev.* D **80** 013003
[4] Paschos E A, Kartavtsev A and Gounaris G J 2006 *Phys. Rev.* D **74** 054007
[5] Berger Ch and Sehgal L M 2009 *Phys. Rev.* D **79** 053003
[6] Paschos E A and Schalla D 2009 *Phys. Rev.* D **80** 033005
[7] Kelkar N G, Oset E and Fernandez de Cordoba P 1997 *Phys. Rev.* C **55** 1964
[8] Singh S K, Sajjad Athar M and Ahmad S 2006 *Phys. Rev. Lett.* **96** 241801
[9] Alvarez-Ruso L, Geng L S, Hirenzaki S and Vicente Vacas M J 2007 *Phys. Rev.* C **75** 055501
[10] Alvarez-Ruso L, Geng L S and Vicente Vacas M J 2007 *Phys. Rev.* C **76** 068501
[11] Amaro J E, Hernandez E, Nieves J and Valverde M 2009 *Phys. Rev.* D **79** 013002
[12] Martini M, Ericson M, Chanfray G and Marteau J 2009 *Phys. Rev.* C **80** 065501
[13] Nakamura S X, Sato T, Lee T-S H, Szczerbinska B and Kubodera K 2010 *Phys. Rev.* C **81** 035502
[14] Hernandez E, Nieves J and Valverde M 2010 *Phys. Rev.* D **82** 077303
[15] P. Vilain *et al.* (CHARM II Collaboration) 1993 *Phys. Lett.* B **313** 267
[16] M. Hasegawa *et al.* (K2K Collaboration) 2005 *Phys. Rev. Lett.* **95** 252301
[17] K. Hiraide *et al.* (SciBooNE Collaboration) 2008 *Phys. Rev.* D **78** 112004
[18] Bell J S 1964 *Phys. Rev. Lett.* **13** 57
[19] Hernández E, Nieves J and Valverde M 2007 *Phys. Rev.* D **76** 033005;
Hernández E, Nieves J, Valverde M and Vicente Vacas M J 2010 *Phys. Rev.* D **81** 085046
[20] Sato T, Uno D and Lee T-S H 2003 *Phys. Rev.* C **67** 065201
[21] Oset E and Salcedo L L 1987 *Nucl. Phys.* A **468** 631; Carrasco R C and Oset E 1992 *Nucl. Phys.* A **536** 445
[22] Leitner T, Mosel U and Winkelmann S 2009 *Phys. Rev.* C **79** 057601
[23] Nieves J, Oset E and Garcia-Recio 1993 *Nucl. Phys.* A **554** 509;
Nieves J, Oset E and Garcia-Recio 1993 *Nucl. Phys.* A **554** 554
[24] Boyd S *et al.* 2009 *AIP Conf. Proc.* **1189** 60
[25] Kurimoto Y *et al.* (SciBooNE Collaboration) 2010 *Phys. Rev.* D **81** 111102

Measurement of π-nuclei interactions PIAνO-Harpsichord

Motoyasu Ikeda

Kyoto University, Department of Physics, Kyoto, Japan

E-mail: motoyasu@scphys.kyoto-u.ac.jp

Abstract. We introduce a new experiment to measure cross sections of interactions between charged pions and nuclei (^{12}C and ^{16}O). The current analysis status and future plans are also explained.

1. Introduction

The latest T2K result [1] shows an indication of $\nu_\mu \to \nu_e$ oscillation via non-zero θ_{13}. If θ_{13} is relatively large, it is possible to measure θ_{13} precisely and discover δ_{CP} in the near future. In this case, the precision on the θ_{13} measurement by T2K will be systematics dominated, where one of the major sources of systematic error is the uncertainty on pion-nucleus interaction cross sections [1]. Thus, we performed a new experiment to precisely measure these cross sections.

In interactions of low energy (sub-GeV) neutrinos, predominant in the T2K beam, the pion production probability through the delta resonance is relatively high. Thus, the understanding of interactions of these pions with matter is important to reduce the systematic error for the T2K measurement. The uncertainty is present at two levels: within the target nucleus of the primary neutrino interaction (FSI: final state interaction), and at the detector level with secondary interactions of pions that emerge from the target nucleus. However, data from past experiments that are used to constrain the π-nuclei transport model in the T2K simulation have 20-30% uncertainties in the absorption cross section around 100-400 MeV/c pion momentum. Therefore, our first goal is to measure the π-nuclei absorption and charge exchange cross sections using a newly developed scintillator tracker with an uncertainty less than 10%. The results of our measurement will improve the uncertainty on the neutrino interaction simulation significantly. For example, the error contribution from π FSI, to the expected number of T2K ν_e events is currently 10%. This uncertainty will be improved to 5% if our goals are achieved. In addition, better understanding of pion propagation in our detector materials will improve both the near and far detector systematic uncertainties for T2K.

2. Experimental setup

Our data was taken at the M11 beamline in TRIUMF, fed by a 500 MeV proton cyclotron. At M11, we can obtain secondary particles such as pions, muons and electrons from collisions between the proton beam and target (secondary protons are also present in higher beam momentum settings). Pions can be selected with purity of more than 99% in our analysis by using information of TOF(time-of-flight) and the Cherenkov detector (Fig. 1). The momentum

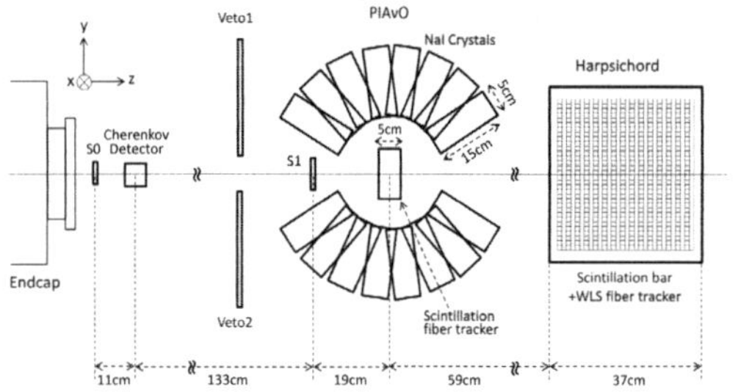

Figure 1. Experimental setup (configuration A)

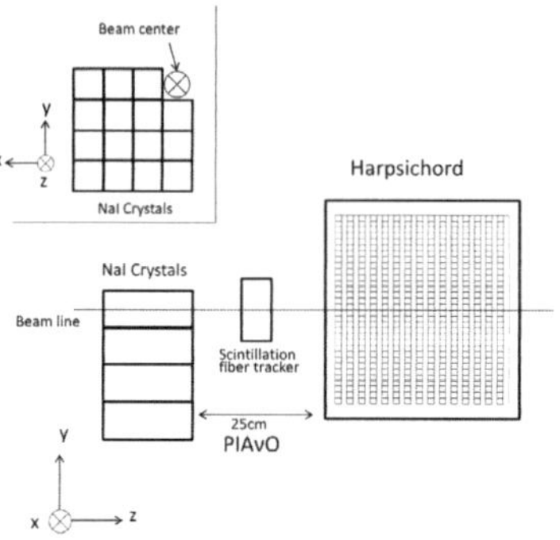

Figure 2. PIAνO and HARPSICHORD in configuration B

of beam particles can be selected by changing the current of magnets in the beamline. In this measurement, we took mainly π^+ data from 150 MeV/c to 375 MeV/c in steps of about 25 MeV/c during a period from October to December, 2010.

Figure 1 shows the overall setup. There are two main components. One is PIAνO (PIon detector for Analysis of neutrino Oscillation), and the other one is HARPSICHORD (HAdron Reconstruction Performance Studies In CH On Reduced Detector). A trigger is generated by the coincidence of a beam particle passing through the S0 and S1 scintillators. The particle then enters the PIAνO scintillation fiber tracker. This tracker is a fully active target. While most π^+s penetrate the tracker, about 10% of π^+ interact via elastic and inelastic scattering, absorption or charge exchange. The final state for each interaction is as follows: there is only a π^+ after elastic scattering, π^+ and nucleons after inelastic scattering, no π^+ and zero or more nucleons after absorption, and a π^0 and zero or more nucleons after charge exchange. Therefore, the tracker is required to observe all particle tracks and identify the particle types to categorize the interaction types. Since π^0 cannot be observed in the tracker, we installed NaI detectors surrounding the tracker to detect γ from the decay of π^0 for separation of absorption and charge exchange events. HARPSICHORD is downstream of PIAνO and consists of removable lead plates sandwiched between scintillator tracking planes. When the lead plates are installed, HARPSICHORD acts as another π^0 detector together with the NaI detectors.

2.1. Configurations of PIAνO and HARPSICHORD

There are two configurations as shown in Fig. 1. In configuration A, 16 NaI detectors are installed around the scintillator fiber tracker, and lead plates are not installed in HARPSICHORD. From this configuration, the angular distribution of γ of π^0 from charge exchange interactions can be measured. The measured angular distribution can be cross-checked with configuration B as shown in Fig. 2. Since it is expected that γ are mostly emitted in the forward and backward direction, configuration B can give more statistics for charge exchange events due to the larger solid angle of HARPSICHORD with lead plates.

2.2. PIAνO

In order to make a super fine grained tracker to observe all tracks, the PIAνO tracker is composed of 1024 of 1.5 mm square scintillation fibers. The tracker has 16 tracking planes (x and y layers in

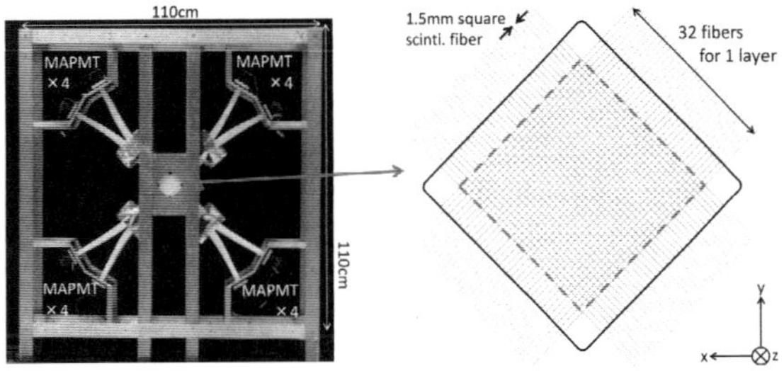

Figure 3. Photo of the PIAνO tracker and drawing of the detector volume (red dashed line).

one plane). One layer consists of 32 fibers as shown in Fig 3. Each fiber is coated by about 25 μm thickness reflective paint (EJ-510). One end-face of a fiber is coated by evaporated aluminum, and the other end-face is connected to one channel of a 64 channel multi-anode PMT (MAPMT: Hamamatsu H8804). We use 16 MAPMTs in total. Signals are recorded by DAQ electronics recycled from the SciBar detector [2]. The detector volume of the tracker is 5 cm $\times$ 5 cm $\times$ 5 cm. Each NaI detector is a 5 cm $\times$ 5 cm $\times$ 15 cm NaI crystal connected to a 38mm (1.5 inch) PMT (Hamamatsu:R580).

2.3. HARPSICHORD

HARPSICHORD is composed of 960 1 cm x 1 cm x 30 cm scintillator bars. A wave length shifting fiber, inserted in each bar, is coupled to a Multi Pixel Photon Counter (MPPC). The same DAQ system as that of Fine Grained Detector (FGD), which is one of near detectors for T2K, is used [3]. HARPSICHORD has 15 tracking planes and one layer consists 32 bars. Removable lead plates ($\sim$1.4 mm thickness) are installed between each tracking plane in configuration B to increase the conversion rate of γ.

3. Analysis status

Event reconstruction is applied to find an incident track, an interaction point (vertex), and secondary tracks. Then, particle identification of each track is determined by dE/dx of the track. The algorithm of the reconstruction is as follows. The first step is to search for an incident track. Assuming a horizontal incidence, a straight-line trajectory which includes the maximum number of hits is chosen as an incident track. The angle of the incident track is fitted by a least squares method using the selected hits. A secondary track is found by searching for a track which starts from near the end point of the incident track and has the most number of hits along it. The cross point of the incident track and the secondary track is assigned as the event vertex.

After reconstruction, absorption or charge exchange events can be selected as follows. 1) Events which have a good incident track are selected. The criteria of a good incident track is that the incident position is within the tracker's fiducial volume and the incident angle is within $\pm 4^\circ$ for both the x and y directions (Fig. 4). 2) Events which have a vertex within the fiducial volume are selected. 3) Events which have no π^+ secondary track are selected. At this stage, both absorption and charge exchange like events are selected. 4) To isolate charge exchange events, events with large NaI hits or shower like tracks in HARPSICHORD are selected.

The left plot in Fig. 5 shows the vertex z distribution in the 250 MeV/c beam momentum run after the fiducial volume cut. The fiducial volume is indicated by the arrows in the figure. The right plot in Fig. 5 shows the fraction of events after the fiducial volume cut to the total number of events which have a good incident track as a function of the beam momentum. A

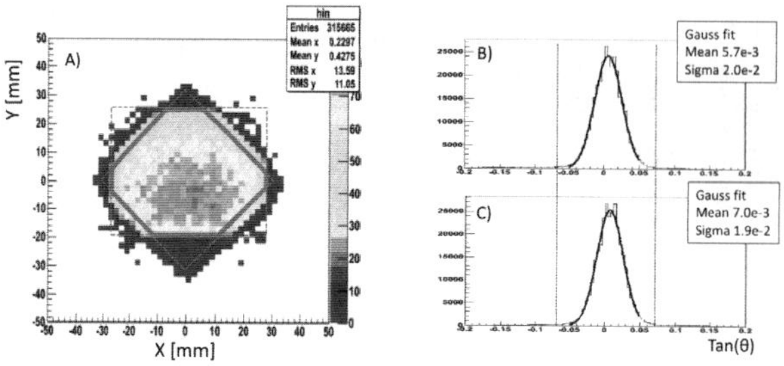

Figure 4. The distributions of incident position (A), x direction (B), and y direction (C) (250 MeV/c, data). The red solid area in (A) shows the fiducial volume in x-y plane

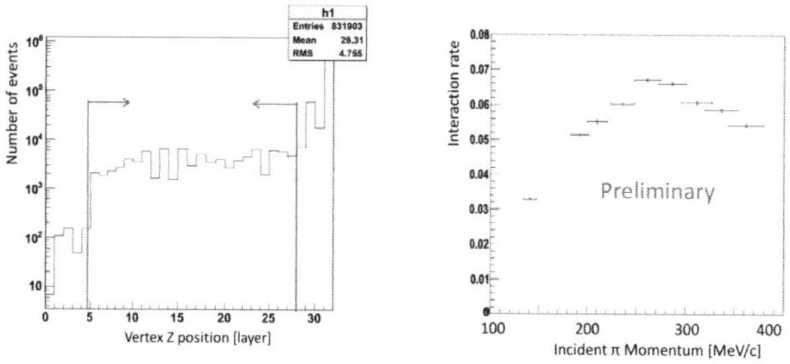

Figure 5. *Left*: The distribution of z position in 250 MeV/c (data). *Right*: The interaction rate as a function of beam momentum (data).

clear Δ resonance peak can be seen.

4. Future plan

We plan to complete this data analysis and measure the cross sections by the end of 2011. Following that, we will evaluate how much our result can constrain T2K's neutrino interaction model, and update the systematic error contribution from the uncertainty in pion interaction cross sections. We hope that the updated model and systematic error will be ready for T2K analysis in 2012. In addition to the data taken in the last year, we took data with a water target in August of this year. The analysis of the water target will be released as soon as possible.

5. Conclusion

Precise measurements of the interaction of charged pions in a nuclear medium are very important for reducing systematic uncertainties in neutrino experiments. Therefore, we started a new experiment using the pion beam at TRIUMF. Our goal is to measure the absorption and charge exchange cross sections separately within 10% uncertainty. In order to achieve this goal, we developed two new detectors: PIAνO and HARPSICHORD. The results from both the scintillator target and the water target will be completed soon to improve the T2K analysis in 2012.

References

[1] Abe K *et al.* 2011 *Phys.Rev.Lett.* **107** 041801
[2] Nitta K *et al.* 2004 *Nucl.Instrum.Meth. A* **535** 147
[3] Abe K *et al.* 2011 *Nucl.Instrum.Meth. A* **659** 106

Neutrino interaction measurements using the T2K near detectors

Daniel G Brook-Roberge for the T2K Collaboration
University of British Columbia, Vancouver, BC, Canada
E-mail: `droberge@triumf.ca`

Abstract. The T2K near detectors provide a rich facility for measuring neutrino interactions in a high-flux environment. This talk will discuss the near detector CC-inclusive normalization analysis for the T2K oscillation result in detail, along with the present result, and describe the plan for its extension to more sophisticated measurements. Selection criteria for CCQE interactions will be presented, as will a strategy for calculating cross-section difference between plastic scintillator and water. The unique capacities of the near detectors to measure other exclusive CC and NC channels in a narrow-band off-axis beam will also be explored.

1. Introduction

The T2K (Tokai to Kamioka) long-baseline neutrino oscillation experiment[1] aims to make high-precision measurements of oscillation parameters using a beam of muon neutrinos. In particular, the appearance of electron neutrinos in the beam is studied in an effort to measure a non-zero value of the mixing parameter θ_{13}. Electron neutrino appearance provides a particular challenge for the near detector, as the electron neutrino fraction in the muon neutrino beam is the most significant background to this measurement.

The T2K experiment has had two experimental runs; one at beam power up to 50 kW from January to June 2010, and one at beam power up to 150 kW from November 2010 to March 2011, when it was cut short by the Tohoku earthquake. 1.43×10^{20} protons on target were accumulated by T2K prior to the earthquake.

Recently, T2K has published an electron appearance result with suggestion of large θ_{13} ($\sin^2 2\theta_{13} \approx 0.1$)[2]. The significance of this result has driven further interest in improving background estimates using the near detector.

2. The T2K Experiment

The T2K experiment has three main components: the beam source, the near detector complex, and the far detector, Super-Kamiokande. The ND280 near detector is the focus of this paper.

2.1. Beamline

The T2K beam is produced from the 30 GeV proton beam of the J-PARC Main Ring. These protons are extracted in eight bunches every 3 seconds and collided with a long graphite target. The positively charged secondary particles are focussed down a 96 m decay volume using three magnetic horns. The predominant contribution is from pions, which decay $\pi^+ \rightarrow \mu^+\nu_\mu$ to produce muon neutrinos. All charged decay products are stopped in a beam dump while the

Figure 1. An exploded view of the ND280 detector.

neutrinos pass through freely. The electron neutrino component is produced by other particles decaying in the beam pipe, such as muons and kaons.

The T2K beamline is angled 2.5° away from Super-Kamiokande; this off-axis beam technique provides a narrower energy spectrum peaking near the oscillation maximum, and dramatically reduces the high-energy tail. T2K is the first oscillation experiment to use the off-axis configuration.

2.2. Far Detector (Super-Kamiokande)

The 50 kt water Čerenkov detector Super-Kamiokande (Super-K) is the far detector for T2K, sitting 295 km from the production target at J-PARC. Super-K images the Čerenkov rings produced by charged particles from neutrino charge-current interactions to detect incident neutrinos. The ring imaging allows a high-purity separation of electrons and muons, as the two particles produce Čerenkov rings with much differently defined edges.

Many of the Super-K backgrounds are intended to be well constrained by ND280. In addition to the intrinsic electron neutrino fraction of the T2K beam, we have substantial background from neutral pions. A neutral pion decays to two photons, which are indistinguishable in Super-K from electrons. If one of the decay photons escapes or is otherwise missed, the remaining photon will mimic a CCQE electron neutrino interaction, i.e. our appearance signal.

2.3. Near Detectors

Two near detectors are located in a pit 280 m from the production target at J-PARC. The INGRID detector is a cross-shaped arrangement of 14 iron/scintillator neutrino detector modules, centred on the beam axis. The event rates in these modules are used to measure the stability of the beam flux and direction. At the T2K design intensity, the interaction rate should be sufficient to monitor these on a day-to-day basis.

At the off-axis angle to the neutrino beam is the ND280 spectrometer. The ND280 detector consists of two central detectors, the P0D (Pi-Zero Detector) and the Tracker, surrounded by an electromagnetic calorimeter (ECal) and placed within the 0.2 T magnetic field of the UA1 magnet. Interleaved in the magnet yokes is an additional detector, the Side Muon Range Detector (SMRD). A full diagram of ND280 can be found in Figure 1.

The Tracker consists of three large-volume Time Projection Chambers (TPCs) and two Fine Grained Detectors (FGDs). The TPC gas detectors allow high precision 3D tracking of charged particles and measurements of the energy loss, while the plastic scintillator FGDs provide an active target mass ($\approx$ 1.1 tons per FGD) allowing tracking back to the neutrino interaction vertex. The TPC can measure the momentum and charge of particles from their curvature in

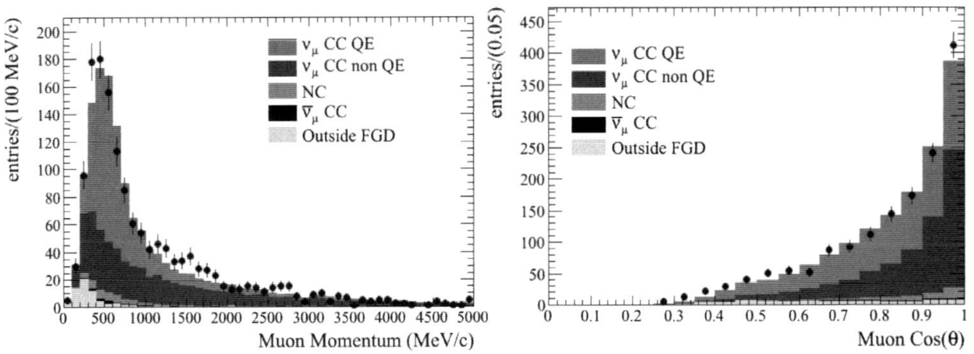

Figure 2. Distribution of muon momentum and angle in data from January-June 2010 and in Monte Carlo.

the magnetic field, and provides a measure of dE/dx for particle identification. Tracks from the TPC can then be traced back into the FGD and associated with FGD activity, allowing the vertex to be localized and the fiducial volume established. FGD-only tracking can then be used to find short-range particles such as protons and pions.

The P0D aims to measure neutral current π^0 production and takes the form of a hybrid calorimeter/tracking detector. The central region of the P0D alternates scintillator planes, water targets, and brass sheets for observation of tracks and showers, while the ends use a scintillator/lead combination to contain the full showers within the P0D.

The lead-scintillator ECal aims to identify particles escaping from the central detector and to detect photons produced in the inner detector. For the first T2K run in early 2010, only the ECal module at the downstream end of the Tracker was operational. The remainder of the ECal was installed and ready for the second run starting November 2010.

3. Current Results

Two neutrino interaction results are available from ND280, both from a Tracker-based analysis: a measurement of the ν_μ charge-current interaction rate and a measurement of the ν_e fraction. These analyses attempt to use the Tracker to find neutrino interactions in the FGD. Presently only inclusive charge-current measurements are complete; work is ongoing to measure specific channels.

3.1. Tracker ν_μ Rate Measurement

The current ν_μ-based analysis is a comparison of the Tracker muon neutrino interaction rate with a NEUT Monte Carlo prediction. The data/Monte Carlo ratio is used to scale the flux prediction at Super-K and provides a useful contribution to the oscillation measurement. We select CC ν_μ events by selecting tracks that begin in the FGD fiducial volume and make it into a downstream TPC. Incoming events are removed by requiring no tracks in upstream TPCs, and the TPC PID is used on the highest-momentum negative track to isolate muons. This selection has a 91% purity for selecting ν_μ charge-current interactions.

The resulting data/Monte Carlo ratio, using the 2.88×10^{19} POT from the first T2K run, is

$$R = 1.036 \pm 0.028(\text{stat})^{+0.044}_{-0.037}(\text{det syst}) \pm 0.038(\text{phys model syst})$$

This normalization reduces the flux uncertainty in the ν_e appearance measurement by 50%.

Figure 2 shows the measured momentum and angular distributions of the selected muons. In each plot, the points are the data with statistical error bars, and the coloured histogram is the Monte Carlo broken down by interaction type.

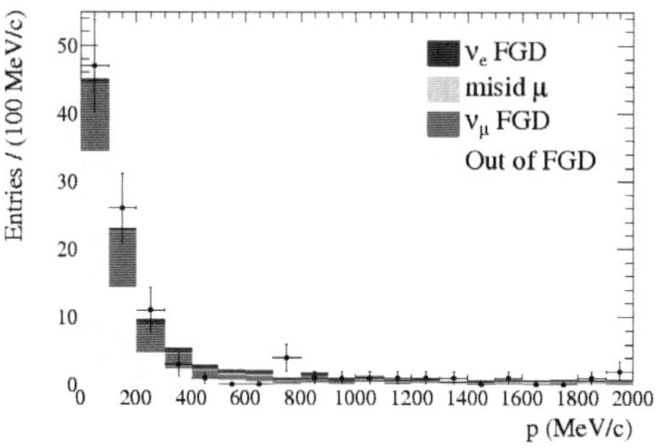

Figure 3. Momentum distribution of electron neutrino candidate events.

3.2. Electron Neutrino Fraction Measurement

The second measurement is of the electron neutrino fraction present in the beam at its origin. This is a more difficult measurement; charge-current muon neutrino interactions are a large fraction of the beam events in ND280, while electron neutrinos have the large muon contribution as background.

While the selection for the electron neutrino candidates is quite similar to that for the muon neutrinos, the extraction of the ν_e fraction must account for the large backgrounds. A likelihood fit with empirical PDFs as a function of momentum for the signal and three categories of background was used to extract the number of electron neutrino interactions.

Figure 3 shows the result of this fit. The resulting ν_e/ν_μ ratio is $(1.0 \pm 0.7(\text{stat}) \pm 0.3(\text{syst}))\%$. To check the validity of the beam simulation, a double ratio can be formed between data and Monte Carlo:

$$\frac{(\nu_e/\nu_\mu)_{data}}{(\nu_e/\nu_\mu)_{MC}} = 0.6 \pm 0.4(\text{stat}) \pm 0.2(\text{syst})$$

While this measurement is dominated by statistical uncertainties, it does show a rough correspondence between the data and Monte Carlo. The uncertainties are too large to constrain the background at Super-K, but this provides a useful cross-check of our Monte Carlo simulation.

4. Future Prospects

Work is ongoing to extend the ND280 ν_μ analysis to include a selection of muon neutrino CCQE interactions. This, in turn, permits a measurement of the neutrino spectrum, rather than just an overall normalization. In addition, the additional data from Run 2 is being incorporated into the analyses, pending consideration of the differences in detector systematics.

The π^0 analysis is proceeding well. Reconstruction of decay photons in the P0D is now possible to the precision necessary for selection of π^0 decays. Incorporation of the Run 2 data and its full complement of ECal modules will improve the π^0 analysis.

5. Conclusion

The T2K ND280 near detector is operational and is making significant contributions to the oscillation measurements. Completed analyses normalize the ν_μ flux and measure the ν_e fraction of the beam at its source. Improvements to these measurements are ongoing, and work is underway to add a NCπ^0 analysis to constrain that important background.

References

[1] K Abe *et al.* 2011 *Nucl. Instr. Meth.* A 10.1016/j.nima.2011.06.067
[2] K Abe *et al.* 2011 *Phys. Rev. Lett.* **107** 041801

Elastic scattering with the MINERνA experiment

Benjamin P Ziemer, for the MINERνA Collaboration
Department of Physics and Astronomy
University of California Irvine
Irvine, CA 92697-4575

E-mail: ziemer@fnal.gov

Abstract. The Main Injector Experiment ν-A (MINERνA) located at Fermi National Laboratory will measure neutrino cross sections, nuclear effects from a broad range of nuclear targets and a variety of other neutrino interactions. Neutrino elastic scattering will be one of the first focuses of the MINERA collaboration; these measurements will be an important input to current and future neutrino oscillation experiments. Results of the charged current quasi-elastic channel exposure in anti-neutrino NuMI running are presented. Future elastic scattering results, both charged current and neutral current, in anti-neutrino and neutrino exposures are also discussed.

1. Introduction

The future of neutrino physics involves the pursuit of CP violation and precision measurements of the neutrino mixing parameters. Precision measurements of the neutrino's interaction with a range of nuclei are needed before the most accurate physics neutrino factories and beta beams can be extracted. Interactions with anti-neutrinos are becoming increasingly more important. Also, broad physics that can be realized when new, precision neutrino and anti-neutrino data are combined. Clearly, future neutrino endeavours can be aided by a careful study of $\nu_\mu/\overline{\nu}_\mu$ interactions.

2. Neutrino Elastic Scattering

Neutrino nucleon elastic scattering is an important class of interactions that is used extensively in the neutrino community. Hence, it is essential to understand this interaction to a high precision. Presently, there is tension between the higher energy charged current quasi-elastic (CCQE) cross sections measured from the NOMAD experiment [1] and the lower energy BooNE experiments [2]; in addition, there are 20% differences among other experiments' cross sections.
The (quasi) elastic cross section can be written in terms of the well-known leptonic current and a total of six form factors of the hadronic current. Using isospin symmetry, neglecting second class currents and ignoring the factor multiplied by $(m_\mu/m_N)^2$, leaves the only one unknown, the axial form factor. The axial factor can be parameterized via a dipole form:

$$F_A(Q^2) = \frac{g_A(Q^2\!=\!0)}{(1+\frac{Q^2}{M_A^2})^2}$$

where $g_A(Q^2\!=\!0)$ is determined from neutron decay to be roughly -1.26 and M_A is the experimentally determined axial mass parameter. Recent results for the experimentally extracted axial mass parameter have been much higher than measurements obtained in the past (see Figure 1). There are a large range of different targets for these experiments and nuclear effects are still relatively unknown for ν's (as well as for $\overline{\nu}$'s). In addition, there is interest in using theories beyond the impulse approximation, such as meson exchange currents and 2p2h contributions, to describe current CCQE data. Conflicting data make precision physics in this interaction channel difficult at the moment.

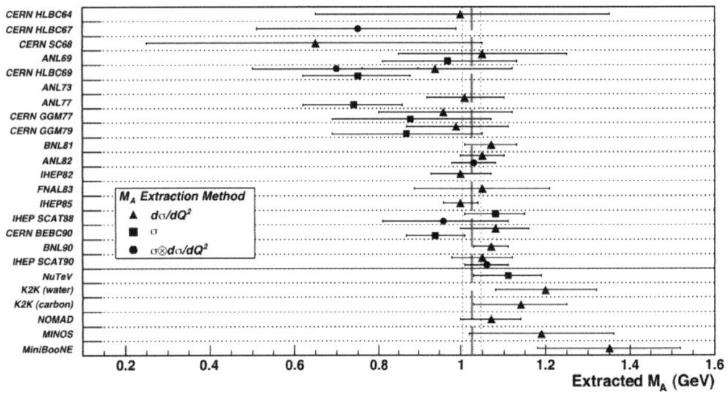

Figure 1. M_A parameters extracted from neutrino experiments with a wide range of targets; data spans 1964-2010. The line represents an average M_A value of 1.026±0.021 GeV/c.

3. MINERνA Detector and Physics Capabilities

The basic element of the MINERνA detector is a hexagon with an apothem of ~1.1m made of varying length, repeating scintillator bars read out by fibers connected to a photomultiplier tube. Scintillator planes have three different orientations for 3D track reconstruction. Two scintillator views are stacked together in a tracker module, a 0.2mm sheet of Pb is attached to each plane in an electromagnetic calorimeter (ECAL) module and one plane view is joined with 2.54cm of Fe in a hadronic calorimeter (HCAL) module. A module is surrounded by steel outer detector. The full detector has 60 tracker modules (~6.5 tons scint.), 10 ECAL modules and 20 HCAL modules. The upstream end of MINERνA has six modules with varying shape solid graphite (C), Pb and Fe targets. A liquid He cryogenic vessel sits in front of MINERνA and there are plans to install a water target module as well. MINERνA lacks a magnetic field; exiting particles are momentum analyzed by the downstream MINOS detector. Position and timing resolutions for the detector are measured to be 2.65mm and 3.0ns, respectively. A more detailed description of the detector can be found in [3].

The MINERνA experiment will collect the largest neutrino data set to date. Charged current events originating the fiducial region are fully contained, except for the muon, for neutrino energies up to 10 GeV. The position and timing resolution will enable studies of busy vertex regions and disentangling of possible multiple neutrino events in the detector. The variety of nuclear targets allows studies of $\nu_\mu/\overline{\nu}_\mu$ interactions in the same beam.

4. MINERνA $\nu_\mu/\overline{\nu}_\mu$ Data Sets

After partial installation of the 20 HCAL modules, the 10 ECAL modules and 30 tracker modules, the detector was fully instrumented and started taking NuMI LE $\overline{\nu}_\mu$ data. After the full detector installation was completed, the NuMI beam switched to the LE ν_μ configuration. The detector has $>98\%$ live time since data taking started (with the partial detector) on 11/2009. The protons on target (POT), NuMI beam mode and the detector configuration for various MINERνA data sets are listed in Table 1.

Table 1. Detector configuration, POT and NuMI beam for MINERνA data sets.

Detector	POT, NuMI mode
partial	$0.8e20\ \overline{\nu}_\mu\ LE$
full	$1.28e20\ \nu_\mu\ LE$
full	$1.5e20\ \overline{\nu}_\mu\ LE$
full (total run plan)	$4.9e20\ \nu_\mu\ LE$
full (total run plan)	$12e20\ \nu_\mu\ ME$

5. MINERνA Elastic Physics

First MINERνA physics results will be Pb/Fe interaction ratios from the most downstream heavy target module and $\overline{\nu}_\mu$ CCQE interactions. The former are reported at this conference [4] and the latter are described here.

Neutrino elastic scattering channels are $\overline{\nu}_\mu \rightarrow \mu^+ n$ and $\nu_\mu \rightarrow \mu^- p$. The first MINER$\nu$A elastic results are $\overline{\nu}_\mu$ scattering with the partial detector. The event selection is a μ^+ momentum analyzed track originating in $\sim$3 ton fiducial volume. Kinematic quantities are derived from the muon alone; the final state neutron is not required to be identified, but a Q^2-dependent vertex recoil energy cut is applied to enhance the CCQE fraction in the sample, see Figure 2.

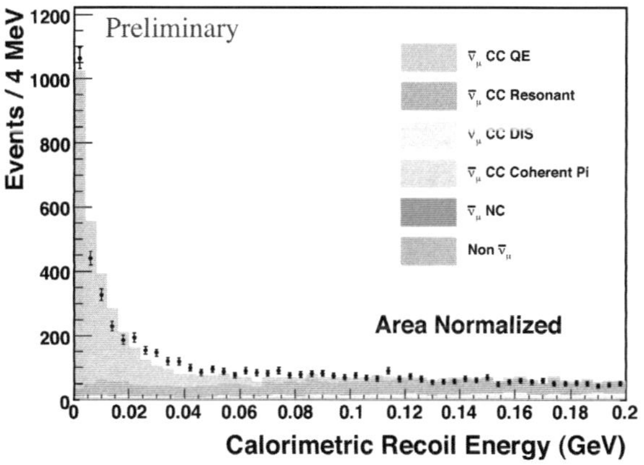

Figure 2. Distribution of the recoil energy of $\overline{\nu}_\mu$ CCQE-like interactions for data and monte carlo.

The area normalized Q^2 and E_ν distributions of data compared to GENIE-generated monte carlo [5] using this event selection are shown in Figures 3 and 4.

Analysis of elastic $\nu_\mu \rightarrow \mu^- p$ scattering is currently underway. Factors from the $\nu_\mu/\overline{\nu}_\mu$ cross section ratio and increased fiducial volume of the full detector will lead to approximately four times the statistics of the $\overline{\nu}_\mu$ partial detector sample for a similar POT exposure. The presence of the highly ionizing proton introduces another signature for the ν_μ CCQE interaction. Using the charge, momentum/range information from the MINOS experiment and knowledge from MINERνA Test Beam detector for proton identification, MINERνA will collect one of the largest samples of this channel to date (with the total POT in the run plan). There is also interest to use knowledge gained from ν_μ CCQE channel analysis to identify single proton tracks originating in the detector; this is the signal for simple neutral current elastic scattering. The $\nu_\mu p \rightarrow \nu_\mu p$ interaction can be studied in ratios and combinations to yield a measurement of g_a^s, the strangeness contribution the nucleon spin [6]. The backgrounds in this channel will

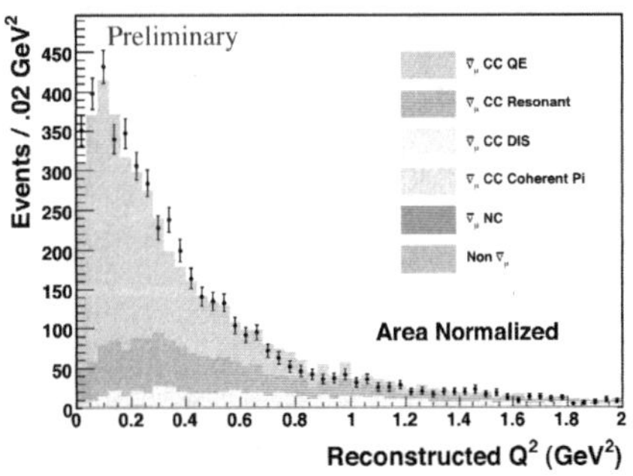

Figure 3. Preliminary Q^2 distribution for CCQE-like $\overline{\nu}_\mu$ candidates.

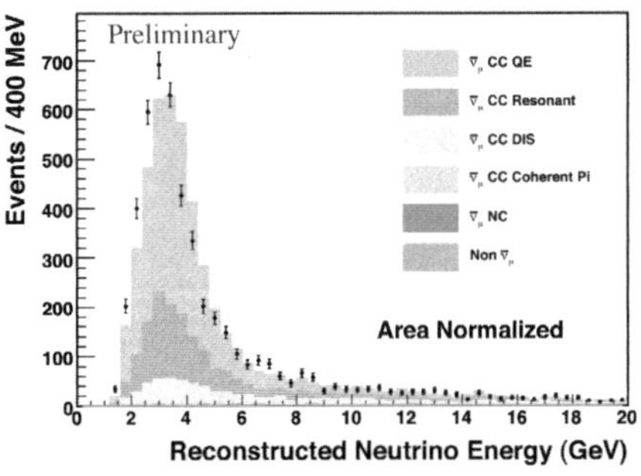

Figure 4. Preliminary E_ν distribution for CCQE-like $\overline{\nu}_\mu$ candidates.

complicate the analysis.

Acknowledgments

The author is very grateful to the organizers for the invitation to speak at the NuFact 2011 and for the financial assistance that, in part, made it possible to attend the venue. The author would also like to extend gratitude, on the behalf of the entire MINERνA Collaboration, to both Alain Blondel and Ilias Efthymiopoulos for their efforts in hosting NuFact 2011.

Appendix A

Kinematic quantities assuming two body $\overline{\nu}_\mu$ CCQE interaction:

$$E_{\overline{\nu}_\mu} = \frac{m_n^2 - (m_p - \epsilon_B)^2 - m_\mu^2 + 2(m_p - \epsilon_B)E_\mu}{2((m_p - \epsilon_B) - E_\mu + p_\mu cos\theta_\mu)}, \quad Q^2 = 2E_{\overline{\nu}}(E_\mu - p_\mu cos\theta_\mu) - m_\mu^2$$

References

[1] Lyubushkin V 2009 A study of quasi-elastic muon neutrino and antineutrino scattering in the NOMAD experiment *AIP Conf. Proc.* **1189** p157-162
[2] Aguilar-Arevalo A 2008 Measurement of muon neutrino quasi-elastic scattering on carbon *Phys. Rev. Lett.* **100** 032301
[3] Budd H 2011 Design of the MINERνA detector *Topics in Part. Phys. Proc.* (Chicago, Elsevier)
[4] Devan J 2012 Charge current inclusive measurements with MINERνA *NuFact Int. Conf. Proc.* (Geneva, JPCS)
[5] Andreopoulos C, *et al.* 2010 The GENIE neutrino monte carlo generator *Nuc. Inst. Meth.* **614** p87-104
[6] Praet C 2009 Modeling quasi-free neutrino-nucleus reactions for accelerator-based experiments PhD. Dissertation (Belgium, Ghent University)

Journal of Physics: Conference Series **408** (2013) 012047 doi:10.1088/1742-6596/408/1/012047

Inclusive measurements with MINERνA

Joshua D Devan (on behalf of the MINERνA collaboration)

Department of Physics, College of William and Mary,
P.O. Box 8795, Williamsburg, VA 23187, USA

E-mail: jddevan@email.wm.edu

Abstract. MINERνA is a neutrino scattering experiment in the NuMI beamline at Fermilab designed to measure neutrino cross-sections, final states and nuclear effects on a variety of targets in the few-GeV region. MINERνA is currently running in the NuMI low energy configuration and will continue in medium energy. We present a preliminary neutrino energy spectra in three beam configurations and a preliminary comparison of iron and lead event rates.

1. Introduction

MINERνA [1] is a neutrino scattering experiment in the NuMI beamline at Fermilab, designed to measure neutrino cross-sections, final states and nuclear effects on a variety of targets. NuMI is an intense, broad band spectrum, flexible beamline originally developed for the MINOS oscillation experiment. MINERνA is currently running in the low energy configuration with a peak neutrino energy of $\sim$3 GeV and will continue in the NOνA era as the energy is raised to $\sim$6 GeV.

2. MINERνA detector

The MINERνA detector is constructed of strips of triangular, extruded polystyrene scintillator assembled into 17 mm thick, 2.5 m point-to-point hexagonal planes. A wavelength-shifting optical fiber located in the center of the strips transmits light to multi-channel photomultiplier tubes for readout. The planes are stacked with a 60° rotation between strips in adjacent planes to form three stereoscopic views. The downstream planes are alternated with sheets of 2 mm lead to form an electromagnetic calorimeter and 1 in steel to form a hadronic calorimeter. The MINOS near detector acts as a spectrometer for rear-exiting muons, measuring momentum by either range or curvature.

The upstream planes of MINERνA contain targets of iron, lead and carbon. A cryostat target is currently filled with liquid helium with a proposal to run hydrogen or deuterium in the NOνA era. A water target is under development.

3. MINERνA test beam

To calibrate the absolute energy scale and validate monte-carlo simulations of the MINERνA detector, a small, reconfigurable version was constructed and exposed to a measured beam of low energy hadrons. The test beam detector consists of 40 planes of scintillator of $\sim$1 m^2 active area. These planes can be interleaved with sheets of lead and steel absorber to emulate the downstream calorimeters of the MINERνA detector.

The project required the development of a new tertiary beamline at the Fermilab Test Beam Facility to produce, identify and momentum-analyze low energy hadrons. The beamline consists

Journal of Physics: Conference Series **408** (2013) 012047 doi:10.1088/1742-6596/408/1/012047

of a target and collimator, two pairs of wire chambers upstream and downstream of a pair of dipole magnets and a time of flight system. In addition to low energy hadrons, the detector was exposed to a broad spectrum, unmeasured muon beam and cosmics. The project ran during the Summer of 2010; calibrations and analysis are in progress. The spectrum of beamline particles is shown in Figures 1 & 2.

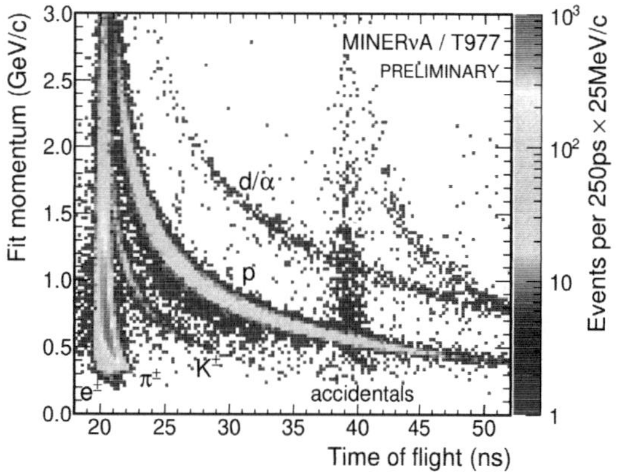

Figure 1. Spectrum of particles observed in the test beam beamline. Accidental triggers from particles in adjacent accelerator buckets produce a shadow at +19 ns.

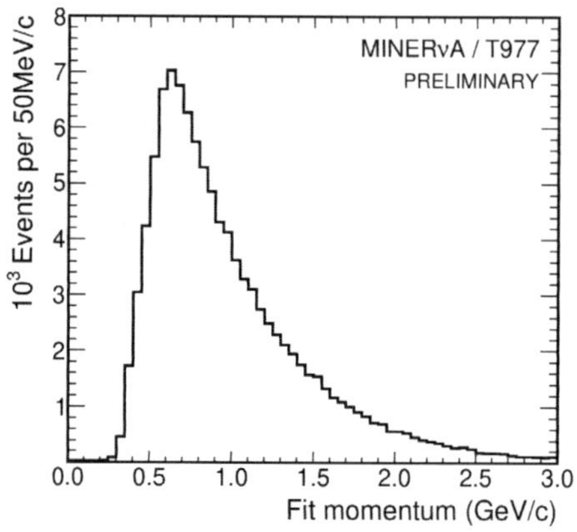

Figure 2. Projection of Figure 1 onto the vertical (momentum) axis. Data is 107×10^3 events from the Summer 2010 run.

4. Calorimetric energy resolution

The calorimetric energy resolution of the detector is determined by a monte-carlo study of neutral current events with a vertex in the upstream, fully active tracker region. Events are generated with Genie 2.6 [2] and simulated in Geant4 [3] with the QGSP BERT model. Visible energy is summed in the tracker and downstream electromagnetic and hadronic calorimeters. The energy in the calorimeters is weighted to account for the additional passive material. An overall energy correction is then applied. The true recoil energy is defined as $E_{recoil} \equiv E_\nu^{in} - E_\nu^{out}$. Using this simple calorimetry, the energy resolution is $\sigma/E = 0.12 \oplus 0.27/\sqrt{E\,(\mathrm{GeV})}$ (see Figure 3). The energy resolution will improve with shower reconstruction algorithms and EM/hadronic compensation.

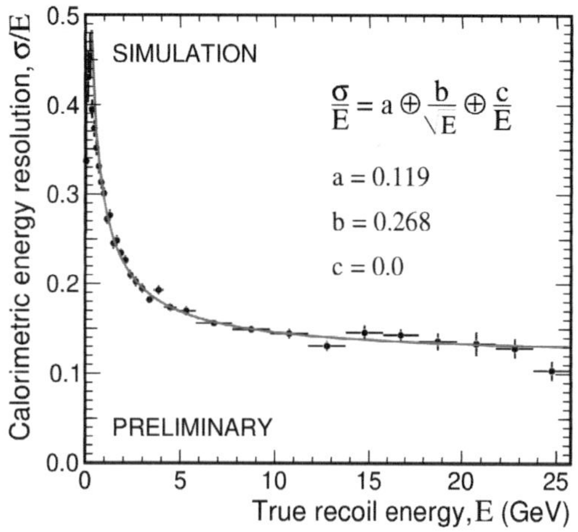

Figure 3. Calorimetric energy resolution for neutral current monte-carlo events. True recoil energy is defined as $E_{recoil} \equiv E_\nu^{in} - E_\nu^{out}$.

5. Towards charged current inclusive cross-sections

MINERνA is actively working towards producing charged current inclusive cross-sections on plastic scintillator and the iron, lead and carbon targets. To that end, we present the following two preliminary analyses.

5.1. Charged current neutrino energy spectra

The NuMI beamline is reconfigurable; the positions of the target and horns and the horn current can be varied to select the neutrino energy spectra. Figures 4 & 5 show the observed spectra of neutrinos producing a charged current interaction in the MINERνA detector in three beamline configurations: low energy (LE), pseudo medium energy (pME) and pseudo high energy (pHE). Events are restricted to a primary vertex in the upstream, fully active tracker region with a muon track analyzed in the MINOS spectrometer. Tracks identified in the final state products are fit to a dE/dx profile for particle identification. The remaining energy is summed calorimetrically as described in Section 4.

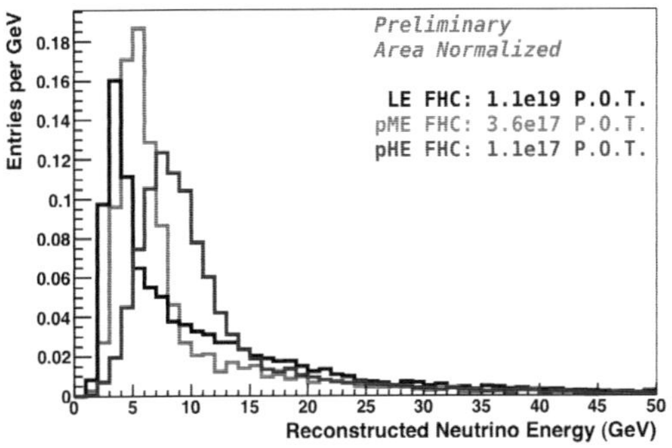

Figure 4. Spectra of neutrinos producing a charged current interaction in the low energy (LE), pseudo medium energy (pME) and pseudo high energy (pHE) beamline configurations in forward horn current (FHC, neutrino-focusing) mode.

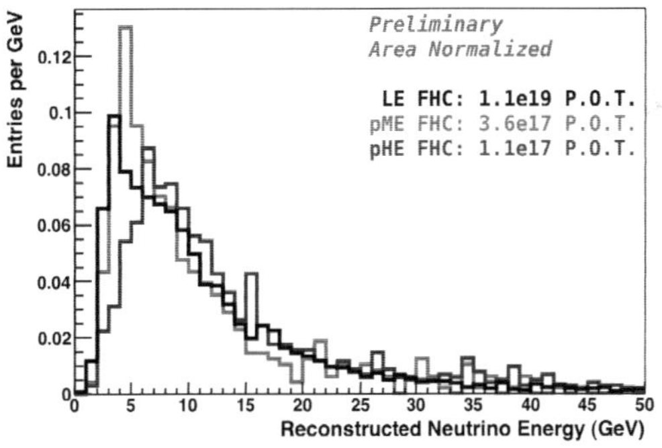

Figure 5. Spectra of anti-neutrinos producing a charged current interaction (from the ~6% $\bar{\nu}$ content of the ν beam).

5.2. Iron/lead charged current event rates

MINERνA is developing analysis techniques for events originating in the passive nuclear targets by studying the most downstream target of iron and lead. Passive target planes are located between active scintillator planes; by requiring muon activity in the first plane downstream of a target, with no activity upstream, a charged current interaction vertex can be localized to the target. The transverse position is determined by projecting the muon track to the center of the target plane. This study requires that the muon is tracked in the MINOS spectrometer.

Backgrounds are introduced from events originating in the scintillator planes upstream and downstream of the target (see Figure 6). Backgrounds and acceptance effects can be studied in data with a scintillator reference target. Eight active planes are declared as a passive target, event selection cuts are applied, then selection purity can be quantified by analyzing the data from within the reference target. The reference target is divided into two regions in the same manner as the passive iron/lead target. A ratio of events originating in these two regions is shown in Figure 7. A data to monte-carlo comparison of events reconstructed in the most downstream target of iron and lead is provided in Figures 8 & 9.

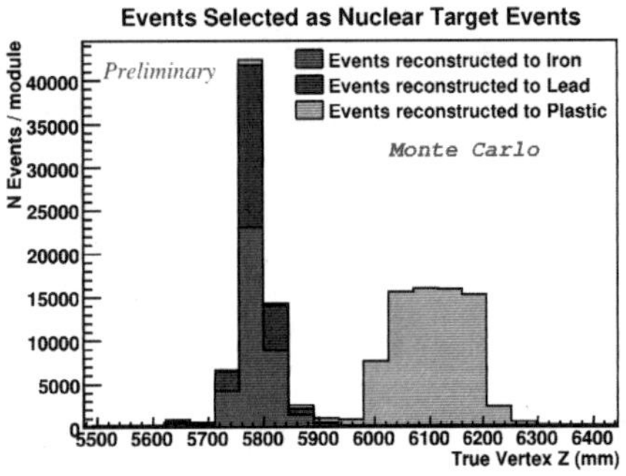

Figure 6. True vertex distribution of charged current interactions reconstructed in the passive and reference targets. Passive target is located at 5760–5800 mm, reference target is 6020–6200 mm.

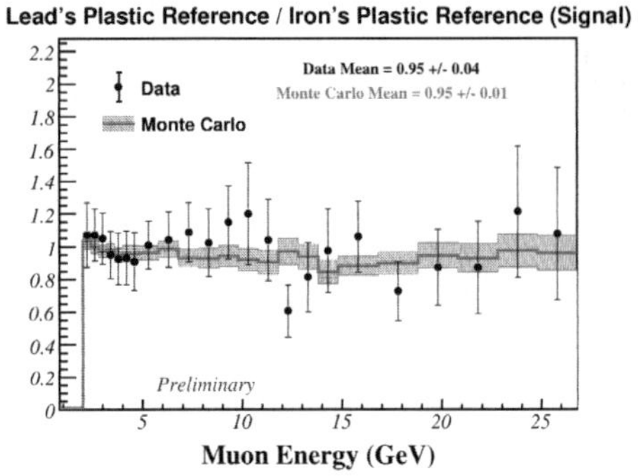

Figure 7. Mass-corrected ratio of events reconstructed in the lead region of the reference target to those in the iron region. Deviations from unity result from acceptance effects.[1]

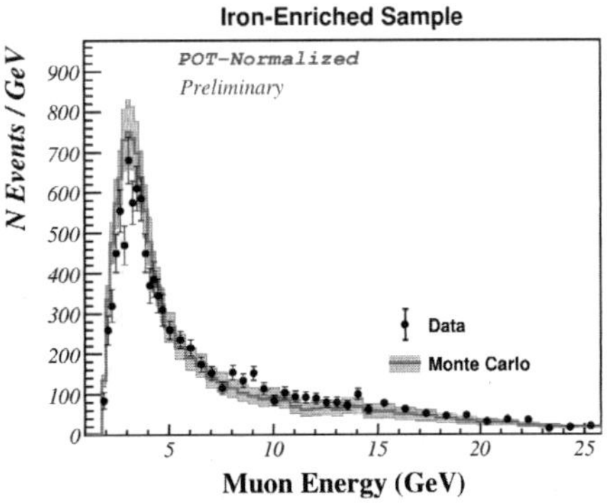

Figure 8. Data/monte-carlo comparison of events reconstructed in the iron region of the most downstream passive target.[1]

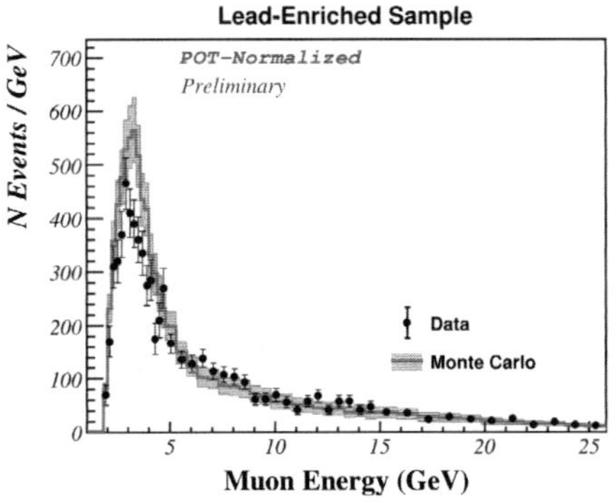

Figure 9. Data/monte-carlo comparison of events reconstructed in the lead region of the most downstream passive target.[1]

6. Conclusion

MINERνA will measure neutrino cross-sections, final states and nuclear effects on a variety of targets in the few-GeV region to reduce systematic uncertainties in oscillation experiments and provide new understanding of the nucleus.

References

[1] http://minerva.fnal.gov
[2] http://www.genie-mc.org
[3] http://geant4.cern.ch

[1] Discrepancies between data and monte-carlo result from incorrect flux modeling. Flux is Fluka08, prior to tuning, with the Genie 2.6 [2] event generator.

Journal of Physics: Conference Series **408** (2013) 012048 doi:10.1088/1742-6596/408/1/012048

Charged pion spectra in proton–carbon interactions at 31 GeV/c

Magdalena Zofia Posiadała
on behalf of the NA61/SHINE Collaboration
Faculty of Physics, University of Warsaw, Hoża 69, 00-681 Warsaw, Poland
E-mail: magdalena.posiadala@fuw.edu.pl

Abstract. The NA61/SHINE experiment at CERN SPS measured charged pion spectra in p+C interactions at 31 GeV/c. These measurements are necessary to improve predictions of the neutrino flux for the T2K long baseline neutrino oscillation experiment in Japan. Presented analysis was based on the data collected during the first NA61/SHINE run in 2007 with an isotropic graphite target with a thickness of 4% of nuclear interaction length. Three different methods which were used in order to obtain π^+ and π^- spectra are introduced. Differential cross sections for negatively and positively charged pions are presented as a function of laboratory momentum in ten intervals of the laboratory polar angle up to 420 mrad.

1. Introduction

The NA61/SHINE (SHINE = SPS Heavy Ion and Neutrino Experiment) is a large acceptance hadron spectrometer located in the North Area H2 beam line of the European Centre for Nuclear Research (CERN) in Geneva. The experiment uses beams from the Super Proton Synchrotron (SPS). The NA61/SHINE experiment combines a rich physics program in three different fields: auxiliary measurements for neutrino experiments, cosmic-ray simulations, and the behavior of strongly interacting matter at high density (see: [1, 2, 3] for details). In its first stage of data taking (years 2007, 2009 and 2010) NA61/SHINE performed measurements of hadron production in hadron-nucleus interactions. Such data are needed for neutrino (T2K) and cosmic-ray (Pierre Auger and KASCADE) experiments.

This article reports on the measured differential cross sections for negatively and positively charged pions obtained with 31 GeV/c p+C interactions taken during first NA61 pilot run in 2007. All measurements presented here were taken with a 2 cm long target with dimensions 2.5(W)x2.5(H) cm (about 4% of nuclear interaction length) with density $\rho = 1.84 \frac{g}{cm^3}$, so-called thin target. In total, during 2007 data taking we managed to register about 678 k events on thin carbon target.

2. Overview of the NA61/SHINE detector

The main components of the detector have been inherited from its ancestor NA49 [4] which has been designed to handle the most complex hadronic final states like high multiplicity (around 1800) charged particles produced in lead-lead collisions. At the same time it is able to deal with charged multiplicity as low as $\geq 3 - 10$, in elementary hadronic (p+p) as well as proton-nucleus collisions. The layout of the NA61/SHINE setup is shown in Fig. 1. The main tracking

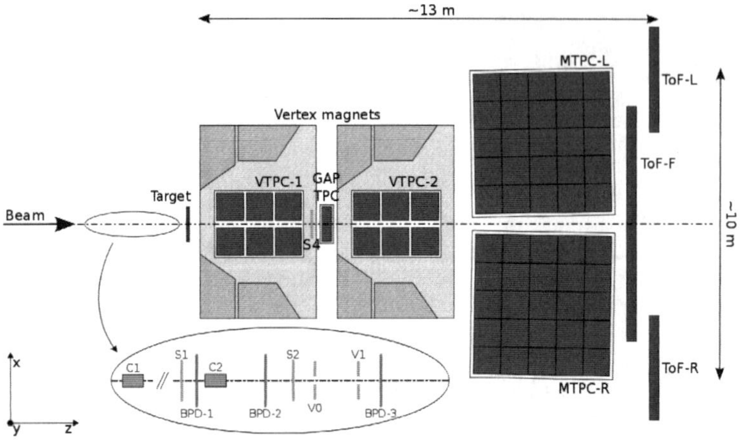

Figure 1. The layout of the NA61/SHINE experiment at the CERN SPS (top view, not to scale). The incoming beam direction is along the z axis. The magnetic field bends charged particle trajectories in the $x - z$ (horizontal) plane. The drift direction in the TPCs is along the y (vertical) axis.

devices are four large volume Time Projection Chambers (TPCs). Two of them, the vertex TPCs (VTPC1 and VTPC2), are located in the magnetic field of two superconducting dipole magnets. Two others MTPC-L and MTPC-R (blue boxes in Fig. 1) are positioned downstream. Two time-of-flight detectors (ToF-L/R) were inherited from NA49 and are able to provide time measurement resolution of $\sigma \approx 60$ ps. In 2007 the experiment has been updated with a new forward time-of-flight detector (ToF-F) in order to extend the acceptance for pion and kaon identification as required for T2K measurements [5]. The ToF-F wall is installed downstream of the main TPCs (black line in Fig. 1), closing the gap between the ToF-L and ToF-R walls. The ToF-F time resolution is $\sigma \approx 115$ ps. More details of the NA61 set-up may be found elsewhere [1, 4].

3. Methods of particle identification

Particle identification in NA61 was possible using energy loss measurements (dE/dx) in active volume of the TPCs and time-of-flight information from the ToF-F detector. If dE/dx is plotted versus a particle's momentum, a clear mass dependence can be seen, as shown in left plot in Fig. 2. When the resolution of the ionization measurement is sufficient, the separation of different species is possible, excluding of course regions where Bethe-Bloch curves cross each others like in momentum range of $1 < p$ $[GeV/c] < 3$ for positively charged particles. For negative particles situation in this region of phase space is better as we do not see antiprotons and antideuterons in the data. This momentum range of $1 < p$ $[GeV/c] < 3$ is very important for the T2K neutrino predictions which is the reason why measurements were also performed with addition of specially designed ToF-F detector (see middle plot in Fig. 2). As a result, combined analysis with dE/dx and ToF-F measurements was carried out. Example of two-dimensional m^2–dE/dx illustration for positively charged particles for momentum interval $2 < p$ $[GeV/c] < 3$ is shown in right plot in Fig. 2.

To conclude, in order to obtain π^+ and π^- yields in broad momentum range, three different analyses were developed. They are as follows:

- energy loss measurements (dE/dx) in the active volume of the TPCs; analysis applicable

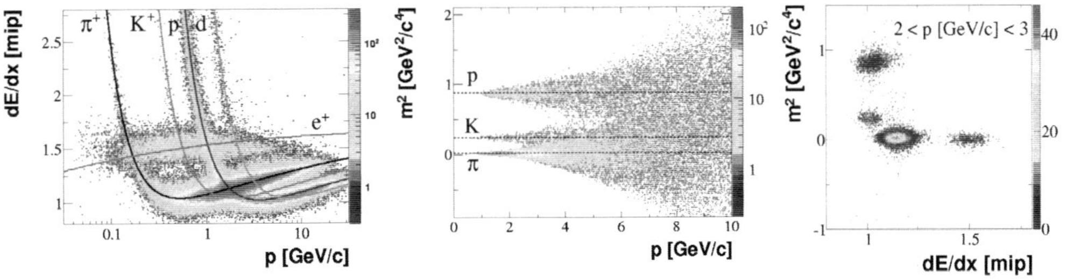

Figure 2. Left: dE/dx as a function of momentum for positively charged particles together with the parametrized Bethe-Bloch curves for positrons (electrons), pions, kaons, protons and deuterons. Middle: Measured mass squared, derived from the ToF-F measurement, versus momentum p. The lines show the expected mass squared values for different particles. Right: Example of two-dimensional m^2–dE/dx plot for positively charged particles for momentum interval $2 < p\ [GeV/c] < 3$. 2σ red contour around fitted pion peak is also shown. Each point in all three plots represents one track detected during 2007 data taking on the thin carbon target.

for low momentum particles *i.e.* $0.2 < p\ [GeV/c] < 1$ for positive and $0.2 < p\ [GeV/c] < 3$ for negative tracks. The topic of this analysis was the subject of the Ph.D. thesis [8, 11].

- combined information from the dE/dx and time-of-flight measurements ($tof - dE/dx$); method used for tracks with the ToF detector acceptance i.e with p >0.8 GeV/c. For more details concerning the analysis the reader is referred to [9, 11].
- analysis of negatively charged particles further referred to as h^-. This method is based on the theoretical and experimental premises that negative particles produced by 31 GeV/c protons consist mainly of negative pion mesons with an admixture of electrons, negative kaons and a negligible fraction of antiprotons. The Monte Carlo simulation [6] was used to calculate corrections for the contribution of electrons, primary K^- and $\overline{p}$. Extensive description of the analysis method can be found in [10, 11].

4. Differential cross sections for π^+ and π^- mesons

In Fig. 3 we present differential cross sections for π^+ (left) and π^- (right) meson production obtained in p+C interactions at 31 GeV/c obtained by three methods of analysis. Details of the normalization procedure may be found elsewhere [7].

5. Conclusions

In the 2007 data taking the first physics data on interactions of 31 GeV/c protons on the thin and T2K replica carbon target were registered. The spectra of π^+ and π^- mesons were obtained using different methods: dE/dx, $tof - dE/dx$ and h^-.

The numerical values of the final charged pion spectra as well as details concerning the analyses, cross section normalization, systematic errors, model comparisons may be found in [11]. These measurements have already been used by the T2K collaboration in order to tune simulations of the neutrino beam which are very significant for searching of electron neutrino appearance in muon neutrino beam, as stated in [12].

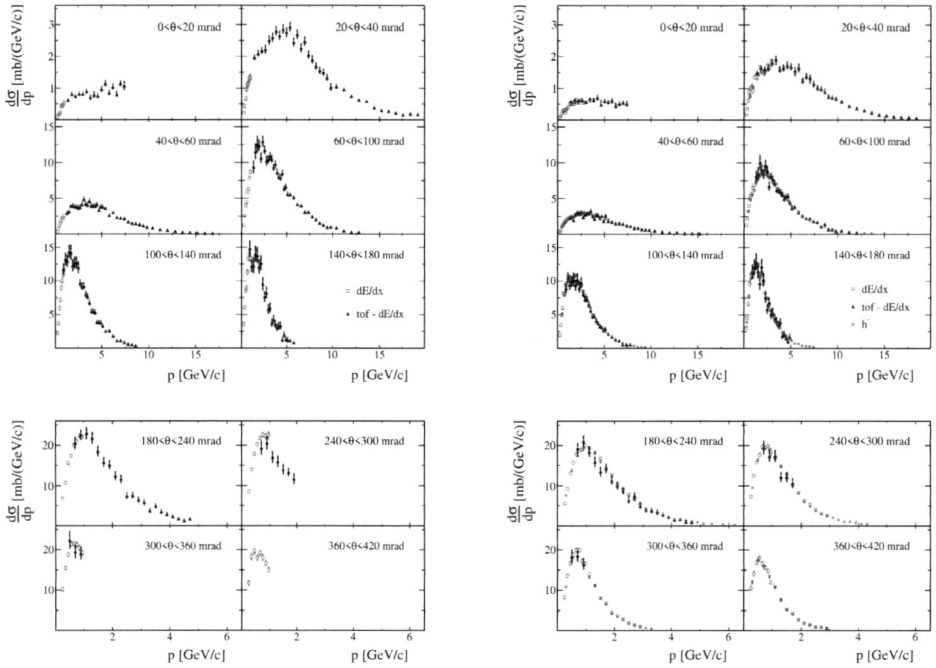

Figure 3. Differential cross sections for π^+ (left) and π^- (right) meson production in p+C interactions at 31 GeV/c. The spectra are presented as a function of laboratory momentum (p) in ten intervals of polar angle (θ). Results obtained using: two analysis methods for π^+ and three for π^- are presented by different symbols: □ - dE/dx analysis, ○ - h^- analysis and ▲ - $tof - dE/dx$ analysis. Error bars indicate only statistical uncertainties.

6. Acknowledgments

The author would like to thank to the Polish Ministry of Science and Higher Education for support with Grant No. N N202 1267 36.

References

[1] Antoniou N *et al.* 2006 CERN-SPSC-2006-034
[2] Antoniou N *et al.* 2007 CERN-SPSC-2007-004
[3] Abgrall N *et al.* 2007 CERN-SPSC-2007-019
[4] Afanasev S *et al.* 1999 *Nucl. Instrum. Meth. A* **430** 210
[5] Abgrall N *et al.* 2008 CERN-SPSC-2008-018
[6] Abgrall N 2011 *Constraining neutrino flux predictions with hadron production data: the NA61/SHINE measurements for the T2K experiment* (Geneva: University of Geneva)
[7] Strabel C 2011 *Evaluation of Particle Yields in 30 GeV Proton-Carbon Inelastic Interactions for Estimating the T2K Neutrino Flux* (Zurich: ETHZ)
[8] Posiadała M 2012 *Measurements of Hadronic Cross Section for Precise Determination of Neutrino Beam Properties in T2K Oscillation Experiment* (Warsaw: University of Warsaw)
[9] Murphy S 2012 *Measurement of charged Pion and Kaon production cross sections with NA61/SHINE for T2K* (Geneva: University of Geneva)
[10] Palczewski T 2012 *Ph.D thesis in preparation* (Warsaw: Soltan Institute for Nuclear Studies)
[11] Abgrall N *et al.* 2011 *Phys. Rev. C* **84** 034604
[12] Abe K *et al.* 2011 *Phys. Rev. Lett.* **107** 041801

Strange particle production in proton-carbon interactions at 31 GeV/c

Silvestro Di Luise
on behalf of the NA61/SHINE collaboration
ETH Zurich, Switzerland
E-mail: Silvestro.Di.Luise@cern.ch

Abstract. Spectra of positively charged kaons in p+C interactions at 31 GeV/c have been measured with the NA61/SHINE spectrometer at CERN SPS. The analysis is performed on the dataset collected during the 2007 pilot run (about 678×10^3 events). The target is a 2 cm long graphite specimen (4% of a nuclear interaction length λ_I). Charged pion spectra have been already measured with the same dataset [1]. Pion and kaon differential production cross sections are required to improve predictions of the neutrino flux for the T2K long baseline neutrino oscillation experiment in Japan. The knowledge of kaon production is especially important for the precise prediction of the intrinsic electron neutrino component of the T2K beam. The results are presented as a function of laboratory momentum in 2 intervals of the laboratory polar angle covering the range from 20 up to 240 mrad. Kaon spectra as well as K^+/π^+ ratios are compared with predictions of several hadron production models. In addition, first results from the analysis of the high statistics 2009 data taking are shown.

The NA61/SHINE (SPS Heavy Ion and Neutrino Experiment) experiment at the CERN SPS pursues a broad physics program in various fields [2, 3, 4]: hadroproduction measurements for neutrino oscillation experiments (T2K [9]), cosmic ray air-showers simulation (Pierre Auger and KASCADE experiments), study of the properties of the onset of deconfinement and search for the critical point of strongly interacting matter.

First NA61/SHINE physics paper was devoted to measurements of charged pion spectra in p+C interactions at 31 GeV/c [1]. These data have already been used for an accurate neutrino flux prediction in the T2K experiment [10]. The set of measurements on kaon production presented here provide another fundamental ingredient towards a complete flux characterization, in particular in the high energy region where most of the electron neutrino to muon neutrino contamination originates from kaon decays. Hadroproduction measurements are particularly suitable in NA61 thanks to the wide kinematic acceptance and the excellent particle identification based on the information from both drift chambers and time-of-flight detectors.

The NA61/SHINE apparatus ([1]-Sec. II) consists of four large volume Time Projection Chambers (TPCs): two Vertex-TPCs (VTPC-1 and VTPC-2) placed in the magnetic field produced by two super-conducting dipole magnets and two Main-TPC (MTPC) located downstream symmetrically with respect to the beam line (Fig. 2). The setup is complemented with a time-of-flight (ToF) detector array horizontally segmented in 64 scintillator bars read out at both ends by photomultipliers. The time resolution is 115 ps. The magnetic field was set to a bending power of 1.14 Tm (the estimated average momentum resolution $\sigma(p)/p^2$ is about 5×10^{-3} (GeV/c)$^{-1}$). The adopted binning scheme, represented in Fig. 1, was mainly driven

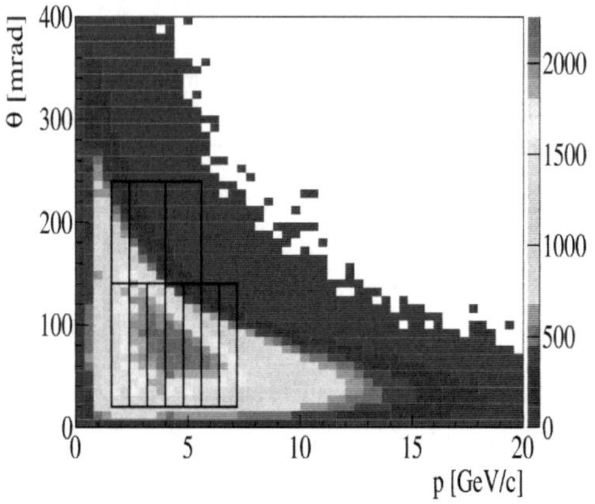

Figure 1. T2K beam simulation: the $\{p, \theta\}$ distribution for K^+ weighted by the probability that their decay produces a neutrino in the acceptance. The binning used in the analysis is superimposed: the maximum kinematic range considered is $1.6 < p < 7.2$ GeV/c and $20 < \theta < 240$ mrad.

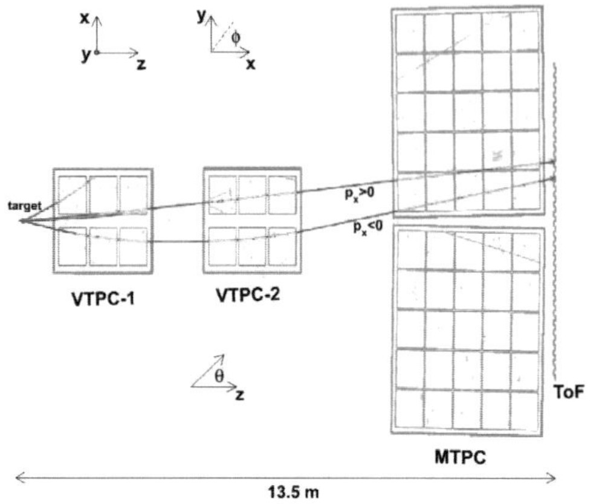

Figure 2. Schematic drawing of the experimental apparatus and a reconstructed p+C interaction event. Yellow (green) points indicate TPC clusters (not) associated to reconstructed primary tracks. Stars correspond to hits reconstructed in the ToF. Red lines are the fitted particle trajectories.

by the limited available statistics and by the necessity to cover most of the phase space relevant for T2K. Two polar angle θ (cfr. Fig. 2 for the definition) intervals are defined: $20 < \theta < 140$ mrad and $140 < \theta < 240$ mrad. The combination with a 0.8(1.6) GeV/c momentum (p) bin yields a few thousands of selected tracks per $\{p, \theta\}$ interval. Measurements have been performed until 7.2 GeV/c. This choice comes from the fact that in the relativistic rise region (above 4-5 GeV/c) the particle identification is a matter of extracting the vanishing kaon signal over the predominant proton one: with the available statistics of 2007 data the applied procedure turned out to be robust only until about 7 GeV/c. Track selections have been optimized to maximize both acceptance and reconstrucion efficiency. In order to minimize biases in the determination of the kaon ionization energy loss dE/dx (in the TPC's) and the time-of-flight ToF measurement due to the kaon decay in flight, only tracks reconstructed until the downstream edge of the MTPC (close to the ToF detector) are accepted. Such a requirement, given the good efficiency of the reconstruction algorithms in the identification of kink topologies, provide a high purity ($\sim 97\%$) sample of stable kaons. It also reduces the calculation of the corrections to the identified kaon yields essentially to a track-by-track reweighting for the survival probability calculated from the measured momentum and path length (from the primary vertex to the ToF wall); minor corrections ($\sim$ 2-3%) to be applied to take into account the contamination from non recognized kaon decays (i.e. when the decay vertex occurs in between the MTPC and the ToF, cfr.Fig. 2). Kaon identification based only on the dE/dx information is troublesome at the lower considered momenta, until 3-4 GeV/c, since protons and kaons have close values of the specific energy loss in this kinematic range (cfr. Fig. 3). Another approach is to identify particle through the measurement of their mass squared m^2. This method is based on the measurement of the time-of-flight as, along with the track length and momentum, it allows for an indirect measurement of the particle mass. Identification based only on the m^2 information is effective at low momenta, below 3-4 GeV/c, where the separation power is higher. The uncertainty on the mass measurement is in fact dominated by the time resolution, it means the m^2 resolution worsens quadratically with increasing momentum. Given the complementarity of the two approaches the particle identification capability can be improved, over the whole kinematic range, by looking at the combined dE/dx and m^2 information.

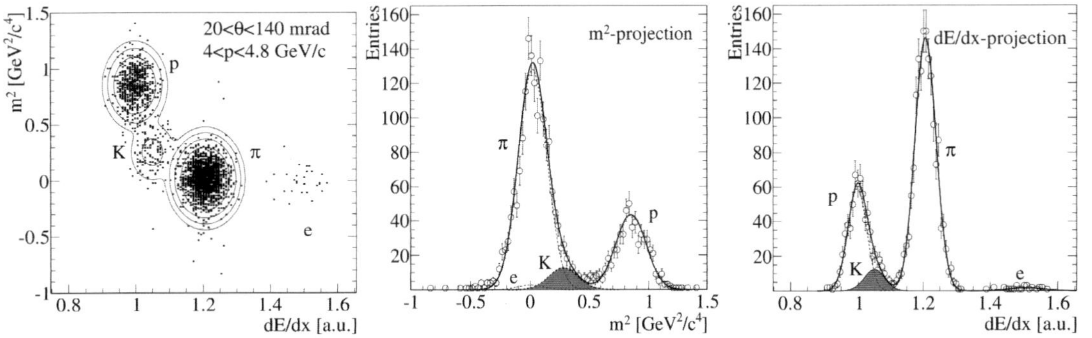

Figure 3. Example of a Maximum Likelihood bidimensional fit to the dE/dx-m^2 distribution. The four contour lines correspond to the 1-σ and the 2-σ levels calculated from the proton (the two innermost) and the kaon (the two outermost) distribution functions. The individual m^2 and dE/dx distributions are also shown superimposed with the results of the fitted functions.

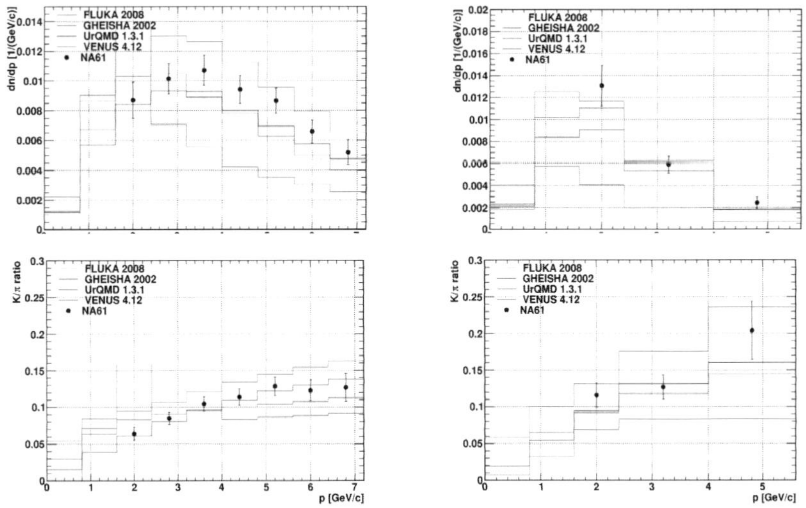

Figure 4. *Top*: Differential cross sections for K^+ production in p+C interactions at 31 GeV/c as a function of the laboratory momentum, p (left: 20< θ <140 mrad, right: 140< θ <240 mrad). Errors are statistical and systematic uncertainties added in quadrature. The overall uncertainty (2.5%) due to the normalization procedure is not included. *Bottom*: Ratio of K^+ over π^+. Errors are statistical only. Curves from several model prediction are superimposed.

Fig. 3 shows the correlation between the measured dE/dx and m^2 for one of the analysed $\{p, \theta\}$ bin: accumulations corresponding to the different particle types are resolvable if compared to the individual dE/dx and m^2 distributions. The functional model used to fit the yields of protons (p), kaons (K^+), pions (π^+) and positrons (e^+) to the data is a superposition of four bidimensional gaussians. A dedicated Monte Carlo simulation has been used to validate the accuracy of the simple gaussian shape approximation so as to avoid biases related to the definition of the particle distribution functions. Given the small K/p and K/π production rate and the signals overlapping at higher momenta, a dedicated procedure has been conceived for the precise determination, directly from the data, of the expected dE/dx and m^2 mean value and of the m^2 variance as well. This allows to narrow the variability range of the fitted peak positions and widths on both dE/dx and m^2 axis. Moreover, considering the $\beta\gamma{=}p/m$ dependence of the dE/dx expected value, the relative peak distances between the different particle accumulations have been fixed: this further reduce the number of degrees of freedom and improve the stability of the fit. Measured K^+ spectra and ratio of K^+ over π^+ cross sections are presented in Fig. 4 compared with the prediction of different simulation model (e.g. VENUS 4.12 [5, 6] and FLUKA 2008 [7, 8]). Errors are dominated by the statistical uncertainty ($\sim 10-15\%$) except for the highest momentum interval where the systematic error associated to the fit procedure rises up to $6-7\%$ After the 2009 data taking campaign an order of magnitude more events have

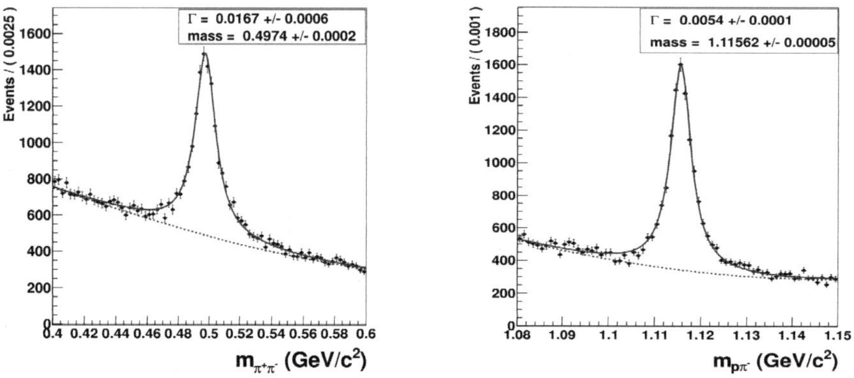

Figure 5. Data quality assessment for the 2009 dataset: invariant mass distribution in the regions of K^0_S (left, $p_T < 0.2$ GeV/c, $1 < y < 1.5$) and Λ (right, $0.4 < p_T < 0.6$ GeV/c, $-1.25 < y < 1$) peak.

been recorded, also thanks to important upgrades in the TPC readout and DAQ system. The data have been calibrated and the physics analysis is started. Data quality plots based on the invariant mass distribution of V^0-like topologies are shown in Fig. 5.

References

[1] Abgrall N *et al.* 2011 (NA61/SHINE Collaboration) *Phys. Rev.* C **84** 034604.
[2] Antoniou N *et al.* 2006 (NA49-future Collaboration) CERN-SPSC-2006-034.
[3] Antoniou N *et al.* 2007 (NA61/SHINE Collaboration) CERN-SPSC-2007-004/019.
[4] Abgrall N *et al.* 2008 (NA61/SHINE Collaboration) CERN-SPSC-2008-018.
[5] Werner K 1991 *Nucl. Phys.* A **525**, 501.
[6] Werner K 1993 *Phys. Rep.* **232**, 87.
[7] Fasso A *et al.* 2005 CERN-2005-10.
[8] Battistoni G *et al.* 2007 *AIP Conf. Proc.* **896**, 31.
[9] Abe K *et al.* 2011 (T2K Collaboration) *Nucl. Instrum. Methods* A **659**, 106-135.
[10] Abe K *et al.* 2011 (T2K Collaboration) *Phys. Rev. Lett.* **107**, 041801.

The NA61/SHINE long target pilot analysis for T2K

Nicolas Abgrall
on behalf on the NA61/SHINE collaboration

Département de Physique Nucléaire et Corpusculaire (DPNC), University of Geneva, 24 quai Ernest Ansermet, 1205, Geneva, Switzerland

E-mail: `nicolas.abgrall@cern.ch`

Abstract. The NA61/SHINE collaboration performed measurements of pC interactions at 31 GeV/c beam momentum with a full size replica of the T2K target (1.9 interaction length) during a pilot run in 2007. Larger statistics runs were also conducted in 2009 and 2010. The NA61/SHINE setup consists in a large acceptance spectrometer located on the H2 beamline of the SPS at CERN. For the first time, the kinematical phase space of interest for an accelerator based neutrino experiment (i.e. kinematical phase space of pions/kaons exiting the target and producing neutrinos in the direction of the near and far detectors) is fully covered by a single hadron production experiment. In a first stage, yields of positively charged pions were measured at the surface of the target. The analysis of the 2007 data set presented here demonstrates that **a)** high quality long target data were successfully taken with the NA61/SHINE apparatus, and **b)** for the first time, neutrino flux predictions can effectively be re-weighted with long target data.

1. Motivations for long target measurements for T2K

A prediction of the neutrino beam content at the T2K ([1], [2]) far detector is given in Table 1. The ν_μ flux is predominantly ($\sim 95\%$) produced by the decay in flight of positively charged pions focused by the magnetic horns of the beamline. While $\sim 30\%$ of the ν_e flux is produced from the decay of positively charged kaons, $\sim 50\%$ of it is due to muons produced in the decay of the same charged pions that generate the ν_μ flux. Thus, in a first stage, pion production data will constrain both ν_μ and ν_e fluxes.

Table 1. Flavour content of the neutrino beam at the far detector. Values integrated up to 10 GeV neutrino energy for the contribution of the different parent particles are quoted for each flavour. Abundances are normalised to the ν_μ content of the beam. Predictions (based on FLUKA2011.2 [7]) with horn currents set to 250 kA each.

			Source									
ν	Flux		π^+ or π^-		K$^+$ or K$^-$ (K2)		K$^+$ or K$^-$ (K3)		K^0_L		μ^+ or μ^-	
species	Abund.	$\langle E_\nu \rangle$	%	$\langle E_\nu \rangle$	%	$\langle E_\nu \rangle$	%	$\langle E_\nu \rangle$	%	$\langle E_\nu \rangle$	%	$\langle E_\nu \rangle$
ν_μ	1.0	0.60	95.1	0.59	4.5	2.39	0.2	1.06	0.1	0.96	0.02	0.57
$\bar{\nu}_\mu$	0.071	0.61	84.3	0.59	5.7	1.38	0.3	0.78	1.3	0.96	8.4	0.61
ν_e	0.011	0.71	1.0	1.08	–	–	30.8	1.13	12.3	1.04	55.8	0.58
$\bar{\nu}_e$	0.002	0.92	0.4	1.06	–	–	16.5	0.82	76.4	1.04	6.6	0.44

In terms of hadron production measurements, the pion contribution to the ν_μ flux at the far detector can be decomposed into a *secondary* component and a *secondary+tertiary* component. The first contribution originates from pions directly produced in the proton primary interaction (secondary pions) or in the decay of a secondary particle. The *secondary+tertiary* contribution refers to all pions exiting the target or pions produced in the decay of an outgoing particle. The dependence of these contributions on the neutrino energy is depicted in Figure 1. While the *secondary* component contributes to $\sim$ 60% of the ν_μ flux at the beam peak energy of $\sim$ 600 MeV, the *secondary+tertiary* component contributes to $\sim$ 90% of the flux. Thus, thin target measurements ($p + C \rightarrow \pi^+ + X$ at 30 GeV) for T2K constrain up to 60% of the ν_μ flux prediction at the far detector, the remaining 40% being produced in secondary interactions in the target and elements of the beamline. Long target measurements would constrain up to 90% of the flux prediction. In this case, only 10% of the flux due to secondary interactions out of the target would require constraints from other data.

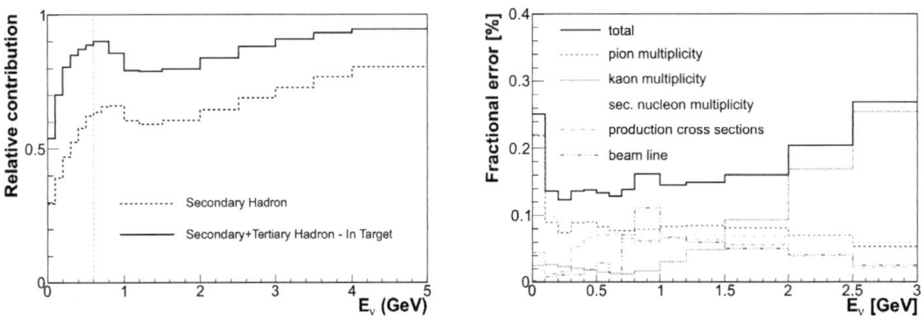

Figure 1. Ratio of *secondary* (dashed) and *secondary+tertiary* (solid) [left] to total contribution for the ν_μ flux at the far detector. Systematic errors for the ν_μ flux at the far detector [right] (for the analysis described in [4]).

As depicted in Figure 1 (right), the total fractional error on the T2K ν_μ flux prediction ([3], [4]) at the far detector is $\sim$ 15% at the beam peak energy. Uncertainties arise from two main contributions: uncertainties related to the beamline setting (e.g. beam optics, beam direction, alignment and currents of the focusing horns) and those related to the re-weighting procedure based on the NA61/SHINE thin target data (within 5 to 10 %)[5]. The fractional error at the beam peak energy is dominated by uncertainties from the second contribution (i.e. pion/kaon multiplicity, secondary nucleon multiplicity and production cross sections). In particular, uncertainties on the re-weighting of secondary nucleon and production cross sections for secondary interactions are applied to $\sim$ 40% ($\sim$ 10%) of the flux in the case of the thin (long) target based re-weighting. These uncertainties rely on interpolation between sparse data sets or comparison to models and are often poorly known. In the case where flux predictions are re-weighted by long target data, these contributions are taken into account at once and there is no need for error prone estimates of secondary interactions in the target. Long target data are therefore needed to obtain a precision of 5% or better on the absolute flux prediction.

2. The NA61/SHINE long target measurements for T2K

The 2007 NA61/SHINE pilot data were taken with a replica of the T2K target at 30 GeV beam energy. The NA61/SHINE spectrometer consists in a set of 5 time projection chambers (TPCs) complemented by an array of time-of-flight (ToF) detectors located downstream of the TPCs, allowing for a full coverage of the forward pion production of interest for T2K (i.e. phase space of pions exiting the target and producing neutrinos in the direction of the near and far detectors). Details of the experimental setup are given elsewhere [5]. The trajectory of each beam proton track is reconstructed with a set of beam position detectors that allow to determine the position of the beam impact on the upstream face of the target. The usage of the ToF in the analysis assures that selected tracks in the TPCs originate from single proton interactions in the target. Data can therefore be normalized to the total number of protons on target (POT).

Tracks reconstructed in the TPCs are extrapolated back through the magnetic field from their first measured point to the surface of the target. A point-of-closest-approach is found between the trajectory of the track and the surface of the target. This requires a precise knowledge of the relative alignment of the beam and target, which is determined by using both beam and TPC tracks. For the 2007 pilot run, the target was tilted with respect to the beam axis (z axis) in both horizontal ($\sim$ 5 mrad) and vertical ($\sim$ 2.8 mrad) directions. The misalignment was taken into account in the extrapolation procedure and effects on the outgoing pion yields (additional systematic uncertainty) studied with dedicated Monte-Carlo simulations. The target is split into 5 bins of 18 cm each along the beam (z axis), and a last bin corresponding to the dowstream face of the target that covers the very forward (below 40 mrad) pion production. Analysis cuts are optimized to improve the momentum and polar angle resolutions at the first measured point on track that determine the achievable resolutions on target, estimated to $\sigma_z = 5$ cm, $\sigma_p/p = 3\%$ and $\sigma_\theta/\theta = 6\%$ for the longitudinal, momentum and polar angle resolutions respectively. As depicted in Figure 2, tracks can effectively be reconstructed on the surface of the target. The analysis coverage extends up to 20 GeV/c in the forward production region (< 40 mrad) and covers angles up to 280 mrad at lower momentum.

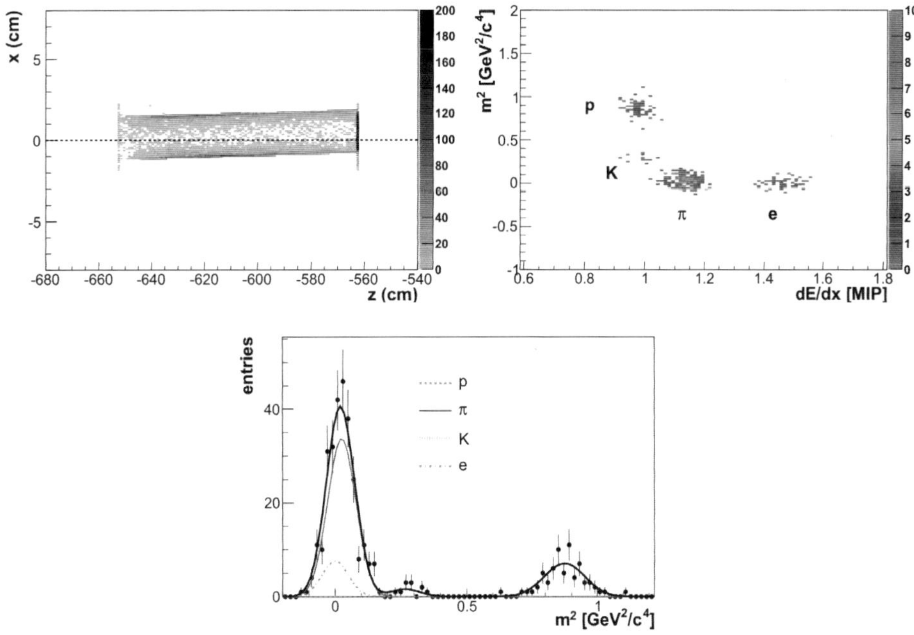

Figure 2. Distribution of the point-of-closest-approach on target in the $x - z$ projection [top left], data distribution in the $\{ToF, dE/dx\}$ phase space for $40 < \theta < 100$ mrad and $2.4 < p < 3.2$ GeV/c [top right], and projection of the fit result along the ToF axis [bottom].

The particle identification (PID) relies on the energy loss (dE/dx) measurement in the TPCs and the time-of-flight that is used to compute the particle mass squared. The combination of the two signals provides a powerful PID over a large momentum range. Yields of positively charged pions exiting the target are extracted in bins of (p, θ, z) using a 2d-likelihood minimization (see

Figure 2). Figure 3 shows that such yields can effectively be measured at the surface of the target.

The FLUKA simulation package [7] used in the T2K beam simulation was interfaced to the NA61/SHINE simulation chain for which a PID based on data parametrizations was furthermore implemented. Data and Monte-Carlo are processed identically with the same reconstruction chain and PID analysis. Thus, reconstructed yields of positively charged pions can be compared to reconstructed Monte-Carlo predictions prior to any corrections necessary to get absolute yields (e.g. all Monte-Carlo based corrections such as acceptance and decay corrections, reconstruction efficiency, etc). Such a comparison is depicted in Figure 3 with the corresponding re-weighting factors defined as the ratio of data to Monte-Carlo yields. Current uncertainties on the re-weighting factors are dominated by the statistical uncertainty which is typically 10% due to the very poor statistics of the 2007 pilot data. Systematic uncertainties are within 5-10%, the dominant source being PID (1-5%), followed by target misalignment (3%) and normalization (1.4%). Systematics from ToF efficiency, beam momentum and target density are all below 3%. More details on the NA61/SHINE long target data analysis and re-weighting procedure are given in [6].

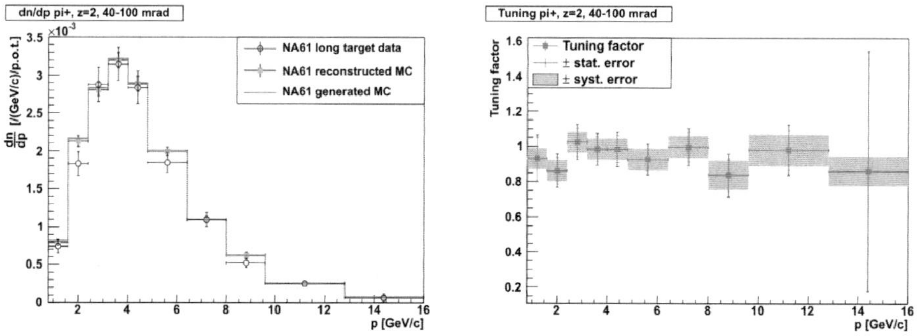

Figure 3. Data and Monte-Carlo yields of positively charged pions normalized to the bin size and to the number of protons on target over the second longitudinal bin of the target for $40 < \theta < 100$ mrad [left], and corresponding re-weighting factors [right].

Statistical uncertainties will be reduced with the 2009/2010 long target data sets, and further improvements are expected (in particular for the target alignment) to reduce the systematic uncertainties down to 5% or better. The pilot analysis of the 2007 NA61/SHINE long target data demonstrates that the T2K neutrino flux predictions can effectively be re-weighted with long target data. A first preliminary re-weighting of the ν_μ flux prediction at the far detector was already implemented and compared to the prediction based on the 2007 NA61/SHINE thin target data ([5], [6]).

References

[1] Itow Y *et al.* (T2K Collaboration) 2001 *Preprint* arXiv:0106019v1 [hep-ex]
[2] Abe K *et al.* (T2K Collaboration) 2011 *Nucl. Instrum. Meth.* A **659** 106-135
[3] Galymov V 2011 these proceedings
[4] Abe K *et al.* (T2K Collaboration) 2011 *Phys. Rev. Lett.* **107** 041801
[5] Abgrall N *et al.* (NA61/SHINE Collaboration) 2001 *Phys. Rev.* C **84** 034604
[6] Abgrall N 2011 PhD thesis CERN Document Server CERN-THESIS-2011-165 http://weblib.cern.ch/
[7] Fasso A, Ferrari A, Ranft J and Sala P R 2005 CERN-2005-10 INFN/TC_05/11 SLAC-R-773

Journal of Physics: Conference Series **408** (2013) 012051 doi:10.1088/1742-6596/408/1/012051

Hadron production simulation by FLUKA

G Battistoni[1], F Cerutti[2], A Ferrari[2], J Ranft[3], S Roesler[2] and P R Sala[1]

[1] INFN, Milan, Italy
[2] CERN, Geneva, Switzerland
[3] University of Siegen, Germany

E-mail: Francesco.Cerutti@cern.ch

Abstract. For the purposes of accelerator based neutrino experiments, the simulation of parent hadron production plays a key role. In this paper a quick overview of the main ingredients of the PEANUT event generator implemented in the FLUKA Monte Carlo code is given, together with some benchmarking examples.

1. Introduction

FLUKA [1, 2] is a general purpose tool for calculations of particle transport and interactions with matter, covering an extended range of applications spanning from proton and electron accelerator shielding to target design, calorimetry, activation, dosimetry, detector design, Accelerator Driven Systems, cosmic rays, neutrino physics, and radiotherapy. 60 different particles plus heavy ions can be transported by the code. The energy range covered for hadron-hadron and hadron-nucleus interaction is from threshold up to 10000 TeV, while electromagnetic and μ interactions can be dealt with from 1 keV (100 eV for photons) up to 10000 TeV. Nucleus-nucleus interactions are also supported up to 10000 TeV/n, thanks to the interfaces with a modified version of rQMD-2.4 [3, 4] (covering the intermediate energy range above $\sim$100 MeV/n) and with DPMJET-III [5] (invoked at high energies, that is above 5 GeV/n). At low energies, the BME event generator [6], which is part of the FLUKA code, applies. Neutron transport and interactions below 20 MeV and down to thermal energies are treated in the framework of a multi-group approach, with cross section data sets developed for FLUKA starting from standard evaluated databases (mostly ENDF/B, JENDL and JEFF).

Transport in arbitrarily complex geometries, including magnetic fields, can be accomplished using the FLUKA combinatorial geometry, supporting object repetition (lattice capability). A suitable voxel geometry module allows to model properly CT scans or other detailed 3D representations, e.g. of human beings, typically for dosimetry or therapy planning purposes.

The code has the ability to run either in fully analogue mode or in biased mode exploiting a rich variety of variance reduction techniques.

FLUKA is jointly developed by the European Organization for Nuclear Research (CERN) and the Italian National Institute for Nuclear Physics (INFN).

The approach to hadronic interaction modelling adopted in FLUKA has been detailed in several papers [7, 8, 9, 10, 11]. Hadron-nucleon inelastic collisions are described in terms of resonance production and decay up to a few GeV. At higher energies, a model [9] based on the Dual Parton Model (DPM) [12], that is a particular quark/parton string model providing reliable results up to several tens of TeV, takes over. In DPM, hadron-hadron interactions result in the creation of two or more QCD color strings, from which hadrons have to be generated.

Hadron-nucleus (h-A) interactions as modelled in the PEANUT (PreEquilibrium-Approach-to-NUclear-Thermalization) event generator of FLUKA, can be schematically described as a sequence of three main steps: i) Glauber-Gribov cascade and high energy collisions, ii) Generalized IntraNuclear cascade and iii) Preequilibrium stage, which are eventually followed by the Evaporation/Fragmentation/Fission phase ending through γ deexcitation.

The next section will briefly recall the interpretation of the Glauber approach in terms of quark chain and particle production diagrams, as well as the essential features of the PEANUT "nuclear environment", in particular the crucial formation zone concept. Its effect is manifestly displayed in the comparison with experimental data, together with the role of the Glauber multiple collision model.

2. Hadron-nucleus reaction modelling

The Glauber formalism [13, 14] allows to compute the scattering amplitude as well as all relevant cross sections for hadron–nucleus interactions, using the knowledge of elementary hadron–nucleon scattering and of the nuclear ground state only.

The Glauber multiple collision model can be formulated as a field theory (the so called Glauber-Gribov model [15, 16, 17]) based on general principles such as unitarity and analyticity [18]. The various multiple collision terms can be shown to be in a one-to-one correspondence with the various Feynman graphs describing the interaction.

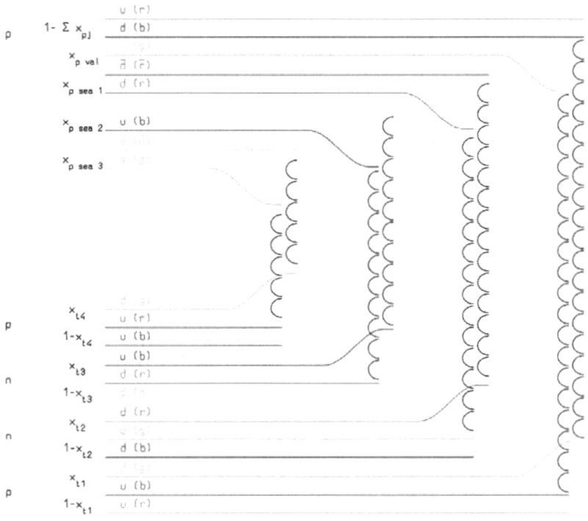

Figure 1. Leading two-chain diagrams in DPM for $p - A$ Glauber scattering with 4 collisions. The color (red, blue, and green) and quark combination shown in the figure is just one of the allowed possibilities.

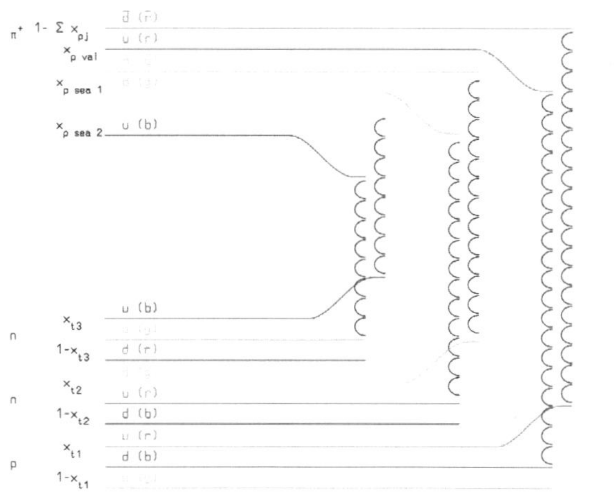

Figure 2. Leading two-chain diagrams in DPM for $\pi^+ - A$ Glauber scattering with 3 collisions. The color (red, blue, and green) and quark combination shown in the figure is just one of the allowed possibilities.

The Glauber-Gribov calculus allows to determine the probability distribution for the number n of target nucleons the projectile is going to interact with. In the DPM language these collisions are described by $2n$ chains, neglecting the contributions from higher order diagrams (referred to as multiple soft collisions), that is considering at most one pomeron exchange with each target nucleon. Two of these chains originate from the projectile valence quarks together with the valence quarks of one target nucleon (resulting in 2 valence-valence chains), whereas the other $2(n-1)$ chains stem from $(n-1)$ $q\bar{q}$ sea quarks of the projectile combined with the valence quarks of the other target nucleons ($2(n-1)$ sea-valence chains). Examples of this chain building process are reported in figures 1 and 2.

At energies high enough to consider coherent effects as corrections, h-A reactions are described as a cascade of two-body interactions, concerning the projectile and the reaction products. This mechanism is called IntraNuclearCascade (INC).

In PEANUT, particle trajectories in the nuclear medium are treated classically (curved in the mean nuclear+Coulomb field), with the nucleus divided into 16 radial zones of different density, plus 6 outside the nucleus to account for nuclear potential, plus 10 for charged particles. Respective nuclear densities and Fermi energies are adopted for neutrons and protons (the shell model ones for A<16). The interaction probability is calculated from σ_{free} taking into account the (density dependent) Fermi motion. Nonetheless, many significant quantum effects (Pauli blocking, formation zone, correlations...) are incorporated. Exact conservation of energy, momenta and all additive quantum numbers is assured on an event-by-event basis, including nuclear recoil.

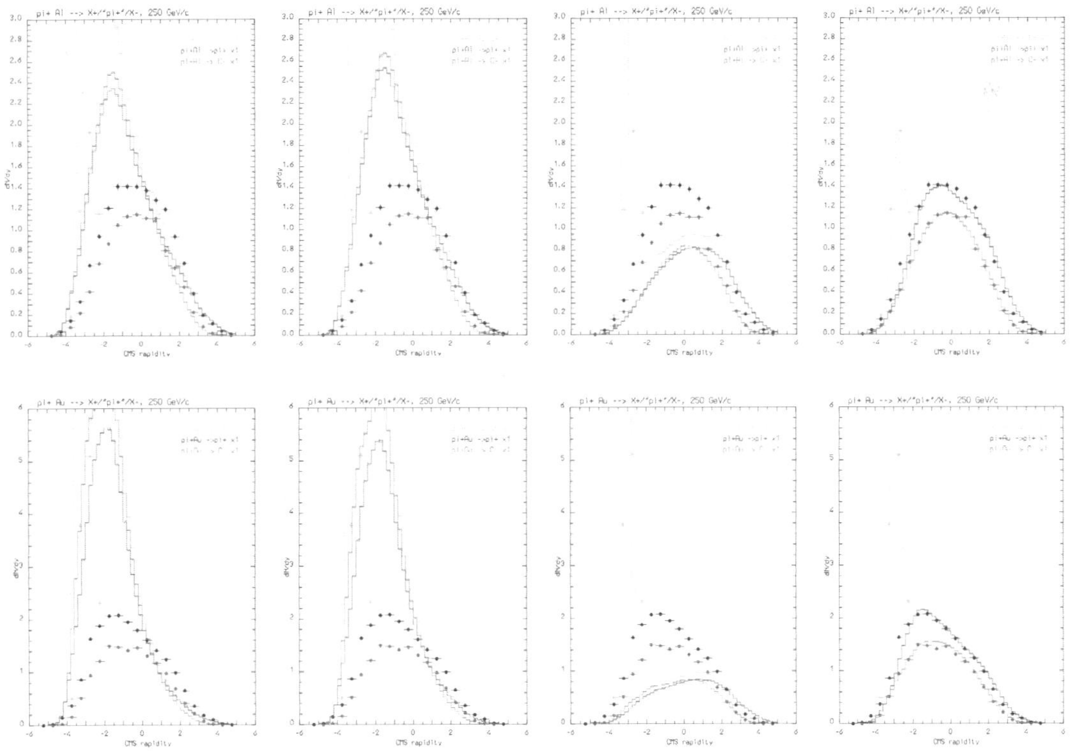

Figure 3. Computed (histograms) rapidity distributions of positive, negative and π^+ particles for 250 GeV/c π^+ on Aluminium (top) and Gold (bottom), compared with experimental data [19]. No formation zone and no Glauber multiple collisions are taken into account in the first column. In the second column formation zone is not taken into account, while Glauber multiple collisions are. In the third column formation zone is taken into account, while Glauber multiple collisions are not. Eventually, in the last column both formation zone and Glauber multiple collisions are taken into account.

The INC goes on in PEANUT until all nucleons are below a smooth threshold around 50 MeV, *and* all particles but nucleons (typically pions) have been emitted or absorbed. The nuclear configuration at this point is used as the starting condition for the preequilibrium stage, that follows the exciton formalism of M. Blann and coworkers [20], with some modifications [9].

The formation zone (or formation time) concept [21, 22] has been proposed since several years as a key factor in modelling high energy nuclear interactions. Formation zone is indeed essential in order to understand the observed reduction of the re-interaction probability with respect to the naive free cross section assumption. It can be understood as a "materialization" time. At high energies, the "fast" (from the emulsion language) particles produced in the Glauber cascade have a high probability to materialize already outside the nucleus without triggering a secondary cascade. Only a small fraction of the projectile energy is thus left available for the INC and the further stages.

Comparisons of calculated and experimental charged particle rapidity distributions, for $\pi^+ Al$ and $\pi^+ Au$ collisions at 250 GeV/c, are shown in figure 3, with Glauber multiple collisions and formation zone either activated or not.

3. Conclusion

We presented the FLUKA modeling of hadron-nucleus reactions relevant for the yield characterization of neutrino parents. The role of multiple Glauber collisions and formation zone was highlighted in comparison with experimental data.

References

[1] Ferrari A, Sala P R, Fassò A and Ranft J 2005 *FLUKA: a multi-particle transport code* (CERN 2005-10, INFN/TC_05/11, SLAC-R-773)
[2] Battistoni G, Muraro S, Sala P R, Cerutti F, Ferrari A, Roesler S, Fassò A and Ranft J 2007 *Proc. of the Hadronic Shower Simulation Workshop 2006 (Fermilab 6–8 September 2006)* ed M Albrow and R Raja (AIP Conf. Proc. 896) 31–49
[3] Sorge H, Stöcker H and Greiner W 1989 *Annals of Physics* **192** 266
[4] Andersen V *et al* 2004 *Advances in Space Research* **34** 1302
[5] Roesler S, Engel R and Ranft J 2001 *Proc. of the MonteCarlo 2000 Conf. (Lisbon, October 23-26 2000)* ed A Kling, F Barão, M Nakagawa, L Távora and P Vaz (Springer-Verlag Berlin) 1033–1038
[6] Cerutti F, Battistoni G, Capezzali G, Colleoni P, Ferrari A, Gadioli E, Mairani A and Pepe A 2006 *Proc. of the 11th International Conf. on Nuclear Reaction Mechanisms (Varenna)* ed E Gadioli (Ric. Scient. ed Educ. Perm., Univ. degli Studi di Milano S126) 507
[7] Fassò A, Ferrari A and Sala P R 2001 *Proc. of the MonteCarlo 2000 Conf. (Lisbon, October 23-26 2000)* ed A Kling, F Barão, M Nakagawa, L Távora and P Vaz (Springer-Verlag Berlin) 159
[8] Fassò A, Ferrari A, Ranft J and Sala P R 1995 *Proc. of the Specialists' Meeting on Shielding Aspects of Accelerators, Targets and Irradiation Facilities (Arlington, April 28-29 1994)* (OECD/NEA) 287–304
[9] Ferrari A and Sala P R 1998 *Proc. of Workshop on Nuclear Reaction Data and Nuclear Reactors Physics, Design and Safety (Trieste, April 1996)* ed A Gandini and G Reffo **2** 424
[10] Fassò A, Ferrari A, Ranft J and Sala P R 1997 *Proc. of SARE-3 (KEK-Tsukuba)* ed H Hirayama (KEK report Proceedings 97-5) 32
[11] Ballarini F *et al* 2003 *Proc. of the 10th International Conf. on Nuclear Reaction Mechanisms (Varenna)* ed E Gadioli (Ric. Scient. ed Educ. Perm., Univ. degli Studi di Milano S122) 579
[12] Capella A, Sukhatme U, Tan C I and Tran Thanh Van J 1994 *Phys. Rep.* **236** 225
[13] Glauber R J and Matthiae G 1970 *Nucl. Phys.* B **21** 135
[14] Glauber R J 1959 *High-energy collision theory (Lectures in Theoretical Physics)* ed A O Barut and W E Brittin (Interscience, NewYork)
[15] Gribov V N 1969 *Sov. Phys. JETP* **29** 483
[16] Gribov V N 1970 *Sov. Phys. JETP* **30** 709
[17] Bertocchi L 1972 *Nuovo Cimento* **11A** 45
[18] Weis J H 1976 *Acta Phys. Pol.* B **7** 851
[19] Agababyan N M *et al* 1991 *Z. Phys.* C **50** 361
[20] Blann M 1971 *Phys. Rev. Lett.* **27** 337; Blann M 1972 *Phys. Rev. Lett.* **28** 757; Blann M and Vonach H K 1983 *Phys. Rev.* C **28** 1475; Blann M 1983 *Phys. Rev.* C **28** 1648
[21] Stodolski L 1975 *Proc. of the 5th International Colloquium on Multiparticle Reactions (Oxford)* 577
[22] Ranft J 1988 *Phys. Rev.* D **37** 1842

Journal of Physics: Conference Series **408** (2013) 012052 doi:10.1088/1742-6596/408/1/012052

Predicting neutrino flux for the T2K experiment

Vyacheslav Galymov, for the T2K collaboration
Department of Physics, York University, Toronto, Canada, M3J 1P3
E-mail: slavic@yorku.ca

Abstract. The T2K experiment is a long-baseline neutrino oscillation experiment. One of its main goals is the discovery of the ν_μ to ν_e oscillations governed by the last unknown mixing angle θ_{13}. Additionally T2K aims to make precision measurements of Δm^2_{23} and θ_{23} oscillation parameters. The experiment uses the two detector approach where a near detector measures the neutrino flux close to the ν source, while a far detector measures the oscillated flux. To determine the oscillation parameters the extrapolation of the measurements at the near detector to the far detector has to be done. Such extrapolation relies on the predictions of the neutrino fluxes for the two detectors. In this paper the estimation of the neutrino flux predictions for the T2K experiment is briefly described.

1. Introduction

The T2K (Tokai to Kamioka) experiment [1] utilizes a 30 GeV proton beam from the Japan Proton Accelerator Research Complex (J-PARC) to conventionally produce an intense muon neutrino beam to study the neutrino oscillation phenomenon. One of its main goals is to discover the ν_μ to ν_e oscillations governed by the yet unknown θ_{13} mixing angle. The possibility that this parameter has a non-zero value has been suggested by the recent T2K result [2]. In addition, the experiment aims to make precise measurements of Δm^2_{23} and θ_{23} oscillation parameters. T2K is the first experiment to use an off-axis configuration [3] to get a narrow band ν_μ beam. The unoscillated neutrino flux is measured by a near detector, ND280, which intercepts neutrinos at $2 - 2.5^\circ$ relative to the beam axis (off-axis angle) and is located approximately 280 m away from the production site. The ν beam direction is measured by the INGRID detector placed on the beam axis at the same distance from the target as ND280. The T2K far detector, Super-Kamiokande (SK), located 295 km away from J-PARC and 2.5° off-axis, measures the flux after a significant fraction of ν_μ have undergone oscillations.

To ensure the stability of the off-axis neutrino beam, the proton beam position and angle at the target are controlled to better than 1 mm and 0.5 mrad, respectively. This translates to approximately 1 mrad uncertainty on the neutrino beam direction where a 1 mrad shift results in a 2% shift in the peak energy of the off-axis beam. The stability of the beam direction is further monitored by a muon monitor and the INGRID detector. The measurements of both detectors are consistent and indicate the stability of the ν beam direction is better than 1 mrad.

The analysis strategy of the T2K experiment is to determine the oscillation parameters by comparing the flux measurements at ND280 to the observations at SK. Due to its proximity to the neutrino source the energy distribution of neutrinos seen at the ND280 detector is not exactly the same as the one at SK. Flux predictions derived from beam Monte Carlo simulation are used to relate the two measurements to each other.

2. T2K neutrino beam Monte Carlo and flux predictions

The primary interactions of the proton beam in the graphite T2K target are treated by FLUKA 2008 (latest version at the time of the analysis) [4, 5]. The particles generated in these interactions are then exported for the subsequent transport through the beam-line to a GEANT3 [6] based simulation which uses GCALOR model [7] for hadronic interactions.

The expected fluxes from different parent types are shown in figure 1 for ν_μ at ND280 and SK as well as ν_e at SK. The ν_μ flux sharply peaks in the region 0.5 – 0.7 GeV, where the first oscillation maximum is located. As can be seen from figure 1 the majority of ν_μ come from pion parents. The same pions are responsible for production of muons that in turn produce a significant fraction of low energy ν_e, which are one of the most important backgrounds for the ν_e appearance signal. Understanding the pion production is therefore an essential first step for both the good sensitivity for the ν_e appearance searches and the precision measurement of the ν_μ disappearance signal.

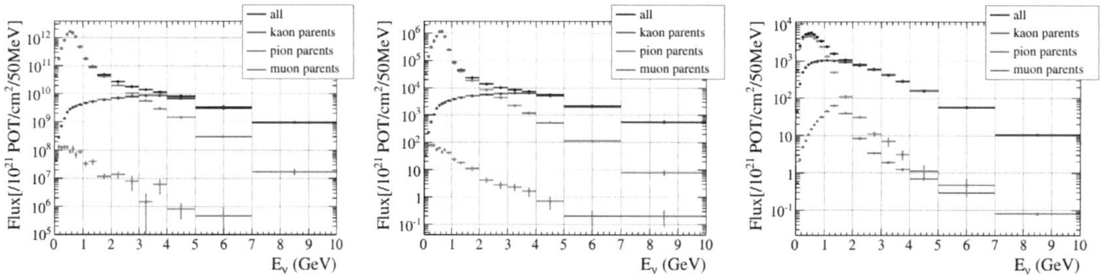

Figure 1. The contributions to the neutrino fluxes and near and far detectors from different neutrino parents. From left to right: ND280 ν_μ, SK ν_μ, and SK ν_e. The fluxes for SK are without the oscillation effects

One of the physics goals of the NA61/SHINE experiment is to provide precise measurements of the hadron production from a graphite target at the T2K beam energy. The experiment has collected data in 2007 and 2009-2010 running with a thin target (0.04 interaction lengths) as well as a replica of the T2K target (1.9 interaction lengths long target). The NA61 pion production results from the analysis of the 2007 run with the thin target [8] have been included in the analysis of the T2K data.

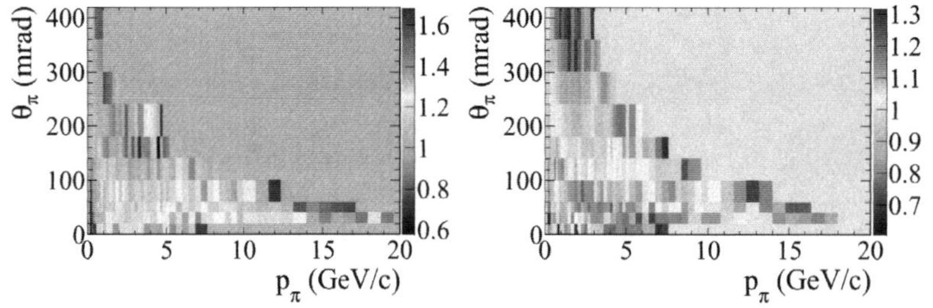

Figure 2. Data/FLUKA ratios of the production multiplicity for π^+ on the left and π^- on the right. Nominal FLUKA values are used outside of the data phase-space coverage

The T2K flux predictions are tuned using the NA61 thin target data by weighting the production of secondary pions—pions produced in the interactions of beam protons—inside

Journal of Physics: Conference Series **408** (2013) 012052 doi:10.1088/1742-6596/408/1/012052

the graphite target according to the ratio of the measured to FLUKA predicted multiplicity. Figure 2 shows such multiplicity ratios for π^+ and π^- for different values of pion momenta and production angles. The NA61 pion data is also used to tune the production multiplicity of tertiary pions. Since these pions are produced by protons of lower momentum, momentum scaling is needed. This is done by converting the multiplicity ratios in figure 2 from $p - \theta$ to $x_F - p_T$ phase-space and then determining appropriate weights according to the pion's x_F and p_T rather than p and θ.

In addition to tuning the pion phase-space distribution, the interaction rates of particles in a given material are also corrected. The particles produced in the forward direction may undergo absorption inside the long target. Escaping particles can also be absorbed in the interactions with the surrounding material which mostly consist of horn aluminum. The rate of these interactions is determined by the magnitude of the production cross-section for a given particle in a given material. While the agreement between FLUKA and existing production cross-section data was found to be adequate, significant discrepancy was observed between GCALOR and data. The particle interaction rates with aluminum outside of the target are therefore corrected.

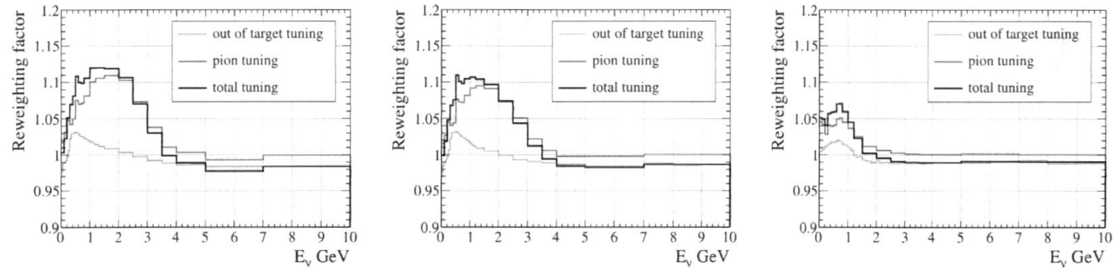

Figure 3. Re-weighting factors for tuning the beam MC as a function of the true neutrino energy. From left to right: ND280 ν_μ, SK ν_μ, and SK ν_e

The weights applied to different neutrino energy bins for ND280 ν_μ, SK ν_μ, and SK ν_e fluxes as a result of tuning the flux predictions based on the NA61 pion measurement and production cross-section data are displayed in figure 3.

3. Systematic uncertainties

Hadron production related uncertainties are the largest contribution to the overall systematic uncertainty on the flux predictions. The experimental errors on the NA61 measurement determines the uncertainty on the pion production in most of the T2K relevant pion phase-space. The uncertainties on kaon and secondary nucleon multiplicities are evaluated from comparison of FLUKA to the measurements with 24 GeV/c protons on beryllium target [9].

In addition to the hadron production uncertainties, other dominant sources of uncertainty are the off-axis angle uncertainty estimated from the INGRID measurement of the neutrino beam direction, proton beam parameters evaluated from the proton monitor measurements, uncertainty in the absolute value of the horn currents, and uncertainty in the alignment of the target and the horns. The fractional error envelopes for each error source are shown in figure 4 as a function of the neutrino energy.

The flux uncertainty on the predicted rate of ν_μ charged current events at ND280, R^{MC}_{ND280}, is 15.4%. The flux uncertainty on the predicted number of electron like events for SK, N^{MC}_{SK}, under $\theta_{13} = 0$ hypothesis (background only) is 16.1%. Given R^{Data}_{ND280}, the measured event rate at ND280, the flux uncertainty on the expected number of e-like events at SK

$$N^{exp}_{SK} = (N^{MC}_{SK} / R^{\mu,MC}_{ND280}) \times R^{Data}_{ND280}$$

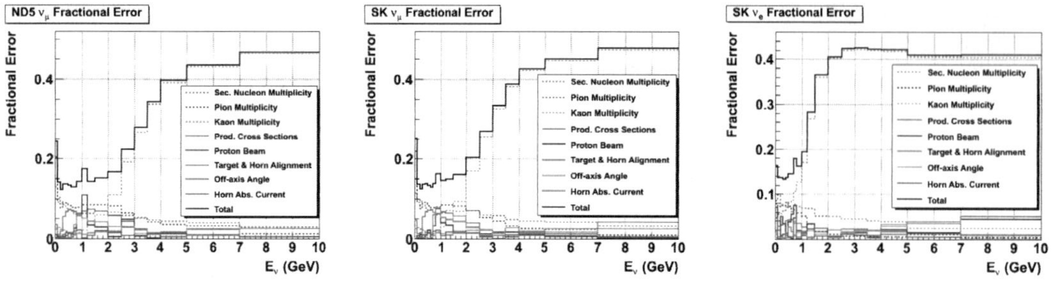

Figure 4. Fractional error as a function of the true neutrino energy. From left to right: ND280 ν_μ, SK ν_μ, and SK ν_e

is the uncertainty on the $N_{SK}^{MC}/R_{ND280}^{MC}$ ratio which leads to a cancellation of some of the flux systematic uncertainties between the two detectors. This results in a smaller flux related uncertainty on N_{SK}^{exp} of only 8.5%.

4. Consistency with ND280 and INGRID measurements

The ratio of the measured to the predicted charged current ν_μ event rate for ND280 is

$$R_{ND280}^{Data}/R_{ND280}^{MC} = 1.036 \pm 0.028(\text{stat.})^{+0.044}_{-0.037}(\text{det. syst.}) \pm 0.038(\text{phys. syst.})$$

The ratio of the event rates for INGRID detector is

$$R_{INGRID}^{Data}/R_{INGRID}^{MC} = 1.057 \pm 0.001(\text{stat.}) \pm 0.040(\text{syst.})$$

where the systematic uncertainties quoted do not include the flux and neutrino cross-section uncertainties. Both near detectors show a good agreement with the expectations from the tuned flux.

5. Conclusions

The T2K flux predictions are based on and validated by external hadron production data. As the majority of the neutrinos come from pion parents, the pion production measurements from the NA61 experiment are of particular importance for the reliable flux calculations. The current T2K flux estimations make use of these measurements as well as and existing cross-section data to tune model predictions. A good agreement is found between the event rates predicted from the tuned flux and the measurements performed at the ND280 and INGRID detectors. The uncertainties on the predicted flux are evaluated resulting in an 8.5% uncertainty on the number of the expected ν_e events at SK from background sources. The kaon production results and the measurements of the particle yields from the T2K replica target by the NA61 experiment are expected to further reduce the flux uncertainties.

References

[1] Abe K *et al.* 2011 *Nucl. Instr. Meth.* A **659** 106
[2] Abe K *et al.* 2011 *Phys. Rev. Lett.* **107** 041801
[3] Beavis D *et al.* 1995 BNL-52459
[4] Battistoni G *et al.* 2007 *AIP Conf. Proc.* **896** 31
[5] Fasso A *et al.* 2005 CERN-2005-10 and SLAC-R-773
[6] Brun R *et al.* 1994 CERN-W5013
[7] Zeitnitz C and Gabriel T A 1994 *Nucl. Instr. Meth.* A **349** 106
[8] Abgrall N *et al.* 2011 *Phys. Rev.* C **84**, 034604
[9] Eichten T *et al.* 1972 *Nucl. Phys.* B **44** 333

Neutrino nucleus reactions within the GiBUU model

O Lalakulich, K Gallmeister, U Mosel

Institut für Theoretische Physik, Universität Giessen, Germany

E-mail: Olga.Lalakulich@theo.physik.uni-giessen.de

Abstract. The GiBUU model, which implements all reaction channels relevant at medium neutrino energies, is used to investigate the neutrino and antineutrino scattering on iron. Results for integrated cross sections are compared with NOMAD and MINOS collaborations data. It is shown, that final state interaction can noticeably change the spectra of the outgoing hadrons. Predictions for the MINERνA experiment are made for pion spectra, averaged over NuMI neutrino and antineutrino fluxes.

1. Introduction

Neutrino and antineutrino scattering on nuclei for neutrino energies above 30 GeV was studied in several experiments starting from the 80s. Theoretically they were successfully described within the quark parton model as Deep Inelastic Scattering (DIS) processes. Recent measurements by MINOS and NOMAD collaborations also covered the intermediate energy region. Here the neutrino reactions are not so easy to model, because of the overlapping contributions from QE scattering, resonance production and background processes. This requires complex approaches that take all of the relevant channels into account.

Nuclear effects in neutrino reactions can also be studied in detail nowadays. The MINERνA experiment intends to perform measurements on Plastic (CH), Iron, Lead, Carbon, Water and liquid Helium targets in the NuMI beam, which would directly allow to compare nuclear effects on various nuclei. Besides muon detection, this experiment will also be able to resolve various final states by identifying the tracks of the outgoing hadrons. Theoretical modelling of such exclusive reactions requires realistic approaches to the desctiption of the initial and final state interactions in target nuclei.

Both these requirements are satisfied by the GiBUU transport model [1, 2], which we use here to study neutrino and antineutrino scattering on iron. Our results are compared with the recent MINOS and NOMAD collaborations data. Predictions are also made for the spectra of the outgoing particles. The calculations are done without any fine tuning to the data covered here with the default parameters as used in the GiBUU framework.

2. GiBUU transport model

The GiBUU model was initially developed as a transport model for nucleon-, nucleus-, pion-, and electron- induced reactions from some hundreds MeV up to tens of GeV. Several years ago neutrino-induced interactions were also implemented for the energies up to a few GeV. Recently the code was extended to describe also the DIS processes in neutrino reactions. Thus, we can study all kind of elementary collisions on all kind of nuclei within a unified framework.

The model is based on well-founded theoretical ingredients and has been tested against various nuclear reactions. For a detailed review of the GiBUU model see [2].

GiBUU describes all processes relevant at medium energies, the cross section is calculated as $\sigma_{\rm tot} = \sigma_{QE} + \sigma_{RES} + \sigma_{BG} + \sigma_{DIS}$. Our approach to quasielastic (QE) scattering, resonance (RES) production and background (BG) processes is described in [3, 4]. The DIS scattering is included as PYTHIA simulation.

In the region of the shallow inelastic scattering, that is at moderate invariant masses, 1.6 GeV $< W <$ 2.0 GeV, there is a potential problem of double counting. Here the same physical events can be considered as originating from decays of high mass baryonic resonances or from DIS. In the GiBUU code we use the ansatz, that both RES, BG and DIS processes contribute in this region. While the RES and BG contributions are smoothly switched off in this region and DIS contribution is switched on. In essence, the DIS processes at $W <$ 2 GeV account for the resonances whose electromagnetic properties are not known and for the non-resonant processes giving several mesons in the final state beyond the one pion background. With this choice, the DIS events become noticeable at neutrino energies around 3 GeV.

3. Integrated cross sections.

As it was measured in 80s, the DIS cross section grows linearly with energy. So at high neutrino energies the data are conveniently presented as cross section per energy $\sigma_{\rm tot}/E_\nu$. Despite the measurements being made on nuclear targets, the world average values are given [5] for isoscalar target: $\sigma_{\rm tot}/E_\nu(\nu) = 0.667 \pm 0.014 \cdot 10^{-38}$ cm^2/ GeV for neutrinos and $\sigma_{\rm tot}/E_{\bar\nu}(\bar\nu) = 0.334 \pm 0.008 \cdot 10^{-38}$ cm^2/ GeV for antineutrinos. Such an approach is meaningful, only if nuclear corrections are neglected.

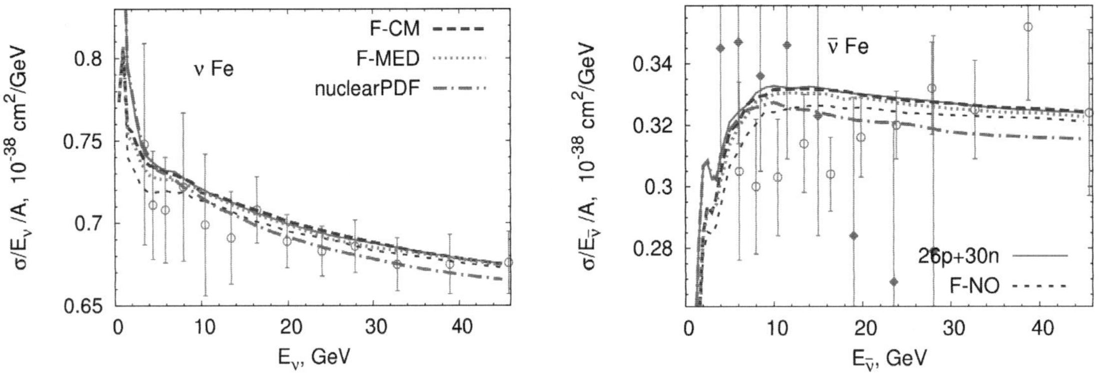

Figure 1. Total cross section per nucleon for neutrino (left) and antineutrino (right) induced reactions on iron target. Various prescriptions to include nuclear effects in DIS are compared. MINOS collaboration data (○) are from [6, 7], IHEP-JINR collaboration data (♦) are from [8].

The actual value of such corrections for neutrino reactions is not known so far, because of both experimental inaccuracies and difficulties in the theoretical description. On one hand, nuclear parton distributions, based on electromagnetic scattering data and intended for description of both charged lepton and neutrino reactions, were introduced. For a review and a list of recent parametrizations see, for example, [9]. On the other hand, recent investigations [10, 11] showed, that in neutrino reactions nuclear corrections to parton distributions are at the same level as for electrons, but have a very different dependence on the Bjorken x variable. The topic remains controversial, with the hope that future precise MINERνA collaboration results on various targets will clarify the situation.

As we already mentioned, the GiBUU code uses PYTHIA for the simulation of DIS processes. In the GiBUU simulation the neutrino interacts with one initial nucleon, bound in the hadronic potential and having nonzero Fermi momentum. In order to be able to use the PYTHIA event simulator, we have to provide some quasi-free kinematics as inputs to PYTHIA. Various prescriptions to do this result in a $3-4\%$ difference in the results. The corresponding cross sections (denoted as "F-NO", "F-CM", "F-THRE") are shown in Fig 1. Nuclear parton distribution functions from [12] are also implemented as one of the options to use (to avoid double counting, nuclear potential and Fermi motion in such calculations are switched off). The result ("nuclearPDF") as well as the free cross section for iron composition ("26p+30n") are also shown in Fig 1. At the moment we consider the various prescriptions mentioned above as intrinsic uncertainty of the GiBUU code, reflecting the lack of our understanding of the nuclear effects. No other event generator, as far as we know, accounts for nuclear corrections in high–energy neutrino reactions.

Fig. 1 shows, that our calculations are in a good agreement with the recent neutrino data, which are also consistent with each other. This figure shows that the decreasing slope of our curves for the neutrino cross section is in agreement with that of the data. This slope is not taken into account in deriving the world-average value, where it was assumed to be negligible. For antineutrino the agreement is good for $E_{\bar{\nu}} > 25$ GeV. For lower energies our curve is above the recent MINOS collaboration data, but below the IHEP-JINR results [8]. The overall agreement of our calculations with the data is therefore better than the agreement of the data with each other.

4. Final state interactions and change of the final hadronic spectra.

After being produced in the initial interaction, outgoing hadrons propagate throughout the nucleus. In GiBUU this process of final state interactions (FSI) is modeled by solving the semi-classical Boltzmann-Uehling-Uhlenbeck equation. It describes the dynamical evolution of the phase space density for each particle species under the influence of the mean field potential, introduced in the description of the initial nucleus state. Equations for various particle species are coupled through this mean field and also through the collision term. This term explicitly accounts for changes in the phase space density caused by elastic and inelastic collisions between particles. FSI decrease the cross sections as well as significantly modify the shapes of the final particle spectra. Such change was seen, for example, in photo-pion production [13] and is described by the GiBUU with a good accuracy. A similar change should be observed in neutrino reactions.

Fig. 2 shows the π^+ (left panels), π^0 (middle panels) and π^- (right panels) spectra for neutrino (upper panels) and antineutrino (lower panels) NuMI fluxes for 1-pion events. Since MINOS [14] collaboration considered only events with the energy of the outgoing muon $E_\mu > 1.5$ GeV, we apply the same cut in our calculations. The green dashed lines show the kinetic energy (T_π) distributions before FSI, i.e. of pions produced in the initial neutrino vertex. The solid red lines show the distributions after FSI, i.e. of pions that made it out of the nucleus. Such spectra can also be calculated for any other outgoing particles (protons, kaons, eta) for any predefined final state and should be measurable in MINERνA experiment.

For dominant channels (π^+ production for neutrino reactions and π^- in antineutrino ones), the FSI decrease the cross section at $T_\pi > 0.2$ GeV. This is mainly explained by pion absorption through $\pi N \to \Delta$ following by $\Delta N \to NN$. Pion elastic scattering in the FSI also decreases the pion energy, thus depleting the spectra at higher energies and accumulating strength at lower energies. Thus, an increase of the cross sections is observed at $T_\pi < 0.15$ GeV; altogether this leads to a significant change of the shape of the spectra.

Scattering can also lead to pion charge exchange. For neutrino-induced reactions, the $\pi^+ n \to \pi^0 p$ scattering in the FSI is the main source of side–feeding for the π^0 channel, leading

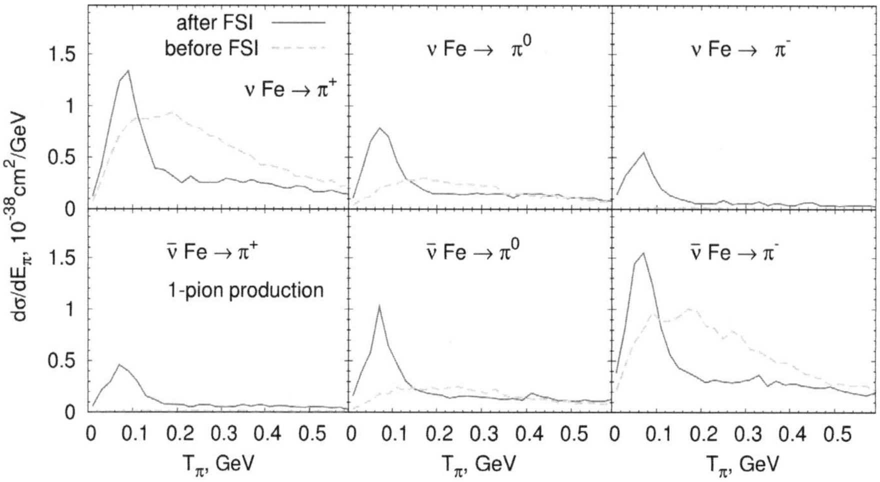

Figure 2. Pion kinetic energy distributions for neutrino and antineutrino induced reactions for 1-pion production (one pion of a given charge and no other pions are produced). Calculations are for the NuMI low-energy mode neutrino/antineutrino fluxes.

to a noticeable increase of the π^0 cross section at low T_π. The inverse feeding is suppressed, because less π^0 than π^+ are produced at the initial vertex. The same mechanism of side feeding from dominant to sub-dominant channel through $\pi^- p \to \pi^0 n$ is working for antineutrino induced reactions.

For the least dominant channel (π^- production for neutrino reactions and π^+ in antineutrino ones), the FSI (in particularly side feeding) represent the main source of the events observed; thus a dramatic FSI effect.

Acknowledgment

This work is supported by DFG. O.L. is grateful to Ivan Lappo-Danilevski for programming assistance.

Reference

[1] `http://gibuu.physik.uni-giessen.de/GiBUU`
[2] Buss O *et al.* 2012 *Phys. Rept.* **512** 1–124 (*Preprint* 1106.1344 [hep-ph])
[3] Leitner T, Alvarez-Ruso L and Mosel U 2006 *Phys. Rev. C* **73** 065502 (*Preprint* nucl-th/0601103)
[4] Leitner T, Buss O, Alvarez-Ruso L and Mosel U 2009 *Phys. Rev. C* **79** 034601 (*Preprint* 0812.0587 [nucl-th])
[5] Amsler C *et al.* (Particle Data Group) 2008 *Phys. Lett. B* **667** 1
[6] Wu Q *et al.* (NOMAD) 2008 *Phys. Lett. B* **660** 19–25 (*Preprint* 0711.1183 [hep-ex])
[7] Adamson P *et al.* (MINOS) 2010 *Phys. Rev. D* **81** 072002 (*Preprint* 0910.2201 [hep-ex])
[8] Anikeev V B *et al.* 1996 *Z. Phys. C* **70** 39–46
[9] Hirai M, Kumano S and Saito K 2009 *AIP Conf. Proc.* **1189** 269–275 (*Preprint* 0909.2329 [hep-ph])
[10] Schienbein I *et al.* 2008 *Phys. Rev. D* **77** 054013 (*Preprint* 0710.4897 [hep-ph])
[11] Kovarik K *et al.* 2011 *Phys. Rev. Lett.* **106** 122301 (*Preprint* 1012.0286 [hep-ph])
[12] Eskola K J, Kolhinen V J and Salgado C A 1999 *Eur. Phys. J. C* **9** 61–68 (*Preprint* hep-ph/9807297)
[13] Krusche B *et al.* 2004 *Eur. Phys. J. A* **22** 277–291 (*Preprint* nucl-ex/0406002)
[14] Aguilar-Arevalo A *et al.* (MiniBooNE Collaboration) 2008 *Phys. Rev. Lett.* **100** 032301 (*Preprint* 0706.0926 [hep-ex])

XIII International Workshop on Neutrino Factories, Super beams and Beta beams (NUFACT11) IOP Publishing
Journal of Physics: Conference Series **408** (2013) 012054 doi:10.1088/1742-6596/408/1/012054

How can we use neutrino nucleus interactions as a probe of the strong interaction?

Jorge G Morfin

Particle Physics Division/Neutrino Department, Fermi National Accelerator Laboratory, Batavia, IL 60510, USA

E-mail: morfin@fnal.gov

Abstract. Neutrino scattering experiments have been studying QCD for around 40 years. An example of the more recent studies of QCD with neutrinos is the NuTeV ν-Fe experiment at Fermilab. The problem the community faces in trying to study QCD with neutrino data is that there is no experimentally verified way to convert neutrino-nucleus (for example, Fe) results to the equivalent neutrino-nucleon values making it difficult to combine neutrino nucleus scattering data in QCD global fits to extract parton distribution functions. This is particularly significant since there are now indications that nuclear effects in neutrino nucleus interactions might be different than those measured in charged-lepton nucleus scattering. To better understand this situation, the MINERνA neutrino-nucleus scattering experiment at Fermilab, a collaboration of elementary-particle and nuclear physicists, is systematically studying neutrino nuclear effects off of He, C, O, Fe and Pb for a more thorough A-dependent study of nuclear PDFs and these correction factors.

1. Introduction

Neutrino scattering plays a crucial role in extraction of fundamental parton distribution functions (PDFs). These PDFs describe parton constituents of protons and other hadrons, and (in the $\overline{MS}$ convention) are precisely defined in terms of operator matrix elements. The necessity of neutrino measurements is obvious, because only neutrinos can resolve the flavor of the nucleon's constituents: the ν interacts with d, s, $\overline{u}$ and $\overline{c}$ while the $\overline{\nu}$ interacts with u, c, $\overline{d}$ and $\overline{s}$. The weak current's unique ability to "taste" only particular quark flavors significantly enhances the study of parton distribution functions.

Large samples, and dedicated effort to minimizing beam-related systematics will allow neutrino experiments to independently isolate all the structure functions $F_1^{\nu N}(x,Q^2)$, $F_1^{\overline{\nu} N}(x,Q^2)$, $F_2^{\nu N}(x,Q^2)$, $F_2^{\overline{\nu} N}(x,Q^2)$, $xF_3^{\nu N}(x,Q^2)$ and $F_3^{\overline{\nu} N}(x,Q^2)$ for the first time. By taking differences and sums of these structure functions, specific parton distribution functions in a given (x,Q^2) bin can in turn be determined. With the manageable systematic uncertainties expected, neutrino experiments can dramatically improve the isolation of individual PDFs by measuring the full set of ν and $\overline{\nu}$ structure functions. Extracting this full set of structure functions will rely on the y-variation of the structure function coefficients in the expression for the cross-section.

In the helicity representation, for example:

$$\frac{d^2\sigma^\nu}{dxdQ^2} = \frac{G_F^2}{2\pi x}[\frac{1}{2}(F_2^\nu(x,Q^2) + xF_3^\nu(x,Q^2)) + \frac{(1-y^2)}{2}(F_2^\nu(x,Q^2) - xF_3^\nu(x,Q^2)) - 2y^2 F_L^\nu(x,Q^2)] \quad (1)$$

Journal of Physics: Conference Series **408** (2013) 012054 doi:10.1088/1742-6596/408/1/012054

By analyzing the data as a function of (1-y^2) in a given (x,Q^2) bin, all six structure functions can be extracted[1].

2. Neutrino iron scattering results

To acquire significant statistics, the use of heavy nuclear targets is unavoidable due to the weak nature of the neutrino interactions. This complicates the extraction of free nucleon PDFs because corrections must be applied to the data to convert from the nucleus A to a nucleon. The results of the latest study of QCD using neutrino scattering comes from the NuTeV experiment [1]. The NuTeV experiment accumulated over 3 million ν and $\overline{\nu}$ events in the energy range of 20 to 400 GeV off a manly Fe target. The main points are that the NuTeV cross section agrees with the CCFR values (obtained using the same detector) for values of $x_{Bj} \leq 0.4$ but is systematically higher for larger values of x_{Bj} . NuTeV agrees with charged lepton data for $x_{Bj} \leq 0.5$ but there is increasing disagreement for higher values. Although NuTeV F_2 and xF_3 agree with theory for medium x, they find a different Q^2 behavior at small x and are systematically higher than theory at high x. These results can be summarized in four main questions to ask subsequent neutrino experiments:

- At high x, what is the behavior of the valence quarks as x $\rightarrow$ 1.0?
- At low W, what is happening in the transition region between resonance production and the DIS regions?
- At all x and Q^2, what is yet to be learned if we can measure all six ν and $\overline{\nu}$ structure functions to yield maximal information on the parton distribution functions?
- At all x, how do nuclear effects with incoming neutrinos differ from nuclear effects with incoming charged leptons?

This last item highlights an overriding question when trying to get a global view of structure functions from both neutrino and charged-lepton scattering data. How do we compare data off nuclear targets with data off nucleons and, the associated question, how do we scale nuclear target data to the comparable nucleon data.
A recent study [1] analyzed the impact of new data sets from the NuTeV [3], CHORUS and E-866 Collaborations on the PDFs. This study found that the NuTeV data set (together with the model then used for the nuclear corrections) pulled against several of the other data sets, notably the E-866, BCDMS and NMC sets. Reducing the nuclear corrections at large values of x reduced the severity of this pull and resulted in improved χ^2 values. These results suggested on a purely phenomenological level that the appropriate nuclear corrections for ν-DIS may well be smaller than the assumed $l^{\pm}$-nucleus corrections.
To investigate this question further, the data from the high-statistics ν-DIS experiment, NuTeV, was used to perform a dedicated PDF fit to neutrino-iron data [2]. The methodology for this fit is parallel to that of the previous global analysis [1], but with the difference that only Fe data has been used and no nuclear corrections have been applied to the analyzed data; hence, the resulting PDFs are for a bound proton in an iron nucleus. Specifically, the iron PDFs were determined using the recent NuTeV differential neutrino (1371 data points) and anti-neutrino (1146 data points) DIS cross section data [3], as well as the NuTeV/CCFR dimuon data (174 points) which are sensitive to the strange quark content of the nucleon. Kinematic cuts of $Q^2 \geq 2$ GeV^2 and W$\geq$ 3.5 GeV were imposed, and a good fit with a χ^2 of 1.35 per data point was obtained [2].

[1] Note that for this type of parton distribution function study, anti-neutrino running will be essential.

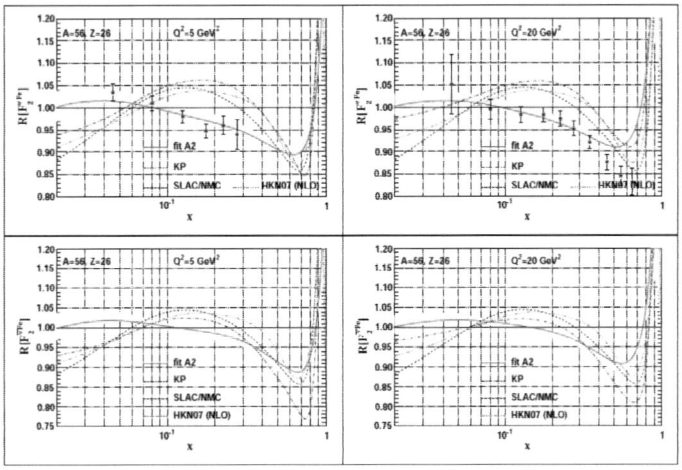

Figure 1. Nuclear correction factor R for the structure function F_2 in neutrino and anti-neutrino scattering from Fe for $Q^2 = \{5,20\}$ GeV^2. The solid curve shows the result of the analysis of NuTeV data; the uncertainty from the fit is represented by the shaded (yellow) band. For comparison, the correction factor from the Kulagin-Petti model (— · —) KP, HKN07 (— · · —) Kumano, and the SLAC/NMC parametrization (– – –) are shown.

3. Nuclear correction factor

By comparing these iron PDFs with the free-proton PDFs (appropriately scaled) a neutrino-specific heavy target nuclear correction factor R can be obtained which should be applied to relate these two quantities. In addition to kinematic variables, R can depend on the observable under consideration simply because different observables may be sensitive to different combinations of PDFs. For example, the nuclear correction factor R for F_2^A and F_3^A will, in general, be different. The nuclear correction factors for $F_2^{\nu Fe}$ and $F_2^{\overline{\nu} Fe}$ at $Q^2 = 5$ GeV^2 and 20 GeV^2 derived in this analysis and labeled A2 are shown in Fig 1.

The SLAC/NMC curve in the figures has been obtained from an A and Q^2-independent parametrization of calcium and iron charged-lepton DIS data [1]. The curves labeled "KP" are from the Kulagin-Petti model [4] constructed specifically to compute the nuclear effects in neutrino-nucleus interactions. Due to the neutron excess in iron, both the A2 and the KP curves differ when comparing scattering for neutrinos and anti-neutrinos.

Although the results of this analysis have general features in common with the KP model and the SLAC/NMC parametrization, the magnitude of the effects and the x-region where they apply are quite different. The present results are noticeably flatter than the KP and SLAC/NMC curves, especially at moderate-x where the differences are significant. The general trend we see when examining these nuclear correction factors is that the anti-shadowing region is shifted to smaller x values, and any turn-over at low x is minimal given the PDF uncertainties. More specifically, there is no indication of "shadowing" in the NuTeV neutrino results. In general, these plots suggest that the size of the nuclear corrections extracted from the NuTeV data are smaller than those obtained from charged lepton scattering (SLAC/NMC) or from the set of data used in the KP model.

4. Continuing the study of neutrino nucleus interactions

While the nuclear corrections extracted from charged current ν-Fe scattering have similar characteristics as the $l^{\pm}$-Fe charged-lepton results, the detailed x and Q^2 behavior is quite different. This observation raises the deeper question as to whether the neutrino and charged-

lepton nuclear correction factors may be substantially different. The CTEQ study of the iron PDFs provides a foundation for a general investigation (involving a variable A parameter) that can definitively address this topic. Resolving these questions is essential if we are to reliably use nuclear data to obtaining free-proton PDFs which form the basis of the LHC analyses.
Experimentally, the MINERνA experiment, a collaboration of elementary-particle and nuclear physicists, is performing a hight-statistics, systematic study of neutrino nucleus interactions. The overall goals of the experiment are to measure absolute exclusive cross-sections and study nuclear effects in ν-A interactions with He, C, O, Fe and Pb nuclear targets. For QCD oriented studies, they are planning systematic studies of the resonance-DIS transition region and the low Q^2 DIS region including the extraction of high-x_{Bj} parton distribution functions.

5. Conclusions

Using the results from the NuTeV neutrino Fe experiment, nuclear effects of charged current deep inelastic neutrino-iron scattering have been studied in the frame-work of a χ^2 analysis of parton distribution functions (PDFs). A set of iron PDFs has been extracted which is then used to compute x_{Bj}-dependent and Q^2-dependent nuclear correction factors for iron structure functions which are required in global analyses of free nucleon PDFs. Upon comparing our results with nuclear correction factors from neutrino-nucleus scattering models and correction factors for $l^{\pm}$-iron scattering we find that, except for very high x_{Bj}, our correction factors differ in both shape and magnitude from the correction factors of the models and charged-lepton scattering. There is no a priori reason to expect them to be the same. On the contrary, with the introduction of the axial-vector current in neutrino scattering, it would be difficult to understand if they were exactly the same. What was unexpected was the degree to which the R factors differ between ν-Fe scattering compared to $l^{\pm}$-Fe charged-lepton results. In particular the lack of evidence for shadowing in neutrino scattering, confirmed by [5] for NuTeV data, is quite surprising.
The nuclear correction factors R are being measured over a wider range of A and with reduced errors by the MINERνA experiment in the NuMI beam in the near future.

References

[1] Owens J F *et al.* 2007 *Phys. Rev.* D **75** 054030 (*Preprint* hep-ph/0702159)
[2] Schienbein I, Yu Y, Keppel C, Morfin J G, Olness F and Owens J F 2008 *Phys. Rev.* D **77** 054013 (*Preprint* arXiv:0710.4897[hep-ph])
[3] Tzanov M *et al.* (NuTeV Collaboration) 2006 *Phys. Rev.* D **74** 012008
[4] Kulagin S A and Petti R 2006 *Nucl. Phys.* A **765** 126 (*Preprint* hep-ph/0412425)
[5] Serge Kulagin, private communication

γ-ray production in neutral-current neutrino-oxygen interactions at energies above 200 MeV

A M Ankowski[1,2], O Benhar[1], T Mori[3], R Yamaguchi[3] and M Sakuda[3]

[1] INFN and Department of Physics,"Sapienza" Università di Roma, I-00185 Roma, Italy
[2] On leave from Institute of Theoretical Physics, University of Wrocław, Wrocław, Poland
[3] Department of Physics, Okayama University, Okayama 700-8530, Japan

E-mail: Artur.Ankowski@roma1.infn.it

Abstract. We report the results of a calculation of the neutrino- and antineutrino-induced γ-ray production cross section for oxygen target. Our analysis is focused on the kinematical region of neutrino energy larger than ~200 MeV, in which single-nucleon knockout is known to be the dominant reaction mechanism. The numerical results have been obtained using a realistic model of the target spectral function, extensively tested against electron-nucleus scattering data. We find that at neutrino energy 600 MeV the fraction of neutral-current interactions leading to emission of γ-rays of energy larger than 6 MeV is ~41%, and that the contribution of the $p_{3/2}$ state is overwhelming.

The observation of γ rays originating from nuclear deexcitation can be exploited to identify neutral-current (NC) neutrino-nucleus interactions in a broad energy range. Following the pioneering studies of nuclear excitations by neutral weak currents of Refs. [1, 2], theoretical calculations of the cross section of γ-ray production from NC neutrino-oxygen interactions have been carried out in the neutrino energy range $E_\nu \sim$ 10–500 MeV [3, 4, 5].

Neutrons, while providing ~50% of NC events, do not emit Cherenkov light. As a consequence, the availability of an alternative signal allowing one to identify NC interactions is very important. Events with γ rays of energy above the observational threshold of 5 MeV can be detected in a water Cherenkov detector, like Super-Kamiokande, and contribute up to ~5% of the total event number [6, 7], independent of neutrino oscillations. Note that in water ~90% (16 out of 18) of the NC interactions take place in oxygen.

At low energy, elastic scattering and inelastic excitation of discrete nuclear states provide the main contribution to the neutrino-nucleus cross section. However, at $E_\nu \gtrsim$ 200 MeV the cross section associated with these processes tends to saturate, and quasielastic (QE) nucleon knockout becomes the dominant reaction mechanism. If the residual nucleus is left in an excited state, these processes can also lead to γ-ray emission.

In the QE regime, neutrino-nucleus scattering reduces to the incoherent sum of elementary scattering processes involving individual nucleons, the energy and momentum of which are distributed according to the target spectral function [8]. A schematic representation of NC QE neutrino-nucleus scattering is given in Fig. 1, where the dashed line represents the threshold for nucleon emission in the continuum.

In this paper, we discuss the emission of γ rays arising from the decay of the residual nuclei of the reactions $\nu + {}^{16}_{8}\mathrm{O} \rightarrow \nu + p + {}^{15}_{7}\mathrm{N}^*$, or $\nu + {}^{16}_{8}\mathrm{O} \rightarrow \nu + n + {}^{15}_{8}\mathrm{O}^*$, the cross sections of

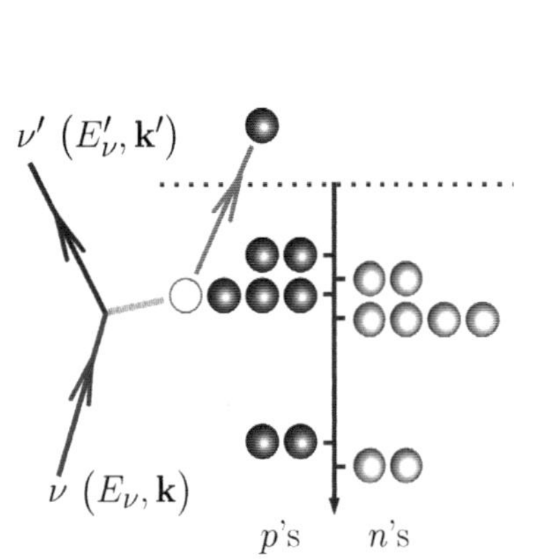

Figure 1. Schematic representation of neutral-current neutrino scattering off $^{16}_{8}$O.

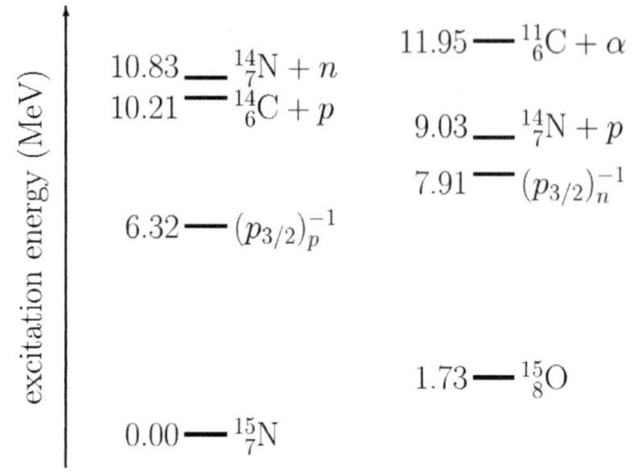

Figure 2. Low-lying excited levels of the residual nuclei produced in $^{16}_{8}\mathrm{O}(\nu, \nu' N)$ scattering.

which have been computed using a realistic model of the oxygen spectral function. Note that the presented formalism applies also to antineutrino-induced γ-ray production.

In our approach, covered in detail in Ref. [9], the cross section of γ-ray production following a NC QE interaction, σ_γ, is written in the form

$$\sigma_\gamma \equiv \sigma(\nu + {}^{16}_{8}\mathrm{O} \to \nu + \gamma + Y + N) = \sum_\alpha \sigma(\nu + {}^{16}_{8}\mathrm{O} \to \nu + X_\alpha + N)\, \mathrm{Br}(X_\alpha \to \gamma + Y), \quad (1)$$

where N is the knocked out nucleon, X_α denotes the residual nucleus in the state α, and Y is the system resulting from the electromagnetic decay of X_α, e.g. $^{15}_{8}$O, $^{15}_{7}$N, $^{14}_{7}$N+n, or $^{14}_{6}$C+p [10, 11]. The energy spectrum of the states of the residual nuclei is schematically illustrated in Fig. 2. The branching ratios $\mathrm{Br}(X_\alpha \to \gamma + Y)$ have been taken from Ref. [16].

The NC QE cross section, $\sigma(\nu + {}^{16}_{8}\mathrm{O} \to \nu + X_\alpha + N)$, has been calculated within the approach discussed in Ref. [15] for the case of charged-current interactions. It may be expressed as

$$\frac{d\sigma_{\nu A}}{d\Omega dE'_\nu} = \sum_{N=p,\,n} \int d^3p\, dE P_N(\mathbf{p}, E) \frac{M}{E_N} \frac{d\sigma_{\nu N}}{d\Omega dE'_\nu}, \quad (2)$$

where $E_N = \sqrt{M^2 + \mathbf{p}^2}$, M being the nucleon mass, $d\sigma_{\nu N}/d\Omega dE'_\nu$ denotes the elementary neutrino-nucleon cross section and the spectral function $P_N(\mathbf{p}, E)$ yields the probability of removing a nucleon of momentum $\mathbf{p}$ from the target leaving the residual nucleus with energy $E + E_0 - M$, E_0 being the target ground-state energy.

According to the shell model, nuclear dynamics can be described by a mean field. In the simplest implementation of this model, protons in the $^{16}_{8}$O nucleus occupy three single-particle states, $1p_{1/2}$, $1p_{3/2}$, and $1s_{1/2}$, with removal energy 12.1, 18.4, and $\sim$42 MeV, respectively [12, 13, 14]. The neutron levels exhibit the same pattern, see Fig. 1, but are more deeply bound by 3.54 MeV [11].

As a consequence of a mean field dynamics, knock out of a target nucleon leaves the residual system in a bound state, and the spectral function can be conveniently written in the form

$$P_N(\mathbf{p}, E) = \sum_{\alpha \in \{F\}} n_\alpha |\phi_\alpha(\mathbf{p})|^2 f_\alpha(E - E_\alpha), \quad (3)$$

where $\phi_\alpha(\mathbf{p})$ is the momentum-space wave function associated with the α-th shell-model state and the sum is extended to all occupied states belonging to the Fermi sea $\{F\}$. The *occupation probability* $n_\alpha \leq 1$ and the (unit-normalized) function $f_\alpha(E - E_\alpha)$, describing the energy width of the α-th state, account for the effects of nucleon-nucleon (NN) correlations, not included in the mean field picture. In the absence of correlations, $n_\alpha \to 1$ and $f_\alpha(E - E_\alpha) \to \delta(E - E_\alpha)$.

A realistic model of the proton spectral function of oxygen has been obtained within the local density approximation (LDA), combining the experimental data of Ref. [12] with the results of theoretical calculations of the correlation contribution in uniform nuclear matter at different densities [15, 17]. The results reported in Ref. [15] show that the LDA spectral function provides an accurate description of the inclusive electron-oxygen cross sections at beam energies around 1 GeV. In addition, it predicts a nucleon momentum distribution in agreement with that obtained from the data of Ref. [18].

As pointed out in Ref. [19], nucleon-knockout experiments measure *spectroscopic strengths*, not occupation probabilities. Spectroscopic strengths are given by the area below the sharp peaks observed in the missing-energy spectra, corresponding to knockout of a nucleon occupying one of the shell-model states, corrected to take into account final-state interactions. On the other hand, occupation probabilities include contributions corresponding to larger removal energy, arising from mixing of the one-hole state with more complex final states [19].

The $p_{1/2}$, $p_{3/2}$ and $s_{1/2}$ spectroscopic strengths have been computed by integrating the oxygen spectral function of Refs. [15, 17] over the energy ranges $11.0 \leq E \leq 14.0$ MeV, $17.25 \leq E \leq 22.75$ MeV, and $22.75 \leq E \leq 62.25$ MeV, respectively. Dividing these numbers by the degeneracy of the shell-model states, one obtains the quantities S_α listed in Table 1. The same spectroscopic strengths have been used for protons and neutrons.

The branching ratios $\mathrm{Br}(X_\alpha \to \gamma + Y)$, necessary to calculate the cross section σ_γ according

α	$p_{1/2}$	$p_{3/2}$	$s_{1/2}$
S_α	0.632	0.703	0.422
$\mathrm{Br}(X_\alpha \to \gamma + Y)$	0%	100%	16 ± 1%

Table 1. Spectroscopic strengths of the $^{16}_{8}$O hole states and their branching ratios for deexcitation by the $E_\gamma > 6$ MeV photon emission.

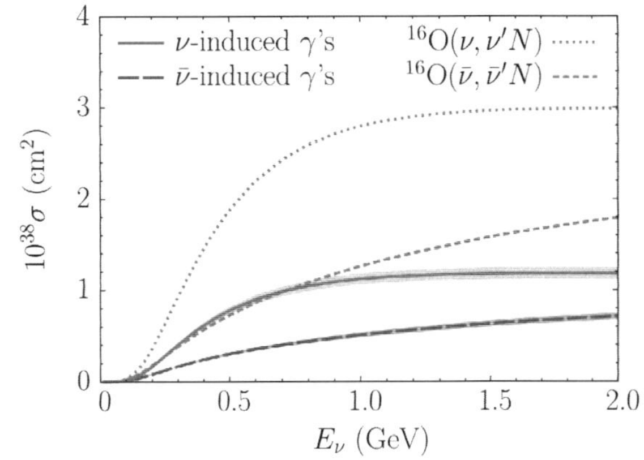

Figure 3. Cross section for γ-ray production following NC QE interaction of neutrino (solid line) and antineutrino (long-dashed line) compared to the NC QE cross section of neutrino (dotted line) and antineutrino (short-dashed line). Only the photons of energy larger that 6 MeV are considered.

to Eq. (1), are collected in Table 1 [16]. In the case of the $p_{1/2}$-proton (neutron) knockout, the residual nucleus is $^{15}_{7}$N ($^{15}_{8}$O) produced in its ground state. Hence, no γ-rays are produced. As the $p_{3/2}$-proton (neutron) hole lies below the nucleon-emission threshold, 10.21 MeV (7.30 MeV), it always deexcites through photon emission with half-life 0.146 ± 0.008 fs (less than 1.74 fs) [11]. When a proton (neutron) is knocked out from the deepest $s_{1/2}$ shell, the excitation energy is high enough for many deexcitation channels to open, of which only two, $^{14}_{6}\mathrm{C} + p$ and $^{14}_{7}\mathrm{N} + n$ ($^{14}_{6}\mathrm{C} + p$ and $^{11}_{6}\mathrm{C} + \alpha$), yield photons of energy higher than 6 MeV [16] (see Fig. 2).

In Fig. 3 our results for the neutrino- and antineutrino-induced γ-ray production cross section are compared to the neutrino and antineutrino NC QE cross sections. The error bands show the uncertainties arising form the determination of the spectroscopic strengths (5.4%), the treatment of Pauli blocking (1%), and the branching ratio of the $s_{1/2}$ state (1%). The σ_γ's dependence on neutrino energy is very similar, although not identical, to that of the NC QE cross section. The discrepancy arises from difference between the average removal energy associated with the whole spectral function and the energy of the $p_{3/2}$ shell, yielding the overwhelming contribution to σ_γ. The neutrino-induced γ-production cross section reaches its maximum at $E_\nu \sim 1.9$ GeV and is slowly decreasing at larger energies. On the other hand, the corresponding antineutrino cross section is an increasing function of E_ν.

In conclusion, we have computed the neutrino- and antineutrino-induced γ-ray production cross section for oxygen target, focussing on the kinematical region in which single-nucleon knockout dominates. Considering photons of energy larger than 6 MeV, we find that the $p_{3/2}$ state provides the overwhelming contribution, and that the ratio $\sigma_\gamma/\sigma_{\mathrm{NC}}$, exhibiting a significant energy-dependence at $E_\nu \lesssim 1$ GeV, is ~41% at $E_\nu = 600$ MeV. Our results, obtained using a realistic model of the target spectral function, provide an accurate estimate of a signal that can be exploited to identify neutral-current events in water-Cherenkov detectors.

Acknowledgments

This work was supported by INFN (Grant MB31), MIUR PRIN (Grant "Many-body theory of nuclear systems and implications on the physics of neutron stars"), and in part by the Grant-In-Aid from the Japan Society for Promotion of Science (Nos. 21224004 and 23340073). A.M.A. was supported by the Polish Ministry of Science and Higher Education under Grant No. 550/MOB/2009/0.

References

[1] Donnelly T W, Hitlin D, Schwartz M, Walecka J D and Wiesner S J 1974 *Phys. Lett.* B **49** 8
[2] Donnelly T W and Peccei R D 1979 *Phys. Rep.* **50** 1
[3] Langanke K, Vogel P and Kolbe E 1996 *Phys. Rev. Lett.* **76** 2629
[4] Kolbe E, Langanke K, Krewald S and Thielemann F K 1992 *Nucl. Phys.* A **540** 599
[5] Kolbe E, Langanke K and Vogel P 2002 *Phys. Rev.* D **66** 013007
[6] Beacom J F and Vogel P 1998 *Phys. Rev.* D **58** 053010
[7] Coulombe J *et al* (T2K Collaboration, ND280 group) 2005 *T2K ND280 Conceptual Design Report*
[8] Benhar O, Day D and Sick I 2008 *Rev. Mod. Phys.* **80** 189
[9] Ankowski A M, Benhar O, Mori T, Yamaguchi R and Sakuda M 2011 *Preprint* arXiv:1110.0679
[10] Ajzenberg-Selove F 1991 *Nucl. Phys.* A **523** 1
[11] Firestone R B, Baglin C M and Chu S Y F 1999 *Table of Isotopes* 8th edition (New York: Wiley)
[12] Bernheim M *et al* 1982 *Nucl. Phys.* A **375**, 381
[13] Leuschner M *et al* 1994 *Phys. Rev.* C **49**, 955
[14] Fissum K G *et al* 2004 *Phys. Rev.* C **70**, 034606
[15] Benhar O, Farina N, Nakamura H, Sakuda M and Seki R 2005 *Phys. Rev.* D **72** 053005
[16] Ejiri H 1993 *Phys. Rev.* C **48** 1442; Kamyshkov Yu and Kolbe E 2003 *Phys. Rev.* D **67** 076007
[17] Benhar O, Fabrocini A, Fantoni S and Sick I 1994 *Nucl. Phys.* A **579** 493
[18] Rohe D *et al* (JLAB E97-006 Collaboration) 2004 *Phys. Rev. Lett.* **93** 182501
[19] Benhar O, Fabrocini A and Fantoni S 1990 *Phys. Rev.* C **41** R24

XIII International Workshop on Neutrino Factories, Super beams and Beta beams (NUFACT11) IOP Publishing
Journal of Physics: Conference Series **408** (2013) 012056 doi:10.1088/1742-6596/408/1/012056

Performance study of the Bucked Coils cooling lattice for the Neutrino Factory

A Alekou[1], J Pasternak[1,2] and C Rogers[3]

[1] Imperial College London, UK
[2] STFC-RAL ISIS, Chilton, Didcot, UK
[3] RAL ASTeC, Chilton, Didcot, UK

E-mail: androula.alekou08@imperial.ac.uk

Abstract. In a Neutrino Factory (NF), neutrinos are produced from the decays of muons. The muon beam itself is produced with large emittance at the NF front-end and in order to be efficiently transported to downstream accelerator systems, the beam emittance needs to be cooled using ionisation cooling. The reference ionisation cooling channel of the NF reduces the transverse emittance by a factor of ~3. However, this lattice has a large magnetic field at the position of the RF cavities and therefore its feasibility has come under question when recent studies indicated high external magnetic field may lower the maximum achievable gradient of the RF cavities. The present work summarizes the performance of a new lattice named "Bucked Coils", designed to mitigate the magnetic field issue of the reference lattice of the NF while aiming to achieve similar cooling performance and transmission. This new lattice offers solution by reducing the longitudinal component of the magnetic field to almost 0 T while at the same time achieving a transmission comparable to the reference lattice within the transverse acceptance of 30 mm. Detailed comparison between the Bucked Coils lattice and the reference lattice of the NF follows with respect to the magnetic field reduction, cooling performance and transmission.

1. Introduction

The Neutrino Factory is the ultimate tool to study neutrino oscillations and search for leptonic CP violation. This future accelerator complex will produce the most intense and high-quality neutrino beam that has ever been achieved, using decays of stored muons [1]. As the muon beam in the Neutrino Factory is produced from pions decaying in flight, its initial transverse emittance is too large to be efficiently transported to further accelerator systems and therefore muon beam cooling is essential. Because of the short muon lifetime (2.2 μs), traditional cooling techniques cannot be applied and hence, the only viable cooling technique is *Ionisation Cooling*: the muons pass through absorbers where they lose momentum in every direction and the lost energy is restored only in the longitudinal direction, when the beam passes through RF cavities.

The reference ionisation cooling channel for the Neutrino Factory, FSIIA (Feasibility Study 2a) [2], performs well with respect to cooling and transmission but at the same time presents a large magnetic field at the end of the RF cavities. Studies have shown [3] that the higher the external magnetic field is, the lower the maximum achievable gradient becomes and therefore the feasibility of FSIIA has come under question.

This paper presents the *Bucked Coils* lattice, a new lattice design created to significantly reduce the magnetic field at the position of the RF cavities while simultaneously achieving a cooling performance and transmission comparable to FSIIA. The geometrical descriptions of FSIIA and the Bucked Coils lattice are given in detail below and the comparison of the achieved magnetic field together with cooling performance and transmission is presented.

2. Methodology and lattice geometry

Using G4MICE version 2-3-0 [4], a beam of 1,000 muons was generated with 10 mm normalized transverse emittance ($\epsilon_\perp$). Different initial transverse beta ($\beta_\perp$) was found for every lattice using the Optics application of G4MICE. The initial momentum followed a Gaussian distribution, centerd at 232 MeV/c with rms of 18 MeV/c.

2.1. FSIIA

A half-cell of the Neutrino Factory reference cooling channel consists of a coil, followed by a 200 MHz RF cavity that has a LiH absorber on each side. The coil's polarity alternates with every half-cell repeat. A full-cell of 1.50 m is shown in fig. 1 (left).

2.2. Bucked Coils

A pair of coils with different radius and opposite polarity are placed homocentrically at the same position along the beam-axis, z. One 200 MHz RF cavity with a LiH absorber on each side, follows the pair of coils, forming a half-cell of the Bucked Coils lattice. The polarity of every coil alternates with every half-cell repeat, as shown in the full-cell depicted in fig. 1 (right). In this paper, three versions of the Bucked Coils lattice, that differ in cell-length and current density, will be presented: BC-I, -II and -III. Table 1 presents the main parameters of FSIIA and the three versions of the Bucked Coils lattice.

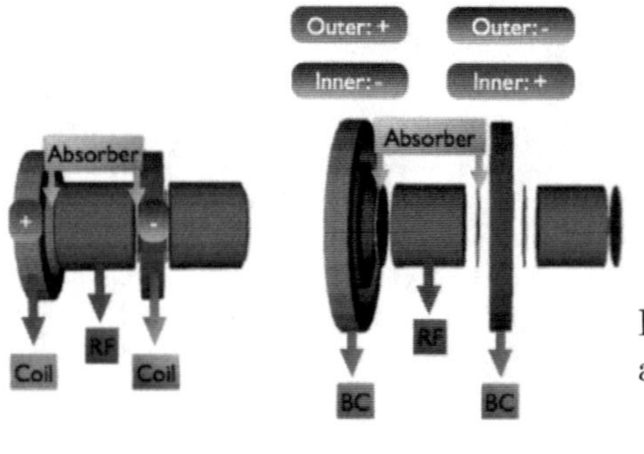

Figure 1: Layout of FSIIA (left) and BC-I (right) full-cell.

3. Results

3.1. Magnetic field

The total magnetic field (B_{tot}), together with the transverse (B_r) and longitudinal (B_z) components of the magnetic field are plotted as a function of the radius in fig. 2a, 2b and 2c respectively for all four lattices. The results for FSIIA are shown in black, whereas the results for BC-I, -II and -III are given in red, green and blue respectively. This colour code is used for all the results shown in this paper. The magnetic field is only plotted at the end of the RF cavities as this is the most sensitive place with respect to RF breakdown [5].

The total magnetic field decrease is obvious in the case of the Bucked Coils lattices. In fig. 2a, FSIIA has a magnetic field that exceeds 4 T at 35 cm radius, whereas BC-I achieves a four times smaller magnetic field. Both BC-II and -III, for the same radius, produce more than two times smaller magnetic field than FSIIA.

In order to gain a better understanding of the magnetic field reduction, B_r and B_z are plotted at the end of the RF cavities, as a function of the radius, R. The transverse component of the

Table 1: Main parameters of the FSIIA and Bucked Coils lattices. "IC" and "OC" correspond to "Inner Coil" and "Outer Coil" respectively.

Lattice	FSIIA	BC-I	BC-II	BC-III
Full-cell Length (m)	1.50	2.10	1.80	1.80
Number of RF cavities	2	2	2	2
Peak Electric Field (MV/m)	15	16	16	16
Phase (degrees)	40	30	30	30
Number of Absorbers	4	4	4	4
Number of Coils	2	4	4	4
IC, Inner Radius (m)	0.35	0.30	0.30	0.30
IC, Outer Radius (m)	0.50	0.45	0.45	0.45
OC, Inner Radius (m)	N/A	0.60	0.60	0.60
OC, Outer Radius (m)	N/A	0.75	0.75	0.75
IC, Current Density (A/mm^2)	106.67	90.24	128.10	128.10
OC, Current Density (A/mm^2)	N/A	120.00	112.80	132.00

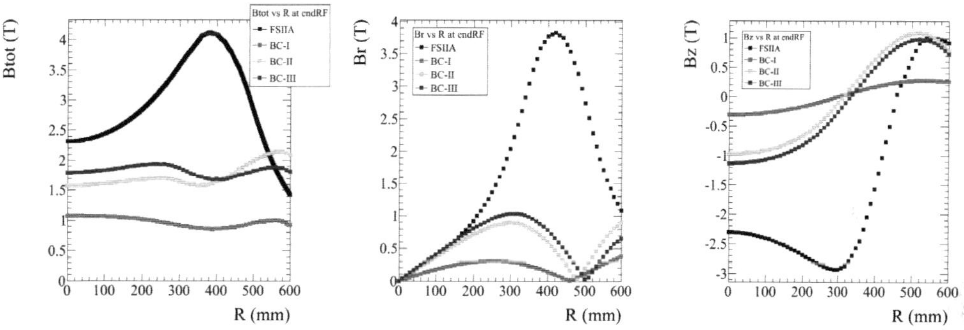

(a) Btot (b) Br (c) Bz

Figure 2: Magnetic field at the end of the RF cavity as a function of radius, R.

magnetic field (see fig. 2b) is as low as 0.1 T for BC-I at 42 cm, i.e. where FSIIA has its maximum (3.83 T). For the same radius, the magnetic field is 38 times smaller in BC-I, and 7 and 6 times smaller in BC-II and -III respectively. The iris of the RF cavity, which is at ~25 cm, is a sensitive location with respect to the RF breakdown, and although the maximum B_r of BC-I is achieved at 24 cm (0.31 T), it is still four times smaller than that of FSIIA.

The magnetic field decrease is apparent in fig. 2c. FSIIA has its maximum B_z (absolute value 3 T) at 30 cm, where BC-I has virtually 0 magnetic field. At this location both, BC-II and -III have three times smaller B_z than FSIIA.

3.2. Optics and cooling performance

Fig. 3 presents the transverse betatron function with respect to z. Although BC-III has the lowest betatron function at the RF position, closely followed by FSIIA, the amplitude of the FSIIA betatron function is smaller than BC-III and therefore the FSIIA cooling performance is expected to be better than that of BC-III. The equilibrium emittance of BC-I and -II will be higher than that of FSIIA and BC-III as can be deduced from their betatron function amplitude.

However, the betatron functions of BC-I and -II are still within the cooling limit.

One of the most important results is the transmission (the number of muons) within the transverse acceptance of 30 mm, the nominal accelerator acceptance defined in the baseline Neutrino Factory design [1]. The best transmission over all is achieved by BC-III at 120 m. What is important to note is that at 70 m, where FSIIA achieves its maximum, all three BC lattices achieve a very similar transmission. Most importantly BC-I, the lattice with the best magnetic field reduction, achieves less than 10% smaller transmission in comparison to the FSIIA lattice at 70 m, and this transmission difference is insignificant.

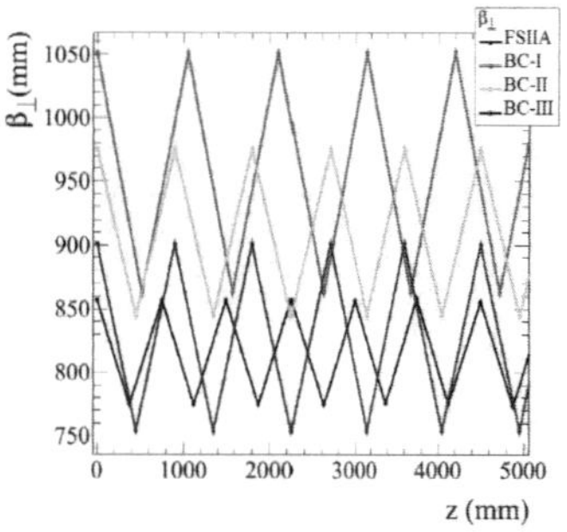

Figure 3: Transverse betatron function with respect to the beam-axis, z.

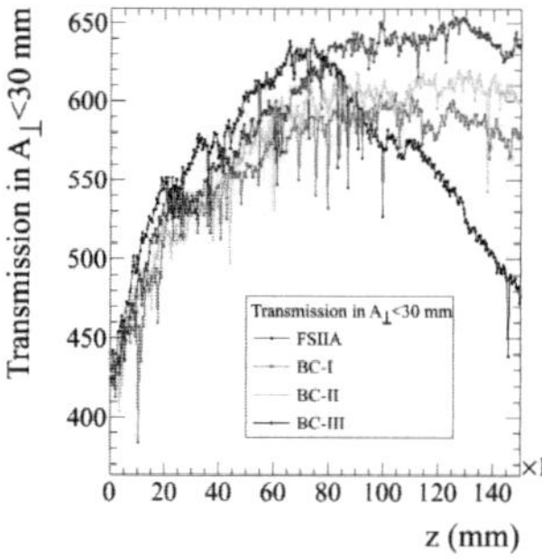

Figure 4: Transmission within the transverse acceptance of 30 mm along z.

4. Conclusions

Several versions of the Bucked Coils lattice were designed with the aim of reducing the magnetic field within the RF cavities. Three versions, BC-I, -II and -III were presented in this paper and compared to the FSIIA lattice with respect to the magnetic field reduction and transmission. When Bucked Coils are used, the magnetic field reduction is clear: BC-I has 4 times smaller B_{tot} and 38 times smaller B_r than FSIIA. At 30 cm the B_z of BC-I is virtually 0 whereas in FSIIA is as high as 3 T. Transmission within the transverse acceptance of 30 mm (fig. 4), shows BC-III to have the best transmission over all the lattices at 120 m. Most importantly, at 70 m, where FSIIA achieves its maximum transmission, BC-I, which achieves the best magnetic field reduction, has a very similar transmission to FSIIA. Bucked Coils lattices not only decreased significantly the magnetic field but also achieved a comparable transmission and cooling performance to the FSIIA lattice.

The Bucked Coils lattice is now one of the three alternative cooling channels for the Neutrino Factory. Optimization of the lattice will be beneficial and further work is planned investigating the feasibility of the lattice with regards to engineering design.

References

[1] The IDS-NF Collaboration 2011 *International Design Study for the Neutrino Factory, IDS-NF-020*
[2] Berg J S *et al.* 2006 *PR-STAB 9.011001*
[3] Moretti A *et al.* (Neutrino Factory and Muon Collider Collaboration) 2005 *Phys. Rev. ST Accel. Beams* **8** 072001
[4] Howlett L and Rogers C 2006 1211–1214 http://micewww.pp.rl.ac.uk/projects/g4mice/wiki (Current October 2011)
[5] Palmer R B *et al.* 2008 (*Preprint* `arxiv:0809.1633v1 [physics.acc-ph]`)

Recent developments on the muon Non-Scaling FFAG for the Neutrino Factory and its subsystems

J Pasternak[1,2], M Aslaninejad[1], J Scott Berg[3], N Bliss[4], C Bontoiu[1], M Cordwell[4], H Witte[5], D Kelliher[6], S Machida[6]

[1]Imperial College London, London, UK
[2]ISIS, Rutherford Appleton Laboratory, STFC, OX11 0QX, UK
[3]BNL, Upton, NY 11973-5000, USA
[4]Technology Department, Daresbury Laboratory, STFC, Daresbury, UK
[5]JAI,Oxford University, Oxford, UK
[6]ASTeC, Rutherford Appleton Laboratory, STFC, OX11 0QX, UK

E-mail: j.pasternak@imperial.ac.uk

Abstract. The current status and recent developments on the muon non-scaling FFAG for the Neutrino Factory studied in the framework of the EUROnu/IDS-NF projects are presented. Beam dynamics studies, including the process of acceleration, are discussed. A first pass at engineering for the layout of the ring cell is described. Progress of studies on the main machine subsystems is discussed. The future plans for the study are described.

1. Introduction

The International Design Study of the Neutrino Factory (IDS-NF) [1], together with the EUROnu Project [2], seek the complete end-to-end conceptual design for the Neutrino Factory, a potential future facility for precision neutrino experiments. The Neutrino Factory (NF) produces a high quality neutrino beam with a well known energy spread and flavour content by injecting muons accelerated to high energy into a decay ring equipped with long straight sections pointing towards far neutrino detectors. This facility aims for an unprecedented precision for a discovery of CP violation in the leptonic sector and is currently recognized as the best facility for this ambitious goal. The current baseline for the NF is described in detail in the recently completed Interim Design Report [3].

Muon acceleration, which is necessary for the neutrino factory to achieve its high performance, represents a challenge for the accelerator system. Firstly, the muon beam produced as tertiary beam has a very large initial transverse emittance and energy spread. Even after applying the RF phase rotation and ionization cooling, assumed in the muon front end, the required accelerator normalized acceptances at the input are as high as 3 π cm rad and 150 mm in transverse and longitudinal phase spaces, respectively. Secondly, a short muon lifetime (2.2 μs at rest) dictates a need for a very fast acceleration. As the first stage of acceleration from 150 MeV to 0.9 GeV, a superconducting (SC) linac is proposed. At higher energy up to 12.6 GeV, the baseline uses recirculating linacs, where the beam passes a few times through the same accelerating RF structure (4-5 times) reducing the cost of the accelerator. In order to further improve the efficiency of acceleration and reduce the cost, a Non-Scaling Fixed Field Alternating Gradient (NS-FFAG) ring is used for the final stage of acceleration to 25 GeV. The NS-FFAG ring allows one to perform more than 11 turns through the same accelerating system using 201 MHz superconducting (SC) RF cavities. This is possible due to the quasi-isochronous optics, which allows the time of flight difference as a function of energy to be minimized by combining the positive and negative bending angles with strong focusing.

2. Non-Scaling FFAG lattice description

The current baseline NS-FFAG lattice [4] is based on FDF triplet lattice using combined SC magnets with superimposed dipole and quadrupole magnetic field components. This magnetic field configuration results in a machine with the so-called "natural" chromaticity, which means that the tunes change as a function of beam energy, as can be seen in Fig. 1. The absence of higher order multipoles simplifies the magnet design and allows for the large dynamic acceptance needed for muon beams, but introduces a time of flight variation as a function of the transverse amplitude. This may result in the longitudinal emittance blow up, which may be prevented by introducing chromaticity correction. A scheme for this correction still needs to be identified in future studies. The time of flight variation as a function of energy has a parabolic behaviour, which is shown in Fig 2.

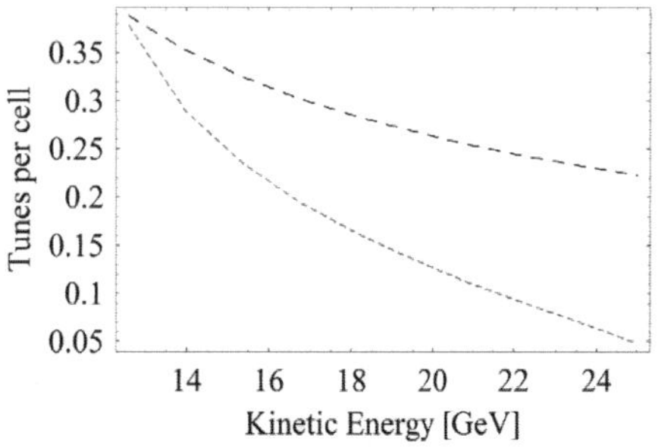

Figure 1. The cell tunes (horizontal-black and vertical-red lines) variation as a function of energy in the NS-FFAG.

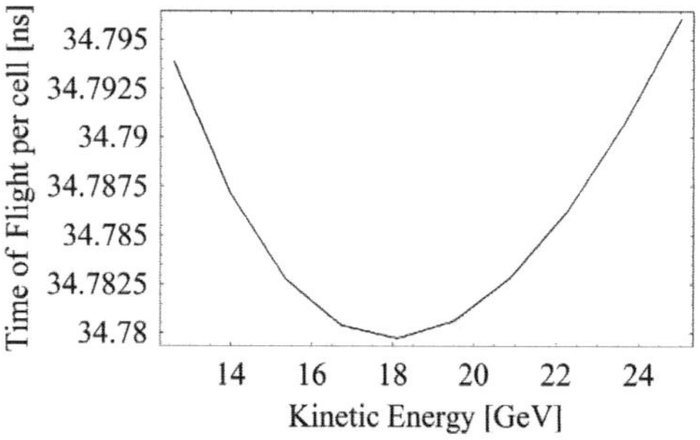

Figure 2. The time of flight per cell as a function of muon energy in the NS-FFAG.

The current lattice configuration has been chosen to allow for sufficient space in the long straight section (5 m) for the extraction septum, whose field is limited to ~2 T to reduce its stray fields. The main NS-FFAG ring parameters are summarized in Table. 1.

Table 1. Main NS-FFAG parameters.

Parameter	Value
Circumference [m]	699
Number of cells	67
Long drift length [m]	5
RF voltage per turn [MV]	1195.6
Max B field [T]	6.2
Max quadrupole gradient [T/m]	18.82
Turns for acceleration	11.8
Muon decay [%]	7.1

3. Preliminary engineering

Engineering studies have been initiated, resulting in a preliminary conceptual layout of the NS-FFAG cell. Attention has been drawn for the space requirements for the injection/extraction magnets, beam diagnostics, RF cavities, etc. The cold-warm transition between the SC magnet block and the long straight section was evaluated, resulting in the reduction of the available space from previously assumed 4.4 to 4 m. The parameters of the preliminary SC magnet design [5] together with the existing SC 201 MHz RF cavity structure were used to create the CAD model of the FFAG cell hosting the RF cavity as shown in Fig. 3.

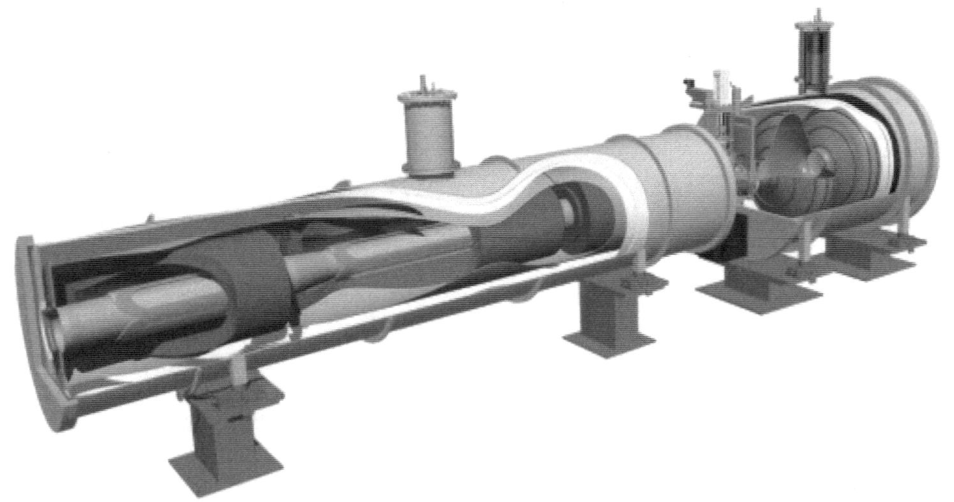

Figure 3. The layout of NS-FFAG machine cell containing the RF cavity. The F and D magnets with coils and yokes (red and blue, respectively) placed in a common cryostat are shown on the left. The RF cavity enclosed in a separate cryostat is shown on the right.

4. Injection/extraction

The injection and extraction systems were designed using distributed kickers deflecting the beam in horizontal plane and placed symmetrically to serve both muon signs. In both systems the kickers are shared for positive and negative muons and are equipped with septum magnets for each polarity on each end of the system. The existing technologies were used as much as possible, but clearly the aperture requirements are beyond those of existing magnets. Also the fields in the extraction septa are higher than in conventional accelerator applications. Currently extraction septa of 2 T are assumed in order to minimize the stray field leakage. The new assumptions on the available space in the straight section, due to the necessary space for cold/warm transitions, resulted in the update of the injection/extraction system parameters, which are summarized in Table 2.

Table 2. Parameters of the injection and extraction systems.

Parameter	Injection	Extraction
Kickers	2	4
Kicker field [T]	0.106	0.8
Septum field [T]	1.2	1.94
Kicker/septum length [m]	3.7/3.4	3.7/4

5. Preliminary beam dynamics simulations with the "serpentine" acceleration.

Particles were tracked in the current baseline NS-FFAG lattice in order to test the performance of the "serpentine" acceleration. First, an initial 6D Gaussian distribution with longitudinal parameters close to the design ones, but with small transverse emittances were launched and 11.8 turns in the machine were tracked. RF cavities were distributed in all cells, neglecting the effects of empty spaces for injection/extraction. Then similar tracking was performed with large (close to design) initial transverse emittances. Both simulations showed a relatively large longitudinal emittance blow up but with almost lossless transport. As can be seen by comparing the Fig.1 and 2, the source of the longitudinal emittance increase is twofold: firstly the initial longitudinal distribution is stretched and filamented in the serpentine channel, secondly the time of flight variation for large transverse amplitude introduces an additional energy spread. The longitudinal emittance blow up in the "serpentine" acceleration has been recognized in previous studies and possible cures were identified [6].

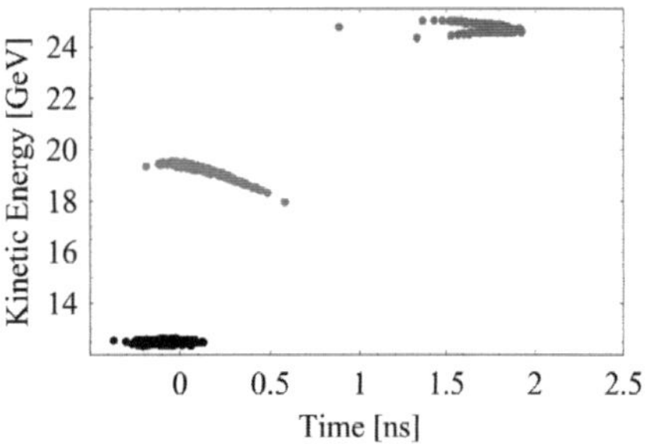

Figure 1. The injected (black), intermediate-after 5 turns (pink) and final (red) longitudinal phase spaces assuming small initial transverse emittances.

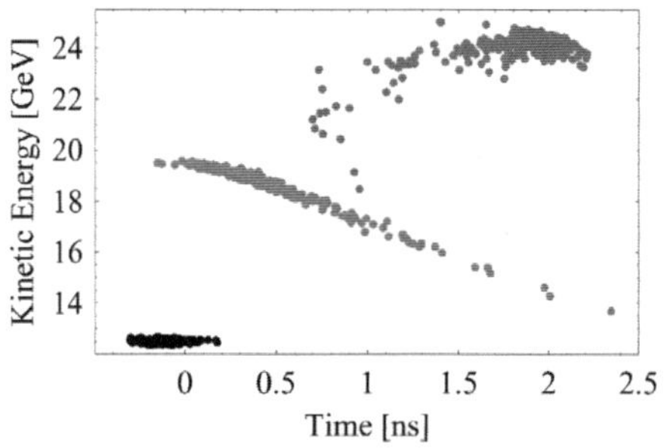

Figure 2. The injected (black), intermediate-after 5 turns (pink) and final (red) longitudinal phase spaces assuming large (design) initial transverse emittances.

6. Summary and future plans

Longitudinal emittance blow up needs to be minimized by optimizing the injection parameters and by altering the machine parameters, which need to be adjusted to give precisely 11.5 turns from injection to extraction. The effect of magnet errors and misalignments on the machine performance will be estimated and used to set the magnet tolerances. The magnet design will be updated to meet those specifications. Various technology solutions for the combined function SC main magnets should be compared including: the current design based on separate multipole conductor layers, a double helix, and a single combined function layer. Chromaticity correction and its effect on the longitudinal and transverse beam dynamics will be studied and the conclusions on the final machine design drawn. Beam loss induced energy deposition from muon decays needs to be studied, and its effect on the feasibility of the cryogenic systems understood, especially the SC extraction septum. More engineering studies of the machine subsystems is needed.

References

[1] www.ids-nf.org
[2] www.euronu.org
[3] Abrams R J 2011 *Interim Design Report* IDS-NF-020 *Preprint:* arXIV:1112.2853v1 [hep-ex]
[4] Scott Berg J *et al.* 2011 in Proc. of IPAC'11San Sebastian Spain p 832
[5] Pasternak J 2011 *et al.* in Proc. of IPAC'11 San Sebastian Span p 829
[6] Scott Berg J 2007 *Nucl. Instrum. and Methods A* **570** Issue 1 p 15

Accelerators for the PS neutrino beam

R Steerenberg, M Calviani, E Gschwendtner, A Pardons, H Vincke

CERN, 1211 Geneva 23, Switzerland

E-mail: Rende.Steerenberg@cern.ch

Abstract. A recent memorandum for an experimental proposal [1] was discussed during the CERN PS and SPS experimental committee (SPSC) of April 2011 and at the Research Board of June 2011. The proposed experiment, with objective to investigate the anomalous $\nu_\mu \rightarrow \nu_e$ oscillations, aims at re-using the discontinued CERN PS Neutrino Facility (PSNF) and experimental zones to install a 150 ton liquid argon time projection chamber (LArTPC) as near detector and a 600 ton LArTPC as far detector. This article will summarize the experimental needs, the proposed facility layout, a primary beam production scheme and the requirements for the reconstruction of the PSNF.

1. Introduction

The proposed experiment aims at re-using the discontinued CERN PSNF. It foresees to install a 150 ton liquid argon time projection chamber (LArTPC) as near detector at 127 m from the production target and the 600 ton ICARUS LArTPC [2] that is presently operational at the Gran Sasso National Laboratory in Italy, as far detector at 850 m distance from the production target, as illustrated in figure1. The CERN-PS will have to produce the required number of protons.

Figure 1. The foreseen layout for the neutrino production and the neutrino beam path on the CERN Meyrin site.

2. The principal experimental requirements

The proposed experiment requires 2.5×10^{20} protons at 20 GeV/c on target to produce the required number of events in the far detector in 2 years time, resulting in 1.25×10^{20} protons per year. This amount is based on the neutrino flux as produced by the target and horn configuration used in 1983. The required neutrino beam is a low energy ν_μ beam, centred around ~1.5 GeV.

3. The PS neutrino facility

The initial PS TT7 tunnel and beam line were designed and constructed in the late 70's, early 80's and exploited for 2 years, starting in 1983, to provide beam to different neutrino experiments (PS169 [3], PS181 [4] and PS180 [5]) in the neutrino pit of building 181 and in the buildings 191 and 36, all situated on the CERN Meyrin site.

Protons were provided by the CERN-PS and extracted via the TT2 line through part of the TT1 line[1], into the TT7 line that directed them towards the 80 cm long beryllium target, which was followed by a magnetic horn, designed to focus the secondary mesons of a momentum around 2 GeV/c. Figure 2 and figure 3 provide an overview of the tunnels that could house the new PSNF.

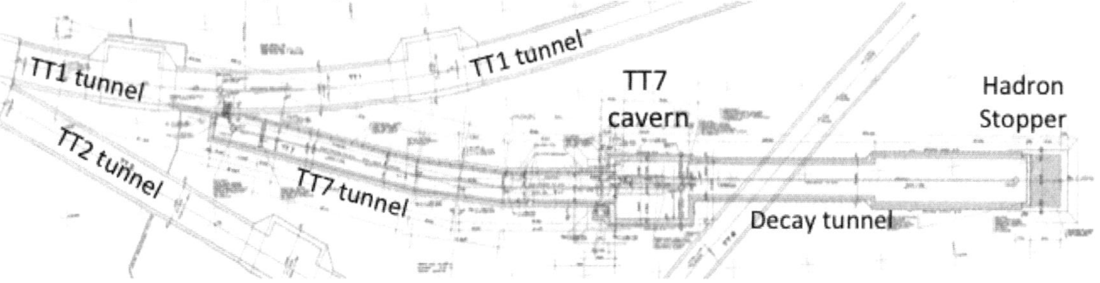

Figure 2. Top view of the transfer tunnels, target cavern, decay tunnel and hadron stopper.

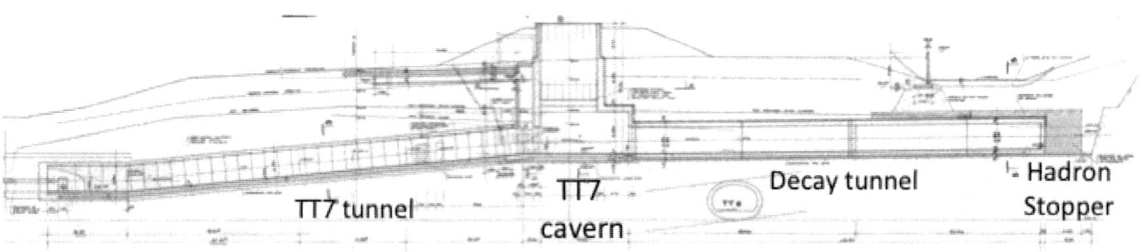

Figure 3. Side view of the TT7 tunnel, target cavern, decay tunnel, and hadron stopper.

4. PS proton beam production

The first experiments (PS169 and PS181) received during the first two months of the 1983 run (February and March) pulses of 1.25×10^{13} protons from the PS at a momentum of 19.2 GeV/c every 1.2 s, representing an average beam power of 32 kW. These experiments integrated in these two months slightly more than 10^{19} protons on target. Later that same year the BEBC experiment (PS180) received the same instantaneous intensity, but at a much lower duty cycle of about 1 pulse every 14.4 s on average, representing an average beam power of slightly more than 2.5 kW. These values together with the potentially maximum CERN-PS production capabilities today are summarized in Table 1.
The CERN-PS provides beam to different experiments by means of time-sharing. A dedicated cycle is allocated to each user and a series of these different cycles compose a super cycle, which is repeated continuously outside the 2 hours per day of dedicated filling of the LHC. Presently the CERN-PS can provide a beam with 3×10^{13} protons/pulse at 20 GeV/c with a maximum repetition rate of 1.2 s,

[1] At that time the TT1 line was supplying beam to ring II of ISR

provided the beam losses can be kept within defined limits. Assuming the above quoted beam together with a yearly physics run of 200 days of 22 hours per day and an average beam availability of 85%, a 37% duty cycle in the super cycle would be required to provide the required 1.25×10^{20} protons per year on the PSNF target. At present this required share of duty cycle in the super cycle is not available.

Table 1. Summary of past and potential PS beam production capabilities.

		Old Neutrino Facility		New Neutrino Facility	
		PS dedicated Feb-Mar 1983	**PS parallel 1983 - 1984**	**PS dedicated**	**PS parasitic[1)]**
Proton Momentum	[GeV/c]	19.2	19.2	20	20
Protons per pulse	[x 10^{13}]	1.2	1.2	3	2.6
Maximum rep. rate	[s]	1.2	14.4	1.2	1.2
Beam energy	[kJ]	38	38	96	84
Average Beam Power	[kW]	32	2.5	80[2)]	70[2)]

[1)] The parasitic beam is in case the PSNF beam and the n_TOF beam will be produced on a single cycle.

[2)] This average beam power is potentially available from the PS provided the losses are within the specified limits and that no other users are present in the PS super cycle.

However, combining the beam production of different experiments on a single cycle and thus making more efficient use of the CERN-PS, could alleviate the situation. The primary proton beam for the CERN neutron Time Of Flight (n_TOF) [6] experiments consists of a single bunch that is accelerated using an RF frequency for the accelerating cavities that is the 8th harmonic of the revolution frequency, resulting in a single bucket occupied by 1 proton bunch and 7 empty buckets that are potentially available to accelerate additional protons for the PSNF. Accelerating 3×10^{13} protons, using the 8th harmonic RF system results in a bunch intensity of 3.75×10^{12} protons. Provided the necessary systems are in place, 7 bunches, with a total intensity of 2.63×10^{13} protons, can be extracted and transferred to the PSNF target and the remaining single bunch, with 3.75×10^{12} protons can be extracted to n_TOF. Based on assumed super cycles, 200 days of 22 hours per day physics run per year and a machine availability of 85%, an estimate of the possible yearly proton production was made recently [7] and is summarized in table 2. About 73% of the yearly requested amount of protons could be accumulated on the target, providing the total requested amount of protons, 2.5×10^{20}, within less than 3 years.

Table 2: Overview of achievable PS primary proton beam intensity and beam power for the PSNF, taking 2 hours per 24 hours into account for dedicated LHC filling.

		Day time	**Night time**	**Average**	**Total**
Duration	[hrs/day]	9.2	12.8	-	**22**
Super cycle length	[s]	38.4	28.2	32.5	-
PSNF/TOF duty cycle in super cycle	[%]	28.1	33.3	33.1	-
Repetition rate	[s]	4.3	3.6	3.9	-
Beam Power	[kW]	19.7	23.3	21.8	-
Integrated proton intensity per day	[x10^{17}]	1.73	2.86	-	4.6
Integrated proton intensity per run	[x10^{19}]	-	-	-	**9.2[1)]**

[1)] This figure is subject to changes depending on the final super cycle configuration.

5. (Re-)Construction of the PS neutrino facility

The existing tunnels and cavern can, after consolidation and some adaptations, be re-used to house the new facility. Since the facility is relatively close to the surface, simulations, addressing the radiological situation for the proposed beam power as given in table 1 and table 2, as well as the energy deposition in the various elements of the secondary beam production zone have been and are still being made [8]. The 150 m long primary proton beam line, the target and the focusing horn will have to be designed and constructed. The activation of the air in the decay tunnel is also an important point that needs to be addressed. A choice between a high performance ventilation system, running at lower than atmospheric pressure, or a helium filled or partial vacuum decay tube needs to be made. Further studies are required to estimate possible dose rates to which maintenance personnel could be exposed as well as the total induced radioactivity in the structure and its surroundings together with the releases to the environment via air and water pathways. The exact composition and status of the present hadron stopper are not known, apart from that it is made of steel blocks, positioned on a concrete slab, extending from the tunnel floor. It is unknown if the steel blocks are sealed to avoid infiltration of water. Putting back into operation the PSNF would require replacing the existing hadron stopper with a new design, taking into account the proposed beam power, radiological aspects and its efficiency stop charged particles and thus minimize the background for the near detector. At the same time it should be investigated if muon detectors could be integrated so that the target/horn and beam alignment can be done independently from the experimental detectors.

6. Present status and concluding remarks

CERN has expressed a genuine interest in the proposed experiment and the experimental collaboration is editing a concrete proposal for the CERN experimental committee. A preliminary feasibility identifying areas that deserve more detailed study has been made [9]. However, in order to address all items, to obtain a detailed cost estimate and a realistic project planning, a full design study must be undertaken. It is worth noting that the PSNF can also be used for other experiment and/or studies, such as neutrino cross-section measurements at very low energies, target R&D and detector studies.

Acknowledgements

The authors of this article would like to thank Y. Body, J. Borburgh, O. Boettcher, P. Chiggiato, P. Costa-Pinto, T. Dobers, I. Efthymiopoulos, A. Ferrari, P. Fessia, M. Giovannozzi, A. Guglielmi, K. Gunnar-Lindell, D. Missiaen, D. Nisbet, F. Pietropaolo, M. Poehler, H. Prin, Y. Robert, P. Sala, D. Tommasini, B. Vandorpe, S. Zeler for fruitful discussions and work done so far.

References

[1] Rubbia C *et al.* 2009 *A New Search for Anomalous Neutrino Oscillations at the CERN PS Preprint* arXiv:0909.0355v2[hep-ex]
[2] Rubbia C *et al.* 2002 *ICARUS: a Second Generation Protons Decay Experiment and Neutrino Observatory at the Gran Sasso Laboratory*
[3] Steinberger J *et al.* 1980 *Proposal: Search for Neutrino Oscillations* CERN/PCC/80-104
[4] Jonker M *et al.* (CHARM Collaboration) 1981 *Contribution of the CHARM Collaboration to the CERN Neutrino Oscillation Program* CERN/PCC/81-8
[5] Baldo-Ceolin M *et al.* 1980 *Proposal to Search for Neutrino Oscillations* CERN/PSCC/80-130
[6] Andriamonje S *et al.* (n_TOF Collaboration) 2000 *The n_TOF Collaboration Technical Report* CERN/INTC-2000-018
[7] Steerenberg R, Gschwendtner E 2011 *PS Primary Protons Beam Produciton Scheme for the Proposed 2LAr@PSNF Experiment* CERN-ATS-2011-193
[8] Calviani M 2011 *Preliminary FLUKA study of the proposed PSNF installation* CERN-ATS-Note-2011-088 TECH1
[9] Steerenberg R *et al.* 2011 *Putting Back in Operation the TT7 PS Neutrino Facility: Critical Issues and Design Study Proposal* CERN-EDMS:11008369

Beta Beams in the CERN complex: PS studies

E Benedetto

NTU-Athens and CERN, 1211 Geneva, Switzerland

E-mail: elena.benedetto@cern.ch

Abstract. For the Beta Beams to become a solid option for a future neutrino-physics facility at CERN, it is fundamental to demonstrate the feasibility of storing and accelerating the Beta Beams ions in the PS and SPS. This work identifies the key issues that should be assessed for the PS, presents the status of the ongoing studies and, in particular, discusses the choice of the injection energy and recent space-charge measurements.

1. Introduction

Beta Beams aims at the production of a pure electron neutrino and anti-neutrino beam from the β–decay of radioactive ions, for oscillation experiments. The baseline design considers ^{6}He and ^{18}Ne as respectively antineutrino and neutrino emitters and it is based as much as possible on existing technology and on CERN facilities [1]. In particular, it relies on ISOLDE experience and infrastructure for the isotopes production, while the ions accelerator chain includes the PS and SPS. The main motivation of this approach is to reduce the cost of the facility and to profit of the experience in machine operation. However, PS and SPS are not optimized for the Beta Beams, therefore it is necessary a careful study on how to accommodate our radioactive ions in these existing machines.

This work is focused on the CERN-PS, assuming the baseline parameters [2], although an optimization of the entire beam transfer chain is ongoing [3]. Activities are done in close collaboration with the LHC Injectors Upgrade (LIU) Project [4], as several issues are in common between Beta Beams and the Upgrade proposal to increase the PS injection energy to 2 GeV. Studies in the CERN-PS include: radiation protection issues, vacuum, injection scheme and collective effects such as space-charge on the injection long flat-bottom and fast head-tail instabilities at transition crossing.

2. PS Injection scheme

The baseline scheme [5] foresees the injection in the PS at 3.5 GeV proton-equivalent kinetic energy. A batch of 20 bunches, arriving from the RCS with a repetition rate of 10 Hz, are accumulated on a long flat bottom (1.9 s), then accelerated up to 26 GeV proton-equivalent and sent to the CERN-SPS.

2.1. Injection energy

The 3.5 GeV injection energy, to be compared with the CERN operational beams energy of 1.4 GeV, has been chosen from space charge consideration and it is based on a maximum Laslett tune shift of 0.22 [6], obtained by considering the bunch parameters summarized in Table 1. This

Table 1: Parameters at PS injection for the Beta-Beams baseline ions [2]

Ion species			^{6}He	^{18}Ne	p-equivalent
Kinetic energy per nucleon	E_k/A	(GeV)	0.787	1.65	3.5
Relativistic gamma	γ		1.84	2.77	4.73
Half life	$\tau_{1/2}$	(mm mrad)	1.49	4.63	–
Normalized horiz. emittance (rms)	ε_H^*	(mm mrad)	10.3	10.3	10.3
Normalized vert. emittance (rms)	ε_H^*	(mm mrad)	5.5	5.5	5.5
Harmonic number	h		21	21	–
Bunch length (rms)	τ	(ns)	21.5	21.5	–
Momentum spread (rms)	dp/p	(10^{-3})	0.23	0.15	–

is a safe margin, as some high intensity operational beams in the PS have a space-charge tune spread larger than 0.3 (although they are immediately accelerated after injection).

Considerations of septum technology, kicker strength and straight-sections length in the PS, indicate that injection at 3.5 GeV p-equivalent is challenging, as also confirmed by PS and injection experts [7, 8].

It is therefore here proposed to lower the Beta-Beams injection energy to 2 GeV p-equivalent. This value corresponds to what is studied for the LIU Project and will allow Beta Beams to profit of synergies with this project and eventually to use the same equipments.

A recent study of a 10 Hz RCS as a PS pre-injector, to replace the CERN-PS Booster [9], proposes the Eddy-Current technology for the septum and the bumper, plus an extra kicker in addition to what already installed in the PS, to provide the 30% increase in kick strength for the 2 GeV injection energy. This work, which will be the starting point when a detailed study of the Beta-Beams injection scheme is required, also identifies issues that need careful study in case of multiple bunches injection from the RCS, namely losses due to the injection bump, the lifetime of eddy-current devices and a vertical acceptance reduction at the septum, compared to what the PS can accept nowadays.

2.2. Long flat bottom

The injection flat-bottom is 1.9 s long, since the PS is receiving 20 injections at 10 Hz from the RCS. This has two consequences. First of all, due to decay losses, on the long flat bottom there is a 43% reduction of the ^{6}He intensity and a 18% for ^{18}Ne. Studies [10] of dose deposition in the magnets indicates that the air activation in the tunnel is well within the limits and that the PS tunnel would remain accessible as a limited-stay controlled area after a few days, depending on the position along the ring and on the operation mode. The report does not take into account beam losses in "hot" regions (e.g. at the septum) and operation losses (injection, transition, extraction). More specific Fluka studies are ongoing now.

The second issue is represented by collective effects and in particular space-charge, which may induce beam blow-up and losses on the long injection flat bottom.

3. Space charge

Space charge induces tune spread in the beam and can make particles in high-intensity and high-brightness beams cross betatron resonances and either be lost, or contribute to emittance growth. Usually the value of $|\Delta Q_{SC,max}| = 0.22$ is assumed for the determination of the maximum allowed Laslett tune-shift. This is safe margin, useful in the design process, but it can be relaxed.

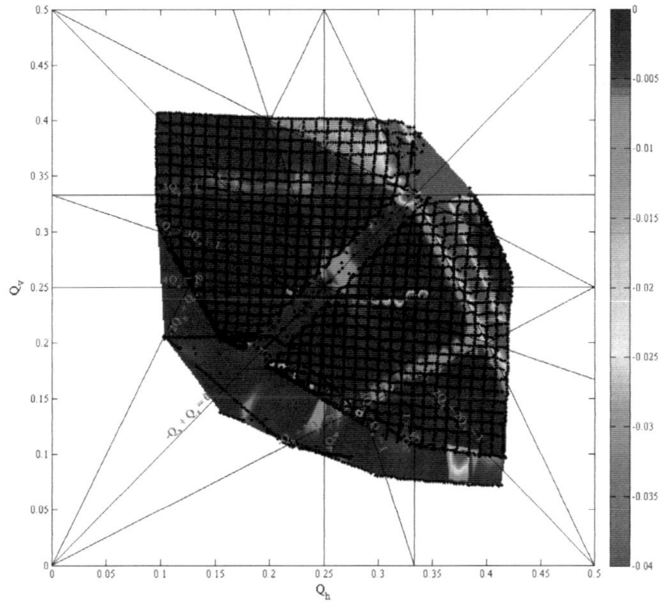

Figure 1: Tune diagram resulting from the loss measurement at 2 GeV. The color scale indicates the derivative of the BCT signal, i.e., the red lines indicate the larger losses.

The identification of the destructive resonance lines and the determination of the optimum working point and the maximum Laslett tune-shift, via beam-based measurements, is fundamental to fully exploit the PS intensity performances for the Beta-Beams, especially if, as discussed in the previous paragraph, an injection at a lower energy (2 GeV p-equivalent) has to be implemented. For the following measurements, the proton beams available in the PS are used, but the results can be easily scaled to the Beta-Beams ions.

3.1. Tune scan for destructive resonance-lines identification

The identification of the most dangerous resonances is realized by applying the method proposed in [11]: a large emittance, moderate intensity, not space-charge dominated beam is kept on a 1.2 s magnetic flat bottom at constant energy of 2 GeV. The emittance of the beam is chosen to be large enough to fill at maximum the vacuum chamber, in such a way that even a weak resonance produces a limited increase of the betatronic amplitude for which losses should become visible. During the scan, the tune is slowly varied with a predefined function and the beam intensity is recorded: if an excited resonance is crossed, beam losses are observed as a slope in the Beam Current transformer (BCT) acquisition.

Figure 1 shows the tune scan for the bare machine (without non-linear elements such as sectupoles, octupoles or Pole Face Windings). Only part of the tune diagram could be explored, because the working point was controlled via two families of quadrupoles which are limited in strength. The black dots indicate the points at which the tune and losses were measured. The colorplot is then obtained by a Matlab© interpolation function. The losses are quoted in terms of the derivative of the BCT signal dN/N.

The plot identify the lines from which the working point should be kept far from. It indicates that the most destructive line is the $2q_x + q_y = 1$ and that the 4th order line $4q_y = 1$ is not excited, but it is the 3th order $3q_y = 1$.

3.2. Space-charge dominated beam measurements

A first attempt was made to measure the losses and the transverse emittance increase for space-charge dominated beams. Two different beams have been used: a) a high-brightness "LHC–like" beam, characterized by a transverse normalized emittance of about 2.5μm mrad and b) a high-intensity "ToF–like" beam, more similar to Beta Beams, with large transverse emittances, to fill as much as possible the vacuum pipe.

For the first, high–brightness low–emittance beam, different regimes have been identified, depending on the working point and on which resonance line is touched. As expected from studies done in the past [12], crossing the integer induces losses-free emittance blow-up on this kind of beam. A different effect was identified for a vertical tune of $Q_y = 6.30$, namely losses with bunch shape degradation. The cause may be the large chromaticity of the beam and particles crossing the $3q_y = 1$ line. Measurements should be repeated after chromaticity correction.

Results for high–intensity, large–emittance beams are more difficult to interpret as any beam blow-up will immediately translate into losses.

4. Conclusions

Efforts are ongoing to assess the feasibility of using CERN-PS to accelerate Beta Beams and possible issues have been identified. Injection at 3.5 GeV p-equivalent is a challenge with existing equipments and therefore we propose to reduce it to 2 GeV, which is the value chosen for the PS Upgrade, within the LIU Project.

The main issue may be space-charge on the long injection flat-bottom, therefore a campaign of measurement has been started, in collaboration with the LIU colleagues. Tune scans have identified the destructive resonance that should be avoided when choosing the working point. The aim of the ongoing space-charge dominated beams is to identify the different mechanisms of beam degradation that may take place and try to find a cure, e.g. chromaticity correction or resonance compensation, in order to accommodate the largest possible Laslett tune spread (i.e. the highest beam intensity at a given energy) for the optimum value of working point.

Acknowledgments

Financial support of the European Community under the European Commission Framework Programme 7 Design Study: EUROnu, Project Number 212372 is acknowledged. The EC is not liable for any use that may be made of the information contained herein.

I am grateful to S. Gilardoni, E. Metral, E. Wildner and the LIU colleagues for discussion, to A. Huschauer and PS-OP crew for help with measurements and the good-quality beam.

References

[1] Zucchelli S 2002 A novel concept for a Neutrino Factory: the Beta-Beam *Phys. Let. B* **532** 166
[2] Wildner E *et al.* 2011 Beta beams database
[3] Wildner E 2011 private communication
[4] Garoby R *et al.* 2011 *Proc. IPAC11*
[5] Benedikt M *et al.* 2008 Conceptual design report for a Beta-Beam facility *The European Phys. J. A - Hadrons and Nuclei* **47** 1-32
[6] Benedikt M *et al.* 2007 Parameter and intensity values, v.3 EURISOL DS Task note 12-25-2009-0014
[7] Gilardoni S 2011 private communication
[8] Goddard B 2011 private communication
[9] Hanke K *et al.* 2011 EDMS-1154705
[10] Magistris M, Trovati S, Kirk M and Delahaye P 2009 CERN-SC-2009-051-RP-TN
[11] Schutt P, Franchetti G, Franczak B 2004 GSI-Acc-Note-2004-05-001
[12] Giovannozzi M *et al.* 2003 *Proc. PAC03*

Collection device for the production of ^{8}Li and ^{8}B radioactive ions.

S Mitrofanov[§,£], T Delbar[§], T Keutgen[§], M Loiselet[§]
[§] Cyclotron Research Centre, Catholic University of Louvain,
Louvain-la-Neuve 1348 Belgium
[£] Flerov Laboratory of Nuclear Reactions, Joint Institute for Nuclear Research,
Dubna, 141980 Russia

E-mail: semen.mitrofanov@uclouvain.be

Abstract. The goal of the European project EUROnu is the study of different ways to produce the future neutrino beams. One of them, proposed by C. Rubbia, is based on the production of ^{8}Li and ^{8}B nuclei and their post-acceleration. Within the "beta beams" work package of this project, our task is to study experimentally the production and collection efficiency of these 2 nuclei. Our R&D work is organised in two phases: first - design and construction of a collection device which will be validated with ^{8}Li; second - experimental study of the possible ways to extract ^{8}B from the collection device.

According to ideas described in [1, 2] to produce radioactive isotopes for neutrino beam creation, the collection device is part of the program set up within the EUROnu collaboration. We describe here the progress made during the latest reporting period for EUROnu [3]. The problems initially encountered with the welding of Tantalum, which were at the origin of some delay in the project, have been successfully solved. Mainly, the reported period was dedicated to test on-line our setup with ^{8}Li produced by the reaction ^{2}H (^{7}Li, ^{8}Li) p. For this purpose, a ^{8}Li beam accelerated to 30 MeV by the cyclotron is sent on a gas cell filled with D_2 at about 200 mbar[1]. After the energy loss in the entrance foil of the cell the energy of the ^{7}Li beam is 24.9 MeV.

The recoiling ^{8}Li are collected in a tantalum tube (d = 28 mm, l = 112 mm) in which they are slowed down by a set of tantalum foils. A diffusion pipe (d = 8 mm, l = 118 mm) brings the ^{8}Li atoms to a cold plate in front of a telescope formed by 2 plastic scintillators to detect the beta decay (see Fig. 1). A set of power supplies allows heating this collection device in order to favour the diffusion of the ^{8}Li. The beam is pulsed, and ^{8}Li is detected during the beam off period. The beam intensity is monitored by the scattering of the beam on a gold foil

[1] Initially a gold plated CD_2 foil (~200 μg/cm^2) was used as a target but after several tests this solution was abandoned, due to the insufficient thermal resistivity strength.

installed just before the gas cell. The backscattered ions are detected by a Silicon detector (PIPS detector from Canberra with 300 μm thickness) which is mounted at an angle of 166^0.

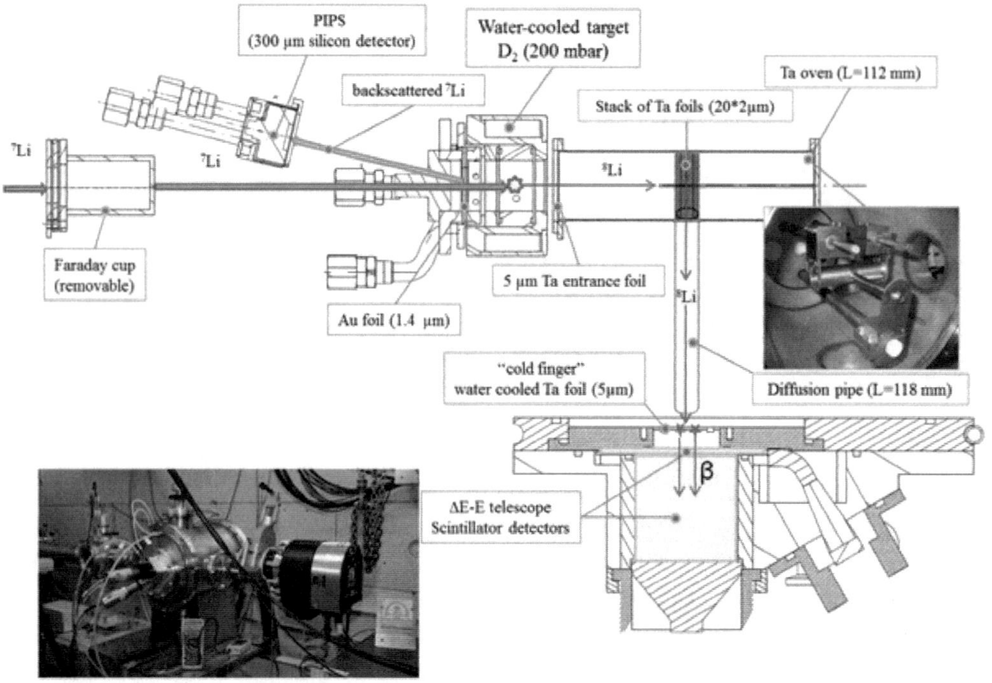

Figure1. Experimental setup (sketch).

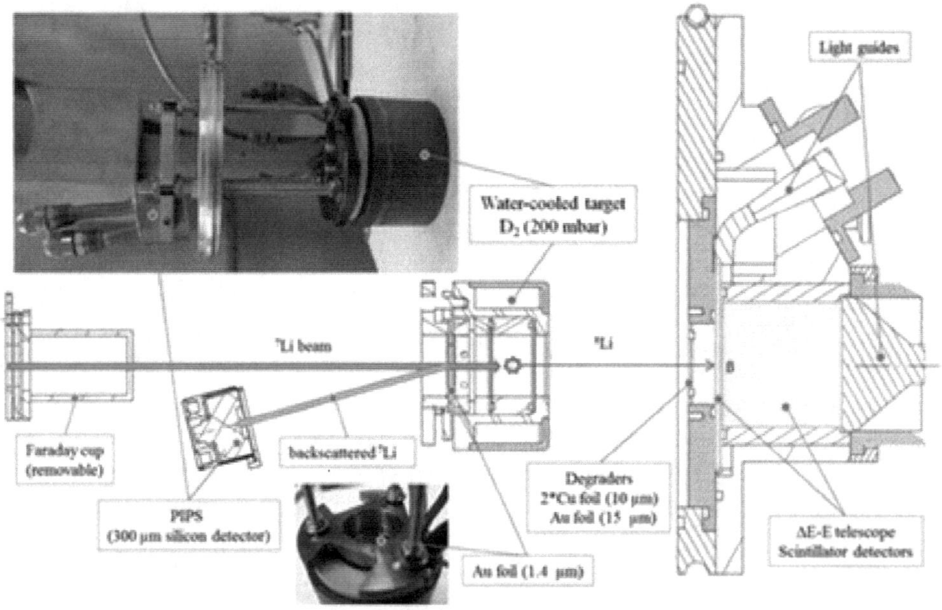

Figure 2. Experimental setup without oven (sketch).

In parallel, we made a second setup (without collection device/oven) to measure the overall production of ^{8}Li because the first one is giving the efficiency only in arbitrary units ("integral measurement").

To decrease the production rate of secondary products of ^{7}Li interaction before the telescope we use degraders (Cu and Al foils), see Fig. 2. To avoid any normalisation factors and to keep the same geometry (dimensions, distances, angles) we use the same target unit in both cases.

During the 2010-2011 period we had 10 beam runs, two of them in June 2011 were cancelled (due to cyclotron problems). Because of it we had to postpone the final ^{8}Li experiments to Fall 2011. Our test runs with the oven during 2010 have shown that the extraction of ^{8}Li will still improve if the temperature of the collection device and of the diffusion pipe is increased see test trends on Fig. 3.

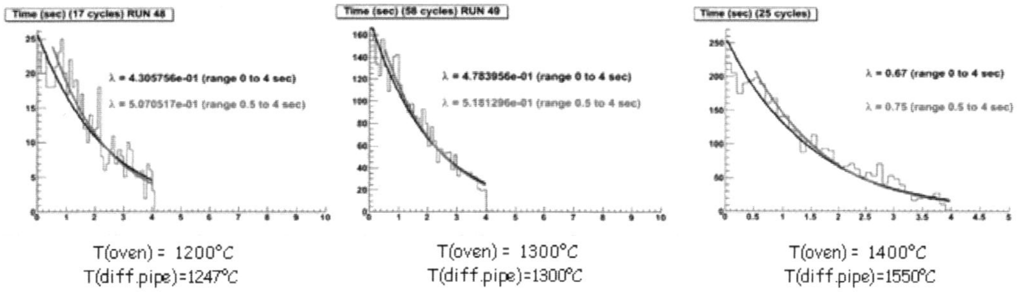

Figure 3. Decay curves depending on the oven temperature. (Decay constant for ^{8}Li: λ 0.83 s^{-1}).

We made also on-line tests for the integral measurement (the total amount of ^{8}Li produced in the target), the ^{8}Li decay constant is obtained (see Fig. 4). From the comparison of these results, it is clearly seen that we need to go higher in temperature; therefore we have ordered new power supplies.

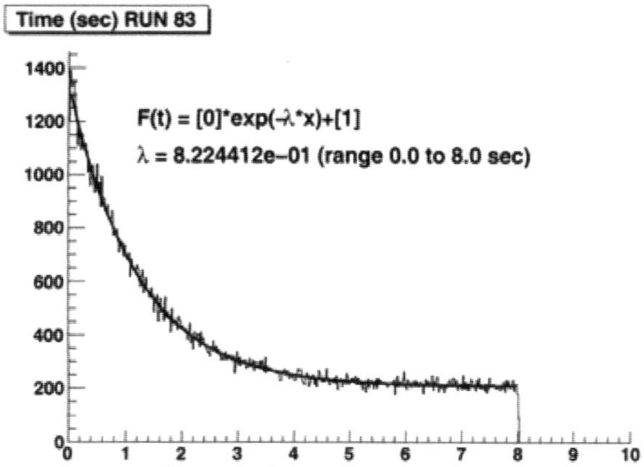

Figure 4. ^{8}Li decay curve. Integral measurement. Preliminary results.

The next step will be the measurement of the amount of ^{8}Li extracted from the target and reaching the end of a heated diffusion pipe of about 120 mm as a function of the temperature up to 2000 ^{0}C. This measurement should be finished during Fall 2011. Preliminary schedule of beam runs for the Fall 2011:

1. 2011/09/23 – integral measurement for ^{8}Li
2. 2011/10/14 - oven measurement for ^{8}Li
3. 2011/11/11 – technical run. To test ^{7}Li beam for ^{8}B
4. 2011/12/05 – reserved for oven measurement for ^{8}Li
5. 2011/12/23 – integral measurement for ^{8}B

The following progress has been made in 2010-2011:

- the preliminary results on the production and release of ^{8}Li have been obtained;
- the behaviour of the setup as a function of the temperature, up to about 1400 °C has been tested;
- the problems related to the presence of the detectors in the vicinity of the collection device at 1400°C have been solved.

In parallel, we are preparing the extraction setup for the production of ^{8}B ions. In this case, the diffusion of Boron ions will be negligible; hence Boron should be embedded in a molecule. We will slow down and stop the ^{8}B ions in AlF_3 and we expect the production of BF_3 which can diffuse far enough to reach the detection setup and, in the final project, to reach the ECR source. For the moment we are in the design study of the ^{8}B setup. These measurements must be finished in spring 2012.

References

[1] Rubbia C *et al.* 2006 *NIM* A **568** 475–487
[2] Zucchelli 2002 *Physics Letters* B **532** 166–172
[3] http://www.euronu.org/ annual report WP1-001/002/003

Target, magnetic horn and safety studies for the CERN to Fréjus Super Beam

E Baussan, M Dracos, G Gaudiot, B Lepers, F Osswald, P Poussot, N Vassilopoulos, J Wurtz and V Zeter on behalf of EUROnu WP2 group

IPHC, Université de Strasbourg, CNRS/IN2P3 F-67037, Strasbourg, France

E-mail: nikolaos.vassilopoulos@iphc.cnrs.fr corresponding author

Abstract. In the framework of the EUROnu design study, a new design for the CERN to Fréjus neutrino beam based on the SPL is under development by the WP2 group. The main challenge of this project lies with the design of a multi-MW neutrino beam facility. The horn and the decay tunnel parameters have been optimized to maximize any potential discovery. The target design, thermo-mechanical analysis, and power supply design of the horn system as well as any safety issues are being studied to meet the MW power requirements for the proton-beam.

1. Introduction

The summary of the recent target and horn studies for the CERN to Fréjus neutrino beam is presented in this paper. The main design and the physics reach of the Super Beam project are described in [1]. The optimization procedure for the horn shape and layout-geometry to achieve optimum physics, the study of a target able to withstand a multi-MW proton-beam power, multi-physics simulation to investigate heat transfer, cooling and mechanical stress for the horn and safety aspects are discussed here.

2. The proton-beam and horn/target station

A 4-MW proton-beam from CERN's SPL is foreseen to be separated by a series of kicker magnets into four beam lines. Then each beam will be focused by a series of quadruples and correctors to a four horn/target assembly (figure 1). In this way, each horn/target assembly is able to accommodate better the multi-MW power and thus increasing its lifetime [2].

A 0.25 mm thick beryllium beam window has been studied as the interface between each 1 MW proton-beam line and each horn/target assembly. Maximum temperatures as high as 180 °C and (109 °C) and Von Mises stresses as high as 50 MPa and (39 MPa) are developed respectively for water and helium cooling: these are well below the beryllium strength limit.

3. Target studies

A packed-bed target with Ti6Al4V-spheres and helium transverse cooling has been chosen as the baseline target option [3, 4]. It is placed inside the upstream part of horn's inner conductor. The advantages of the packed-bed target are among others a) large surface area for heat transfer with coolant able to access areas with highest energy deposition b) minimal thermo-mechanical and inertial stresses and c) potential heat removal rates at the hundreds kilowatt level with high helium flow rate. Advantages of the helium transverse cooling are a) almost beam neutral b) no generation of stress wave in coolant and c) low activation of coolant with no corrosion problems.

Alternatively, a pencil-like geometry of solid beryllium has been studied [3]. This pencil-like geometry gives steady-state thermal stress within acceptable range for Beryllium. Pressurized helium cooling appears feasible but center proton-beam effects could be problematic because of the stress induced: this point needs further thermo-mechanical studies.

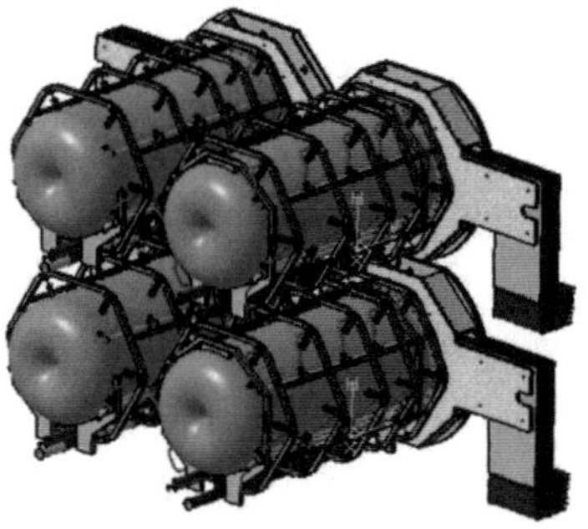

Figure 1. Four-horn assembly.

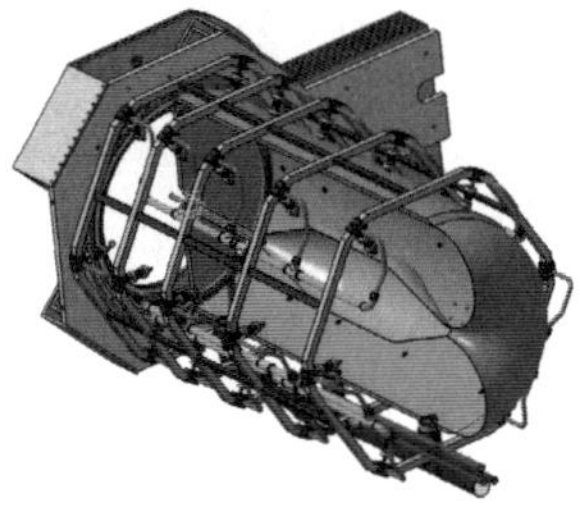

Figure 2. Horn with cooling system.

4. Horn studies

4.1. Horn shape and layout-geometry optimization

The end-design consists of an inner conductor with a cylindrically shaped upstream part to decrease the transverse momentum of the low-energy charged mesons, followed by a trapezoidal shaped middle part to select a special particle energy spectrum (for optimum physics) and finally a convex downstream plate to de-focus wrong-sign mesons that contribute to the background neutrino spectra. This configuration has been selected as the best compromise between physics performance and reliability under 1 to 1.3 MW proton-beam power [3]. The detailed drawing for a horn is shown in figure 2.

The horn parameters as well as the geometrical parameters of the decay tunnel (length and radii) are optimized for the best achievable sensitivity limit on $\sin^2 2\theta_{13}$. The beam parameters are initially scanned broadly and then restrictly in three iterations in order to minimize the CP-violation averaged 99% C.L. sensitivity limit on $\sin^2 2\theta_{13}$ [1, 5].

4.2. Horn thermo-mechanical and dynamical stress studies

The Aluminum alloy Al-6061T6 horn is subjected to cyclic deformation due to a pulsed magnetic pressure load. In addition, the temperature field creates a thermal static stress. For a non uniform cooling with maximal temperature of 60 °C, the maximal static thermal stress is calculated about 60 MPa and is located in the upstream corner and downstream top part of the horn. If a uniform temperature is achieved everywhere, the horn is expanding, and the maximum thermal static stress is 6 MPa [6]. The combined stresses in the inner conductor due to the magnetic pulses and the thermal stress for a uniform achieved temperature of 60 °C are around 20 MPa. For the inner conductor horn, the magnetic pressure pulse creates a peak Von Mises stress of about 16 MPa. The stress and the deformation are shown in figures 3 and 4.

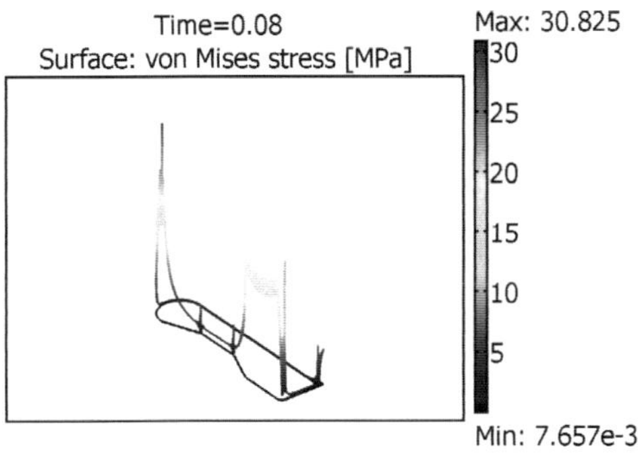

Figure 3. Horn stress for uniform cooling, s_{max} = 30 MPa.

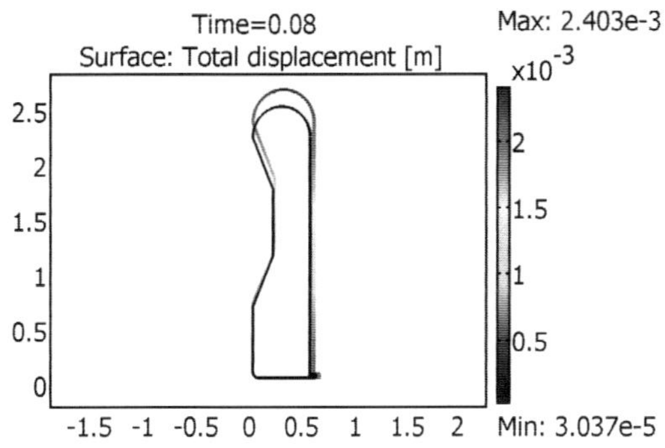

Figure 4. Horn deformation for uniform cooling, u_{max} = 2.4 mm.

4.3. Cooling system

To remove a total power of about 60 kW and maintain a temperature of about 60 ° C a water-jet cooling system is being studied. This system will be made of 5 rows of 6 nozzles (figure 2) and to spray water toward the inner conductor of the horn. The estimated water flow rate is calculated between 60 to 120 l/min per horn depending on the design. To minimize the thermal static stress, the nozzle size and disposition should be properly located in order to achieve the most uniform temperature everywhere inside the horn [7].

4.4. Fatigue

There is no fatigue limit available for Aluminum alloy, which means that fatigue data can only give a probability of failure for a determined level of stress and number of cycles. According to [8], the fatigue strength limit is 20 MPa for 10^9 pulses with a mean static stress due to thermal dilatation. For the weld junction a limit of 10 MPa should be respected to maximize horn lifetime [6].

4.5. Power supply

A one-half sinusoid current waveform with a 350 kA maximum current and pulse-length of 100 μs at 12.5 Hz frequency is needed for each horn. In order to compensate for these extreme conditions a power supply with 10 modules branched in parallel with 35 kA current each running at 50 Hz frequency is being studied. Also, the power supply components lifetime is very important as 0.216 10^9 pulses for each horn's switch and a total of 0.864 10^9 pulses per 200 days for the power supply are expected. Thus, the following designs are being investigated: a power supply with energy recovery made by a) a self that implies huge stresses on the switches and the capacitor due to high reverse voltage and current and b) a diode that implies less stress on the switches and the capacitor, but increase the electric consumption and energy thermal dissipation. Their comparison in terms of manufacturing, reliability and electric costs will define the final solution.

5. Safety

The future design of Multi-Mega Watt sources facility has to take to account the significant amount of radiation produced during beam operation and the radio-activation of the surrounding environment. The design of the shielding should reduce the dose equivalent rate to a minimal level. In order to reach these dosimetry objectives the ALARA (As Low As Reasonably Achievable) approach will be used in the design of the facility. ALARA consists of an iterative process between three phases: a) Preparation, design of the facility, dose equivalent rate map, study the intervention procedures for workers b) execution, engineering phase check/improve the dosimetry objectives and c) analyse and feedback on safety from previous experiments.

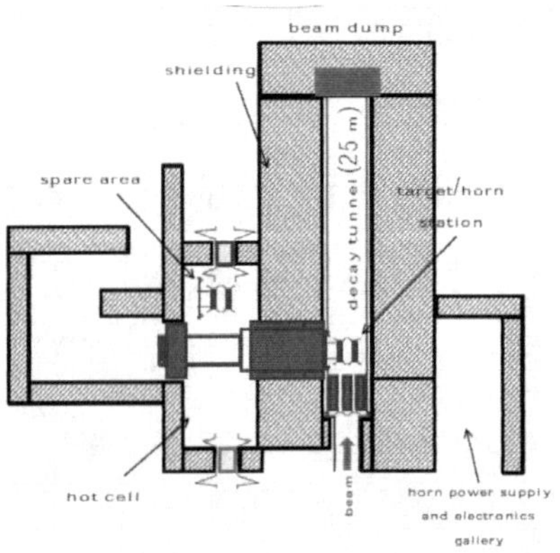

Figure 5. Super Beam infrastructure.

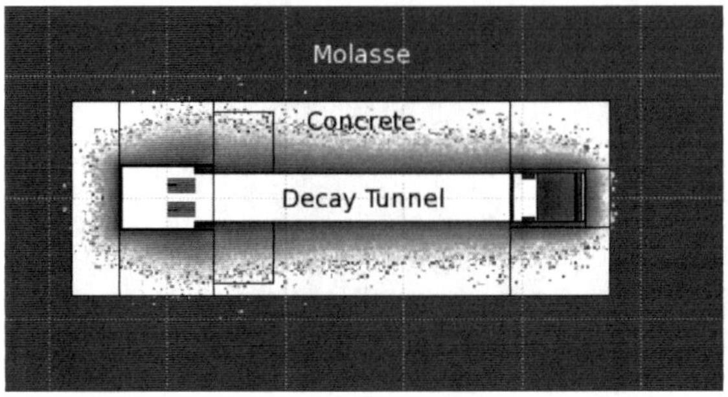

Figure 6. Activity distribution in facility.

The Super Beam infrastructure consists of: a) Proton Driver line, b) Experimental Hall (Target Station, Decay Tunnel, Beam Dump), c) Spare Area Room, d) Hot Cell, e) Service Galleries (Power supply, Cooling system) f) Air-Ventilation system and g) Waste Area (figure 5).

As example of safety Monte-Carlo studies with FLUKA, activation of the whole structure and of the molasse rock surroundings have been evaluated after 200 days of irradiation: there is minimum rock (molasse) activation surrounding the Super Beam facility due to concrete shielding with 5.5 m of thickness as shown in figure 6.

6. Conclusion

Monte-Carlo, thermo-mechanical and dynamical stress finite-elements analysis studies show that the four-horn/target system can be feasible under the extreme 4 MW proton-beam power conditions. Furthermore, R&D is needed for the target and the horn in order to study the fatigue, cooling, power supply designs and radiation degradation.

Acknowledgments

We acknowledge the financial support of the European Community under the European Commission Framework Programme 7 Design Study: EUROnu, Project Number 212372. The EC is not liable for any use that may be made of the information contained herein.

References

[1] Longin A 2010 *A new design for the CERN-Fréjus Super Beam* (EUROnu-WP2-10-04) http://www.euronu.org
[2] Bobeth C *et al* 2009 *The proton driver for the neutrino Super Beam: status and issues* (EUROnu-WP2-09-11) http://www.euronu.org
[3] Bobeth C *et al* 2011 *The target and horn for the SPL-based Super Beam: preliminary design report* (EUROnu-WP2-11-01) http://www.euronu.org
[4] Densham C 2011 *(M)MW Target and horn for neutrino super-beams* (in these proceedings)
[5] Bobeth C and Longhin A 2010 *Optimization of hadron focusing* (EUROnu-WP2-10-02) http://www.euronu.org
[6] Lepers B 2011 *Horn studies* (EUROnu-WP2-11-xy, note to be published) http://www.euronu.org
[7] Lepers B 2010 *Water jet cooling system for an electromagnetic horn* (EUROnu-WP2-10-06) http://www.euronu.org
[8] Yahr G 1993 *Proc. Conf. on Codes and Standards in a Global Environment (Colorado)* ed J Staffiera (American Society of Mechanical Engineers) p 43

Recent progress of RF cavity study at Mucool Test Area

Katsuya Yonehara on behalf of MTA working group

Fermilab, MS 221, P.O. Box 500, Batavia, IL 60510, USA

E-mail: yonehara@fnal.gov

Abstract. In order to develop an RF cavity that is applicable for a muon beam cooling channel, a new facility, called Mucool Test Area (MTA) has been built at Fermilab. MTA is a unique facility whose purpose is to test RF cavities in various conditions. There are 201 and 805 MHz high power sources, a 4-Tesla solenoid magnet, a cryogenic system including a Helium liquifier, an explosion proof apparatus to operate gaseous/liquid Hydrogen, and a beam transport line to send an intense H^- beam from the Fermilab Linac accelerator to the MTA hall. Recent activities at MTA will be discussed in this document.

1. Introduction

A muon collider is an attractive possibility for the next generation of colliders at the energy frontier. The muon provides clean collisions and due to its mass, there is ignorable energy loss from bremsstrahlung compared to an electron. Since the muon has a short lifetime, one of greatest challenges in a muon acceleration is how quick a muon beam is accelerated to achieve the desired beam energy. This requires muon phase space cooling in the acceleration process. Due to the characteristics of muon, an ionization cooling method is the most effective way. From ionization cooling simulations, the muon beam six-dimensional phase space can be cooled more than 10^6 of magnitude down within the muon lifetime [1,2]. However, a muon ionization cooling channel has never been built since the beam lattice is more complicated than a conventional beam channel. Because a beam dynamics in the ionization cooling process is perturbed by energy loss and stochastic interactions the beam lattice of cooling channel needs to have a larger acceptance than usual. The Muon Ionization Cooling Experiment (MICE) has begun to investigate the ionization cooling theory and its practicality [3].

A conceptual muon cooling channel consists of a focusing magnet, ionization cooling absorber, and RF cavity. Muon beam is focused at the absorber by the magnet to maximize the transverse momentum distribution. The amplitude of momentum is isotropically reduced by the ionization energy loss process, and an RF cavity restores the energy lost in longitudinal direction. As a result, only the transverse muon beam phase space is cooled. The cooling absorber should be a low z material to minimize a multiple scattering effect. For this reason, gaseous or liquid Hydrogen, solid Beryllium, or solid LiH prefers to be used as an ionization cooling absorber. The energy restore RF cavity needs to be located in a strong magnetic field to make a compact ionization cooling channel. However, the available RF field gradient is strongly limited by the strength of an external magnetic field. A dark current in the cavity is focused by the external magnetic field and accelerated by an RF electric field. This creates larger damage on the cavity surface by dumping its energy to smaller spot and induces RF breakdown at lower electric field in stronger magnetic fields. The goal electric field is 25 MV/m at 3

Tesla. But, the present available electric field is about a half of the goal value [4]. In order to investigate this problem, we have built a new test facility, Mucool Test Area (MTA) and have tested several conceptually new RF cavities. This document will introduce a unique feature of MTA and summarize a recent result of RF cavity tests.

2. MTA

A main purpose of MTA is to observe a characteristic of an RF cavity under the same condition as a muon ionization cooling channel. MTA has many unique instruments to accomplish this purpose. There are 201 and 805 MHz high power RF sources and waveguides. Peak powers of RF sources are 4.5 and 12 MW for 201 and 805 MHz, respectively. There is an RF switch on the 805 MHz waveguide to share the RF power into two RF cavities. There is a 4-Tesla superconducting solenoid magnet that has a 44 cm bore hole to assemble an 805 MHz cavity in the magnet. There is a liquid helium recovery system that has a large cooling power, 250 W enough to cool down the magnet as well as other cryogenic system. The apparatus in the MTA experimental hall is explosion proofed to operate gaseous and liquid Hydrogen. The MTA experimental hall is kept cleaning (Class 100) to handle a precision instrument. Recently, the commission of MTA beam line has been completed [5]. Now a 400 MeV full Linac H^- beam (7×10^{12} H^-/pulse) is used for the RF cavity test although the maximum beam repetition rate is 60 pulses per hour due to the limit of radiation hazard.

3. Test special RF cavity

3.1. 201 MHz RF cavity

A key issue to improve the RF cavity performance in a strong magnetic field is how to remove a dark current or reduce its intensity in the cavity. A possible source of dark current is a surface emission electron. This mechanism is well known as a Fowler-Nordheim model. According to the model, a probability of electron emission depends on a field enhancement factor related with a surface roughness, a work function of a cavity material, and the strength of an applied electric field. By treating a cavity wall surface with a special rinsing technique that is used for a superconducting RF cavity, the surface roughness of a normal conducting cavity becomes drastically small. Besides, using a Beryllium RF window that has a high work function makes lower probability of surface emission of electron than Copper. The special surface-treated 201 MHz pillbox cavity with a Beryllium RF window has been tested under a magnetic field by using a fringe field of the 4-Tesla solenoid. The observed electric fields were 21 and 12 MV/m at 0 and 0.75 Tesla at the center of the cavity. The cavity will be tested with a real ionization cooling magnet to measure the available electric field in 2012.

3.2. 805 MHz button RF cavity

In order to investigate the property of cavity material, an 805 MHz cavity that has a replaceable "button" electrode has been tested under a strong magnetic field. We have tested Copper, Molybdenum, Tungsten, and Tin coated Copper materials. But, none of these reached to the goal value in a strong magnetic field [6]. Unfortunately, the RF power line needs to access from longitudinal direction of the cavity in the present cavity configuration. This structure makes a high electric field spot near the RF power port. In fact, there were many pits that were made during the breakdown process around the port. This can cause an uncertainty of the measurement to find the available RF gradient. The cavity has been refurbished to eliminate such an unwanted hot spot. We plan to repeat the experiment in fall 2011.

3.3. 805 MHz box RF cavity

Another unique 805 MHz cavity test has been made. It is a box cavity to configure the RF electric field direction perpendicular to the external magnetic field. In this field configuration, a dark current is diverged from a high electric field region to the low one by an E×B force [7]. As a result, a surface emission electron is not accelerated by the electric field and absorbed on the cavity wall without heating up the surface. Experiment has been made and the observed available electric fields were 50 and 33 MV/m in 0 and 3 Tesla at the E×B angle 90 degrees, respectively. However, an allowable angle between electric and magnetic fields is only 90 ± 2 degrees. It requires a special shape RF cavity that has a half of shunt impedance as a normal pillbox cavity [8]. It will be expensive.

3.4. Gas-filled 805 MHz RF cavity

A gas filled 805 MHz RF cavity contains a dense buffer gas in the cavity to stop electron acceleration by an RF field. The gas even avoids focusing of the dark current in a strong magnetic field. Besides, the gas can be used as the ionization cooling absorber when gaseous hydrogen is filled in the cavity. A high pressure hydrogen gas filled RF cavity has been demonstrated in a strong magnetic field and no RF field degradation has been observed [9]. A crucial issue on this cavity is a beam-induced plasma loading effect. Ionized electrons in the cavity, which is produced by an interaction between an incident charged particle and a dense Hydrogen gas, can consume a stored RF power. Even ionized electrons may induce the RF breakdown. The beam test has been made at MTA to measure the beam-induced plasma loading effect. First of all, no RF breakdown has been observed. The observed RF power consumption rate is good agreement with the expected value that comes from a plasma dynamics model in a DC field. For example, an RF power consumption of single electron is 5×10^{-17} Joules/cm/RF cycle/electron at applied RF electric field 20 MV/m. An electronegative dopant gas effect has also been tested. A 0.01 % of SF6 was doped in a pure hydrogen gas. As a result, the electron capture rate in the doped hydrogen gas was two orders of magnitudes better than that in a pure hydrogen gas. Figure 1 shows the plasma loading effect in a pure Hydrogen gas (red) and Hydrogen gas with 0.01 % SF6 dopant gas (blue). It is clear that the plasma loading is significantly mitigated by SF6. The detail analysis is under study.

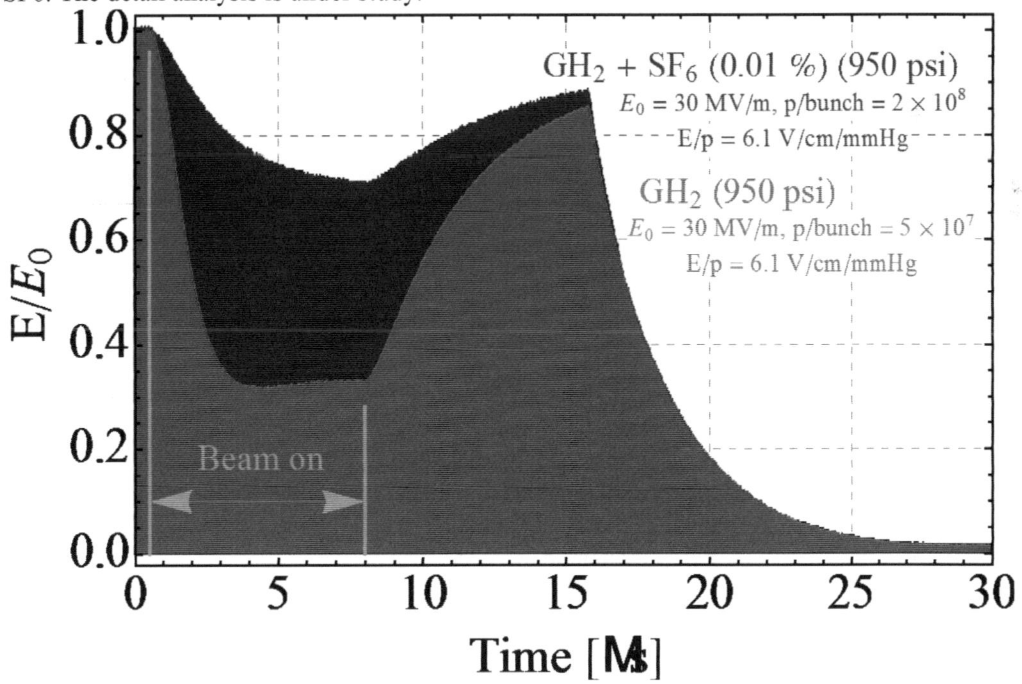

Figure 1. Observed RF pickup signals in a pure hydrogen gas and hydrogen gas with 0.01 % SF6 dopant gas.

4. Conclusion

Within the first five years of the MTA, development of RF technology for an ionization cooling channel is progressing at a rapid pace. Gas-filled RF cavities have been demonstrated to drastically mitigate the effects of RF breakdown, and are therefore a good candidate for a cooling channel. A total RF power consumption of the plasma loading in a gas-filled RF cavity with a real muon beam for a collider can be estimated from the experimental result. The number of ionized electrons in one muon beam bunch will be 10^{14}~10^{16} electrons. Thus, from the observed RF power consumption per single electron (5×10^{-17} Joules/RF cycle/electron/cm), a total RF power consumption is 0.005~0.5 Joules/RF cycle/cm. It corresponds to 0.1~10 % of stored energy in a 200 MHz pillbox cavity in a unit length. These values can be overestimated at dense plasma condition in a real cooling channel since this simple power loss estimation does not include the recombination process. We need more beam test.

We also demonstrated that the beam plasma density was significantly reduced by using a small amount of electronegative dopant gas. 0.01 % of SF6 dopant reduces factor 100 of the electron density. However, SF6 gas may not be practical or the ionization cooling channel since the channel will be operated at cryogenic temperature. Oxygen will be a good candidate of capturing electrons since the lowest flammable level (LFL) of Oxygen molecule in Hydrogen gas is 7 % that is two orders of magnitude higher than the required amount of doping rate to remove the electron swarm. Researching electronegative gas should be made.

Once we find the possible RF cavity for an ionization cooling channel, we will integrate all muon beam cooling elements including with an infrastructure and demonstrate an entire system in MTA.

References

[1] Yonehara K *et al.* 2010 *Proceedings of IPAC 2010* MOPD076 pp 870
[2] Palmer R *et al.* 2005 *Phys. Rev. STAB.* **8** 061003
[3] *Proposal to the Rutherford Appleton Laboratory* http://mice.iit.edu/mnp/MICE0021.pdf
[4] Moretti A *et al.* 2004 *Proceedings of LINAC 2004* TU204 pp271
[5] Moore C D *et al.* 2010 *Proceedings of IPAC 2010* THPEB038 pp 3966
[6] Huang D *et al.* 2009 *Proceedings of PAC 2009* TU5PFP032 pp 888
[7] Torun Y *et al.* 2010 *Proceedings of IPAC 2010* THPEA054 pp 3795
[8] Stratakis D *et al.* 2011 *Proceedings of PAC 2011* THOAS4
[9] Hanlet P *et al.* 2006 *Proceedings of EPAC 2006* TUPCH147 pp 1364

MICE Step 1: First measurements of emittance using particle physics detectors

David Adey on behalf of the MICE collaboration
Physics Department, University of Warwick, Coventry, UK, CV47AL
E-mail: d.adey.1@warwick.ac.uk

Abstract. The muon ionization cooling experiment (MICE) is a strategic R&D project intended to demonstrate the only practical solution to prepare high brilliance beams necessary for a neutrino factory or muon colliders. MICE is under development at the Rutherford Appleton Laboratory (UK). It comprises a dedicated beam line to generate a range of input emittance and momentum, with time-of-flight and Cherenkov detectors to ensure a pure muon beam. The emittance of the incoming beam is measured in the upstream magnetic spectrometer with a fibre tracker. A cooling cell will then follow, alternating energy loss in LH_2 absorbers and RF acceleration. A second spectrometer identical to the first and a second muon identification system measure the outgoing emittance. In the 2010 run the beam and most detectors have been fully commissioned and a first measurement of the emittance of a beam with particle physics (time-of-flight) detectors has been performed. The next steps of more precise measurements, of emittance and emittance reduction (cooling), that will follow in 2011 and later, will also be outlined.

Introduction

In order for a muon beam produced by the target of a neutrino factory to meet the downstream acceptance of the accelerator chain it must undergo a reduction in its emittance. Current methods of emittance reduction, or cooling, are too slow given the short lifetime of the muon. A new method is proposed in ionisation cooling [1]. An absorber material is used to reduce the total momentum of a muon, maximising energy loss against multiple scattering, and RF cavities restore momentum longitudinally. The net loss of transverse momentum results in a reduction in the transverse emittance.

MICE

The Muon Ionisation Cooling Experiment (MICE) [2] based at the Rutherford Appleton Laboratory (UK), aims to demonstrate ionisation cooling with a 10 % reduction in transverse emittance measured to within 1 %. A mechanical target dips into the ISIS proton synchrotron producing pions which decay into muons in the MICE beamline, formed from nine quadrupoles, two dipoles and a superconducting solenoid. The emittance of the beam will then be deliberately inflated with a diffuser of varying thicknesses; MICE will in this way be able to measure the emittance reduction for various emittance and momentum beams, ranging from 3 to 10 mm transverse emittance and 140 MeV/c to 240 MeV/c momentum.

Two scintillating fibre trackers [3] placed within 4 T superconducting solenoids will make precision measurements of the position and momentum of each muon, allowing for a measurement of emittance based on an ensemble of single particles. Time of flight detectors provide an accurate timing measurement for use in particle identification based on the known momentum and time of flight between detectors. A calorimeter-like detector, the electron-muon ranger (EMR), will add sensitivity to muons which have decayed in the MICE beamline. An additional KLOE-like (KL) scintillating fibre detector is placed in front of the EMR. Currently, all three TOFs, KL, Cerenkov PID detectors and luminosity monitors are installed. The trackers are due to be installed in 2012.

The absorber material chosen in MICE is liquid hydrogen, and some solid absorbers such as lithium hydride are under consideration. The acceleration will be performed by 8 MV 201 MHz RF cavities. The arrangement of the detectors, absorbers and RF cavities is shown in Figure 1.

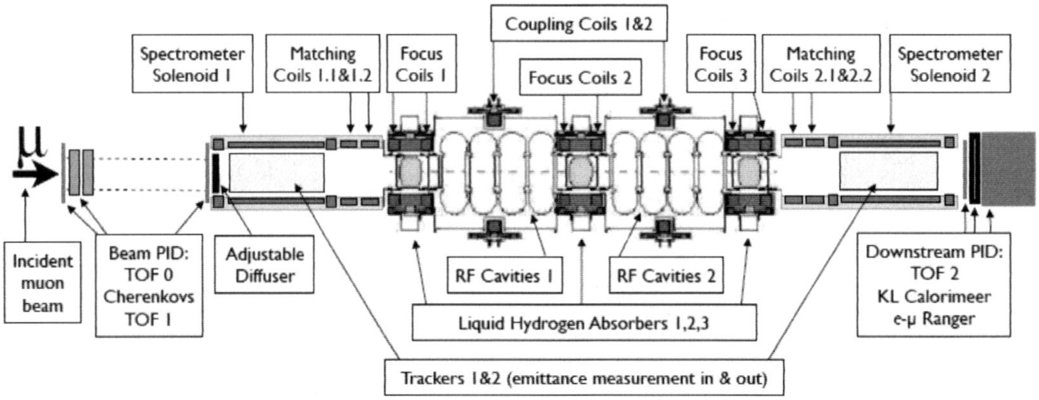

Figure 1. Layout of the Muon Ionisation Cooling Experiment, showing the full Step 6 MICE cooling channel.

Beam Rates

The rate of muon tracks seen in MICE is a consequence of the pions produced by the interactions of protons on the target. To maximise the muon rates the target would present maximum material to the proton beam. However, the resulting losses in the ISIS synchrotron are of concern, and so beam losses must be kept within set limits. A clear relationship has been shown [4] between the beam losses observed by ISIS and the number of tracks reconstructed as muons in the MICE hall. This is shown in Figure 2.

Time of Flight detectors

Three scintillating bar time of flight detectors with high time precision have been constructed for use in MICE [5]. The scintillating bars are arranged in an $x - y$ formation, with 10, 7 and 10 bars in each plane for TOFs 0, 1 and 2 respectively. The bars themselves are between 4 and 6 cm wide and approximately 2.5 cm thick. A calibration procedure accounting for time walk and other factors has produced a resolution of less than 60 ps for each time of flight detector. The resolutions are shown in Figure 3 and the individual resolutions are given in the caption. Using the $x - y$ geometry and the time difference measured between the photomultipliers at each end of the bars, a position measurement is also obtainable.

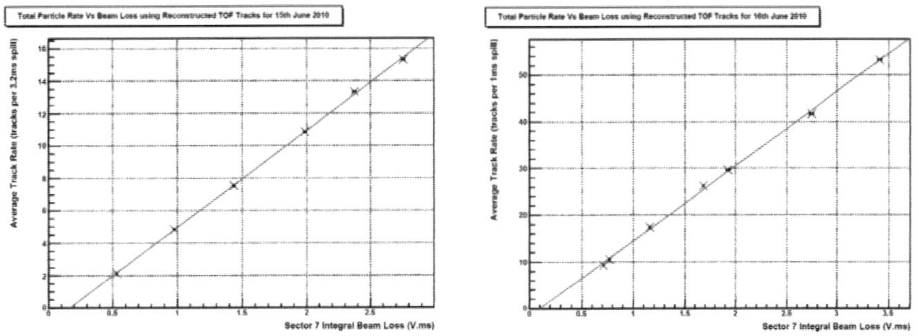

Figure 2. Reconstructed TOF tracks as a function of ISIS beam losses. Rates are shown for 1 ms (left) and 3.2 ms (right) spill gates.

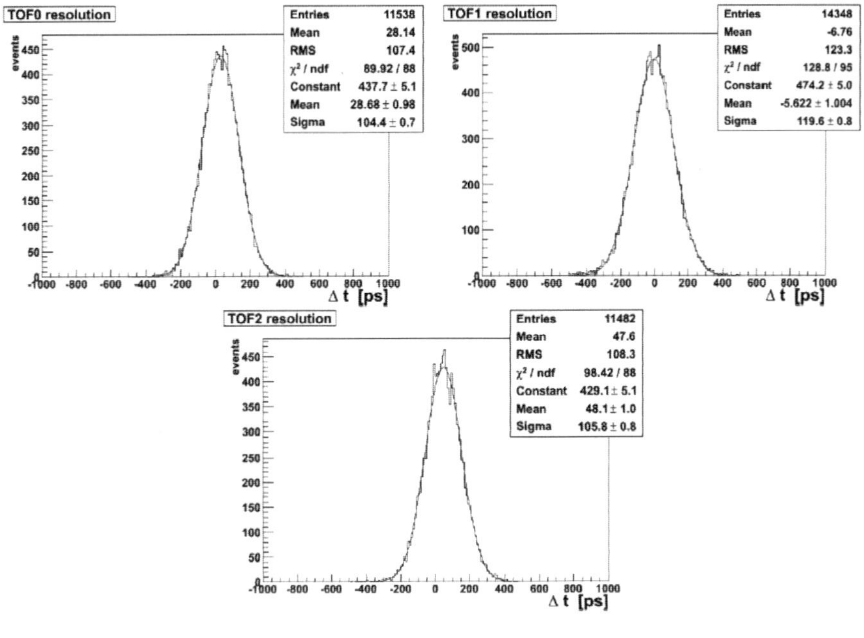

Figure 3. Timing resolution of the MICE time of flight detectors. Resolutions of 51 ps, 58 ps and 52 ps have been shown for TOFs 0, 1 and 2 respectively. The σ shown on the figures represents approximately twice the calibrated resolution, which is found from one plane only in each detector.

Measurements of emittance

Using the position and time information provided by the time of flight detectors, an iterative reconstruction method was developed [6] to calculate the single particle transfer matrix between two detectors, including the intervening quadrupole optics. In this manner, the optical functions of the beam were calculated. Based on an intensive period of running comprising Step 1 of MICE, the full range of operating modes, for both positive and negatively charged muons, were investigated. The reconstructed emittance and β functions for these modes are shown in Figure 4.

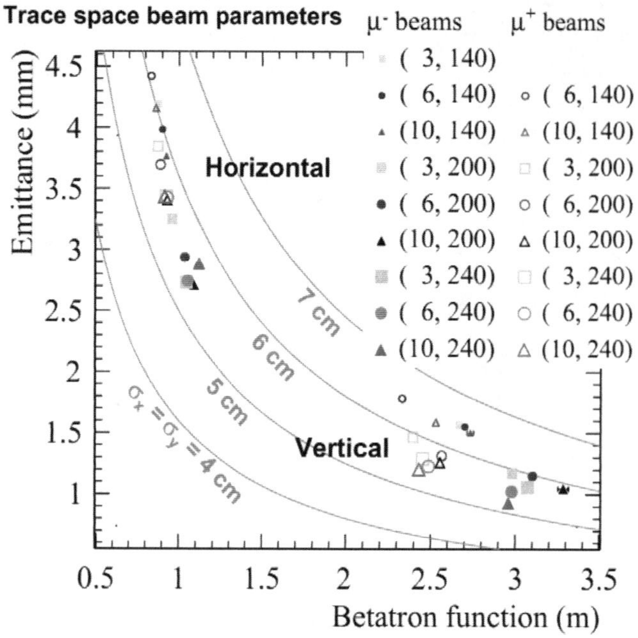

Figure 4. Reconstructed emittance (mm rad) and beta functions for the operating modes of MICE, including positive and negative polarity beamlines. The values in parentheses refer to the emittance (mm) and momentum (MeV/c) for which that beam is intended in the presence of the diffuser.

Conclusion

A run campaign during 2010, marking Step 1 of MICE, has provided data at each of the target operating modes for MICE. Using a new reconstruction method the emittance and betatron function of each of these beams has been calculated, although this measurement is preliminary based on the detectors available in the beamline. The full precision measurement of emittance will await the installation of the trackers. The MICE muon beamline has been commissioned, and the installed detectors shown to be operating to specification.

References

[1] Neuffer D (1983) *Principles and Applications of Muon Cooling Particle Accelerators Vol 14 75*

[2] Gregoire G *et al.* (MICE Collaboration) 2003 *An International Muon Ionization Cooling Experiment* (MICE) MICE Note 21

[3] Ellis M *et al.* 2011 *The design, construction and performance of the MICE scintillating fibre trackers Nucl. Intstr. Meth. A Vol. 659 Iss. 1 136-153*

[4] Dobbs A 2011 *Particle Rate and Host Accelerator Beam Loss on the MICE Experiment* PhD thesis Imperial College London

[5] Bertoni R *et al.* 2010 *The design and commissioning of the MICE upstream time-of-flight system Nucl. Intstr. Meth. A Vol. 615 Iss. 1 14-26*

[6] Rayner M A 2011 *The development of a novel technique for characterizing the MICE muon beam and demonstrating its suitability for a muon cooling measurement* PhD thesis The University of Oxford

Progress in the construction of the MICE cooling channel

Gail G Hanson[1] **(on behalf of the MICE collaboration)**
Department of Physics and Astronomy,
University of California, Riverside, Riverside, California 92521, USA

E-mail: Gail.Hanson@ucr.edu

Abstract. The international Muon Ionization Cooling Experiment (MICE), sited at Rutherford Appleton Laboratory in the UK, aims to build and test one cell of a realistic ionization cooling channel lattice. This comprises three Absorber–Focus-Coil (AFC) modules and two RF–Coupling-Coil (RFCC) modules; both are technically challenging. The Focus Coils are dual-coil superconducting solenoids, in close proximity, wound on a common mandrel. Each pair of coils is run in series, but can be configured with the coil polarities the same ("solenoid mode") or opposite ("gradient mode"). At the center of each FC there is a 20-L liquid-hydrogen absorber, operating at about 14 K, to serve as the energy loss medium for the ionization cooling process. The longitudinal beam momentum is restored in the RFCC modules, each of which houses four 201.25-MHz RF cavities whose irises are closed with 42-cm diameter thin beryllium windows. To contain the muon beam, each RFCC module also has a 1.4-m diameter superconducting coupling solenoid surrounding the cavities. Both types of magnet are cooled with multiple two-stage cryo-coolers, each delivering 1.5 W of cooling at 4 K. Designs for all components are complete and fabrication is under way. Descriptions of the various components, design requirements, and construction status will be described.

1. Introduction

The international Muon Ionization Cooling Experiment (MICE) at the Rutherford Appleton Laboratory (RAL) [1], shown in figure 1, aims to build and test one cell of a realistic ionization cooling lattice and was designed to measure a 10% emittance reduction by ionization cooling to within 1%, an absolute emittance measurement of 0.1%. MICE is being assembled in Steps, illustrated in figure 2, as detector components become available, with data taken at each step. The MICE beam line, Step I detectors, and results from Step I data taking are presented in [2]. (Steps II and III consisted of one spectrometer solenoid and tracker and both spectrometer solenoids and trackers but without the liquid hydrogen absorber, respectively, and will be combined in Step VI.)

Measurements of the initial and final emittances using the spectrometer solenoids, trackers, focus coil, and various absorbers will take place in Step IV. In Step V the re-accelerating RF, coupling coil, and second absorber-focus-coil (AFC) module will be added. The second RF-coupling-coil (RFCC) module and third AFC module will be installed in Step VI to complete the full cell of the cooling channel, and studies of the cooling channel will be carried out.

[1] Supported by the US National Science Foundation PHY-0970178.

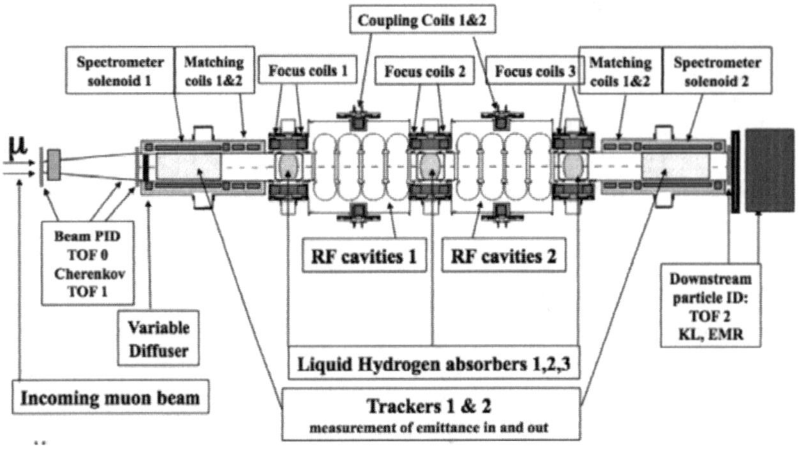

Figure 1. MICE layout.

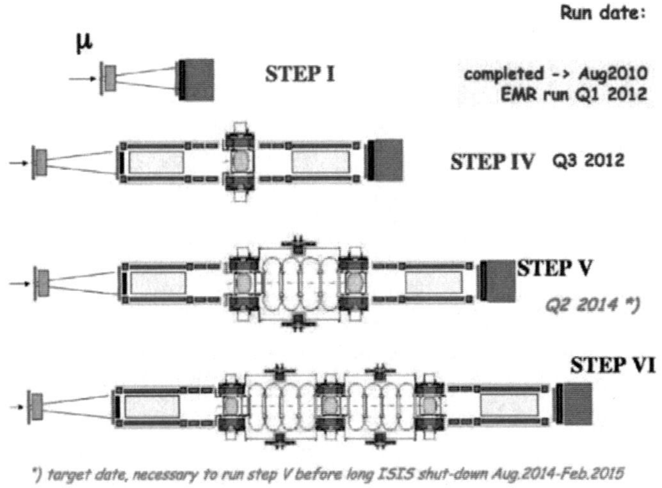

Figure 2. MICE schedule.

2. MICE Step IV

The goal of Step IV is the measurement of the ratio of emittances before and after the absorber. The spectrometer solenoids, trackers, and absorbers will be installed by the third quarter of 2012, with data taking during the fourth quarter of 2012 and first quarter of 2013.

The two scintillating fiber trackers with VLPC readout [3] were constructed with contributions from the Japan, the UK and the US institutions and have been operational since 2009. The trackers are used to measure the emittance using a single-particle technique. They are currently being re-commissioned with cosmic data and upgraded software.

The two spectrometer solenoid magnets are superconducting magnets, each consisting of five coils wound on a common aluminum mandrel. Cooling of the radiation shield and cold mass is provided by a series of fivetwo-stage cryocoolers and one single-stage crycooler. Liquid helium is maintained in the cold mass by means of a recondensation circuit. The magnets, one of which is shown in figure 3, are being built by Wang NMR in Livermore, California, with management now directly under the Lawrence Berkeley National Laboratory (LBNL) Accelerator and Fusion Research Division (AFRD), which has led to a substantial increase in resources. Good MICE magnet coordination has been established. Reconsiderations of the issues of cooling a large magnet with cryocoolers plus electromagnetic coupling between the coils has led to better solutions for thermal balance and quench protection of the magnet and leads. A plan and milestones for Step IV running have been established.

The AFC module consists of a focus coil magnet assembly with two superconducting coils that can be operated with the same ("solenoid mode") or opposite ("flip mode") polarity, a liquid hydrogen absorber body, and absorber windows and safety windows. A liquid hydrogen storage and delivery system is also required. The magnet assembly, shown in figure 4, is being constructed by Tesla in the UK. The first liquid hydrogen absorber was built and cooling-tested at KEK and has now been delivered to RAL. The absorber windows, one of which is shown in figure 5, are being provided by the University of Mississippi (United States).. There is also a system to install solid absorbers.

Data will be taken with vacuum (with only helium and windows for the tracker), liquid hydrogen absorber empty, liquid hydrogen absorber full, and various solid absorbers (lithium hydride disks and a wedge-shaped absorber to study emittance exchange).

There is good progress on the focus coil, the absorbers, and the liquid hydrogen system.

Figure 3. Spectrometer solenoid at Wang NMR.

Figure 4. Focus coil (Tesla Eng., Ltd.).

Figure 5. Absorber window (University of Mississippi).

3. MICE Step V

The goal of Step V is to gain experience with the RFCC module operation. The RFCC module, a second AFC module, and RF power, controls and connections will be installed. Tests of "sustainable cooling," with the momentum lost in the absorbers restored by the RF cavities, begin with Step V. The aim is to take Step V data before the long ISIS shutdown August 2014 to February 2015.

The coupling coils located around the RF cavity assemblies provide additional longitudinal magnetic field to confine the beam between the absorbers. The coupling coil design and construction are being carried out by a collaboration between LBNL and the Harbin Institute of Technology (HIT) Institute of Cryogenics and Superconductivity Technology in China, with management now directly under the LBNL AFRD, which has led to a substantial increase in resources. The first coil has been wound at Qi Huan Corp, China, and shipped to HIT in June 2011. Welding of the cold mass cover plate will take place at HIT. The coil will then be shipped to the US. The plan is to test the coil in a cryostat, probably at Fermilab. The test of the first coil is critical. More conductor will be ordered for the two remaining coils (an additional coil for MICE and one for the Fermilab MuCool Test Area). The design of the quench protection and leads is underway.

RF cavity production is advancing well. Ten (8 + 2) cavity bodies have been delivered to LBNL. Nine titanium-nitride coated beryllium windows have also been delivered and accepted. A single-cavity test module has been designed and will be tested at Fermilab MuCool Test Area (MTA). The problem of RF cavity breakdown in high magnetic fields is under investigation at the MTA , and the first coupling coil will be used for this crucial test.

The RF power distribution system, shown in figure 6, and infrastructure have been designed. RF power units from LBNL and CERN are being refurbished. The LBNL unit has been tested at 1 MW output power with no evidence of significant X-ray production and will now be pushed towards 2 MW. The CERN unit will be reassembled and tested next. The total RF power needed will be 4 MW for Step V and 8 MW for Step VI.

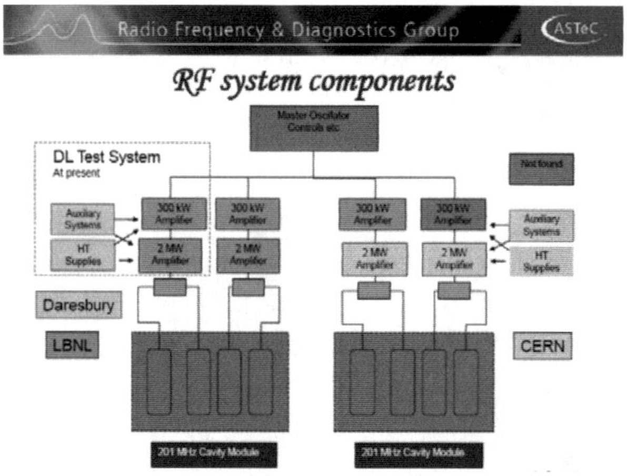

Figure 6. RF power distribution system.

4. MICE Step VI

The addition of another RFCC module and a third AFC module will bring MICE to Step VI, the full cell of the cooling channel, which has the optics and characteristics needed for 10% or more emittance reduction. Data will be taken to study the performance of the cooling channel, the goal of MICE. If the Step V RFCC module is not ready before the long ISIS shutdown, proceeding directly to Step VI will be considered, although there are some risks involved.

5. Conclusions

We have made good progress on the re-baselined MICE schedule. We have a plan and milestones established for the spectrometer solenoids for Step IV running in October 2012. MICE aims to take data in Step V mode before the ISIS long shutdown in August 2014. Completing the coupling coils is critical to this plan, and the schedule depends on the cryostat test of the first coil. MICE is on the way towards demonstrating ionization cooling!

References

[1] Gregoire G *et al.* 2003 *An international muon ionization cooling experiment (MICE)* Proposal to the Rutherford Appleton Laboratory MICE-NOTE-GEN-0021

[2] Adey D 2011 *First measurements of emittance at MICE* these proceedings

[3] Ellis M *et al.* 2011 *The design, construction and performance of the MICE scintillating fibre trackers Nucl. Instrum. Meth.* A **659** pp 136-153

Journal of Physics: Conference Series **408** (2013) 012065 doi:10.1088/1742-6596/408/1/012065

RF Cavity Processing and Testing Plan for MICE

Derun Li*, Allan DeMello†, S P Virostek†, and Michael S. Zisman*

*Accelerator & Fusion Research Division, Lawrence Berkeley National Laboratory, Berkeley, CA 94720 U.S.A.
†Engineering Division, Lawrence Berkeley National Laboratory, Berkeley, CA 94720 U.S.A.

E-mail: dli@lbl.gov

Abstract. Fabrication status of the MICE RF cavities will be described. Design of a single-cavity vessel to test the cavities will be discussed, along with the plans for processing and testing the cavities.

1. Introduction

The international Muon Ionization Cooling Experiment (MICE), sited at Rutherford Appleton Laboratory (RAL), is currently under way. At the present time the muon beam line has been successfully installed and commissioned, and components for the remainder of the experiment are being fabricated. Figure 1 shows a layout of the MICE cooling channel.

Figure 2 shows two views of the RFCC module, indicating the locations of its four RF cavities. Each of the cavity irises is terminated with a curved 42 cm diameter Be window. The power is fed into each cavity via a pair of coaxial loop couplers. Tuning is accomplished by pushing or pulling on the cavity via the six tuner arms. The RF cavities compensate for the muon energy loss in the liquid-hydrogen absorbers (located in the center of each FC module), maintaining the longitudinal momentum.

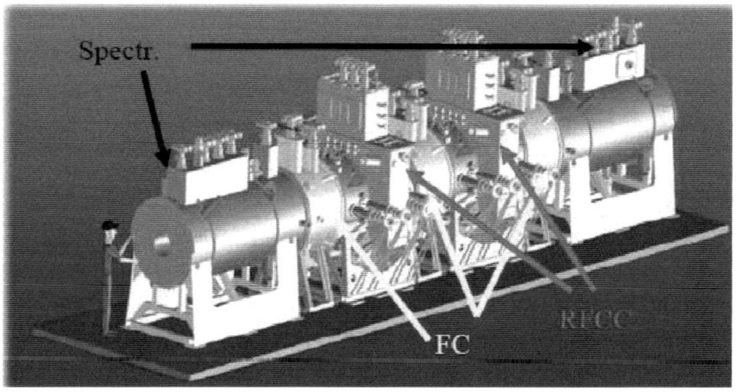

Figure 1. Layout of the MICE cooling channel, showing a cooling cell comprising three Focus Coil (FC) modules and two RF–Coupling-Coil (RFCC) modules. At each end is a spectrometer solenoid magnet containing a scintillating fiber tracker that measures particle coordinates before and after the cooling channel cell.

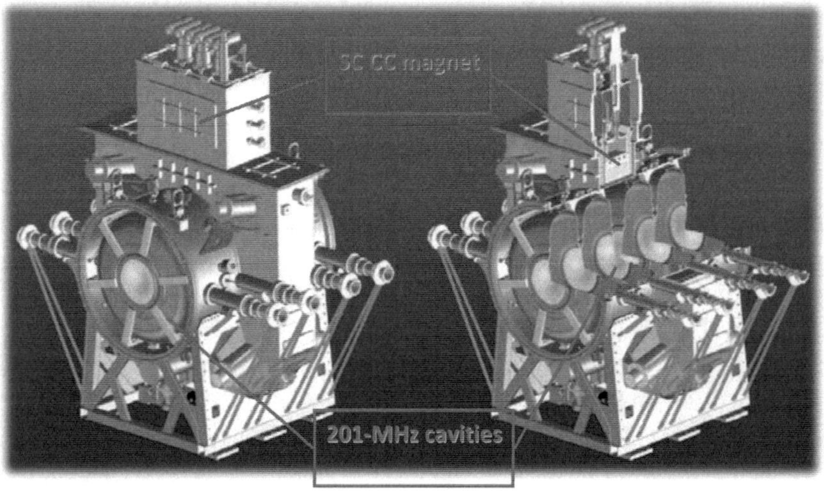

Figure 2. (*left*) drawing of the RFCC module indicating locations of superconducting Coupling Coil magnet and 201 MHz RF cavities; (*right*) cutaway view.

2. Cavity Measurement and Testing

The MICE RF cavities (8 for the two RFCC modules and 2 spares) were delivered to LBNL in December 2010. The actual MICE cavities already have their cooling pipes attached; the cooling pipes are not attached for the two spares so that they can be used for either of the two configurations needed for MICE. To date, five of the cavities have been measured, including both mechanical and RF measurements.

From other vendors, we have received and accepted nine of the 42 cm diameter curved Be windows used to electromagnetically terminate the cavity irises. Ten ceramic RF windows have also been received from a vendor, of which 4 are at LBNL and 6 are with our collaborators at University of Mississippi. Design of the RF input coupler is complete and the device is ready for fabrication.

The tuner design, based on using 6 tuner arms per cavity, operated in a push-pull configuration to tune the cavity frequency via shape changes (similar to the approach used for superconducting cavities), is complete. A prototype of a single tuner arm with its control circuitry has been successfully tested on a dummy load. A full set of six tuner arms for testing the cavities is being fabricated now. The mechanical designs for the cavity supports within the vacuum chamber, and the vacuum chamber itself are also complete.

The remaining step before testing the cavities is to complete their processing, including mechanical polishing, electropolishing, and a high-pressure water rinse. These tasks will be carried out at LBNL under close supervision. To facilitate testing, a single-cavity vacuum vessel has been designed and is now being fabricated.

The measurement program we envision includes the following:

- Physical measurements
  - cavity, tuners, tuning mechanism, support structure
  - vacuum system, cooling water, and integration scheme
- RF measurements (both with and without the Be windows)
  - low-power determinations of RF parameters (f, Q, coupling), tuning range and sensitivity, tuning circuit mechanism and control circuitry
  - high-power processing to high gradient and tests of operational reliability

The high-power measurements will make use of the single-cavity vacuum vessel (see Section 3) to avoid postponing initial testing until the modules are fully assembled in the MICE Hall at RAL.

3. Single-Cavity Vacuum Vessel

3.1. Design
The design goal for this chamber was to keep as many as possible of the features and dimensions of the actual MICE RFCC vacuum vessel, but in a system designed to hold only one cavity at a time. The resultant design is illustrated in Fig. 3.

3.2. Test Plan
The first check will be on the engineering and mechanical design in order to fully define the fabrication process for the real RFCC vacuum chamber. Any required changes in the single cavity vessel would subsequently be incorporated into the RFCC design. Another test will be to make use of the cavity tuning system and its controls to verify the tuning range and reproducibility of the system.

Although the test chamber only accommodates one cavity, it will permit us to develop the techniques for cavity installation, including

- fixturing for installing and aligning cavities using hexapod struts
- Be window installation procedures
- RF coupler connections
- cooling water connections
- vacuum port connections and getter performance
- tuning and actuator circuit connections
- vacuum vessel handling

and to obtain hands-on experience in using them. Finally, the test vessel will serve as a chamber to test and process each of the MICE cavities. For one of the cavities, we intend to explore the effects of LN-temperature operation, as this is a possible option for the cavities at MICE.

4. Preparations for Cavity Processing and Testing
In preparation for RF testing, each cavity will be electropolished at LBNL, followed by a high-pressure water rinse. Figure 4 shows the setup for electropolishing, based on the techniques developed and carried out by JLab staff for the existing MuCool test cavity. Plans call for beginning the electropolishing in October 2011.

The parameters that must be monitored during high-power commissioning and operation include:

- cavity body, water, window, and RF circulator temperatures
- RF amplifier forward and reflected power
- cavity frequency, RF voltage amplitude and phase (referenced to drive power)
- cavity and window region vacuum
- x-ray radiation level
- sparking rate
- status of interlocks (machine protection, personnel protection, RF amplifiers).

Figure 3. Single-cavity vessel: (*left*) end view; (*center*) side view; (*right*) cutaway view.

Figure 4. (*left*) RF cavity electropolishing setup; (*right*) close-up of polishing electrode.

The list of low-level RF control points that we expect to use is:

- peak power limit for a given pulse length
- duty factor
- pulse length
- initial frequency and frequency correction increment
- initial power, power increment, and maximum peak power
- delay from sparking event to restart
- reflected power threshold and detection delay
- vacuum burst limits
- temperature limit
- predetermined actions for various types of detected events

Using the information above, we will carry out cavity conditioning for each of the 8 MICE cavities. We will measure the cavity frequency shift as a function of average power and time, particularly that due to heating of the Be windows. Measurements will then be repeated with the active tuners to measure tuning range and sensitivity. If possible, we would like to repeat these measurements in the presence of a suitable magnetic field, either at the Fermilab MuCool Test Area (MTA) or at RAL.

5. Summary

A test plan for the MICE RF cavities has been developed based on using a specially designed single-cavity test vessel, whose fabrication is now under way. In the near term, the plan includes completing RF measurements on the second group of five cavities at LBNL, carrying out the required cavity cleaning and polishing activities, coarse cavity tuning via cavity deformation to set the central frequencies to a common value, and completion of an initial tuner system (6 arms plus actuators) and other required ancillary items for the testing process.

Acknowledgments

We would like to thank Andy Moss and the Daresbury Laboratory RF group for helpful discussions of the test procedures and instrumentation. We also thank Alan Bross and Yağmur Torun for useful discussions about MTA operations issues.

This work was supported by the Director, Office of Science, Office of High Energy Physics, of the U.S. Department of Energy, under Contract No. DE-AC02-05CH11231.

References

[1] See http://mice.iit.edu/.

Journal of Physics: Conference Series **408** (2013) 012066 doi:10.1088/1742-6596/408/1/012066

Muon RLA - design status and simulations

K B Beard[1], S A Bogacz[2], V S Morozov[2], Y R Roblin[2]

[1]Muons, Inc., Batavia, IL, USA

[2]Jefferson Lab, Newport News, VA, USA

Email: beard@muonsinc.com

Abstract. The Neutrino Factory baseline design involves a complex chain of accelerators beginning with a linac. This first pre-linac follows the capture and bunching section and accelerates the muons from about 244 to 900 MeV and must accept a high emittance beam about 30 cm wide with a 10% energy spread. It uses counterwound, shielded superconducting solenoids and 201 MHz superconducting cavities, and currently consists of 24 3 m and 24 5 m long cryomodules. The next stage is a 1st dogbone-shaped RLA that takes the total energy from 900 MeV to 3.6 GeV in 4.5 passes, followed by a 2nd RLA that takes the energy from 3.6 to 12.6 GeV in 4.5 passes. Simulations are in progress to optimize the optics and determine the radiation loads from beam loss and muon decay.

1. Overview

The baseline Neutrino Factory (NF) acceleration scheme has been evolving for some time. It begins with a pre-linac that takes the muons from about 244 MeV total energy to around 900 MeV; from there, it is injected into the middle of a 4.5 pass dogbone shaped Recirculating Linear Accelerator (RLA) that accelerates the beam to 3.6 GeV; next it is injected into the middle of second dogbone shaped RLA which takes the beam to 12.6 GeV in 4.5 passes (Fig 1.). Muons of opposite sign travel in alternating buckets in the same direction down the linacs and around the arcs in opposite directions.

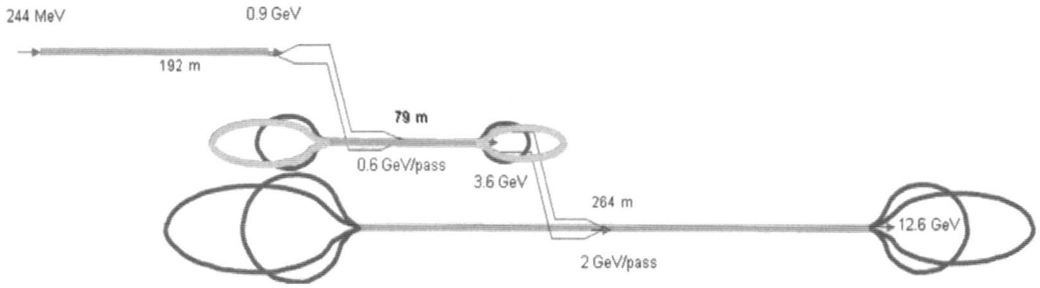

Figure 1. Baseline Neutrino Factory acceleration 0.2-12.6 GeV.

This scheme is evolving as the simulations progress and become more detailed. The very large size of the bunch, $\varepsilon_x/\varepsilon_y = 4.8$ mm rad and $\varepsilon_l = \sigma_{\Delta p}\,\sigma_z/m_\mu c = 24$ mm, is an important consideration.

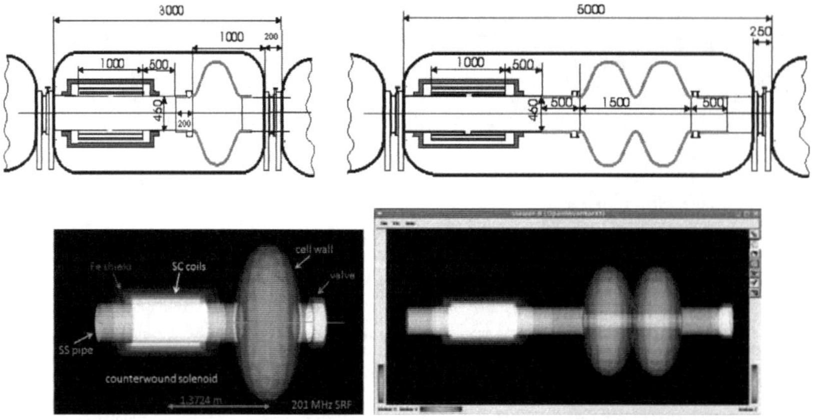

Figure 2. Schematic and G4beamline simulation of the 3m and 5m NF cryomodules.

2. Recent Changes to the Pre-Linac

For some time, the pre-linac consisted of 3 types of cryomodules: a 3 m, a 5 m, and an 8 m.[1] All 3 contained a counterwound and magnetically shielded superconducting solenoid and superconducting 201.25 MHz RF. The 3 m contained one single cell RF cavity, the 5 m one double-cell cavity, and the 8 m two double-cell cavities. G4beamline uses field maps calculated using Superfish for the solenoids and COMSOL for the RF cavities.[4] The beam was estimated to be about 30 cm wide with a 10% energy spread. Simulations revealed that the transverse focusing was insufficient and significant beam losses would occur. It also became clear that the longitudinal beam dynamics was more of a challenge than originally anticipated, so the first cavities are run far off-crest.

To address this[2], the pre-linac now uses 24 3 m and 24 5 m cryomodules (Fig. 2)[3,4], allowing for more rapid acceleration and greater focusing. An improved pillbox routine in G4beamline was used to better tune the cavities and estimate the Transit Factor effect[5].

The transverse optics is very straight forward, with the tune taken from OptiM (Fig. 3).

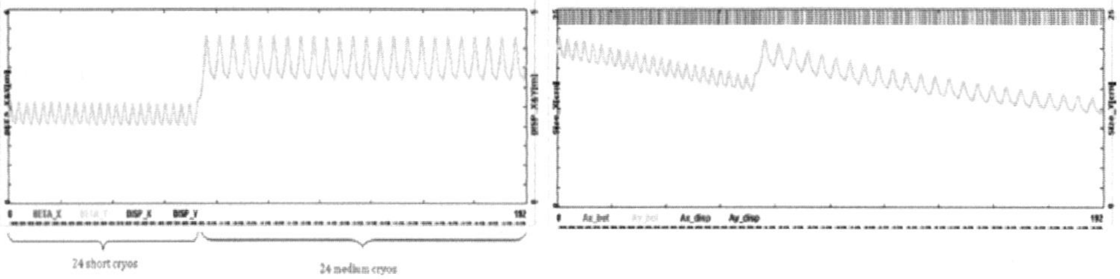

Figure 3. NF pre-linac beta functions (left) and bunch size (right) in OptiM.

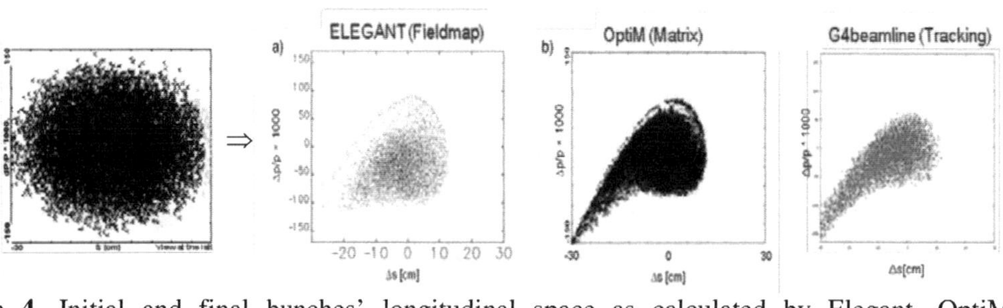

Figure 4. Initial and final bunches' longitudinal space as calculated by Elegant, OptiM, and G4beamline.

However, the longitudinal optics is less simple. To capture the bunch with such a large longitudinal size, the first cells are run 72° off crest and linearly varied with position to reach on-crest by the end of the linac. Simulations in OptiM, Elegant, and G4beamline agree well, but work is ongoing to improve the transport (Fig. 4).

3. Recent Changes to the RLA

3.1. Linac

Upon each pass through the RLA's linac, the focusing gets weaker. The beta-beating technique has the linac's quads strengths symmetrically increasing with distance from the center of the linac, allowing more passes through the linac with acceptable beta functions. While originally a linear dependence was used, a nonlinear strength vs. position yields even better results (Fig. 5).

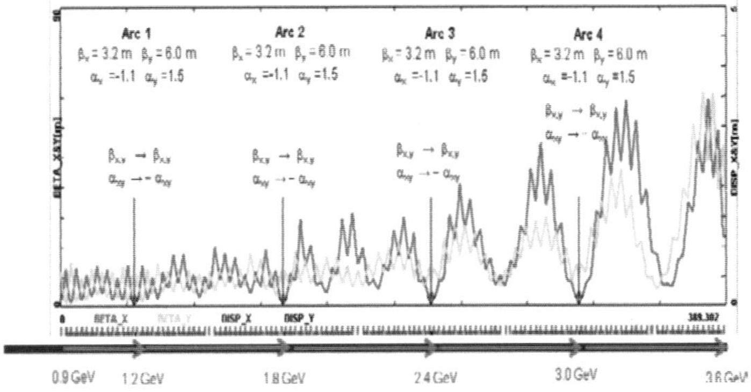

Figure 5. Beta functions in the 1st RLA linac's 4.5 passes; the arcs are not shown.

To use this technique, the beam must be injected into the center of, and extracted from one end of, the linac. It has 4 90° advance cells horizontally and 3 120° advance cells vertically. A small chicane corrects the higher energy passes.

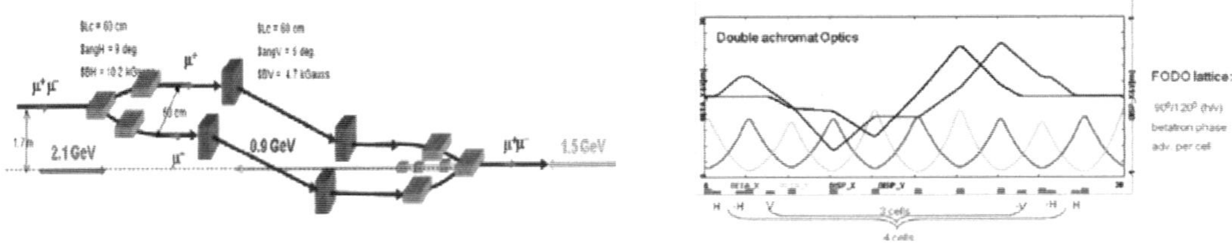

Figure 6. Injection chicane for the first RLA.

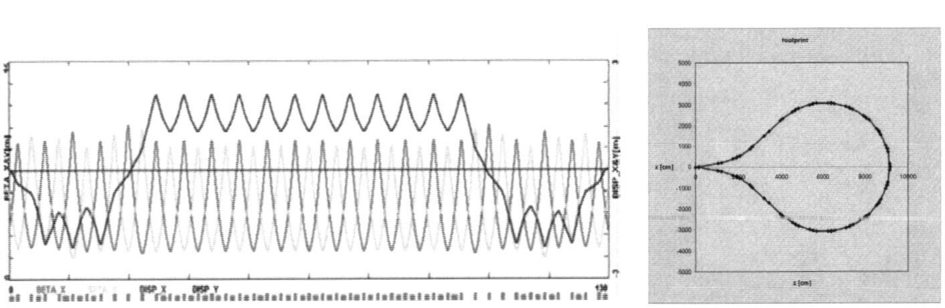

Figure 7. Teardrop arc beta functions and shape.

3.2. *Arcs*

The teardrop shaped arcs use 4 outward bending cells, two 2 cell transition regions, and 10 identical but inward bending cells. The current baseline is for one arc/pass. A vertical bypass based on 4 90° cells has been designed.

We are investigating a new concept, the multi-pass arc, in which two or more beams with greatly differing momenta travel through the same magnets (Fig. 8). This has the promise of greatly simplifying the RLA and reducing costs.

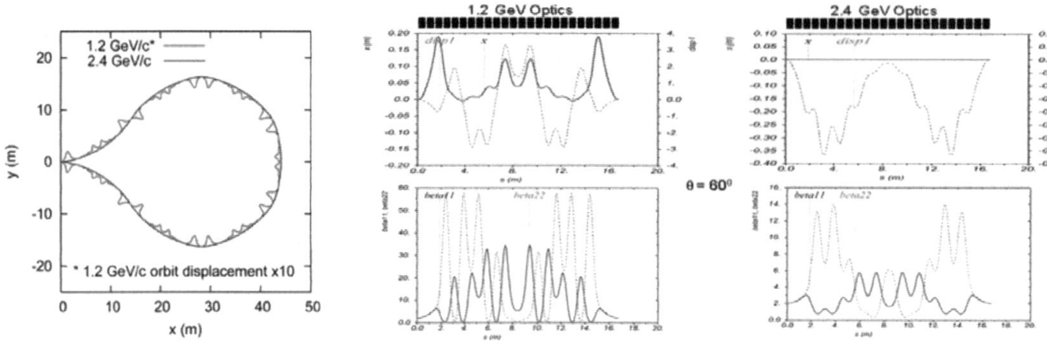

Figure 8. Optics for a 2:1 momentum ratio multi-pass arc.

3.3. *Splitter.*

If a multi-pass arc is used, the μ^+ and μ^- beams of two differing momenta must be directed to the same place. This can be done by removing the quadrupole components in the first magnets (Fig. 9).

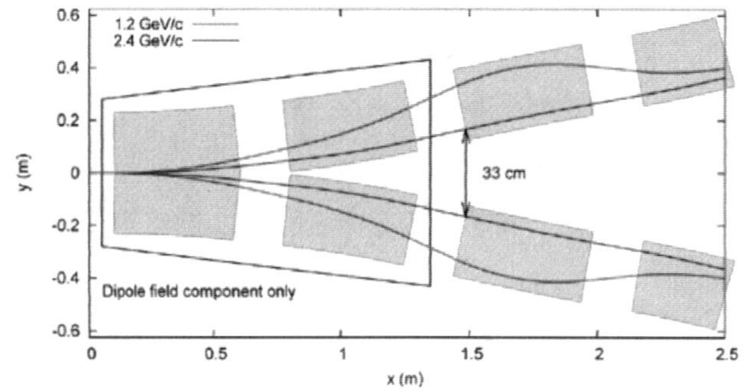

Figure 10. Splitter for a multi-pass arc with trajectories shown to scale.

4. Summary

The evolution of the design of the pre-linac and both RLAs for a NF is continuing. The pre-linac's optics are being optimized and a detailed design of the injection and vertical bypass chicanes has been done. An alternative multi-pass arc scheme is being examined.

References

[1] Beard K B *et al.* 2011 *2011 Particle Accelerator Conference - Proceedings (New York)* p 181
[2] Bogacz S A, Berg J S and Beard K B private communication.
[3] IDS-NF Collaboration 2011 Interim Design Report **IDS-NF 020** (http://www.ids-nf.org)
[4] Bontoiu C *et al.* 2010 *International Particle Accelerator Conference (Kyoto)* p 4590-4592.
[5] Beard K B *et al.* Neutrino Factory Linac RF Cavity Transit Factor **MAP-doc-4308** (http://map-docdb.fnal.gov/cgi-bin/ShowDocument?docid=4308)

Straight Scaling FFAG line

J-B Lagrange[1], T Planche[1,5], E Yamakawa[1], Y Ishi[2], Y Kuriyama[2], Y Mori[2], B Qin[2], T Uesugi[2], K Okabe[3], A Sardet[4] and R Wasef[4]

[1]Graduate School of Engineering, Kyoto University Katsura, Kyoto, Japan
[2]Kyoto University Research Reactor Institute, Kumatori, Japan
[3]University of Fukui, Fukui, Japan
[4]Institut National Polytechnique de Grenoble, Grenoble, France

E-mail: `lagrange@rri.kyoto-u.ac.jp`

Abstract. Recent developments in scaling fixed field alternating gradient (FFAG) accelerators have opened new ways for lattice design, with straight sections. An experiment to study zero-chromatic straight sections is in progress at KURRI.

1. Introduction

Until today, scaling FFAG accelerators were only designed in a ring shape. But new criterion of the magnetic field configuration satisfying the scaling condition even for straight FFAG beam line has been recently found. It is indeed possible to guide particles with no overall bend in scaling FFAGs, with a field law in the mid-plane following [1, 2]

$$B_z = B_0 e^{m(x-x_0)} \mathcal{F}(s), \tag{1}$$

with x and s the horizontal and longitudinal coordinates, respectively, $B_0 = B_z(x_0)$, $\mathcal{F}$ an arbitrary function of s and m the normalized field gradient

$$m = \frac{1}{B}\frac{dB}{dx}. \tag{2}$$

To verify and study this new field law, an experiment is in progress at Kyoto University Research Reactor Institute (KURRI).

2. Goal of the experiment

The goal of the Straight Experiment is to measure the phase advance for different energies in a straight scaling FFAG cell. In order to do it, the Twiss parameters as well as position and angle of the beam will be measured at the exit of the Straight Cell when it is launched off its reference trajectory. The linear transfer matrix of the line is

$$\begin{pmatrix} x_1 \\ x'_1 \end{pmatrix} = \begin{pmatrix} \sqrt{\frac{\beta_1}{\beta_0}}(\cos\psi + \alpha_0 \sin\psi) & a_{12} \\ \frac{(\alpha_0-\alpha_1)\cos\psi-(1+\alpha_0\alpha_1)\sin\psi}{\sqrt{\beta_1\beta_0}} & a_{22} \end{pmatrix} \cdot \begin{pmatrix} x_0 \\ 0 \end{pmatrix},$$

[5] moved to TRIUMF, Vancouver, Canada

with x_1 and x'_1 the distance and the angle to the reference trajectory at the exit of the cell, respectively. x_0 is the distance to the reference trajectory at the entrance of the cell. No angle with the reference trajectory is added at the entrance of the cell. α_0 and β_0 are the Twiss parameters at the entrance of the cell, while α_1 and β_1 are the Twiss parameters at the exit of the cell. ψ is the linear phase advance. The system of equations to be solved is then

$$\begin{cases} \sqrt{\frac{\beta_1}{\beta_0}}(\cos\psi + \alpha_0 \sin\psi) = \frac{x_1}{x_0} \\ \frac{(\alpha_0-\alpha_1)\cos\psi-(1+\alpha_0\alpha_1)\sin\psi}{\sqrt{\beta_1\beta_0}} = \frac{x'_1}{x_0} \end{cases}, \tag{3}$$

and we get

$$\begin{cases} \sin\psi = \sqrt{\frac{\beta_0}{\beta_1}}\frac{1}{x_0(1+\alpha_0^2)}\left((\alpha_0-\alpha_1)x_1 - \beta_1 x'_1\right) \\ \cos\psi = \sqrt{\frac{\beta_0}{\beta_1}}\frac{1}{x_0(1+\alpha_0^2)}\left((1+\alpha_0\alpha_1)x_1 + \alpha_0\beta_1 x'_1\right) \end{cases}. \tag{4}$$

So we can calculate the linear phase advance ψ by:

$$\tan\psi = \frac{(\alpha_0-\alpha_1)x_1 - \beta_1 x'_1}{(1+\alpha_0\alpha_1)x_1 + \alpha_0\beta_1 x'_1}. \tag{5}$$

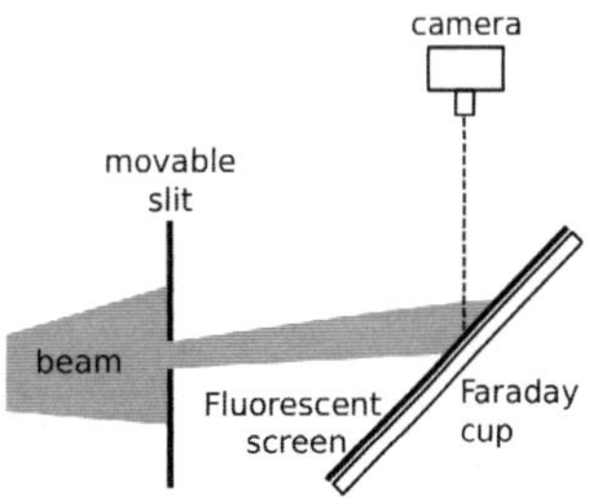

Figure 1. scheme of the emittance measurement system to be installed after the Straight Cell.

3. Layout of the experiment

For the Straight Experiment, the injector of the 150 MeV FFAG complex is used, a Linac delivering H- particles of 7 MeV and 11 MeV in the KURRI FFAG complex. The prototype of the straight scaling FFAG cell is installed in the beam line between the Linac and the FFAG Main Ring. Two collimators will be installed before the Straight Cell to set the Twiss parameters at the entrance. An emittance measurement system will be added after the Straight Cell to determine the exit parameters of the beam. This system is similar to the "slit-grid method" [3], but the grid is replaced by a fluorescent screen (see Fig. 1). This system has been successfully checked both for horizontal [4] and vertical [5] emittance in the H- beam line at the exit of the Linac.

Since each energy has a different reference trajectory (corresponding to a closed orbit in terms of circular machines), the cell is able to move horizontally to match the different trajectories without changing the magnetic field. Bellows before and after the cell allows the vacuum chamber to move along with the magnets without breaking the vacuum.

4. Straight cell design

The prototype for the straight scaling FFAG cell is a FDF triplet. The parameters of the cell are summarized in Tab. 1. The pole shape has been designed with TOSCA code [6] (see Fig. 2 and 3). A "C" type magnet has been chosen to have an easier access to the pole for field measurement.

Table 1. Parameters of the Straight Cell.

Type	FDF
m-value	11. m^{-1}
Total length	2.4 m
Length of F magnet	15. cm
Length of D magnet	30. cm
Max. B Field	0.35 T
Horizontal phase advance	83. deg
Vertical phase advance	80. deg

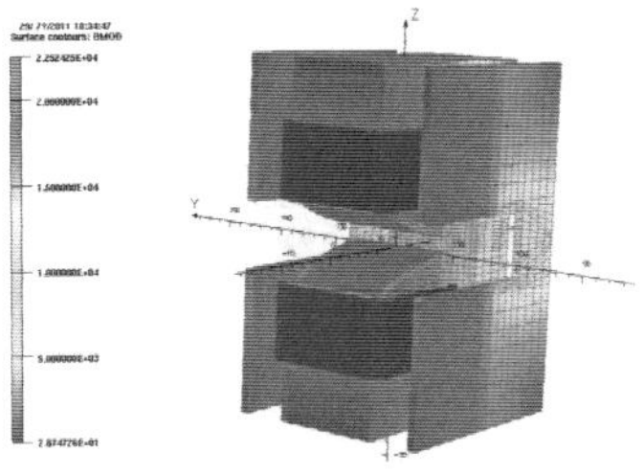

Figure 2. Magnetic field in F magnet (Tosca model).

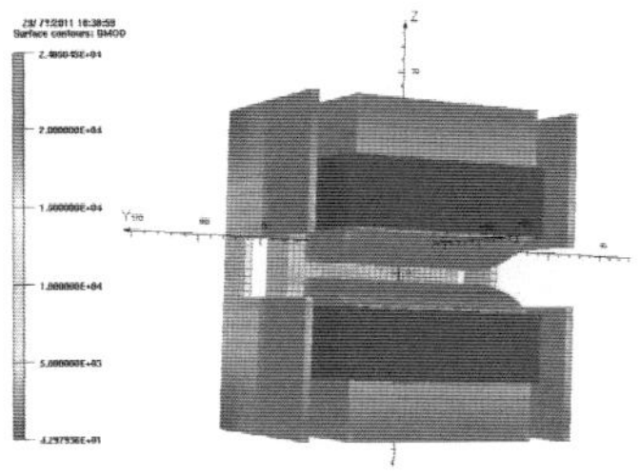

Figure 3. Magnetic field in D magnet (Tosca model).

5. Tracking in field map

Tracking step by step using Runge Kutta integration has been done in field maps from TOSCA [6]. The reference trajectories of 7 MeV and 11 MeV are shown in Fig. 4. The beta functions in the system are shown Fig. 5. The local phase advances in the Straight Cell are plotted in Fig. 6. The results show a constant horizontal phase advance, since the local m-value in the field map is constant (see Fig. 7) and a slightly changing vertical phase advance, typical of the scaling FFAGs, because the fringe fields do not scale due to the pole shape. It would be possible to adjust the length of the fringe field by adjusting the effective field boundaries [7], but rectangular magnets have been kept for economical reasons.

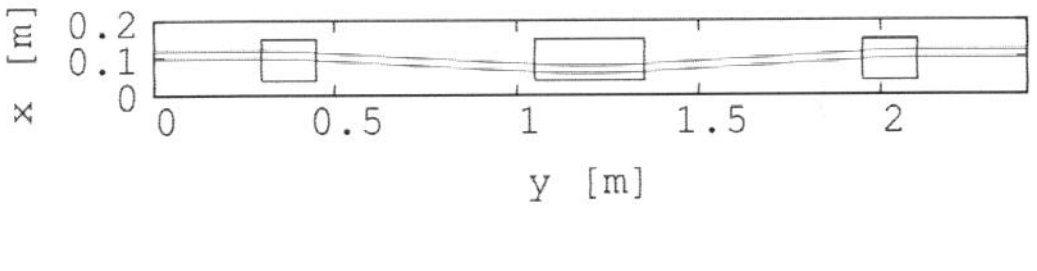

Figure 4. Reference trajectories in the Straight FFAG line.

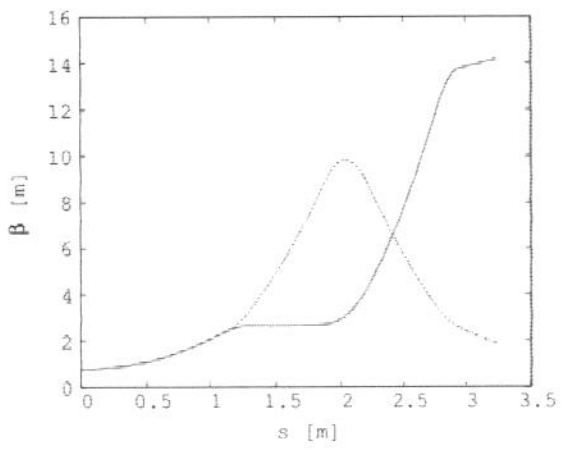

Figure 5. Horizontal (plain red) and vertical (dotted purple) beta functions in the Straight FFAG line.

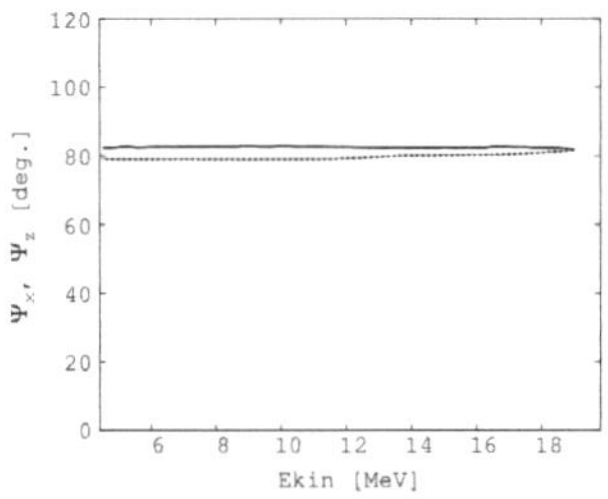

Figure 6. Horizontal (plain) and vertical (dotted) local phase advances versus kinetic energy in the Straight FFAG line.

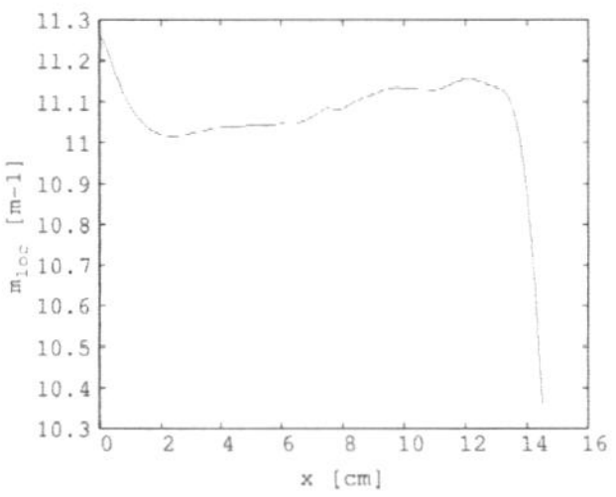

Figure 7. Local m value versus horizontal abscissa in the Straight FFAG line.

Figure 8. Picture of the straight scaling FFAG magnets prototype.

6. Summary

To verify the straight scaling field law, an experiment is in progress at KURRI. Design and tracking in field maps from TOSCA [6] has been done, the magnets have been built and delivered (see Fig. 8). Field measurement of the magnets is underway and the results of the experiment are expected before the end of this year.

References

[1] Lagrange J-B *et al.* 2009 "Straight section in scaling FFAG accelerator" PAC'09 (Vancouver) FR5PFP002.
[2] Lagrange J-B *et al.* 2011 "Zero-chromatic FFAG straight section" *IJMPA* **26** pp. 1785-93.
[3] Fork P "Lecture notes on Beam instrumentation and diagnostics", JUAS course 2007.
[4] Wasef R 2010 "Emittance and Twiss parameters measurement", Internship report for Master degree.
[5] Sardet A 2011 "Vertical emittance and Twiss parameters estimation at the exit of a LINAC", Internship report for Master degree.
[6] TOSCA Magnet Design Code Vector Fields Inc. Aurora, IL.
[7] Qin B *et al.* 2011 "Compact superferric FFAG accelerators for medium energy hadron applications" *Nucl. Instr. Meth.* A **648** pp. 28-34.

Opportunities for neutrino experiments at ISOLDE

Tânia Melo Mendonça[1,2], Rastilav Hodák[3] and Thierry Stora[1]

[1] CERN, 1211 Geneva 23, Switzerland

[2] IFIMUP and IN - Institute of Nanosciences and Nanotechnologies, Rua do Campo Alegre 687, 4169-007 Porto, Portugal

[3] Department of Nuclear Physics and Biophysics, Comenius University, Mlynská dolina F1, SK-842 15 Bratislava, Slovakia

E-mail: taniamel@mail.cern.ch

Abstract. A molten salt target is currently proposed for the production of high rates of ^{18}Ne required for the beta beams project. Intermediate proton energy beams at close to 1 MW power are used on a circulating molten NaF eutectic to produce and extract 1×10^{13} ^{18}Ne/s.

A conceptual design of such a circulating loop is presented. The application of a molten salt target for the monitoring of noble gas production is discussed with respect to the planned experimental program, which will be used to validate the proposed concept. Furthermore, a systematic experimental study on the production and release of gaseous and volatile elements from molten sodium-based targets, in preparation at CERN-ISOLDE, is presented.

1. Introduction

The beta beams concept for the production of highly energetic pure electron neutrino (ν_e) and antineutrino ($\bar{\nu}_e$) beams coming from the β decay of radioactive ions was first proposed in 2002 [1]. Such concept would allow the measurement of ν_e to ν_μ oscillation with exceptional sensitivity for the determination of the θ_{13} mixing angle and CP-violating phase. This type of facility could be advantageously located at CERN making use of the PS and SPS to accelerate the radioactive ions, allowing the production of neutrino beams from their β decay in a dedicated ring with straight sections directed towards an underground detector [1, 2]. Fig. 1 shows the proposed beta beam baseline to be implemented at CERN.

The feasibility of the beta beams has been studied within the EURISOL-DS project, where the production of ^{6}He and ^{18}Ne was investigated [3]. Both radioisotopes, otherwise known as baseline isotopes, can in principle be produced in large quantities and do not have any long-lived daughter products. A top-down approach showed the need for production of about 6×10^{13} ^{6}He and 1×10^{13} ^{18}Ne ions per second which would lead to rates of 2.9×10^{19} $\bar{\nu}_e$ and 1.1×10^{19} ν_e over a running period of 2 and 8 years, respectively [3].

As presented in fig. 1, different production schemes are proposed depending on the ion type. The production of ^{6}He has been successfully validated using the isotope separation online (ISOL) method [4]. ^{6}He ions have been obtained with fast neutrons, produced with a compact tungsten converter, on a thick beryllium oxide target [3] through the ^{9}Be$(n, \alpha)^6$He reaction. Online tests performed at CERN-ISOLDE showed that this figure validated the production of 10^{14} ^{6}He/s with 100 μA, 2 GeV protons and an optimized geometry [6].

The production of the required 10^{13} ^{18}Ne/s was found to be more problematic. A first approach investigated the reaction of 1 GeV proton beam onto oxide targets, alike the standard

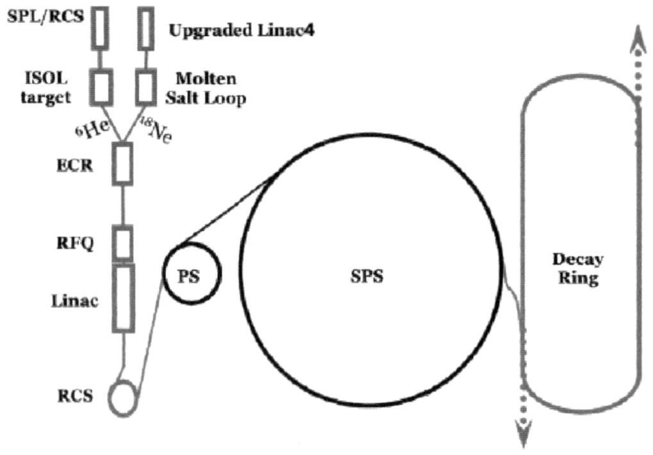

Figure 1. Proposed beta beam baseline implementation at CERN for the production of ^{6}He and ^{18}Ne ions.

procedure at CERN-ISOLDE, and kW beam power [5]. The production cross-sections have been evaluated numerically, which resulted in a production shortfall of more than one order of magnitude at a power of 100 kW showing the need of a more favorable approach. Alternatively, low energy beam interacting with solid oxides such as magnesium oxide have been also investigated [6]. The reaction cross-sections have been determined experimentally and the production yields have been computed for protons and ^{3}He in the 5 to 20 MeV range. An in-target production rate of 1×10^{13} ^{18}Ne/s is expected for 85 mA, 21 MeV beam. The required intensity is expected to decrease when increasing the beam energy to 30 MeV.

Another approach to produce ^{18}Ne beams was inspired by the method applied at Centre de Recherche sur les Cyclotrons, Louvain-La-Neuve by using a 9 kW, 30 MeV proton beam on molten lithium fluoride salt cased in a graphite disk [7]. Extrapolation of the results show that 35 mA proton beam at 50 MeV and multi-megawatts proton driver and target would be needed to deliver the required 1×10^{13} ^{18}Ne/s, assuming 100% isotope extraction efficiencies.

In this paper, an alternative route to produce ^{18}Ne beams is presented. This route consists on a circulating loop of molten salt analogous to the one proposed for a molten metal loop at EURISOL-DS [3]. A schematic representation of the proposed loop is shown in fig. 2. The proton beam energy range is between 50 and 160 MeV, which is also relevant for the production of radioisotopes used for proton emission tomography imaging and cancer treatments.

2. Production of ^{18}Ne using a circulating loop of molten sodium salts

The production of ^{18}Ne can be performed by (p, X) (or (^{3}He,X)) reaction on Na, F or Mg targets [8]. Therefore, the use of targets which contain any of these elements would present direct advantages on the production of Ne. Moreover, the use of sodium molten salts would present several other advantages since these compounds have been extensively studied due to their application as coolants in nuclear reactors [9, 10], and their physical and chemical properties are very well known.

Amongst the extensive list of available molten salt systems, the best candidate to the present application would be sodium fluoride (NaF). Nevertheless, the high melting point of this salt (995°C) limits its applications and the choice of a binary system containing NaF presents even more advantages. In order to choose the most adequate binary system, several physical properties have been compared. Table 1 summarizes some candidates for the present application together

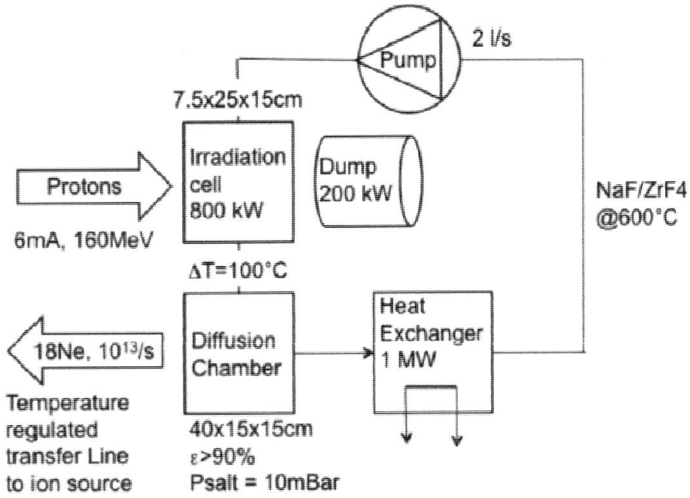

Figure 2. Conceptual layout of a molten salt loop for ^{18}Ne production.

with the respective physical properties at 700°C. The calculated production yields indicates that the composition to obtain the highest ^{18}Ne intensity for a given beam power is the mixture of sodium and zirconium fluorides (NaF-ZrF_4). One shall note that the production yields have been calculated assuming 6 mA, 160 MeV proton beam.

Table 1. Physical properties at 700°C, vapor pressure at 900°C and respective production yields for ^{18}Ne of several sodium salts [9–11].

Salt	Composition (% mol)	Melting Point (°C)	Density (g/cm^3)	Viscosity (cP)	900°C vapor pressure (mmHg)	Production yield ^{18}Ne/s
NaF-BeF_2	57-43	340	2.01	7	1.2	8.8×10^{12}
NaF-ZrF_4	60-40	500	3.14	5.1	5	1×10^{13}
NaF-$NaBF_4$	8-92	385	1.75	0.9	9500	8.4×10^{12}

Accounting for the high reactivity and corrosive nature of the fluoride salts, the choice of the container material is an important issue. It is well known that nickel is one of the most resistant elements to dissolution in these salts . Therefore, the container shall be made from a nickel rich alloy, which is resistant to fluoride environments at high temperatures.

The dimensioning of the irradiation and diffusion chambers has been performed accounting for the materials, heat transfer and diffusion coefficients. Refined calculations and dimensioning of the production unit is still ongoing.

3. Online tests at ISOLDE using a static target unit

A first test on the feasibility of the production of ^{18}Ne using molten fluoride salts will be performed using a static target unit at CERN/ISOLDE with an experiment scheduled in November 2011.

The target material was synthesized at LPSC/Grenoble by heating a mixture of stoichiometric quantities of NaF and ZrF_4 above the eutectic point of this system. The target unit consists of a nickel-rich alloy container equipped with a condensation tube to connect the target to the ion source. The container will be filled up to 3/4 of its volume with the melt, allowing a free surface for the isotopes to diffuse out of the target. The target unit will be connected to a versatile arc discharge ion source (VADIS) via a temperature controlled transfer line.

The foreseen online measurements will allow a systematic study of the production and release of ^{18}Ne. Several tests will be performed with different target temperatures and proton beam intensities. The forthcoming results will give an important contribution for the design and test of the molten salt circulating loop.

4. Conclusions

In this paper, a molten fluoride salt target is proposed for the production of the required rates of ^{18}Ne for the beta beams project. Intermediate proton energy beams at close to 1 MW power are used on a circulating NaF-ZrF_4 eutectic to produce and extract 1×10^{13} ^{18}Ne/s.

The application of a molten fluoride salt target for the monitoring of noble gas production is discussed with respect to the planned experimental program. Furthermore, a systematic experimental study on the production and release of volatile radioisotopes from a static target unit, in preparation at CERN/ISOLDE, is presented.

References

[1] Zuchelli P 2002 *Phys. Lett.* B **532** 166.

[2] Benedikt M, *et al.* 2011, *Eur. Phys. J. A*, **47**, 24.

[3] Final Report of the EURISOL Design Study 2009, J. Cornell Ed, GANIL.

[4] Ravn H R and Alardyce B W 1989 *On-Line Mass Separators*, Treatise on heavy-ion science Edt. D.A. Bromley, Plenum Press, New York.

[5] Autin B, *et al.* 2002, *The acceleration and storage of radioactive ions for a neutrino factory*, **CERN-PS 2002/78**.

[6] Stora T 2010, *Proc. of the Workshop on European Strategy for Neutrino Physics*, CERN Yellow Report 2010, **CERN-2010-003**, 110.

[7] Gaelens M, *et al.* 1998, *Proc. of the 15th International Conference on Application of Accelerators in Research and Industry*, Denton, **ISBN-1-56396-825-8**, 305.

[8] Lagunas-Solar M C 1992, *Nucl. Instr. Methods Phys. Res. B* **69**, 452

[9] Williams D F, Toht L M and Clarno K T 2006, *Assessment of Candidate Molten Salt Coolants for the Advanced High-Temperature Reactor*, **ORNL/TM-2006/12**, Oak Ridge National Laboratory, Oak Ridge.

[10] Williams D F 2005, *Assessment of Candidate Molten Salt Coolants for the NGNP/NHI Heat-Transfer Loop*, **ORNL/TM-2006/69**, Oak Ridge National Laboratory, Oak Ridge.

[11] Benes O, *et al.* 2009, *ALISIA: Review Report on Liquid Salt for Various Application*, Version V4.

Results and prospects on lepton flavor violation at Belle/Belle II

Kiyoshi Hayasaka (for the Belle and Belle II collaborations)

Center for Experimental Studies, Kobayashi-Maskawa Institute, Nagoya University, Furo-cho, Chikusa-ku, Nagoya, Japan

E-mail: hayasaka@hepl.phys.nagoya-u.ac.jp

Abstract. We report the results and prospects on the τ lepton flavor violation at Belle and Belle II.

1. Introduction

Charged-lepton-flavor violation (cLFV) is one of the clear signatures for new physics (NP) since cLFV is forbidden in the SM and its probability is very small even if neutrino oscillations are taken into account. In particular, τ lepton-flavor violation (τ LFV) is promising because the τ could have a strong coupling to the NP due to its heaviest mass among the leptons. In addition, the τ has many possible τ LFV modes: not only merely lepton-flavor-violating process but also lepton-number- or baryon-number violating processes. In order to reveal the structure of the NP, the information from these τ LFV modes, such as a magnitude relation for the branching fractions, will help us. Since a B-factory is also a τ-factory, the KEKB/Belle experiment is a good place to search for τ LFV decays. In addition, since the superKEKB/Belle II experiment is planned, where 40-times larger luminosity (8×10^{35} cm^{-2}s^{-1}) is expected, τ LFV search in the new experiment is also important.

We report recent results using a $\sim$1000 fb^{-1} data sample obtained by Belle [1] and the future prospects evaluated by the Belle II collaboration [2].

2. Method

All searches for LFV τ decays follow a similar procedure. We search for $\tau^+\tau^-$ events in which one τ (signal side) decays into the LFV mode under study, while the other τ (tag side) decays into one charged particle and any number of additional photons and neutrinos (e.g. left plot in Fig. 1). To search for exclusive LFV-decay modes, we select low multiplicity events with zero net charge, and separate an event into two hemispheres (signal and tag) using the thrust axis. The backgrounds in such searches are dominated by $q\bar{q}$ ($q = u, d, s, c$), generic $\tau^+\tau^-$, two-photon, $\mu^+\mu^-$ and Bhabha events. To obtain a good sensitivity, we optimize the event selection using particle identification and kinematic information for each mode separately. Because every τ LFV decay is a neutrinoless one, the information from missing tracks is very powerful to reject the background from generic $\tau^+\tau^-$ events. After signal selection criteria are applied, signal candidates are examined in the two-dimensional space of the invariant mass $M_{\rm inv}$, and the difference between their energy and the beam energy in the CM system ΔE. A signal event

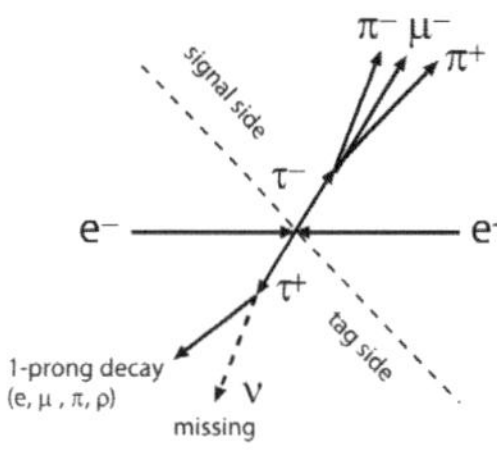

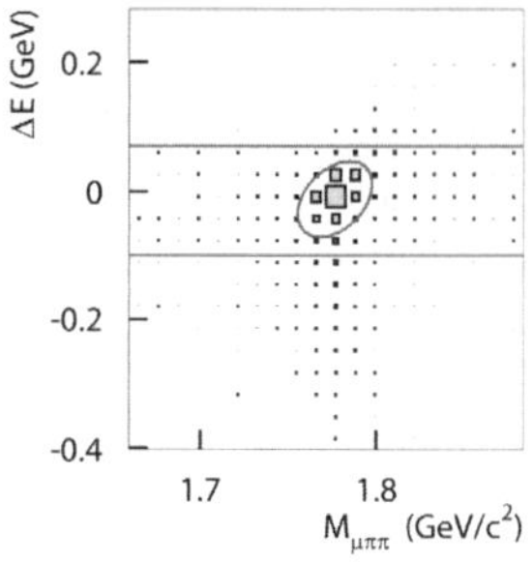

Figure 1. Left: Topology for the signal event, where the decay signal is $\tau \to \mu\pi\pi$. Right: Resulting distribution for the signal events in the $M_{\rm inv}$-ΔE plane. The boxes size shows the magnitude of the density for the signal events, the red ellipse indicates the $\pm 3\sigma$ signal region and the region between the two red lines is defined as the $\pm 5\sigma_{\Delta E}$ side-band region for the background estimation.

should have $M_{\rm inv}$ close to the τ-lepton mass and ΔE close to 0 as shown in Fig. 1 (right). We blind a region around the signal region in the $M_{\rm inv}$-ΔE plane so as not to bias our choice of selection criteria. The expected number of background events in the blind region is first evaluated, and then the blind region is opened and candidate events are counted. By comparing the expected and observed numbers of events, we either observe a τ LFV decay or set an upper limit (UL) using a counting method [3].

3. Results

3.1. $\tau \to \ell h h'$ ($\ell = e, \mu$, $h = \pi, K$)

Although $\tau^- \to \ell^- h^+ h'^-$ violates only the lepton flavor, $\tau^- \to \ell^+ h^- h'^-$ violates the lepton number as well. Here, we search for 8 modes of $\tau^- \to \ell^- h^+ h'^-$ and 6 modes of $\tau^- \to \ell^+ h^- h'^-$. The former ones are expected to be enhanced by the Higgs-mediated model [4] while the latter are motivated by the Majorana neutrino model [5].

We have updated the analysis for these modes with the 854 fb^{-1} data sample. Main backgrounds come from $\tau \to \pi\pi\pi\nu$ for $\ell = \mu$ and $\tau \to \pi\pi^0\nu$ for $\ell = e$, where the photon, that is a daughter of π^0, converts to e^+e^-. To reduce them, $\tau \to \pi\pi\pi\nu$ veto, using $M_{\pi\pi\pi}$, and γ-conversion veto are newly introduced for $\mu\pi K$ modes and $\ell = e$ modes, respectively. As a result, we have found 1 event for $\mu^+\pi^-\pi^-$ and $\mu^-\pi^+K^-$ modes while no events are observed in the other modes. Since this is consistent with the expected number of the backgrounds, we set the UL on the branching fraction. The evaluated numbers to set the UL and the number of observed events in the signal region are summarized in Table 1. The ULs for this mode are the most sensitive (preliminary).

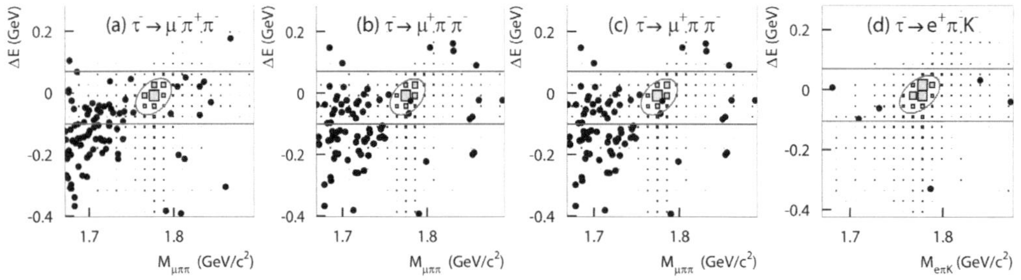

Figure 2. Data ● and signal events density in the $M_{\rm inv}$-ΔE plane for different τ decay modes.

Table 1. ULs for the recently updated modes: efficiency (ε), number of expected background events estimated from the sideband data ($N_{\rm BG}$), total systematic uncertainty ($\sigma_{\rm syst}$), number of observed events in the signal region ($N_{\rm obs}$), 90% CL UL on the number of signal events including the systematic uncertainties (s_{90}) and 90% CL UL on the branching fraction for each individual mode.

Mode	ε (%)	$N_{\rm BG}$	$\sigma_{\rm syst}$ (%)	$N_{\rm obs}$	s_{90}	$\mathcal{B}$ (10^{-8})
$\tau^- \to \mu^-\pi^+\pi^-$	5.83	0.63 ± 0.23	5.3	0	1.87	2.1
$\tau^- \to \mu^+\pi^-\pi^-$	6.55	0.33 ± 0.16	5.3	1	4.02	3.9
$\tau^- \to e^-\pi^+\pi^-$	5.45	0.55 ± 0.23	5.4	0	1.94	2.3
$\tau^- \to e^+\pi^-\pi^-$	6.56	0.37 ± 0.18	5.4	0	2.10	2.0
$\tau^- \to \mu^-K^+K^-$	2.85	0.51 ± 0.18	5.9	0	1.97	4.4
$\tau^- \to \mu^+K^-K^-$	2.98	0.25 ± 0.13	5.9	0	2.21	4.7
$\tau^- \to e^-K^+K^-$	4.29	0.17 ± 0.10	6.0	0	2.28	3.4
$\tau^- \to e^+K^-K^-$	4.64	0.06 ± 0.06	6.0	0	2.38	3.3
$\tau^- \to \mu^-\pi^+K^-$	2.72	0.72 ± 0.27	5.6	1	3.65	8.6
$\tau^- \to e^-\pi^+K^-$	3.97	0.18 ± 0.13	5.7	0	2.27	3.7
$\tau^- \to \mu^-K^+\pi^-$	2.62	0.64 ± 0.23	5.6	0	1.86	4.5
$\tau^- \to e^-K^+\pi^-$	4.07	0.55 ± 0.31	5.7	0	1.97	3.1
$\tau^- \to \mu^+K^-\pi^-$	2.55	0.56 ± 0.21	5.6	0	1.93	4.8
$\tau^- \to e^+K^-\pi^-$	4.00	0.46 ± 0.21	5.7	0	2.02	3.2
$\tau^- \to \bar{\Lambda}\pi^-$	4.80	0.21 ± 0.15	8.2	0	2.26	2.8
$\tau^- \to \Lambda\pi^-$	4.39	0.31 ± 0.18	8.2	0	2.17	3.0
$\tau^- \to \bar{\Lambda}K^-$	4.11	0.31 ± 0.14	8.6	0	2.17	3.1
$\tau^- \to \Lambda K^-$	3.16	0.42 ± 0.19	8.6	0	2.08	4.2

3.2. $\tau \to \Lambda h$ $(h = \pi, K)$

This mode violates the lepton number (L) as well as the baryon number (B): $\tau^- \to \bar{\Lambda}h^-$ conserves $(B - L)$ while $\tau^- \to \Lambda h^-$ violates $(B - L)$. Some GUTs can make $(B - L)$ conserving decays while a more complicated model is necessary to induce $(B - L)$ violating decays.

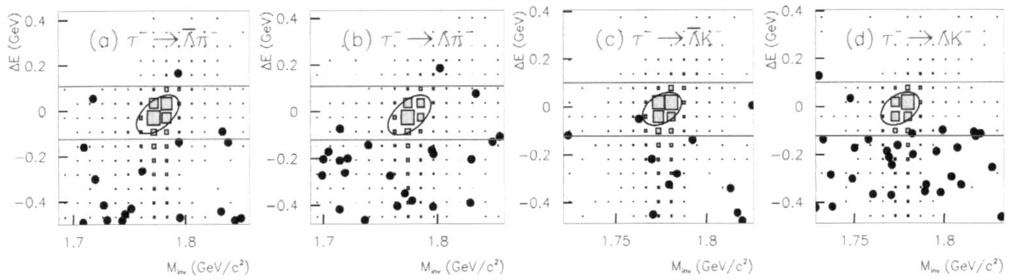

Figure 3. Data ● and signal events density in the $M_{\rm inv}$-ΔE plane for different τ^- decay modes.

We have searched for these decays with the 906 fb^{-1} data sample. Main backgrounds come from $\tau \to \pi K^0_S\nu$ and $q\bar{q}$ $(q = u, d, s)$ events having Λ and π. In the former background, K^0_S is misidentified as Λ. This can be rejected by K^0_S veto using $M_{\pi\pi}$. On the other hand, the latter is likely to have a proton as a charged track on the tag side because of the baryon number conservation. Therefore, we veto the proton for the tag-side track. As a result, no events are observed for each mode; no excess is found for the signal. Thus, we set the UL on the branching fraction. The obtained numbers are summarized in Table 1. We obtain the most stringent ULs (preliminary).

4. Future Prospect

The superKEKB/Belle II experiment is planned to collect 50-times larger integrated luminosity than the KEKB/Belle, i.e., a 50 ab^{-1} data sample will be accumulated. LFV sensitivity depends on the remaining background level. In the case where there is a large (negligibly small) remaining background, such as that in the $\tau \to \mu\gamma$ ($\tau \to \ell\ell\ell$) analysis, the sensitivity is scaled as $1/\sqrt{\mathcal{L}}$ ($1/\mathcal{L}$), where $\mathcal{L}$ means luminosity. Therefore, the accessible branching fractions for $\tau \to \mu\gamma$ and $\tau \to \ell\ell\ell$ at the superKEKB/Belle II are $O(10^{-8/-9})$ and $O(10^{-9/-10})$, respectively.

5. Summary

We have searched for 46 major modes of lepton flavor violating τ decays with the 1000 fb^{-1} data sample. The current status of the τ LFV searches in B-factory experiments is summarized in Fig. 4. No evidence for these decays is observed and we set 90% CL ULs on the branching fractions at the $O(10^{-8})$ level. The sensitivity for the LFV search is 100 times improved in comparison with CLEO's one due to the effective background rejection and increase of the data sample. In the future experiment, i.e., the superKEKB/Belle II experiment, the sensitivity will reach $O(10^{-9/-10})$ with a 50 ab^{-1} data sample.

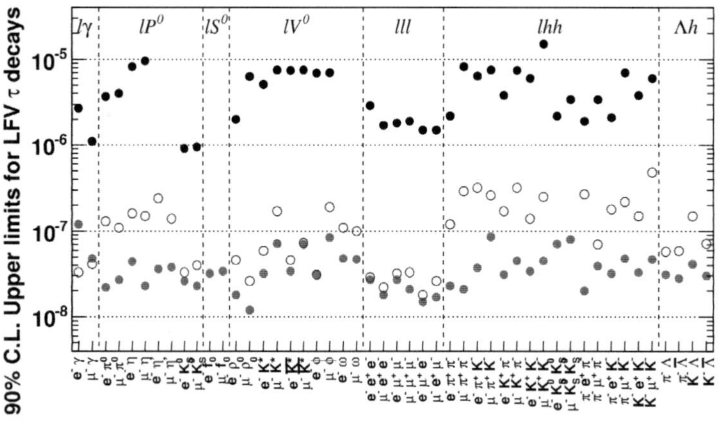

Figure 4. Current 90% CL ULs for the branching fraction of τ LFV mode for the Belle ●, Babar [6] ○ and CLEO [7] ● experiments.

References

[1] Abashian A *et al* (Belle collaboration) 2002 *Nucl. Instrum. Meth. A* **479** 117
[2] Abe T *et al* (Belle II collaboration) 2010 Belle II Technical Design Report *Preprint* arXiv:1011.0352v1 [physics:ins-det]
[3] Feldman G J and Cousins R D 1998 *Phys. Rev. D* **57** 3873
[4] Arganda E, Herrero M J and Portolés J 2008 *JHEP* **06** 079
[5] Atre A, Barger V and Han T 2005 *Phys. Rev. D* **71** 113014
[6] Cervell A *et al* (BaBar collaboration) 2009 *AIP Conf. Proc.* **1182** 656
[7] Weinstein A J *et al* (CLEO collaboration) 2003 *Nucl. Phys. B (Proc. Suppl.)* **123** 107

A novel experiment searching for the lepton flavour violating decay $\mu \rightarrow eee$

Niklaus Berger

Physics Institute, University of Heidelberg, Philosophenweg 12, 69121 Heidelberg, Germany

E-mail: nberger@physi.uni-heidelberg.de

Abstract. Since the discovery of neutrino oscillations it is known that lepton flavour is not conserved. Lepton flavour violating processes in the charged lepton sector have so far however eluded detection; as they are heavily suppressed in the standard model of particle physics, an observation would be a clear signal for new physics and help to understand the source of neutrino masses and CP violation. We propose a novel experiment searching for the decay $\mu \rightarrow eee$ with the aim of ultimately reaching a sensitivity of 10^{-16}, an improvement by four orders of magnitude compared to previous experiments. The technologies enabling this step are thin high-voltage monolithic active pixel sensors for precise tracking at high rates with a minimum of material and scintillating fibres for high resolution time measurements.

1. Introduction

In the *standard model* (SM) of elementary particle physics, the total number of leptons of each flavour is conserved. The discovery of neutrino oscillations [1–3] revealed that this is a broken symmetry and leptons can change their flavour. *Lepton flavour violation* (LFV) has however never been observed in the charged lepton sector. The exact mechanism and size of LFV being unknown, its study is of large interest, as it is linked to neutrino mass generation, *CP* violation and new physics beyond the SM.

Charged lepton flavour violating reactions can be induced by neutrino mixing in a loop diagram (see figure 1 a) which is however strongly suppressed, leading to (unobservable) LFV branching fractions in muon decays of less than 10^{-50}. In many extensions of the SM, such as grand unified models [5–7], supersymmetric models [8] (see figure 1 b), left-right symmetric models [9–11], models with an extended Higgs sector [12] and models where electroweak symmetry is broken dynamically [13], an experimentally accessible amount of LFV is predicted. The observation of LFV in the charged lepton sector would be a sign for new physics, possibly at scales far beyond the reach of direct observation at e.g. the large hadron collider (LHC).

Many experiments have been performed or are in operation to search for LFV in the decays of muons or taus. Most prominent are the search for the radiative muon decay $\mu \rightarrow e\gamma$ [14–17], the decay $\mu \rightarrow eee$ [18], the conversion of captured muons to electrons [19] and LFV tau decays [20].

2. Theory of $\mu \rightarrow eee$

The lepton flavour violating three electron decay of the muon can be mediated either via loops (see figure 1 a) and b)) or at tree level, via the exchange of a (heavy) new particle (see figure 1 c)). The effective couplings of the loop diagram are of tensor type (dipole couplings) if the

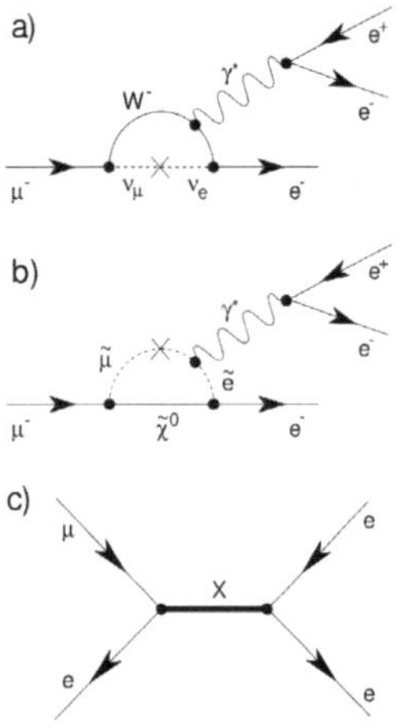

Figure 1. a): Feynman diagram for the $\mu \to eee$ process via neutrino oscillation (indicated by the cross). b): Diagram for lepton flavour violation involving supersymmetric particles. c) Diagram for lepton flavour violation at tree level.

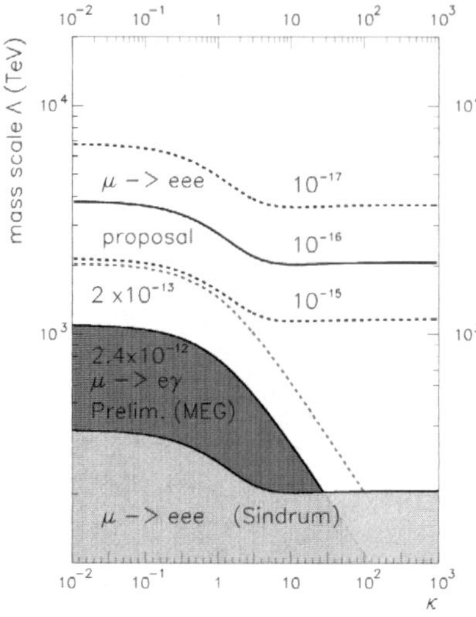

Figure 2. Experimental limits and projected limits on the LFV mass scale Λ as a function of the parameter κ (see equation 1), taken from [4], after a figure by A. de Gouvêa.

photon is real and receive additional vector-type contributions if the photon is virtual. The four fermion coupling of the tree diagram c) can be described as a contact interaction due to the large mass of the intermediate particle. In order to allow comparisons between $\mu \to e\gamma$ and $\mu \to eee$ experiments, the following simplified Lagrangian, introducing a common mass scale Λ is used:

$$\mathcal{L}_{LFV} = \frac{m_\mu}{(\kappa+1)\Lambda^2}\bar{\mu_R}\sigma^{\mu\nu}e_L F_{\mu\nu} + \frac{\kappa}{(\kappa+1)\Lambda^2}(\bar{\mu_L}\gamma^\mu e_L)(\bar{e_L}\gamma_\mu e_L), \tag{1}$$

here the left-left vector coupling is chosen as exemplary for the contact term (right), the parameter κ describes the strength of the contact interaction amplitude relative to the loop amplitude. Figure 2 shows limits on the mass scale Λ as a function of κ for the MEG and SINDRUM experiments, as well as limits expected from the MEG experiment at design sensitivity and for a $\mu \to eee$ experiment with sensitivities from 10^{-15} to 10^{-17}. At the same branching fraction, $\mu \to e\gamma$ experiments are more sensitive at low values of κ (dominance by the loop diagram), whilst $\mu \to eee$ experiments constrain Λ at high values of κ. In the case of a dominant on-shell photon contribution ($\kappa \to 0$), a quasi model-independent relation between the $\mu \to e\gamma$ and $\mu \to eee$ decay rates can be derived:

$$\frac{B(\mu \to eee)}{B(\mu \to e\gamma)} \approx 0.006. \tag{2}$$

In order to set competitive constraints on LFV dipole couplings, a limit on the branching fraction of the decay $\mu \to eee$ thus has to be about two orders of magnitude smaller than the best $\mu \to e\gamma$ limit. In case of a discovery, an $\mu \to eee$ experiment (with a three-body final state) would give access to CP observables and thus allow for a search for CP violation. In $\mu \to e\gamma$ experiments this would require a (difficult) measurement of the photon polarization.

3. Challenges for a $\mu \to eee$ experiment

A $\mu \to eee$ experiment aiming for a sensitivity of 10^{-16} has to be able to suppress accidental backgrounds and the process $\mu \to eee\nu\bar{\nu}$ (*internal conversion*) at least at the same level whilst running at a rate in excess of 10^9 muons on target per second, rates which are achievable with a beam-line upgrade at the Paul Scherrer Institut (PSI) in Switzerland. Accidental background can be controlled by an excellent vertex and timing resolution, whilst the only distinguishing feature of the internal conversion decay is the missing momentum carried away by the neutrinos; thus a very precise track reconstruction is needed.

Conventional tracking detectors either have a limited rate capacity and resolution (gaseous detectors) or induce unacceptable amounts of multiple scattering (hybrid silicon pixel sensors). The development of thin, high-voltage active silicon pixel sensors (HV-MAPS) and scintillating fibre trackers opens up new possibilities which we plan to exploit for a novel experiment searching for the LFV decay $\mu \to eee$.

4. Detector technologies

Recently, thin active silicon pixel detectors have become available [21–26], opening up new possibilities for low-material, high-precision tracking detectors. Produced in a CMOS process, these sensors include amplification and read-out logic inside of pixels with pitches from 20 to 100 μm. The silicon can be thinned to a thickness of 50 μm without affecting the signal collection efficiency. For the application in the $\mu \to eee$ experiment, the high-voltage CMOS monolithic active pixel sensor (MAPS) technology developed at the Institute for Computer Science of the University of Heidelberg (*ZITI*) is particularly interesting [22–24]. Here a process originally developed for driving automotive and industrial devices with high voltage signals is used to create a monolithic pixel sensor, where all the transistors are placed in a deep N-well, which can be reversely biased with regard to the P-substrate with a high voltage, typically more than 50 V. This creates a relatively large depleted area at the transition. Particles passing through this depleted zone will generate a sizeable amount of charge (around 2200 e for a depletion zone depth of 9 μm, leading to a signal-to-noise ratio above 30), which is separated by the strong electric field, leading to a fast current signal. The technology is well suited for tracking low momentum particles: because the signal is collected close to the chip surface, the chip can be thinned without signal loss. It is also resistant to both non-ionizing and (with a suitable design of the electronics) ionizing radiation. As a standard industrial process is used, the sensors are relatively cheap. The process has reticles of up to 2×2 cm size, which can be combined by wire-bonding or stitching to 2×6 cm sensors.

With cooling by gaseous helium only, the sensors can be read out at a maximum frequency of 10 to 20 MHz, leading to up to hundred electron tracks per readout frame. Precise timing information helping to suppress accidental background should come from a scintillating fibre tracker with a time resolution better than 100 ps.

5. Detector concept

The best geometry for a detector providing high resolution tracking for momenta from 15 to 53 MeV/c with a large angular coverage is currently under study. Both for resolution in a multiple scattering environment and ease of reconstruction, a spectrometer like arrangement of two double layers separated by free space in a solenoidal magnetic field is preferred. In order to improve the momentum resolution at high momenta, tracks curling back into the detector are reconstructed and fitted. Various detector concepts are currently being studied using a Geant4 [27] based simulation. Depending on the exact geometry, the detector will produce a data rate of about 50 Gbit/s, which is more than can practically be stored, but low enough for a triggerless operation mode with a farm of GPU assisted PCs performing fast track fits.

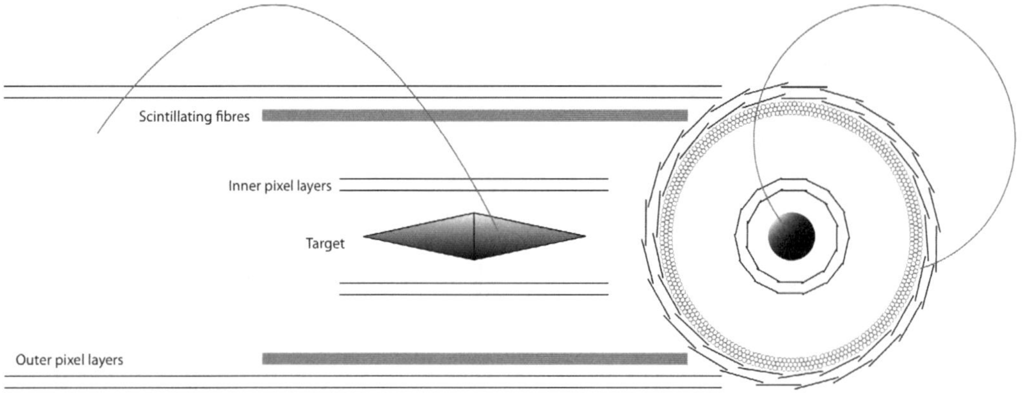

Figure 3. Schematic view of a possible $\mu \to eee$ experiment.

6. Conclusion

New detector technologies (especially HV-MAPS), open the opportunity to build an experiment searching for the LFV decay $\mu \to eee$ with a sensitivity of up to 10^{-16}, and thus the potential of finding physics beyond the standard model or strongly constraining models. A collaboration is currently forming and a letter of intent will be submitted to PSI soon[1].

References

[1] Fukuda Y *et al.* (Super-Kamiokande Collaboration) 1998 *Phys. Rev. Lett.* **81** 1562–1567 (*Preprint* `hep-ex/9807003`)
[2] Ahmad Q R *et al.* (SNO Collaboration) 2001 *Phys. Rev. Lett.* **87** 071301 (*Preprint* `nucl-ex/0106015`)
[3] Eguchi K *et al.* (KamLAND Collaboration) 2003 *Phys. Rev. Lett.* **90** 021802 (*Preprint* `hep-ex/0212021`)
[4] Schöning A Proposal for an Experiment to Search for $\mu \to eee$ CHIPP Plenary Meeting 2011
[5] Pati J C and Salam A 1974 *Phys. Rev. D* **10** 275–289
[6] Georgi H and Glashow S L 1974 *Phys. Rev. Lett.* **32** 438–441
[7] Langacker P 1981 *Phys. Rep.* **72** 185
[8] Haber H E and Kane G L 1985 *Phys. Rep.* **117** 75
[9] Mohapatra R N and Pati J C 1975 *Phys. Rev. D* **11** 566–571
[10] Mohapatra R N and Pati J C 1975 *Phys. Rev. D* **11** 2558
[11] Senjanovic G and Mohapatra R N 1975 *Phys. Rev. D* **12** 1502
[12] Kakizaki M, Ogura Y and Shima F 2003 *Phys. Lett. B* **566** 210–216 (*Preprint* `hep-ph/0304254`)
[13] Hill C T and Simmons E H 2003 *Phys. Rep.* **381** 235–402 (*Preprint* `hep-ph/0203079`)
[14] Brooks M L *et al.* (MEGA Collaboration) 1999 *Phys. Rev. Lett.* **83** 1521–1524 (*Preprint* `hep-ex/9905013`)
[15] Nicolo D (MEG Collaboration) 2003 *Nucl. Instrum. Meth. A* **503** 287–289
[16] Adam J *et al.* (MEG Collaboration) 2010 *Nucl. Phys. B* **834** 1–12 (*Preprint* `hep-ex/0908.2594`)
[17] Adam J *et al.* (MEG Collaboration) 2011 (*Preprint* `hep-ex/1107.5547`)
[18] Bellgardt U *et al.* (SINDRUM Collaboration) 1988 *Nucl. Phys. B* **299** 1
[19] Kaulard J *et al.* (SINDRUM II Collaboration) 1998 *Phys. Lett. B* **422** 334–338
[20] Hayasaka K *these proceedings*
[21] Turchetta R *et al.* 2001 *Nucl. Instrum. Meth. A* **458** 677–689
[22] Peric I 2007 *Nucl. Instrum. Meth. A* **582** 876
[23] Peric I and Takacs C 2010 *Nucl. Instrum. Meth. A* **624** 504
[24] Peric I, Kreidl C and Fischer P 2011 *Nucl. Instrum. Meth. A* **650** 158
[25] Winter M 2010 *Nucl. Instr. Meth. A* **623** 192
[26] Masi R D *et al.* 2011 *Nucl. Instr. Meth. A* **628** 296
[27] Agostinelli S *et al.* 2003 *Nucl. Instr. Meth. A* **506** 250

[1] The letter of intent has been submitted in January 2012 and is available from `http://www.psi.ch/mu3e`

The final measurements of the muon decay parameters from the TWIST experiment

R Bayes, on behalf of the TWIST Collaboration
School of Physics and Astronomy, University of Glasgow, Glasgow G12 8QQ, UK.
E-mail: Ryan.Bayes@glasgow.ac.uk

Abstract. The TWIST (TRIUMF Weak Interaction Symmetry Test) experiment probes the Lorentz structure of the weak interaction using muon decay. This structure has a very well defined form under the Standard Model (SM) which makes precise predictions for the shape of the decay positron spectrum with respect to momentum and angle. The shape of the spectrum may be described under some rather general assumptions using a set of decay parameters whose values according to the SM are $\rho = \delta = 3/4$, $\eta = 0$, and $\xi = 1$. TWIST uses a large sample of muon decays in a large acceptance spectrometer to measure the decay parameters to an order of magnitude greater precision than previous measurements. This experiment saw its last year of data collection in 2007. As TWIST is a systematics dominated experiment, much effort has been spent on refinements of the estimates of the systematic uncertainties over previous TWIST results. These proceedings will discuss the measures taken to achieve the precision goal of parts in 10^4, and the physics implications of the experiment.

1. Introduction

In the standard model (SM) the weak interaction moderates charge current interactions between fermions via the $W^{\pm}$ boson with a $V - A$ coupling. This structure of the SM weak interaction is based on observation, so it is essential to actively test this behaviour. The TRIUMF Weak Interaction Symmetry Test (TWIST) was conceived as a direct test of the weak interaction using the positron spectrum of the muon decay $\mu^+ \to e^+\nu_e\bar{\nu}_\mu$. Prior measurements were completed by TWIST in 2004 [1, 2], 2006 [3], and 2008 [4]. These proceedings describe the recently published final measurement of the TWIST experiment [5, 6].

The generalized, Lorentz invariant, derivative free, four fermion point interaction is described by the matrix element [7]

$$\mathcal{M} = \frac{4G_F}{\sqrt{2}} \sum_{\substack{\gamma=S,V,T \\ \epsilon,\mu=R,L}} g^{\gamma}_{\epsilon\mu} \langle \bar{e}_\epsilon | \Gamma^\gamma | (\nu_e)_n \rangle \langle (\bar{\nu}_\mu)_m | \Gamma_\gamma | \mu_\mu \rangle \tag{1}$$

where a μ handed muon couples to an ϵ handed positron via a scalar (S), vector (V), or tensor (T) interaction with a coupling strength $g^{\gamma}_{\epsilon\mu}$. The neutrino chiralities, n and m are dictated by the type of coupling that occurs. The $g^{\gamma}_{\epsilon\mu}$ represent 19 free parameters in the system as they are complex constants, assuming unitarity and that $g^T_{RR} = g^T_{LL} \equiv 0$.In the SM all $g^{\gamma}_{\epsilon\mu}$ are zero with the exception of $g^V_{LL} = 1$.

The positron spectrum from muon decay is given by the differential distribution [8]

$$\frac{\partial^2\Gamma}{\partial x \partial \cos\theta} = \frac{m_\mu}{4\pi^3} W^4_{e\mu} G^2_F \left(F(x) + |\vec{P}_\mu| \cos\theta \, G(x) \right) \tag{2}$$

where $x = E_e/W_{\epsilon\mu}$ is the reduced momentum when $W_{\epsilon\mu} = 52.83$ MeV/c is the maximum energy of the decay positron, and θ is the angle between the muon polarization and positron direction, $\cos\theta = \vec{P}_\mu \cdot \vec{p}_e/|\vec{P}_\mu||\vec{p}_e|$. $F(x)$ and $G(x)$ are tree level isotropic and anisotropic spectra,

$$F(x) = \sqrt{x^2 - x_0^2}\Big(x(1-x) + \frac{2}{9}\rho(4x^2 - 3x - x_0^2) + \eta x_0(1-x)\Big) + R.C. \quad (3)$$

$$G(x) = \frac{1}{3}\xi(x^2 - x_0^2)\left(1 - x + \frac{2}{3}\delta\left(4x - 3 + \left(\sqrt{1 - x_0^2} - 1\right)\right)\right) + R.C. \quad (4)$$

where ρ, η, δ, and ξ are the muon decay parameters. The decay parameters are bi-linear combinations of the weak coupling constants, and $\rho = \delta = 3/4$, $\xi = 1$ in the SM. TWIST was not sensitive to the η parameter.
The goal of the TWIST experiment was to improve the precision of ρ, δ, and $P_\mu\xi$ by an order of magnitude over previous experiments. To achieve this level of precision all radiative corrections, denoted as R.C. in Eq. 3 and 4, up to the second order leading logarithmic corrections ($\mathcal{O}(\alpha^2 ln^2)$) were included in the analysis [9]. The second order correction $\mathcal{O}(\alpha^2)$ was not included in the analysis and is a systematic uncertainty.

2. The TWIST apparatus

TWIST was conducted at TRIUMF in Vancouver, B.C., using the 500 MeV cyclotron to produce surface muons from a graphite target. Surface muons, with polarization $P_\mu^\pi = 1$ in the SM, were transported with no depolarization down the M13 secondary beamline into the TWIST experimental area. The primary apparatus was a parallel plane spectrometer immersed in a 2 Tesla solenoidal magnetic field which is known to a precision of 0.5 Gauss. The final focus of the beamline was immediately before the spectrometer along the magnetic field axis to minimize muon depolarization. A pair of low mass, time expansion chambers (TEC) were periodically introduced in the beamline at its focal point to monitor the beam position and angle.
The TWIST spectrometer was composed of 44 planer drift chambers (DC) and 12 planer proportional chambers (PC) with a scintillator system upstream of the spectrometer providing a muon event trigger [10]. The chambers are arranged symmetrically around a 99.999% pure Al or Ag muon stopping target foil. The stopping position of the muons was controlled using a variable density gas degrader coupled to a software feedback system which maintained the mean position of the stopping distribution at the micron level in the target foil. The detector dimensions and wire positions were known to a precision of parts in 10^5 in all three dimensions allowing for a high precision measurement of the passage of positron tracks.

3. Analysis

The data used by TWIST for the purpose of this analysis were 10^{10} muon events collected in 14 sets wherein the experimental conditions were held constant. Analysis of single muon decay events begin with pattern recognition using PC hit times and simple fits using DC wire centres. Momenta and angles describing the helical positron path are extracted from a least squares fit using the closest distance of approach to the wires supplied by tabulated DC space time relationships. Cuts and event selections are applied to the data to identify events with low systematic bias. A total of 5×10^8 high quality events were identified after cuts.
The measurement of the muon decay parameters (MDPs) relies on a very good Monte Carlo simulation (MC) of muon decays within the TWIST detector. This MC was generated from pre-defined samples of muon decay spectra input to a GEANT 3 simulation of the detector. These samples were defined using a randomly selected set of MDPs which were hidden from the experimenters. The value of η was fixed to -0.0036 ± 0.0069 [4]. The simulation was generated with 2.7 times the number of events in data. The MCs were subjected to the same analysis as the

Journal of Physics: Conference Series **408** (2013) 012071 doi:10.1088/1742-6596/408/1/012071

data. The data spectrum was fit to a linear combination of the simulated spectrum and a set of "derivative" spectra with a χ^2 statistic to determine the difference in the MDPs ρ, ξ, and $\xi\delta$. This approach is possible because spectrum has a linear dependence on the muon decay parameters. The difference in momentum scales between data and MC is angle dependent and significant relative to the precision goals of TWIST; $\Delta p/W_{e\mu} \sim$ (10 keV/c)/52.828 MeV/c $\sim 2 \times 10^{-4}$. The momentum calibration was corrected in the data by removing the linear interpolation of the differences with respect to $1/|\cos\theta|$ from each event. The nature of this correction as a function of momentum is unknown so its implementation was a systematic uncertainty. Two different calibration methods were used; a momentum shift $p_{ec} = p_{raw} + \Delta p$, and a scaling $p_{ec} = p_{raw}/(1 - \Delta p/W_{e\mu})$. MDPs were taken to be the average value from the two methods.

4. Systematics

TWIST is a systematics limited experiment so a majority of effort in the experiment was spent controlling and understanding systematic effects on the decay spectrum. The systematic uncertainties are shown in Fig.1. Systematics are categorized as target material specific, common to all data sets, and specific to measurements involving the polarization. The positron interaction systematics are examples of the systematic evaluation. The difference between data and MC is less than 1% for delta ray production and 2.4% for bremsstrahlung production. The uncertainties in the muon decay parameters were measured from special MCs which exaggerate the effect of these positron interactions fit to a nominal MC. MDP differences from this fit, scaled relative to the interaction difference between data and MC, are the systematic errors.

5. Physics results

The measured values of the decay parameters are shown in Table 1. The best measurements completed prior to the TWIST experiment are also shown demonstrating that the experimental goals of the experiment were achieved. The precise data set needed to be selected by the

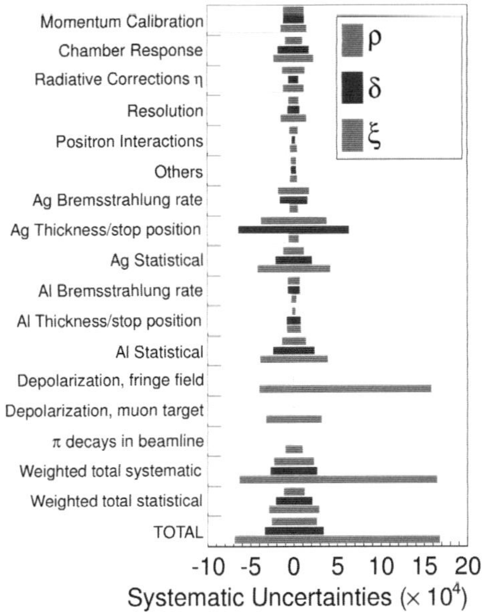

Figure 1: Graphical representation of the systematics for the parameters ρ, δ, and ξ.

Table 1: MDPs values measured from this work in contrast to the previous best measurements.

	Current work	
ρ	$0.74997 \pm 0.00012(\text{st.}) \pm 0.00023(\text{sys.})$	
δ	$0.75049 \pm 0.00021(\text{st.}) \pm 0.00027(\text{sys.})$	
$P^{\pi}_{\mu}\xi$	$1.00084 \pm 0.00029(\text{st.})^{+0.00165}_{-0.00063}(\text{sys.})$	
$P^{\pi}_{\mu}\xi\delta/\rho$	$1.00179^{+0.00156}_{-0.00063}$	
	> 0.99909 (90% C.L.)	
	Previous results	
ρ	0.7518 ± 0.0026	PDG 2003
δ	0.7486 ± 0.0038	[11]
$P_{\mu}\xi$	1.0027 ± 0.0085	[12]
$P_{\mu}\xi\delta/\rho$	0.99787 ± 0.00082	[13]

Table 2: Coupling constants determined from the global analysis of the MDPs.

$\|g^S_{RR}\| < 0.035$	$\|g^V_{RR}\| < 0.017$	$\|g^T_{RR}\| \equiv 0$
$\|g^S_{LR}\| < 0.050$	$\|g^V_{LR}\| < 0.023$	$\|g^T_{LR}\| < 0.015$
$\|g^S_{RL}\| < 0.420$	$\|g^V_{RL}\| < 0.105$	$\|g^T_{LR}\| < 0.105$
$\|g^S_{LL}\| < 0.550$	$\|g^V_{LL}\| < 0.960$	$\|g^T_{LL}\| \equiv 0$

collaboration with an agreed analysis such that the results needed to show good internal consistency and the systematics needed to be similarly agreed upon by all members of the collaboration before the analysis was completed and the hidden parameters were revealed. All criteria were satisfied and the hidden parameters were revealed on January 22, 2010.
The measured value of the combination of parameters $P_\mu\xi\delta/\rho$ is greater than 1 by 2.9σ, implying that the spectrum is not positive definite for a small region at the endpoint outside of the fiducial region. This is not physical, initiating an internal review of the results. The review resulted in the separation of the systematics into stopping target dependent and common, with an associated re-weighting of the average decay parameter to accommodate this change. A systematic uncertainty pertaining to the the mean muon stopping position was identified. There was very little change in the MDPs after the review with a small change in $P_\mu\xi\delta/\rho$. A global analysis was used to evaluate the effect of these new measurements on the weak couplings[14]. The three MDPs measured here are not sufficient on their own to constrain the values of the weak coupling constants so parameters from decay positron polarization measurements [15, 16] were included with current and past measurements of ρ, δ, and $P_\mu\xi$ as shown on Table 1. Table 2 shows the weak coupling constants where there is improvement in the right-right and right-left coupling constants. Further improvements in limits on left-right coupling constants require inverse muon decay measurements. A lower limit $P_\mu\xi\delta/\rho > 0.99909$ was set from the global analysis.

6. Conclusions

The measurement of the MDPs is used as a model independent test of the SM weak interaction. TWIST completed its measurement of the three MDPs ρ, δ, and $P_\mu\xi$ in 2010, having achieved its stated goal of an order of magnitude improvement in precision over previous measurements. The value of the endpoint anisotropy $P_\mu\xi\delta/\rho$ calculated from these results produced a non-physical result and so prompted a review of the experimental analysis of the collaboration after the hidden parameters were revealed, but no errors in the analysis were found that could explain the inconsistency. In spite of this the experiment makes a significant contribution to the limits of the weak interaction couplings.

Acknowledgments

All TWIST collaborators and students deserve profound thanks for their efforts in producing these results. This work was supported in part by the National Science and Engineering Research Council and the National Research Council of Canada, the Russian Ministry of Science, and the U.S. Department of Energy. Computing resources were supplied by WestGrid and Compute/Calcul Canada.

References

[1] Musser J *et al.* (TWIST Collaboration) 2005 *Phys. Rev. Lett.* **94** 101805 (*Preprint* hep-ex/0409063)
[2] Gaponenko A *et al.* (TWIST Collaboration) 2005 *Phys. Rev. D* **71** 071101 (*Preprint* hep-ex/0410045)
[3] Jamieson B *et al.* (TWIST Collaboration) 2006 *Phys. Rev. D* **74** 072007 (*Preprint* hep-ex/0605100)
[4] MacDonald R *et al.* 2008 *Phys. Rev. D* **78** 032010
[5] Bayes R *et al.* (TWIST Collaboration) 2011 *Phys. Rev. Lett.* **106** 041804
[6] Bueno J *et al.* (TWIST Collaboration) 2011 *Phys. Rev. D* **84** 032005 (*Preprint* arXiv:1104.3632[hep-ex])
[7] Fetscher W, Gerber H J and Johnson K 1986 *Phys. Lett. B* **173** 102–106
[8] Amsler C *et al.* (Particle Data Group) 2008 *Phys. Lett. B* **667** 1
[9] Arbuzov A, Czarnecki A and Gaponenko A 2002 *Phy. Rev. D* **65** 113006
[10] Henderson R S *et al.* 2005 *Nucl. Instrum. Meth. A* **548** 306–335 (*Preprint* hep-ex/0409066)
[11] Balke B *et al.* 1988 *Phys. Rev. D* **37** 587–617
[12] Beltrami I *et al.* 1987 *Phys. Lett. B* **194** 326–330
[13] Jodidio A *et al.* 1986 *Phys. Rev. D* **34** 1967–1990
[14] Gagliardi C A, Tribble R E and Williams N J 2005 *Phys. Rev. D* **72** 073002 (*Preprint* hep-ph/0509069)
[15] Burkard H *et al.* 1985 *Phys. Lett. B* **150** 242–246
[16] Danneberg N *et al.* 2005 *Phys. Rev. Lett.* **94** 021802

The H line: a brand new beam line for fundamental physics at the J-PARC muon facility

N Kawamura[1], A Toyoda[2], M Aoki[3], N Saito[2], S Mihara[2], K Shimomura[1] and Y Miyake[1]
[1] Muon Science Laboratory, Institute of Materials Structure Science, KEK
Tsukuba, Ibaraki 305-0801, Japan
[2] Institute of Particle and Nuclear Studies, KEK
Tsukuba, Ibaraki 305-0801, Japan
[3] Department of Physics, Graduate School of Science, Osaka University
Toyonaka, Osaka 560-0043, Japan

E-mail: nari.kawamura@kek.jp

Abstract. The muon facility, J-PARC (Muon Science Establishment; MUSE), has been operated since first beam in 2008. Starting with a 200 kW proton beam, the beam intensity has reached 3×10^{6}/ muons/s, the most intense pulsed muon beam in the world. A 2 cm thick graphite target permits the extraction of four secondary muon beams. A brand new beam line, the H line, is planned to be constructed. The new beam line is designed to have a large acceptance, will provide the ability to tune the momentum, and use a kicker magnet and/or Wien filter. This beam line will provide an intense beam for experiments that require high statistics and must occupy the experimental areas for a relatively long period.

1. Introduction

The Japan Proton Accelerator Research Complex (J-PARC) project was proposed jointly by the Japan Atomic Energy Agency (JAEA) and the High Energy Accelerator Research Organization (KEK) [1]. The new facility consists of a 400 MeV linac and 3 GeV and 50 GeV synchrotron rings. These provide an intense proton beam to pursue particle and nuclear physics, materials and life science and nuclear technology. A 1 MW proton beam is transported from the 3 GeV synchrotron ring to the Materials and Life Science Facility (MLF), which consists of the muon and the spallation neutron facilities. As shown in Fig. 1 and 2, the proton-beam-line tunnel runs through the center of the MLF building from the north to the south. The east and the west wings are experimental halls where the neutron and muon beam lines will be constructed. In order to avoid the diffusion of radioactive contamination, and to make the structure of the radiation shield simple, the proton-beam transport tunnel is isolated from the experimental hall. Therefore the beam transport tunnel divides the MLF building into east and west wing. Although this building designs achieves its purpose, the design of the secondary beam lines is under severe geometrical constraints.

The muon-production target is inserted 30 m upstream of the neutron target in the proton beam line. From the muon target, we decided to extract four muon beam lines. Each of them provides the world-strongest pulsed muon beam with an individual design concept to be used for a variety of muon science experiments. Two of the secondary beam lines are extracted at the angle of 60 degrees to the proton beam line (forward), and the others are at 135 degrees (backward) [2].

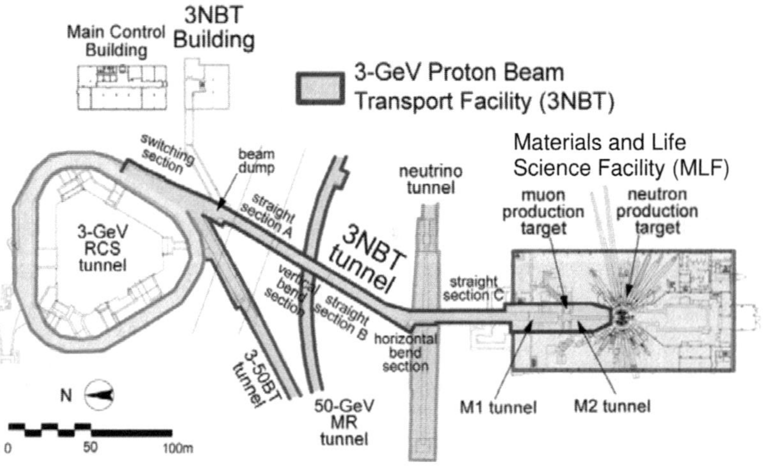

Figure 1. A schematic drawing from 3-GeV rapid cycling synchrotron (RCS) to MLF

The decay muon beam line (D line) was constructed first, and this is the only existing beam line. The D line extracts both negative and positive decay muons up to a momentum of 120 MeV/c, as well as 30-MeV/c surface and cloud muons [3]. This high momentum-tuneability meets the wide demands of user programs. The second beam line to be constructed in JFY2011 is the large acceptance beam line, the U line, which provides the most intense surface-muon beam of the four secondary lines [4]. This intense beam is used to generate an ultra slow muon beam by the laser resonant ionization method [5]. The remaining two beam lines, the S and H line, are located in the east experimental hall, and the construction will start in calendar 2012. The S line is designed to transport surface muons to several experimental areas by using kicker devices to dedicated μSR spectroscopy. Many sample materials can examined simultaneously under a variety of conditions. A typical period of an experiment in the field of material science is a few days to obtain enough statistics for μSR spectrum of sample materials. Therefore, a study requiring much higher statistics [6-8] does not match the design of the of S, D or U lines. The fourth beam line, the H line, aims to answer the demands for experiments [6-8] that need high statistics, *i.e.* the use of a high intensity beam for a long period (more than one year).

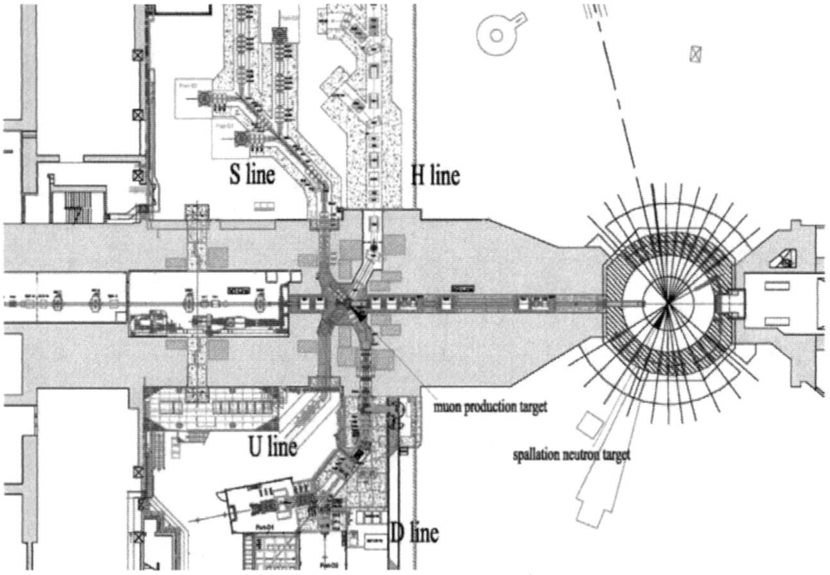

Figure 2. The layout plan of MUSE.

2. Beam line optics

The intended experiments in the H line require high statistics and therefore a high intensity beam. Momentum tuneability is also needed. Both features will be important for potential experiments in the future. The design of the H line will meet these two needs. The basic idea of the beam line optics was proposed to use a large-aperture muon-capture solenoid, a wide-gap bending magnet and a pair of two solenoid magnets that have opposite-direction fields to each other [9]. Conventional matrix calculations are not applicable because the near-axis approximation is not good in a large-aperture solenoid. Thus, G4BEAMLINE [10] is applied for optics calculation, and detailed design work is in progress.

The front-end capture solenoid magnet (HS1) consists of 8 coils, and three kinds of currents are supplied to the first, the second and the remaining coils. The first coil provides the main capture field, and by tuning the second and the remaining coils, the beam profile is minimized near the first bending magnet (HB1) and the gate valve (HGV1) which separates the beam line vacuum of the secondary line from that of the primary line to make the maintenance of the secondary line free from the primary line operation. The second solenoid magnet (HS2) makes either a parallel or weak-focusing beam by a weak field, enabling the installation of a Wien filter or other non-transport magnet like kicker magnets proposed in the experiment [8]. The third solenoid magnet (HS3) provides a field in the opposite direction to HS2 to cancel the beam-profile rotation by HS2. This cancelling function does not depend on the momentum, and thus beam loss due to momentum dispersion at the second bending magnet (HB2) are reduced. The final focusing (FF) magnets are discussed separately [11].

Figure 3 shows the result of a typical transport calculation. The beam is delivered to the first experimental area which is the nearest to the muon source. If HB2 is turned off, the beam will go straight ahead, so that the beam can be transported to the other experimental areas by applying the pair of opposite-direction solenoid field magnets. Thus the H line can provide intense beam to a few experiments that use different apparati and cannot share a common experimental area.

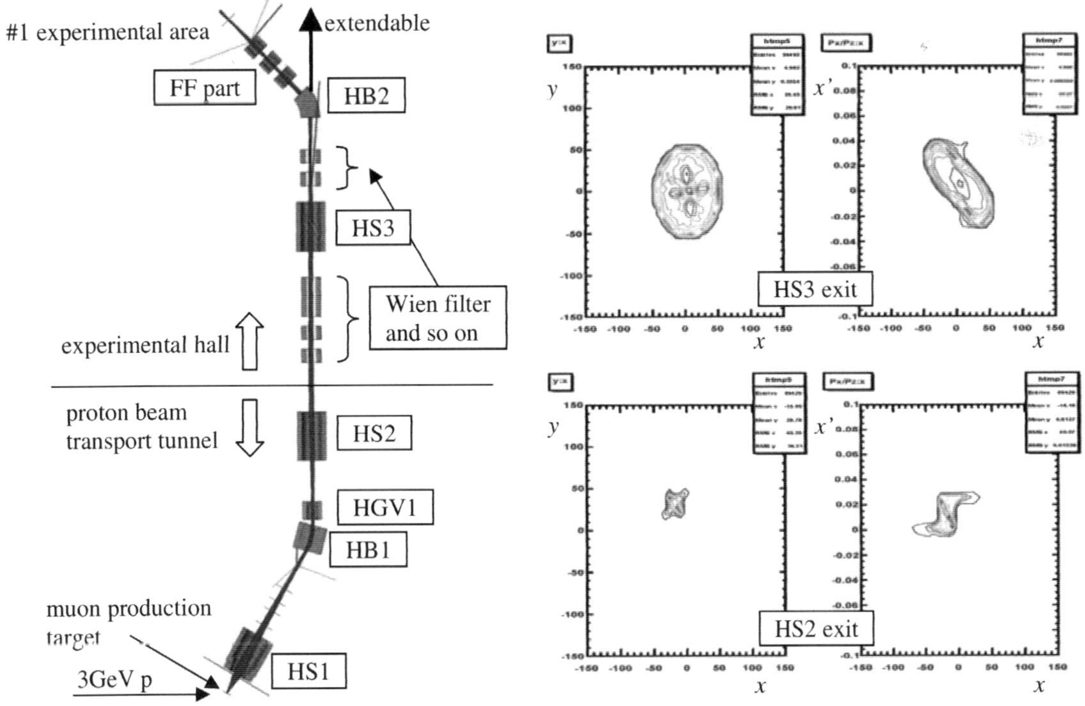

Figure 3. A typical result of a G4BEAMLINE calculation. The blue lines are beam trajectories (left). The surface muon beam from a point source is transported to the first experimental area through three

solenoid magnets and two bending magnets. The Wien filter and the kicker magnets are also shown, but they are not in use in this calculation. The beam profile at the exit of HS3 and HS2 are shown at the upper and lower right.

3. Summary

The beam optics calculation predicts a high transmission efficiency: more than 80% of the captured muons can be transported. The detailed design work of each magnet and other beam-line components is in progress, and the result will be fed back to the optics calculation.
As stated earlier, the muon facility has been under operation since 2008. The temporary radiation shields have been placed the in H line, and they have been activated. The radiation dose from the residual activities is expected to reach the order of 1 Sv/h in the next year. Thus HS1, HB1 and the other front-end devices will need to be installed in the next year. The mechanical design work is in progress along the design strategy for devices in a high radiation field [12].
In summary, the H line will be constructed in the world's strongest pulsed muon facility, MUSE. A high intensity muon beam is provided for the proposed experiments [6-8] in order to observe new physics beyond the standard model.

[1] Joint project team of JAERI and KEK 1999 KEK Report 99-4; JAERI-Tech 99-056; JHF-99-3.
[2] Miyake Y *et al.* 2005 *Nucl. Phys. B* **149** 393.
[3] Shimomura K *et al.* 2009 *Nucl. Inst. Meth.* A **600** 192.
[4] Nakahara K *et al.* 2010 *AIP Conf. Proc.* **1222** 420.
[5] Bakule P *et al.* 2003 *Spectrochimica Acta Part B* **58** 1019.
[6] Shimomura K *et al. these proceedings.*
[7] Saito N *et al. these proceedings.*
[8] Mihara S *et al. these proceedings.*
[9] Doornbos J private communication.
[10] http://www.muonsinc.com/
[11] Toyoda A *et al. these proceedings.*
[12] Kawamura N *et al.*2009 *Nucl. Inst. Meth.* A **600** 114.

J-PARC MUSE H-line optimization for the g-2 and MuHFS experiments

A Toyoda[1], Y Fujiwara[3], Y Fukao[1], O Kamigaito[3], N Kawamura[1], Y Matsuda[2], T Mibe[1], T Ogitsu[1], N Saito[1], K Sasaki[1], K Shimomura[1], M Sugano[1], K Tanaka[2], D Tomono[3], and H A Torii[2]
[1]KEK, 1-1, Oho, Tsukuba, Ibaraki, 305-0801, JAPAN
[2]The University of Tokyo, Institute of Physics, Tokyo, JAPAN
[3]RIKEN, Saitama, JAPAN

E-mail: akihisa.toyoda@j-parc.jp

Abstract. Significant deviation of the anomalous magnetic moment value (g-2) observed by the muon g-2 experiment [1] should be confirmed by the other experiment. This value is experimentally determined by frequency difference observed by the g-2/EDM experiment and muon magnetic moment observed by the muonium hyperfine splitting experiment (MuHFS). Both two experiments are planned to be performed at H-line of the J-PARC/MUSE under construction. We optimized the beamline layout for each experiment with G4beamline [5]. For both experiments, statistics is the most important, thus beamline transmission efficiency should be maximized. Especially for the g-2, the purpose of the present effort is to compromise between small beam size and small leakage field. For the MuHFS, it is crucial to minimize leakage field at around final focus position, and to get all stopped muons within good field region of MuHFS magnet. Conceptual design of the several final focusing systems will be presented.

1. Introduction

The 3.4 sigma deviation of anomalous muonic magnetic moment (g-2) from the standard model (SM) prediction reported by the BNL E821 experiment [1] is expected to be an indication of new physics, and should be confirmed by the other experiment with higher accuracy. Also it is needed to increase experimental accuracy of ratio of muonic magnetic moment to proton magnetic moment (μ_μ/μ_p) to achieve the g-2 precision improvement. This value is determined by the muonium hyperfine splitting measurement (MuHFS). The other topic is that the 3.9 sigma deviation of the positronium hyperfine splitting (PsHFS) measurement from the SM prediction also indicates new physics. Accuracy of the MuHFS is about 40 times worse than that of PsHFS, and thus this should be increased more.

Under the circumstances, two new experiments, g-2/EDM experiment [2] to measure g-2 and MuHFS experiment [3] to measure MuHFS, are proposed and planned to be performed at H line of the J-PARC/MUSE facility as shown in Figure 1. Basic strategy and optimization of the H line is reported in reference [4]. This article mainly focused on the final focusing design of these two experiments.

2. g-2/EDM beamline optimization

In this section we discuss the g-2/EDM beamline optimization done by using G4BEAMLINE program [5]. For this experiment, the following conditions are important to be satisfied.

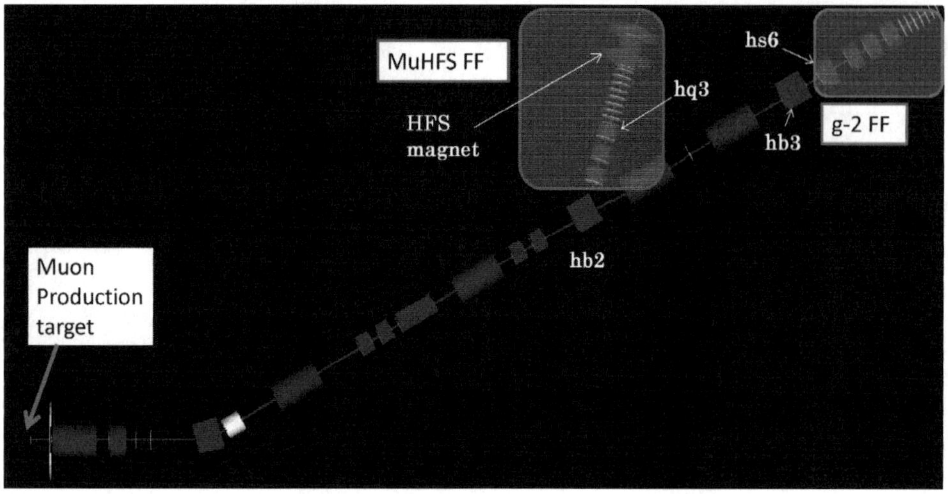

Figure 1. Layout of the J-PARC/MUSE H line. Purple box shows MuHFS beamline and its area. Blue box shows g-2/EDM beamline.

- *Transmission efficiency should be as high as possible.* Statistics is the most important issue of this experiment.
- *Focus point size should be less than 4 cm in diameter.* This experiment installs muonium production target at the beamline final focus point. The muonium emitted from this target is irradiated with a laser in order to produce the ultra slow muon beam. Small beam spot size helps to increase the slow muon intensity.
- *Leakage field at final focus position should be as small as possible.* The slow muon has low momentum at around the final focus, thus the leakage field disturbs the slow muon trajectory.

Possible final focus systems are evaluated in the following subsection.

2.1. Final focus system with 1 solenoid case

The simplest way to make small beam spot size is to use solenoid magnet. A solenoid magnet of hs6 is installed just downstream of the last bending magnet of hb3. Transmission efficiency is as high as 87.5 % inside 40 ϕ at final focus position. Focus point size is small 1.7 cm σ for x, 1.8 cm σ for y. However, leakage field of hs6 magnet is the problem. Figure 2 shows the magnetic field along beam axis. To make small beam spot, maximum field of hs6 is as high as 0.42 T. This leads to short focal length of 0.4 m, thus leakage field at the final focus point is as high as 0.07 T. To use this option, we need to design magnetic shield for the hs6 magnet carefully.

2.2. Final focus system with 3 quadrupole magnets

To solve the leakage field problem mentioned in the above subsection, we have an option to use 3 quadrupole magnets as the focusing system. Last component is quadrupole, thus the leakage field along the beam axis is zero. The maximum field of the quadrupole magnet is lower than that of the solenoid magnet therefore the leakage field at around the beam axis is also expected to be smaller.
The result of the optimization is that the transmission efficiency is 63.7 %, and the beam size is 3.5 cm σ for x, 1.8 cm σ for y. This result is not good enough therefore we cannot use this option.

2.3. Final focus system with 1 solenoid and 3 quadrupole magnets

The smallness of the transmission efficiency of the above 3 quadrupole system comes from its small focusing power. The solenoid system has the large leakage field. To solve these problems, we devise the following method as shown in Figure 3. First of all, we make a small focus spot by using the strong focusing power of an hs6 solenoid magnet. This small focus spot is transported to the final focus by using 3 quadrupole magnets. This method has an advantage of a small focus spot and a small leakage field. The estimated transmission efficiency is as high as 83.4 %, and the final focus size is as small as 2.1 cm σ for x, 1.1 cm σ for y. This method will be applicable to the g-2/EDM experiment.

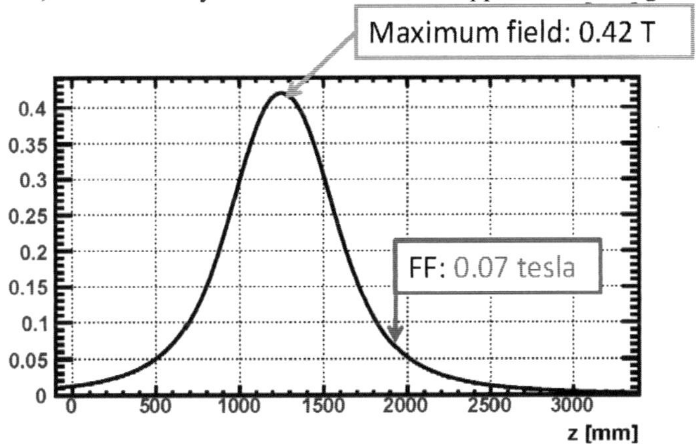

Figure 2. Magnetic field [T] distribution along the beam axis (z [mm]).

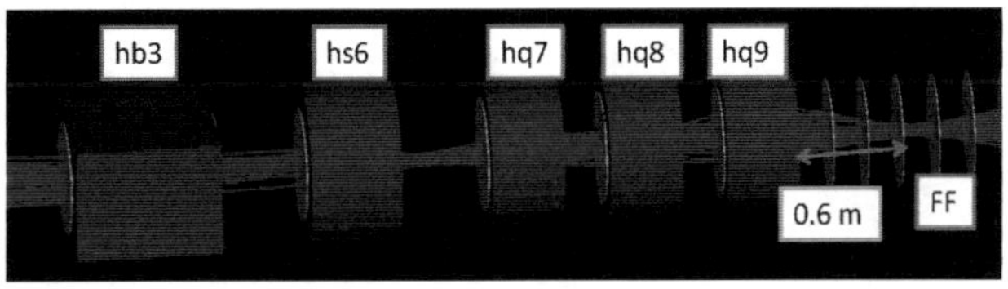

Figure 3. Layout and beam profile with 1 solenoid + 3 quadrupole magnets.

3. MuHFS beamline optimization

In this section we discuss the MuHFS beamline optimization. For this experiment, the following conditions are important to be satisfied.

- *Transmission efficiency should be as high as possible.* Statistics is the most important also for this experiment.
- *Focus point size should be less than 6 cm in diameter.* For this experiment, the beam focusing is not so much serious issue compared to the above g-2/EDM experiment, because we installed the HFS solenoid magnet and the strong magnetic field as high as 1.7 T eventually focuses the surface muon beam. This condition is decided to make sure that all muons stop inside good field region of the HFS magnet (20 cm in diameter).
- *Leakage field at final focus position should be less than 1.7 gauss.* One of the key issues of this experiment is to make a large and homogenious magnetic field region. This limit of 1.7 gauss comes from the limitation of the magnetic field correction power by the shim coil [6].
- *Muon stopping distribution should be inside good field region of the HFS magnet (30 cm long).* This condition will limit the muon momentum spread.

Possible final focus systems are evaluated as follows.

The most effective way to suppress the leakage field of the beamline magnet is to keep the focal length as long as possible. To achieve this long focusing, we devise the following method. At first, we prepare almost parallel beam by using 3 quadrupole magnets as shown in Figure 4.

Next we put the HFS magnet at the final focus point. The created magnetic-field distribution is consistent with OPERA [7] calculation. The HFS magnet design is described in detail in reference [6].

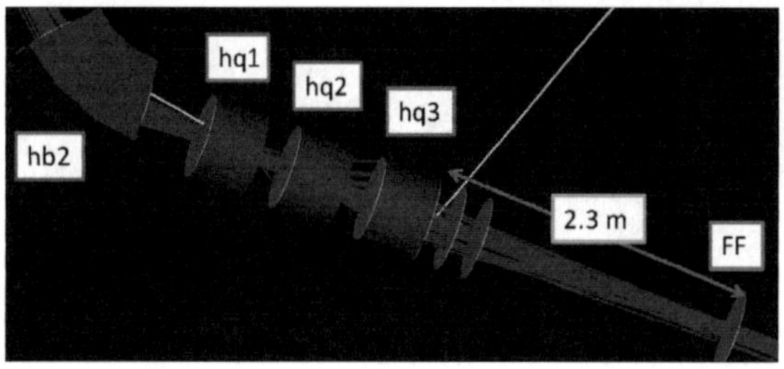

Figure 4. Layout and beam profile of the MuHFS final focus beamline.

One of the possible problems is effect of the leakage field of the strong HFS magnetic field to the muon beam transport. This leakage field is carefully investigated as shown in Figure 5, and the leakage field at a position of the last quadrupole magnet (hq3) is estimated to be as small as 0.005 T. We also checked the MuHFS leakage field effect to the beam profile at the hq3 end. It is confirmed that the MuHFS leakage field does not affect the beam transport seriously. Finally, we estimated the transmission efficiency and the beam profile at the final focus point with the MuHFS magnet. The result is that the transmission efficiency is as high as 93.6 % and final focus spot size is as small as 1.3 cm σ for x, 1.3 cm σ for y. It is concluded that this method will be applicable to the MuHFS final focus system.

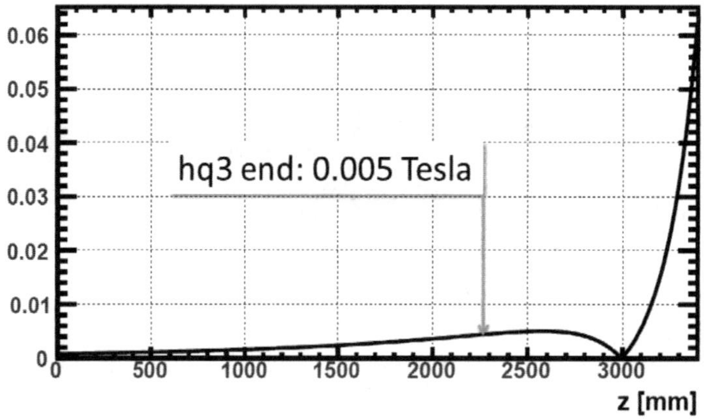

Figure 5. Magnetic field [T] distribution along the beam axis (z [mm]).

4. Summary and future plan

For the g-2/EDM beamline, the final focusing with 1 solenoid and 3 quadrupole magnets is effective, the transmission of 87.5 % is high and the beam size of 2.1 cm σ for x, 1.1 cm σ for y is small enough. For the MuHFS beamline, the final focusing with 3 quadrupole magnets works well, the transmission is as high as 93.6 % and the beam size is as small as 1.3 cm σ for x, 1.3 cm σ for y.

In the future, we need to estimate leakage field from the quadrupole magnets at a final focus point for both beamlines. For the MuHFS beamline, it is necessary to evaluate the muon stopping distribution.

References

[1] Bennett G W *et al.* 2006 *Phys. Rev.* D **73** 072003
[2] Saito N *et al.* these proceedings
[3] Shimomura K *et al.* these proceedings
[4] Kawamura N *et al.* these proceedings
[5] http://www.muonsinc.com/
[6] Sasaki K *et al.* these proceedings
[7] Cobham Technology Services Vector Fields Software http://www.cobham.com/

Development status of superconducting solenoid for the MuHFS experiment at the J-PARC

Ken-ichi Sasaki, Michinaka Sugano, Hiromi Iinuma, Toru Ogitsu, Naohito Saito, Koichiro Shimomura and Akira Yamamoto
KEK, 1-1, Oho, Tsukuba, Ibaraki, 305-0801, JAPAN
E-mail: ken-ichi.sasaki@kek.jp

Abstract. The development of a superconducting solenoid for a new Muonium HyperFine Structure, MuHFS, measurement at J-PARC has been underway since 2010. High homogeneity of magnetic field below 1 ppm is required for this experiment. Superconducting main coils were designed and error fields caused by the main coil misalignment were evaluated for designing superconducting shim coils. The coil deformation by the coil winding, thermal contraction and hoop stress due to the magnetic force was also estimated. A quench protection study was performed to determine the required parameters of the superconducting strand.

1. Introduction

A new measurement of hyperfine transitions in the ground state of muonium, called the HFS experiment, is part of the new J-PARC project [1]. The goal of this experiment is to measure the value of the ground state hyperfine structure interval of muonium down to the level of a few ppb, and also determine the ratio of muon and proton magnetic moment with accuracy of a 10 ppb. For this experiment, a superconducting solenoid system with a magnetic field of 3.4 T is under devlopment [2]. The solenoid requires a high field uniformity with a deviation less than 1 ppm within the spheroid region of 20 cm in diameter and 30 cm in length. This magnet is also considered as a prototype magnet for the g-2/EDM experiment at J-PARC [3], in order to minimize the development time and maximize the synergy between both experiments.

This paper reports the current status of the HFS magnet system development. The development of a field monitoring system to evaluate the field uniformity are also discussed.

2. Magnet design

2.1. Overview

A schematic view of the magnet system for the MuHFS experiment is shown in Figure 1. A NbTi superconductor will be used for the coil to minimize the cost of the strand. The strand diameter and Cu/Sc ratio are 2.0 mm and 6, respectively. The coil is immersed in a liquid helium of 4.2 K, and the re-condensation cryocoolers will be used for long term operation. The magnet will be operated in the persistent current mode to minimize the field fluctuation due to the ripples of the power supply.

2.2. Main coil design

Figure 2 shows the ideal coil block geometry of the MuHFS magnet. The magnet consists of 6 main coils and 2 shielding coils. The inner and outer diameter of the magnet are about 600 mm and 1460 mm, respectively. Contour lines in Figure 2 represent the field difference of ± 0.5 ppm from the center field of about 3.4 T at the current density of 1.18×10^8 A/m^2, corresponding to the strand current of about 411 A. The required uniform region is also plotted in Figure 2 by the dotted line. The Figure indicates that this coil configuration satisfies the required uniformity. The peak field on the coil and the stored energy are about 4.7 T and 4.0 MJ, respectively.

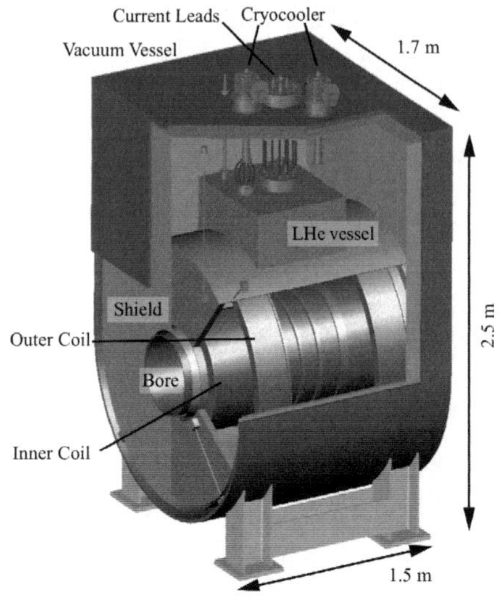

Figure 1. Schematic view of the magnet system for the HFS experiment at J-PARC.

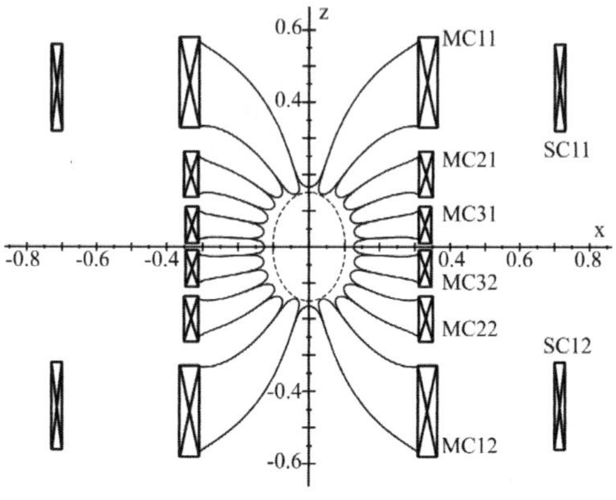

Figure 2. Coil cross sections and calculated field distribution. The contour lines represent the field difference of ± 0.5 ppm from the field at the magnet center. The dotted line represents the required uniform region.

2.3. Quench Protection

The quench protection performance of this coil was simulated. Fig. 3 shows the electrical circuit for the quench protection study in the case of persistent current operation mode. All the coils including the shield coils and the persistent current switch, PCS, are connected in series. The diodes have a forward voltage of 5 V and are connected in parallel with each section as shown in Fig. 3. It is assumed in the simulation that the coils are adiabatic, and that the normal zone does not propagate between coils.

Fig. 4 shows the simulation result during the quench of the MC31 coil. The current decay in the circuit including the quenched coil, is almost completed in about 15 sec after the quench, while the current in the another circuit is only slightly decreased. This is caused by the mutual inductance coupling with the active shield coils, which are wound in the counter direction to the main coils. A quench in the MC31 is the most serious case in terms of the peak temperature of the superconductor. However, the peak temperature after a quench of about 230 K is acceptable. The highest resistive voltage generates in the case of the quench in the SC11 coil. Even in that case, the peak voltage in the coil is allowable, about 2.1 kV. These results indicate that the magnet could be protected without any active protection scheme, such as protection heaters.

2.4. Field Tuning Scheme

In order to achieve a suffciently high homogeneity of the magnetic field, with deviations below 1 ppm, various sources of error field have to be considered beforehand, and the correction

Journal of Physics: Conference Series **408** (2013) 012074 doi:10.1088/1742-6596/408/1/012074

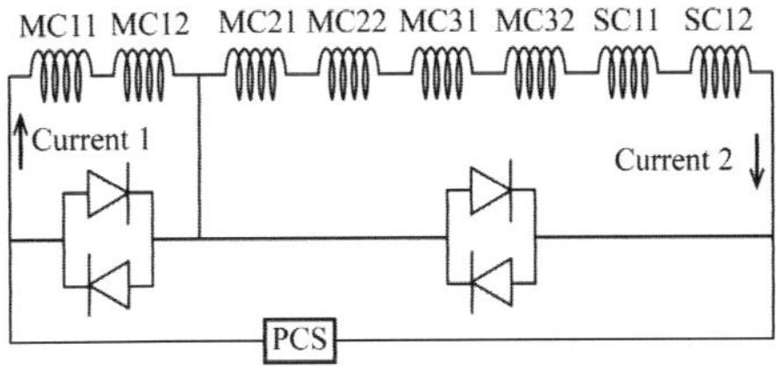

Figure 3. Electrical circuit for the study of quench protection during persistent current operation.

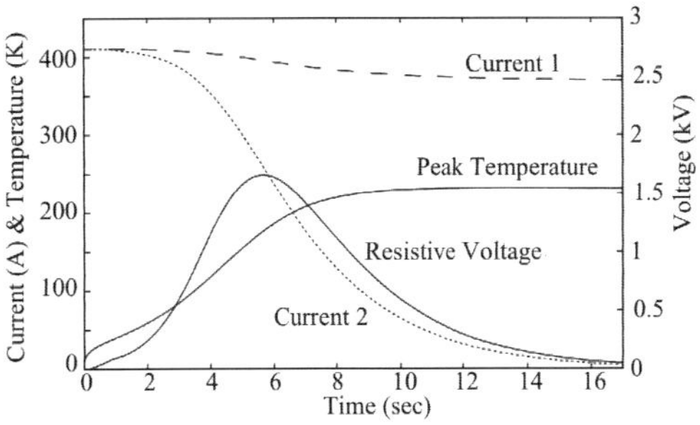

Figure 4. Quench simulation result in the case of the MC31 quench.

scheme must be determined ahead of time as much as possible. The error field due to the coil misalignments were evaluated. We find the error field is less than about 190 ppm, even if the errors in magnet construction were as large as 0.1 mm and 0.1 mrad [2], In order to guard against such a huge error field in the HFS magnet, a combination of steel pieces and superconducting shim coils will be employed; such a method is commonly used in commercial MRI magnets [4] . The error field would be roughly compensated to several ppm by steel pieces mounted on the surface of the magnet bore. For example, when the steel ring with 0.42 m in inner diameter and rectangular cross section (10 $\times$ 20 mm) is inserted in the bore of the coil MC31, the field change at the magnet center is calculated to be about 521 ppm. The complicated error field distribution could be corrected by arranging positions of steel pieces. The detailed design work of the superconducting shim coil is ongoing.

3. Field Monitoring System

3.1. Moving Stage

In order to compensate the error field with the scheme mentioned in the previous Section, a precise field monitoring system is required. An NMR probe will be used to measure the magnetic field; however, the probe size is very small, less than 5 mm. Therefore, a 3-axis moving stage is required. Figure 5 shows the schematic view of the r-θ-z moving stage developed now. The NMR probe is mounted on the slide stage in radial direction, and the slide stage is mounted on a rotating disk. Both the radial slide stage and the rotating disk are driven by ultrasonic motors. The whole stage can move along the z- axis on the rail, and is driven by a stepping motor, which is 2 m away from the magnet end, far away enough so that there will be no significant distortion of the magnetic field. Almost all the parts of this system consist of non-metallic material except for the case of the ultrasonic motor and NMR probe, which are made by aluminum and copper.

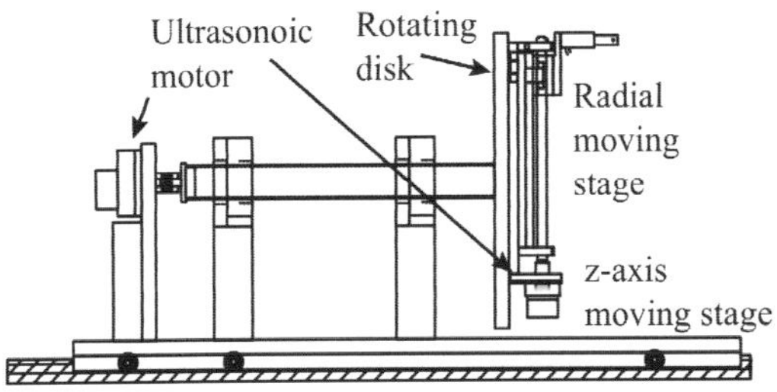

Figure 5. Schematic view of the r-θ-z moving stage.

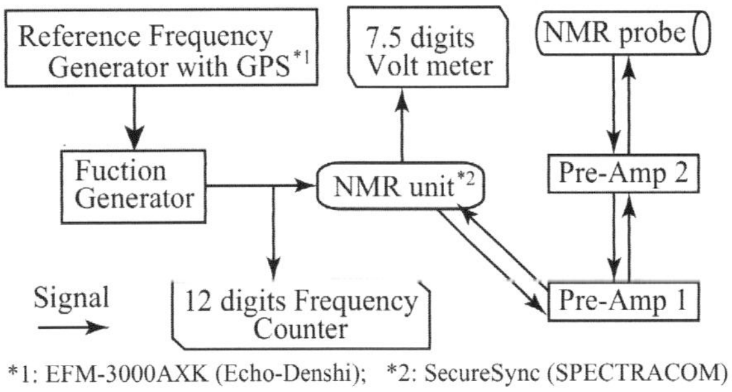

Figure 6. Connection diagram of the NMR system.

3.2. NMR system

Figure 6 shows the connection diagram of the field measurement system. A Continuous Wave type (CW) NMR unit and probe are used for this system. The most effective way to improve the stability and accuracy of the CW-NMR system is to improve the accuracy of the RF signal input to the NMR sample. In our system, a reference frequency generator with GPS, which has an accuracy of 1×10^{-12}, is connected to the function generator. The NMR unit output is a voltage proportional to the measured field.

The stability of this NMR system was checked using 3 T MRI magnet at the NIRS, National Institute of Radiological Sciences. Figure 7(a) shows the error distribution of the RF frequency output from the function generator measured by the frequency counter. The input frequency is 12.771271699 MHz, corresponding to a field strength of 2.9996145 T. The standard deviation is measured to be 1.4 $\times10^{-4}$ ppm. This demonstrates that the output signal of the function generator is highly stabilized thanks to the reference frequency generator, compared to the required accuracy of 1 ppm. However, the output voltage of the NMR unit was not sufficiently stable. Figure 7(b) shows the error distribution of the output voltage of the NMR unit, and the standard deviation is 0.025 ppm. The horizontal axis is the voltage difference between the mean value of the measured voltage. The reason for this result is not clear. However, one suspected source is the pre-amps between NMR probe and unit. In this test system, the frequency of the RF signal is amplified by ten times in the pre-amp circuits. This modulation seems to decrease the stability. A new NMR unit and pre-amps excluding the frequency modulation circuit will be tested in the next round of tests.

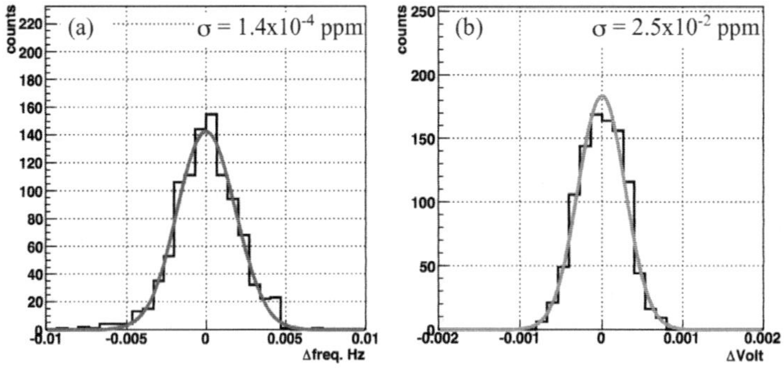

Figure 7. Error field distribution of the NMR system. (a): Frequency of the output signal of the function generator. (b): Output voltage of the NMR unit.

4. Summary

A superconducting solenoid with high field homogeneity for the muonium HFS experiment is being developed. The design of superconducting main coils was completed. The quench protection study showed that the main coils could be safely protected during the magnet quench. The field tuning scheme was discussed and both the steel pieces and the superconducting shim coils were considered. The field monitoring system using the continuous wave type NMR system was tested, and the new system will be tested at the NIRS. The basic design work will be finished soon, and the magnet system including the field monitoring system will be ready for the experiment in mid-2014.

References

[1] Shimomura K 2011 these proceedings
[2] Sasaki K, Sugano M, Ohkubo R, Iinuma H, Ogitsu T, Saito N, Shimomura K and Yamamoto A 2011 *22nd Int. Conf. on Magnet Technology*
[3] Saito N 2011 these proceedings
[4] Robitaille P M and Berliner L 2006 *Ultra High Field Magnetic Resonance Imaging* (New York: Springer-Verlag)

Status of a MIND type Neutrino Factory Far Detector

R Bayes[1], A Bross[3], A Cervera-Villanueva[2], M Ellis[4,5], A Laing[1], F J P Soler[1] and R Wands[3]

[1] School of Physics and Astronomy, University of Glasgow, Glasgow, G12 8QQ, UK
[2] IFIC and Universidad de Valencia, Valencia, Spain
[3] Fermilab, P.O. Box 500, Batavia, IL 60510-5011, USA
[4] Brunell University West London, Uxbridge, Middlesex UB8 3PH, UK
[5] Westpac Institutional Bank, NSW, Australia

E-mail: Ryan.Bayes@glasgow.ac.uk

Abstract. A realistic simulation and analysis of a Magnetized Iron Neutrino Detector (MIND) has been developed for the purpose of understanding the potential sensitivity of such a facility. The status of the MIND simulation and reconstruction as discussed in the interim design report is reviewed here. Priorities for producing a more realistic simulation for a reference design report will be discussed, as will be the steps that have already been taken towards an improved simulation.

1. Introduction

A magnetized iron neutrino detector (MIND) has been proposed as a far detector for a neutrino factory[1]. The detector has a high mass with a simple method of producing a magnetic field for the purpose of charge identification. The purpose of the far detector, in association with a near detector system, is to measure θ_{13} and δ_{CP} through the measurement of "Golden channel" neutrino oscillations, $\nu_e(\bar{\nu}_e) \rightarrow \nu_\mu(\bar{\nu}_\mu)$ [2]. A MIND detector will measure this by identifying the production of muons with the opposite sign of the neutrino source beam. A realistic simulation of a MIND-type detector is required for the optimization of golden channel measurements at the neutrino factory.

2. Conceptual design of a Neutrino Factory MIND

The proposed MIND is composed of alternating iron and scintillator planes. Each scintillator plane consists of twinned arrays of scintillator bars, 1 cm thick and 3 cm wide, arranged to measure the x and y position of an incident particle. The planes are to be built on an octagonal cross-section, 14 m$\times$14 m, with projections on either side of the iron plates to support the detector.

The stresses and distortions of an iron plane have been studied using a finite element simulation. Because the iron cannot be fabricated in a solid sheet, the iron plates will be constructed of iron strips, 2 m in width and 1.5 cm thick, in two perpendicular layers to provide strength and rigidity. The total thickness of an iron plate is 3 cm. The simulation shows the maximum deformation in any direction is less than 2 mm with vertical deformations on the support projections less than 0.2 mm.

The iron plate provides the magnetic field for the detector using a superconducting transmission line [3] (STL) running along the detector's axis of symmetry as a current source.

An STL can carry the 100 kA required to generate the 1 Tesla average magnetic field within a 7.8 cm diameter contained by a 10 cm bore. The magnetic field inside the iron has been simulated showing field deformations because of the octagonal geometry and the gaps between the iron strips used in the fabrication of the plates, as shown in Fig.1. It is thought that these sudden changes in the field direction and intensity will affect scattering in the muon tracks, which will require study in the full detector simulation.

To extract the required physics, a two baseline experiment using two different MINDs has been proposed. A 100 kTon MIND at a distance of 4000 km in coordination with a 50 kTon MIND at a distance of 7500 km is one example. For a cuboid geometry the detectors must be 125 m and 62.5 meters long, respectively. With 429 scintillator bars per plane, there will be 4.28×10^6 readout channels for the 100 kTon detector and 2.14×10^6 channels for the 50 kTon detector. To achieve the same mass using an octagonal geometry the detectors must be 150 m and 75 m long respectively with a proportional increase in the number of scintillator channels.

3. Simulation

A simplified detector was used for the simulation detailed in the neutrino factory interim design report (IDR) [1]. This detector used a square cross section and was immersed in a uniform magnetic field oriented in the +y-direction. GEANT 4 [4] was used to provide the material interactions of charged species while a combination of NUANCE [5] and LEPTO [6] were used to generate the neutrino interactions in the detector. The simulation included a wide range of neutrino interactions including deep inelastic scattering, quasi-elastic scattering, and pion production channels.

Muon tracks resulting from the neutrino interaction events were identified and fit using a Kalman fitter supplied using the RecPack software package [7]. A selection process is applied to the output that Kalman fit using a number of criteria such as the length of the track, the number of hits used in the analysis, the quality of the reconstructed track, the amount and direction of bending of the track, and the likelihood that the track is the result of a charge current event. The effect of these cuts on sets of μ flavoured neutrino charge current events is shown in Fig.2.

The simulated response of the detector in the 25 GeV/c μ beam is used to synthesise a sensitivity of the detector to the measurement of $\sin^2 2\theta_{13}$ and δ_{CP}. The fluxes and oscillation probabilities at a far detector based on the initial beam state was calculated using the Neutrino Tool Suite (NuTS) [8]. For the purpose of this study 2.5×10^{20} μ^+ and μ^- decays are assumed to occur in the storage ring per year as the true relative muon yields are not known and changes

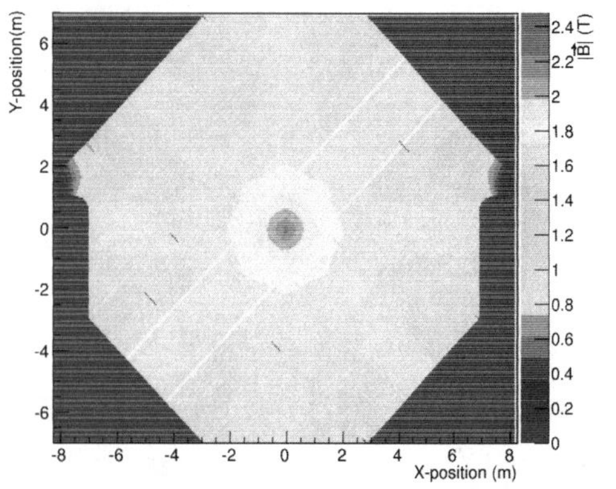

Figure 1. Magnetic field map simulated in the realistic geometry of the iron plate.

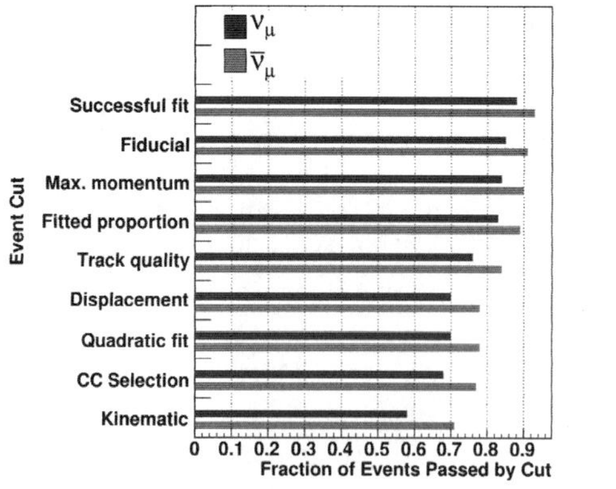

Figure 2. Fraction of events left after the application of the listed cuts applied to isolate charge current events.

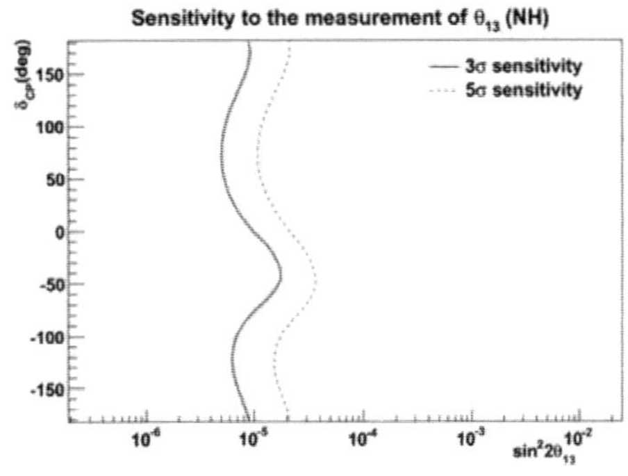

Figure 3. Sensitivity of the neutrino factory to measurement of $\sin^2 2\theta_{13}$ for given the shown values of δ_{CP}.

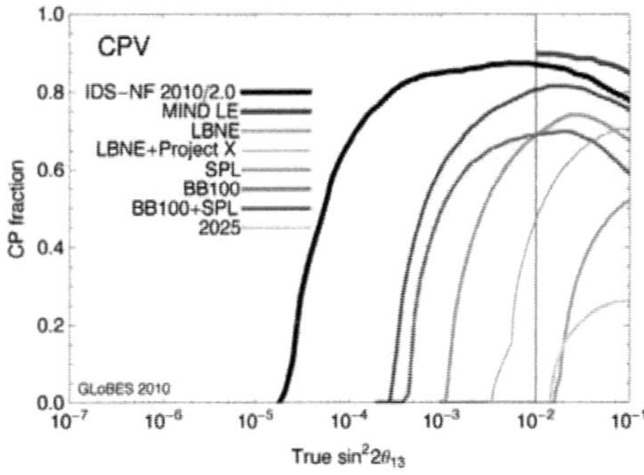

Figure 4. Sensitivity of the neutrino factory to measurement of δ_{CP} given the shown values of $\sin^2 2\theta_{13}$.

in the relative yields do not change the physics results. Assuming a two detector experiment described above, both $\sin^2 2\theta_{13}$ and δ_{CP} are fit simultaneously by minimizing

$$\begin{aligned}\chi^2 &= \sum_j \left\{ 2 \times \sum_e^{E_\mu} \left(A_j x_j N^e_{+,j}(\theta_{13}, \delta_{CP}) - n^e_{+,j} + n^e_{+,j} \log\left(\frac{n^e_{+,j}}{A_j x_j N^e_{+,j}(\theta_{13}, \delta_{CP})}\right)\right.\right. \\ &\left.\left. + A_j N^e_{-,j}(\theta_{13}, \delta_{CP}) - n^e_{-,j} + n^e_{-,j} \log\left(\frac{n^e_{-,j}}{A_j N^e_{-,j}(\theta_{13}, \delta_{CP})}\right)\right) + \frac{(A_j - 1)^2}{\sigma_A} + \frac{(x_j - 1)^2}{\sigma_x} \right\} \quad (1)\end{aligned}$$

where $n^e_{i,j}$ is the "data" for an energy bin e and detector baseline $j \in \{4000, 7500\}$ for μ signals with charge, i, and $N^e_{i,j}$ is the prediction for the same bin. Two systematic uncertainties are explicitly included in the fit. An uncertainty in the number of interactions, A_j, was assessed as $\sigma_A = 0.05$ owing to errors in the fiducial mass of the detectors. The known differences in interactions between neutrinos and anti-neutrinos, x_j, are limited by near detector measurements as $\sigma_x = 0.01$.

The sensitivity of these experiments to θ_{13} was defined as the set of points in $(\theta_{13}, \delta_{CP})$ where, $\chi^2(\theta_{13} = 0) - \chi^2_{min} \geq n^2$. The bands of sensitivity to $\sin^2 2\theta_{13}$ corresponding to $n = 3$ and $n = 5$ are shown in Fig. 3. The fraction of δ_{CP} accessible to a MIND detector at a neutrino factory compared to other experiments, as calculated from GLoBES [9], is shown in Fig. 4.

4. Toward a realistic simulation

The IDR MIND simulation does not represent the intended final geometry, so it is imperative to move to a more realistic model using an octagonal cross-section and corresponding magnetic field. This work has begun and requires the optimization of the analysis to this new geometry. A hadronic reconstruction for the MIND analysis is planned as is the introduction of the muon momentum reconstruction using the muon range. A major systematic study of the background introduced by cosmic rays is planned which will have an impact on choosing the detector site, specifically the depth required to achieve the experimental goals.

The first step towards a realistic simulation has been to use GENIE [10] to sample the neutrino interactions. It is believed that the parton distributions used in GENIE are more realistic than those used in LEPTO because the inelasticity distribution resulting from ν interactions in GENIE as shown in Fig 6 is more flat than the distribution from LEPTO.

5. An Alternative Neutrino Factory Far Detector

A totally active scintillating detector has also been proposed as a far detector for a neutrino factory. This detector would be constructed entirely of scintillator bars, with a square overall

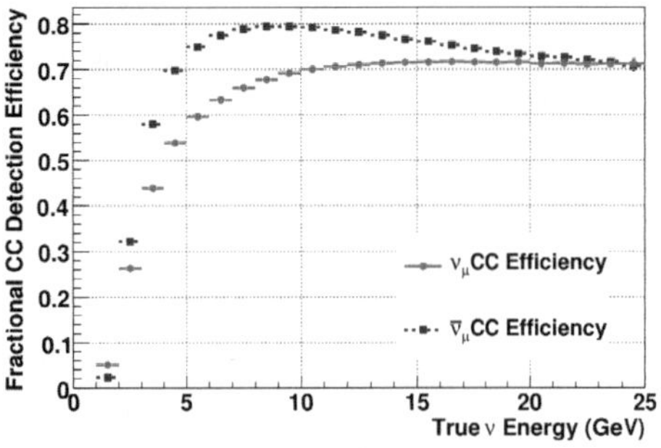

Figure 5. Charge current detection efficiency as a function of neutrino energy using a GENIE neutrino generator.

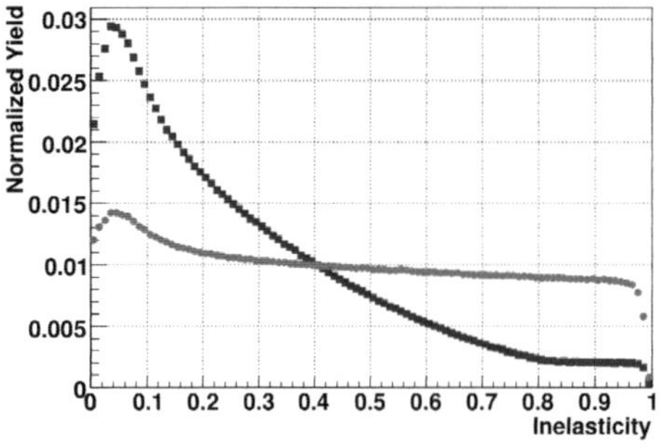

Figure 6. Inelasticity of neutrino interaction events generated using the GENIE neutrino generator.

detector cross-section, 15 m×15 m, and 150 m long. A simulation of a TASD has been produced by Malcolm Ellis for the purpose of studying its response parallel to the MIND simulations.

The benefit of using such a detector is a sensitivity to lower energies than a MIND allowing an effective reconstruction of electron showers generated by $\nu_e(\bar{\nu}_e)$ and neutral current $\nu_\mu(\bar{\nu}_\mu)$ interactions. Thus it can also measure $\nu_\mu(\bar{\nu}_\mu) \rightarrow \nu_e(\bar{\nu}_e)$, platinum channel, oscillations in addition to the golden channel oscillations accessible to a MIND.

The challenge with such a detector is that the magnetic field must be generated for charge identification outside of the detector. The magnetic cavern proposed for this purpose is a technical problem as it is as yet unclear whether the superconducting transmission line[3], again proposed to carry current for the magnetic field generation, can operate when bent into a 7 m radius coil. MIND is still the leading candidate for this reason.

6. Conclusion

These proceedings have reviewed the status of a proposed far detector for the neutrino factory. A great deal of work has been done on the engineering of the detector for the publication of the IDR including structural simulation and field simulation of a realistic design. An interim simulation has demonstrated the feasibility and sensitivity of a MIND type detector using a cuboid geometry and a dipole field. This simulation has shown very high efficiency and good background suppression leading to a sensitivity to θ_{13} and CP violation that surpasses all of the proposed next generation neutrino facilities. The new MIND simulation, currently in progress, uses a more realistic assumptions, with the proposed specifications of the realistic design. The reconstruction and analysis of the simulation output requires optimization for this new detector and magnetic field geometry. It is expected that the results of the full realistic simulation will come within this year.

References

[1] Choubey S *et al.* 2011 Interim Design Report Tech. Rep. IDS-NF-020 IDS-NF
[2] Cervera A *et al.* 2010 *Nucl. Instrum. Meth. A* **624** 601–614 (*Preprint* arXiv: 1004.0358[hep-ex])
[3] Ambrosio G *et al.* (VLHC Design Study Group) 2001
[4] Allison J *et al.* 2006 *Nuclear Science, IEEE Transactions on* **53** 270 –278 ISSN 0018-9499
[5] Casper D 2002 *Nucl. Phys. Proc. Suppl.* **112** 161–170 (*Preprint* hep-ph/0208030)
[6] Ingelman G *et al.* 1997 *Comput. Phys. Commun.* **101** 108–134 (*Preprint* hep-ph/9605286)
[7] Cervera-Villanueva A *et al.* 2004 *Nucl. Instrum. Meth. A* **534** 180–183
[8] Burguet-Castell J *et al.* 2001 *Nucl. Phys. B* **608** 301–318 (*Preprint* hep-ph/0103258)
[9] Huber P *et al.* 2007 *Comput. Phys. Commun.* **177** 432–438 (*Preprint* hep-ph/0701187)
[10] Andreopoulos C *et al.* 2010 *Nucl. Instrum. Meth. A* **614** 87–104 (*Preprint* arXiv: 0905.2517[hep-ph])

Journal of Physics: Conference Series **408** (2013) 012076 doi:10.1088/1742-6596/408/1/012076

NOνA detector technology with initial performance from the surface prototype

Mathew Muether

Fermi National Accelerator Lab, Neutrino Department, P.O. Box 500 MS-220, Batavia, IL 60510 USA

E-mail: muether@fnal.gov

Abstract. NOνA, the NuMI Off-Axis ν_e Appearance experiment, will study $\nu_\mu \rightarrow \nu_e$ oscillations characterized by the mixing angle θ_{13}. Provided θ_{13} is large enough, NOνA may ultimately determine the ordering of the neutrino masses and measure CP violation in neutrino oscillations. A complementary pair of detectors will be constructed ~14 mrad off beam axis to optimize the energy profile of the neutrinos. This system consists of a surface based 14 kTon liquid scintillatior tracking volume located 810 km from the main injector source (NuMI) in Ash River, Minnesota and a smaller underground 222 Ton near detector at the Fermilab. The first neutrino signals at the Ash River Site are expected prior to the 2012 accelerator shutdown. In the meantime, a near detector surface prototype has been completed and neutrinos from two Fermilab sources have been observed using the same highly segmented PVC and liquid scintillator detector system that will be deployed in the full scale experiment. Design and initial performance characteristics of this prototype system are being fed back into the design for the full NOνA program.

1. Introduction

NOνA, the NuMI Off-Axis ν_e Appearance experiment will study $\nu_\mu \rightarrow \nu_e$ oscillations at a baseline of 810 km (L/E of 400 km/GeV). Provided θ_{13} is as large as early indications from the T2K experiment [1], NOνA will have the reach to determine the ordering of the neutrino masses and constrain the CP violating phase, δ [2]. Additionally, NOνA will study the differences in oscillations of neutrinos and antineutrinos, as well as make a precision measurement of θ_{23} by observing the muon neutrino disappearance.

Achieving these goals requires a detector with 14 kTons of material capable of suppressing ν_μ charge current (CC) and neutral current (NC) backgrounds at the 99% level. Additionally, good ν_e detector efficiencies and energy resolution less than 8% of the expected signal width for ν_e CC event are required [2].

2. Experimental design

2.1. Beam

To achieve the goal of the NOνA project, two detectors will be placed 14 mrad off-axis to the primary direction of the NuMI (Neutrinos from the Main Injector) source. As shown in Figure 1, the ν_μ flux near the first $\nu_\mu \rightarrow \nu_e$ oscillation maximum at around 2 GeV is optimized at this angle [2]. The off-axis beam also reduces high energy neutral current background events. To accommodate the needs of the experiment, the NuMI beam power will be upgraded to 700 kW during the shutdown of the accelerator from March to December 2012.

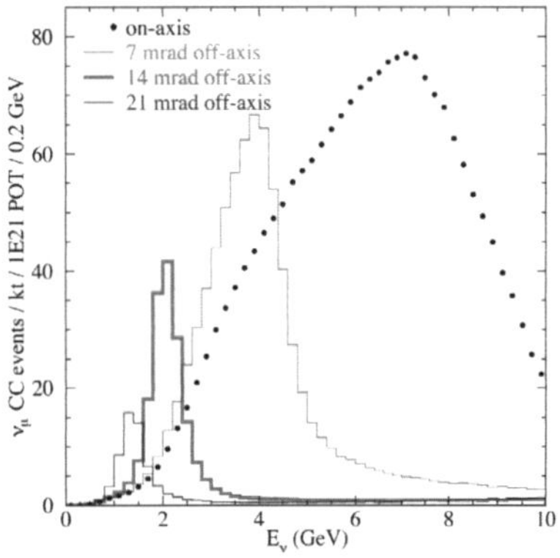

Figure 1. ν_μ CC energy distribution for off-axis angles with medium energy tune [2].

Figure 2. Photograph of the NOνA Near Detector On the Surface.

2.2. Detectors

The NOνA detector system consists of a complementary pair of detectors constructed 14 mrad off-axis to the NuMI source. Both detectors will be highly segmented tracking calorimeters built entirely from low Z ($\sim$0.15 radiation lengths per layer) PVC, glue, and mineral oil based liquid scintillator with a 65% active volume [2]. The far detector will be a surface based 14 kTon volume located 810 km from NuMI in Ash River, Minnesota. A smaller 222 Ton unit will be built 1.1 km from the source at Fermilab in a 105 meters deep underground cavern.

3. Near and far detector status

Beneficial occupancy of the far detector facility was obtained in April 2011. Construction of the far detector at Ash River is scheduled to begin in the first quarter of 2012 with a goal to have a detector segment in place before the Fermilab accelerator shutdown. The full detector is on track for completion in the first half of 2014.

Excavation of the underground cavern for the near detector will also begin following the beam shutdown. In the meantime, a near detector surface prototype (NDOS) has been completed and neutrinos from both the NuMI and Booster sources at Fermilab have been observed using the same highly segmented PVC and liquid scintillator detector system that will be deployed in the full scale experiment. This prototype has been taking data since October 2010. The initial performance characteristics of this prototype system are highlighted in the text below.

3.1. PVC cells

The NOνA detector is built up from extruded TiO_2 loaded PVC cells [2]. Each cell is 3.8 cm by 5.9 cm in cross section with 90% reflectivity for light at 430 nm. Extrusions are joined together to produce a sealed module of 32 cells. In NDOS, the modules are either 4.2 m (vertical orientation) or 2.9 m (horizontal) long while far detector modules are 15.6 m long. These modules are glued together into alternating planes of horizontal or vertical orientation to create self-supporting 32 layer blocks. $\sim$360,000 cells make up the 14 kTon far detector.

Six blocks were constructed along with a 1.7 m muon ranger (Fig. 2) for NDOS. Building the NDOS fully exercised the quality assurance/quality control (QA/QC) techniques in preparation for full production running for the far detector. This process revealed cracks in ~20% of the manifold covers. These covers have been repaired and a new more robust design will be used for future production. 1200 far detector sized sized extrusion have been produced to date.

3.2. Liquid scintillator
PVC blocks are filled in place with a "home brew" of mineral oil containing 5% pseudocumene and wavelength shifters to produce 400-450 nm light. The liquid scintillator makes up 65% of the total detector mass [2]. The scintillator is required to have 80% of the light output of commercial Bicron BC517P at 1 meter. NDOS required ~30,000 gallons of scintillator while the 14 kTon far detector will use over 3 million gallons.

Oil work at NDOS gained us experience in the filling process. Some internal module obstructions were observed during filling; the causes of these has been resolved. NOνA has currently taken possession of around 100,000 gallons of the oil that will be used for the far site.

3.3. Fiber
Internal to each cell is a 0.7 mm diameter looped fiber. The fiber shifts the light collected in the scintillator to 490-550 nm [2]. The fiber ends are routed through the manifold covered to an optical connector where they are available for single sided readout. ~113 km of fiber is used in the near detector design with 13,000 km needed for the far detector.

NDOS fiber handling allowed us to overcome tangling problems related to spooling techniques. We have also learned to measure the fiber performance in real-time as modules are strung. We have received nearly 50% of the required fiber.

3.4. Avalanche photodiodes
The light from the fiber ends is incident on Hamamatsu avalanche photodiodes (APD) which have 85% quantum efficiency for 520-550 nm light [2]. The devices are operated at -15 °C with a gain of 100. For NOνA a 20 photoelectron (pe) signal from a minimum ionizing particle at the far end of a far detector sized module is required with a 10-15 pe threshold applied. Based on initial system verification, we expect 38 pe for such a signal, well above the requirement. 496 APD arrays are required for the near detector and about 12,000 are used in the far detector design.

Surface cleanliness and sealing issues have led to many of the NDOS APDs becoming unusably noisy. 274 installed unit have been removed from the detector for cleaning and study. New surface coatings and installation techniques are under study.

3.5. DAQ
The signals from the APDs are processed by front-end electronics (FEBs) which operate in continuous baseline subtraction digitization mode while sampling each channel every 500 ns [2]. 64 FEBs are fed to a Data Concentrator Module which packages and passes the data in 50 μs blocks to a processing farm. The data is then buffered at the farm for several seconds at which point a software trigger may be issued to record available data in a specified window.

For the near detector, a 500 μs trigger window was used with three separate trigger sources; the 0.4 Hz NuMI spill signal, the 1.2 Hz Booster beam signal, and a 10 Hz cosmic pulser. Real throughput capabilities of the DAQ have doubled since initial running. During stress tests of the system, stable running was achieved with a 96% duty factor (80 ms trigger windows at 12 Hz).

Journal of Physics: Conference Series **408** (2013) 012076 doi:10.1088/1742-6596/408/1/012076

4. Results from the prototype near detector on the surface (NDOS)

Analysis of the data from NDOS is in progress. A full suite of available Monte Carlo (masked to behave like our prototype) together with tracking on real data has allowed us to begin to calibrate and reconstruct.

NDOS has collected 5.6×10^{19} protons on target (POT) worth of data in reverse horn current beam and 8.4×10^{18} POT in forward horn mode from NuMI. Analysis of this sample has yielded 1254 candidate neutrino events with 108 expected cosmic background events. Figure 3(a) shows an excess of tracks pointing back to the NuMI source over the out-of-time cosmic background [4]. Similar distributions have been seen in a Booster neutrino sample of 222 event (with 92 expected background events) from 3×10^{19} POT [5].

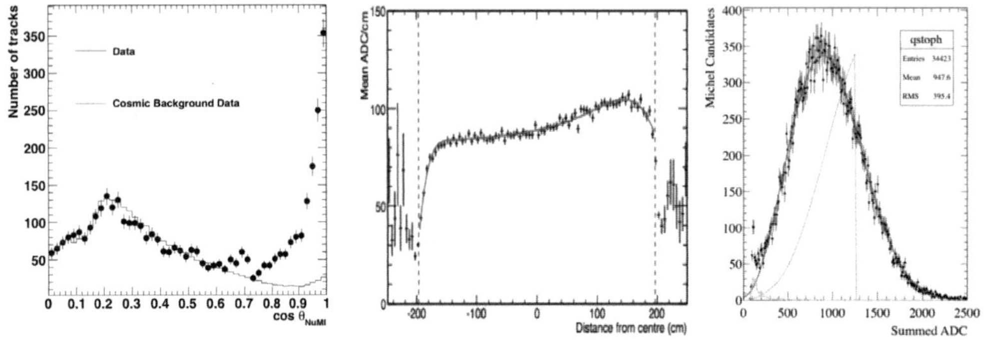

(a) Angular track distribution [4]. (b) Attenuation from muons [6]. (c) Michel spectrum [3].

Figure 3. Preliminary analysis of events from the NuMI source and cosmic data in NDOS.

Additional studies have been performed to understand the energy deposited in the detector and its cell by cell calibrations. Figure 3(b) shows the mean ADC value as a function of the distance from the center of the cell from a cosmic muon sample [6]. A sample fit which could be used to calibrate the detector response is shown. Figure 3(c) shows a sample Michel electron distribution which can be compared against expectation from simulation to provide an electromagnetic energy calibration [3].

5. Conclusion

The NOνA NDOS is taking and analyzing data now. This surface prototype has proved invaluable to all aspects of the experimental program, providing critical feedback for design enhancements and operational experience. Recent results from T2K which hint at a large value for θ_{13}, are very encouraging for the long term physics reach of NOνA and open the opportunity to make real contributions in understanding the neutrino. The far detector is on track to begin data taking in 2012, with the detector hall construction nearly complete and expected beam upgrades running on time. The support for NOνA continues to grow with the collaboration now consisting of 110 physicists from 24 institutions in 4 different countries.

References

[1] Abe K, *et al.* 2011 *Indication of electron neutrino appearance from an accelerator-produced off-axis muon neutrino beam, Preprint* arXiv:1106.2822 [hep-ex]
[2] Feldman G, *et al.* 2007 *NOνA Technical Design Report*
[3] Messier M 2011 *Internal Nova Document NOVA-doc-6023*
[4] Betancourt M 2011 *Internal Nova Document NOVA-doc-5874*
[5] Johnson C 2011 *Internal Nova Document NOVA-doc-5873*
[6] Backhouse C 2011 *Internal Nova Document NOVA-doc-5931*

Journal of Physics: Conference Series **408** (2013) 012077 doi:10.1088/1742-6596/408/1/012077

R&D project for Gd-doped water Cherenkov detector

Takaaki Mori for the Super-Kamiokande Collaboration

Department of Physics, Okayama University, Okayama, 700-8530, Japan

E-mail: takaaki@fphy.hep.okayama-u.ac.jp

Abstract. A dedicated test facility for a proposed Gadolinium doped water Cherenkov detector is being constructed in the Kamioka mine near the Super Kamiokande detector. Anti-electron neutrinos ($\bar{\nu_e}$) from inverse beta decay can be identified with high efficiency by taking advantage of Gd's large cross section on thermal neutron capture and by taking coincidence of a prompt positron and the delayed 8 MeV gamma cascades.

1. Introduction

In the universe, there exists supernova relic neutrinos (SRN) from all past supernovae. Super Kamiokande (SK) [1] search for these SRN events via inverse beta decay interaction ($\bar{\nu_e} + p \rightarrow e^+ + n$) having the largest cross section in visible mode for the SRN interactions. The colored lines in Figure 1 show the theoretical neutrino flux for different models. We have a narrow window of 10-30 MeV to search for the electron anti-neutrino SRN due to the reactor and atmospheric neutrino backgrounds. The expected number of SRN signals in SK is 0.8-5 per year in 10-30 MeV region. Figure 2 shows the expected SRN signals at the SK for different models, and their current SK limits. It is still difficult to observe the SRN signals even with a detector as large as SK.

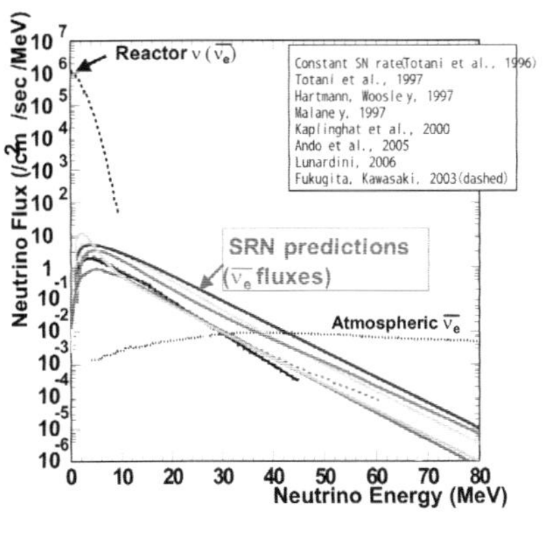

Figure 1. Colored lines show predicted SRN spectrum for different models at SK. Black dotted lines are representing background neutrinos.

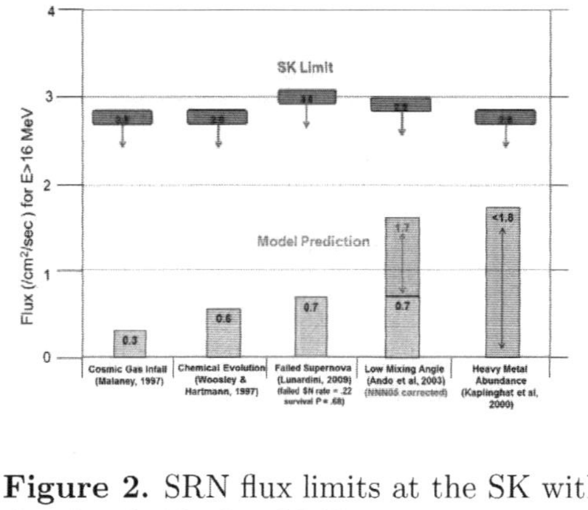

Figure 2. SRN flux limits at the SK with the threshold of 16 MeV.

Thus addition of 0.1% gadolinium (Gd) solution in the SK detector is proposed [3]. The proposed detector is named GADZOOKS! [Gadolinium Antineutrino Detector Zealously Outperforming Old Kamiokande, Super!]. We take advantage of the large cross section of Gd on thermal neutron capture. Neutrons from the inverse beta decays are quickly thermalized in the detector volume and captured by the Gd. Gd subsequently emits about 3-4 γ rays having the total energy of about 8 MeV. By taking the coincidence of the prompt positron and the delayed gamma cascade, we can uniquely identify the antineutrino events and lower the energy threshold down to 10 MeV. This will leads us to the first observation of a SRN signal in SK.

1.1. Gd compound

Soluble Gd is in powdered forms as chemical compounds, so we determined first which Gd compound to be dissolved in the SK water. The candidates of Gd compounds are $GdCl_3$, $Gd(NO_3)_3$, $Gd_2(SO_4)_3$. Corrosion tests show $GdCl_3$ highly corrodes the stainless steel, while no damages were apparent on the another two. However in $Gd(NO_3)_3$ solution, strong absorption by nitrate was evident below 350 nm, in the Cherenkov light wavelength region. Although $Gd_2(SO_4)_3$ is not so easily dissolved into water, we can help dissolving it with the help of H_2SO_4. $Gd_2(SO_4)_3$ is found to be the best candidate thus far.

1.2. Material soak test

We have performed the material soak test for all components (31 elements) of the SK detector to check any corrosions and deteriorations. We put each piece into a polypropylene bottle filled with 500 ml of 0.1% Gd solution or pure water. Only affected material is the PMT support rubber in the Gd solution. Further studies with this rubber show the effects on the SK detector with much greater volume is expected to be small, if water temperature is kept 15^oC. We will further examine the effects of the SK detector materials in the Gd doped water at Gd test facility.

2. Test Gd facility

We are building the Gd test facility called EGADS [Evaluation Gadolinium's Action on Detector Systems]in order to demonstrate the principle of GADZOOKS!. EGADS facility is consisted of 200 ton water Cherenkov detector with the same 20 inch the photomultiplier tubes (PMT) used for SK, a Gd mixing pre-treatment device, a main Gd-water circulation system, and a water transparency measurement device. The following items will be tested at this facility:

(i) Transparency of the Gd doped water. The transparency should be long enough to do as various physics analysis as in SK.

(ii) Gd-water purification system. Gd doped water should be as pure as that of the SK with only Gd compounds.

(iii) Effect of Gd on detector materials. Any corrosions and deteriorations of SK materials by Gd should be checked.

(iv) How to introduce/remove gadolinium. Is the Gd uniformly mixed in the tank? How quickly/economically/completely can Gd be removed?

(v) Ambient neutron level in the tank. Does it cause significant increase in the trigger rate for solar analysis?

(vi) Neutron capture efficiency. We can measure the efficiency of neutron capture by inserting a neutron source in the tank.

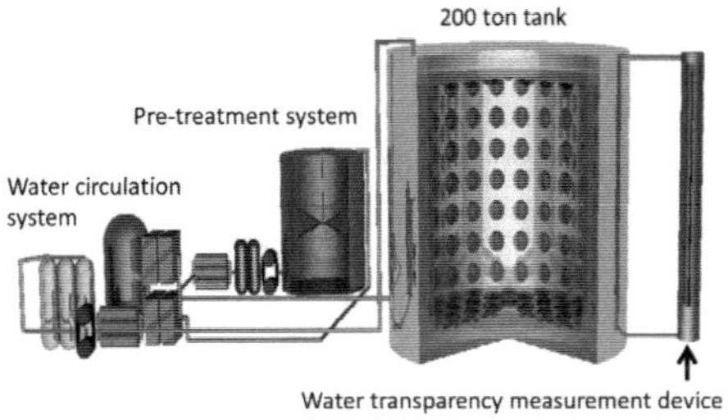

Figure 3. Schematic view of the EGADS system. From left to Right: Gd-water purification system, Gd-water mixing system with 15 ton tank equipped with a sterrer, a 200 ton tank with 20 inch PMTs, and a transparency measurement device.

2.1. Gd water circulation system

The circulation system consists of two parts:pre-treatment system where we mix the Gd compunds into water and main circulation system.

The $Gd_2(SO_4)_3$ powder is first mixed into water at the 15 ton pre-treatment tank to make the 0.1% solution. The tank has a built-in stirrer to completely dissolve $Gd_2(SO_4)_3$. With a small amount of H_2SO_4, $Gd_2(SO_4)_3$ can be dissolved much quickly. The pre-treatment system is equipped with a UV filter to kill bacteria and Amberjet resin to remove uranium and thorium with more than 99% efficiency, along with mechanical filters.

The water at EGADS must be kept ultra pure without removing $Gd_2(SO_4)_3$. We call the system "Selective filtration" [4], since multivalent ions such as Gd^{3+}, and $(SO_4)^{2-}$ should be selected out. Schematic view of of this system can be seen in Figure 4. Gd and SO_4 ions are rejected at the nanofilters (NF), because the pore size of the membrane is smaller. Total Oxidizable Carbon (TOC) can be a food source to bacterial. Ultraviolet photons from TOC destruction system can excite the organic molecules and products ionized carbonic acids which in truth can be removed by deionizer (DI) unit. The Gd-rejected water is then sent through the Reverse Osmosis (RO), where only H_2O molecules can pass through the membrane. The pure water and $Gd_2(SO_4)_3$ are combined at collection buffer tank and circulated to the 200 ton tank. Details of this water system can be found in [4] [5].

The current Gd removal system installed at the EGADS site consists of ion exchange resin. This system can remove Gd in the efficiency of more than 99%. In the near future, we will install a removal system with a mechanical filter press which is currently in development.

2.2. Water transparency measurement device

To evaluate the quality of the water in the EGADS system, a water transparency measurement device is installed. Keep long light attenuation length would be key issue in a Gd-doped water Cherenkov detector. This device can measure light intensity continuously as a function of light travel distance and monitor relevant Cherenkov region using multiple leasers with different wavelength. Details of this device can be found in [5].

2.3. Quality check of PMTs before the installation

We plan to use 240 20 inch PMTs in EGADS tank. Since available PMTs have been left unused for 7 years, we had to check the quality of all candidate PMTs in terms of dark rate, uniform gain, and peak-to-valley ratio, and select good 240 PMTs to be used for installation. This pre-calibration system is schematically shown in Fig.4. We used two light sources for the precalibration: 1) a xenon flash lamp and a scintillation ball to produce a uniform light intensity, and 2) an LED to measure 1 photo-electron (p.e) peak, thus providing a Gain and a peak-to-valley (P/V) ratio. The output charge of a PMT (Q) illuminated by a scintillation ball is expressed as follows:

$$Q = N_{photon} \times QE \times Gain = N_{photon} \times QE \times (\alpha \times V^{\beta}) \quad (1)$$

where N_{photon} is the number of photons hitting cathode of PMT, QE is quantum efficiency. α and β is the parameter for each PMTs. We determined an operational HV for a PMT to produce a charge of 30 photo-electrons within 2%, which should be equivalent to a Gain of 10^7. The absolute Gain at the operational HV has been measured as one photo-electron peak generated by LED. This 1 p.e measurement also provided the P/V ratio. Fig.5 shows a 1 p.e. measurement. With thus determined operational HVs, we measure an average Gain of 2.25 pC/1p.e which is similar to 2.23 pC/1p.e, being the Gain of SK PMTs. Thus, we selected good 240 PMTs which all have good properties for noise ratio (<35 kHz), P/V ratio (>1.2), and similar occupancy (~QE). It is important to evaluate the sensitivity of each PMT to 1 p.e since a signal of 1 p.e or a few p.e.s are expected from γ rays by neutron captured on Gd.

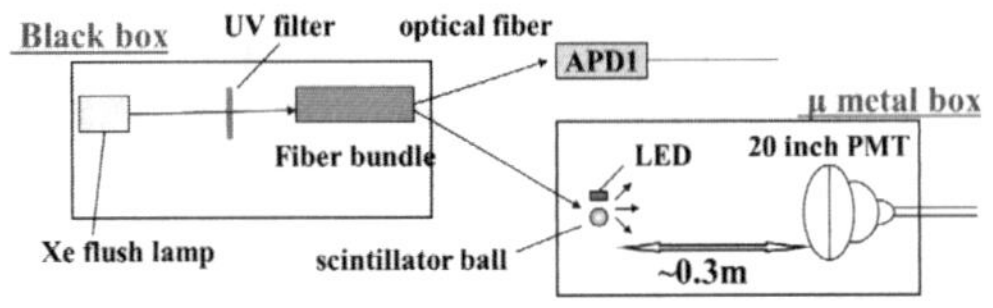

Figure 4. The schematic view of pre-calibration system. A test PMT was set in a μ-metal box to suppress a extra-terrestrial magnetic field down to 40 mG.

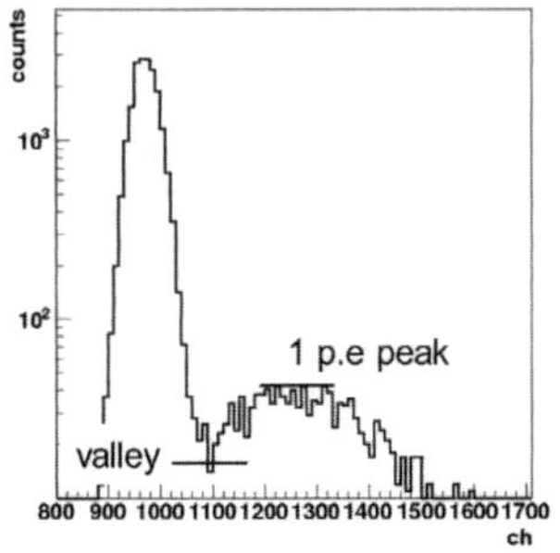

Figure 5. Measurement of Gain of PMT which will be installed in 200 ton tank by 1 photo electron distribution.

3. Current status and future plan

All equipments of EGADS without PMTs are ready in the laboratory in Kamioka mine. We checked water circulation system with pure water for about 6 month. Measurement of water transparency shows EGADS water purification system can keep the pure@water as clean as the SK water. Gd2(SO4)3 has been dissolved successfully in the pre-treatment system. This year, we focus on the operation of the circulation system with Gd, where selective filtration, pressure sensor, and some membrane will be tested to achieve the best efficiency of Gd recovery and transparency of Gd-doped water. After the stable condition can be obtained, we plan to circulate the Gd-doped water through the 200 ton tank. Installation of PMTs follow then.

In the 2012-2013, we plan a complete proof-of-principle test of Gd-loaded water Cherenkov technology using EGADS. The new neutron tagging technology with Gd will have a large impact on future large neutrino experiments exploiting neutrino oscillation and neutrino astronomy.

References

[1] Fukuda Y *et al.* 2003 *Nucl. instrum. Methods Phys. Res.* A **501** 418
[2] Malek M *et al.* 2003 *Phys. Rev. Lett.* **90** 061101
[3] Beacome J F and Vagins M R 2004 *Phys. Rev. Lett.* **93** 171101
[4] Vagins M R 2010 *Detection of Supernova Neutrinos Int. Conf. on Neutrino Physics and Astrophysics*
[5] Renshaw A 2011 *R&D project for a Gadolinium Doped Water Cherenkov Detector Technology and Instrumentation in Particle Physics*

Journal of Physics: Conference Series **408** (2013) 012078 doi:10.1088/1742-6596/408/1/012078

ICARUS and status of liquid argon technology

D Stefan[1]
Laboratori Nazionali del Gran Sasso dell'INFN,
S.S 17 BIS km. 18.910 67010 Assergi L'Aquila, Italy

E-mail: dorota.stefan@lngs.infn.it

Abstract. ICARUS is the largest liquid argon TPC detector ever built (~600 ton LAr mass). It operates underground at the LNGS laboratory in Gran Sasso. It has been smoothly running since summer 2010, collecting data with the CNGS beam and with cosmics. Liquid argon TPCs are really "electronic bubble chamber" providing a completely uniform imaging and calorimetry with unprecedented accuracy on massive volumes. ICARUS is internationally considered as a milestone towards the realization of the next generation of massive detectors (~tens of ktons) for neutrino and rare event physics. Results will be presented on the data collected during 2010 with the detector at LNGS.

1. Introduction

Currently the Liquid Argon Time Projection Chamber (LAr-TPC) technology [1] is of great interest due to its potential in investigation of the neutrino properties and rare events such as proton decay or dark matter. The sensitivity is around 4 times more with respect to the similar measurement performed in the water Cherenkov detector of the same mass. The ICARUS detector is the largest liquid Argon (LAr) detector which has ever been built, with mass of 600 ton (T600). The experience gained from the detector construction and operation and also data analysis should be further used in the future LAr detectors. The advantage of the LAr TPC is its excellent spatial and calorimetric resolution which makes possible a perfect visualization of the charged particles tracks. Such a detection technique is often compared to the bubble chamber filled with heavy liquids, but with the electronic readout. In order to take advantage of the modern experimental techniques, the track reconstruction and data analysis should be fully automatized. This requires studying new reconstruction methods and solving many algorithmic problems, which are similar to those from automatic image processing and pattern recognition.

2. The ICARUS experiment

The ICARUS detector [2] is now installed in the Hall B of the Gran Sasso underground National Laboratory (LNGS) of Istituto Nazionale di Fisica Nucleare (INFN), shielded against cosmic rays by about 1400 meters of rock. The detector is running smoothly under stable conditions starting from October 1st 2010. It is simultaneously collecting self-triggered events of cosmic rays and neutrino interactions associated with the CNGS neutrino beam, focusing on neutrino oscillation search and in particular on $\nu_\mu \rightarrow \nu_\tau$ appearance. The trigger system and liquid Argon purity measurement of the detector are very important issues, described in [2].

The detection principle is following. Charged particles, generated for example by a neutrino interaction in LAr, produce ionization along their path. Thanks to the low transverse diffusion of

[1] On behalf of the ICARUS Collaboration

charge in LAr and to its exceptionally high purity, electrons are drifted along the lines of the electric field applied to the LAr volume and, are projected onto the anode (TPC).

Anode is made of three parallel planes of wires, 3 mm apart and 3 mm of wire pitch, facing the drift path of 1.5 m. By appropriate voltage biasing, the first two planes (Induction-1, Induction-2) provide signals in a non-destructive way, whereas the ionization charge is finally collected by the last one (Collection plane).

Wires are oriented on each plane at a different angle (0°, $+60^{o}$, -60°) with respect to the horizontal direction. Therefore, combining the wire coordinates on each plane at a given drift time, a three-dimensional image of the ionizing event can be reconstructed.

The measurement of the absolute time of the ionizing event is combined with the electron drift velocity information, and provides the position of the track along the drift coordinate. The determination of the absolute time of the ionizing event is accomplished by the prompt detection of the scintillation light produced in LAr by charged particles.

3. Approach to the event reconstruction

In the event reconstruction two types of objects are distinguished: tracks and cascades. The scheme of the reconstruction procedure is presented on the figure 1. First, the signals (hits) on all three independent wire planes have to be found. Next, only in the Collection wire planes, the fitting procedure to the signal shape is applied to obtain the charge deposition on wires. At this stage, the rough estimation of calorimetric measurement can be provided.

3.1. Spatial reconstruction

In order to obtain precise calorimetric information for tracks, the 3D reconstruction is necessary. For this purpose, first, the 2D hit clusters are found separately for two wire planes. Then Polygonal Line Algorithm (PLA) [3] is used to form the 2D trajectories and sort hits along them. Finally, 3D track reconstruction is performed by matching hits from two corresponding 2D trajectories, while third wire plane is used in ambiguous topologies. Matching is based on both the same electron drift time and order of hits sorted with PLA.

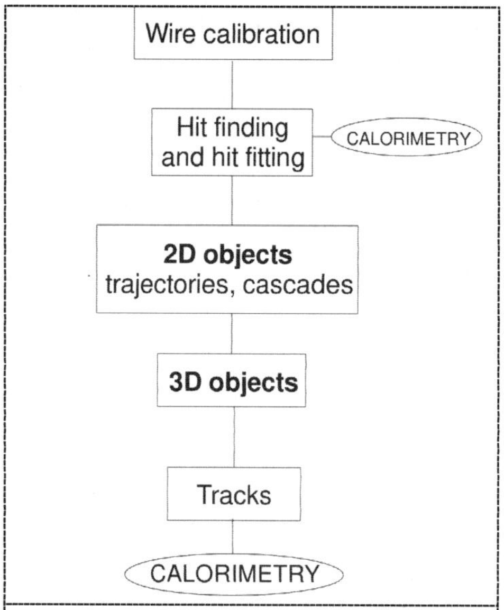

Figure 1. Scheme of the event reconstruction procedure.

3.2. Calorimetric reconstruction

The ionization charge is measured at the collection wire plane. The energy E associated to a given hit is:

$$E = \frac{CW}{R} Q e^{\frac{t-t_0}{\tau}}$$

where: C – the calibration factor; W – the average energy needed for the creation of an electron-ion pair; R – the electron-ion recombination factor; Q – the primary deposited ionization charge; t – the electron drift time; t_0 – time provided by the photomultiplier trigger; τ - the electron lifetime in LAr. The recombination factor R depends on the absorber medium, on the applied electric field and on the density of the released charge. In case of the electromagnetic cascades, the constant mean recombination factor is applied for all hits while for the tracks precise dependence of recombination on dE/dx can be modeled by Birk's law, which has been applied to LAr detectors in [4]:

$$\frac{dE}{dx} = \frac{dQ}{A - \frac{k}{\varepsilon\rho} \times \frac{dQ}{dx}}$$

where: dE/dx – a real particle ionization loss per track length unit; A – the normalization factor (~0.8); k – the Birks coefficient accounting for recombination (~0.05 (kV/cm)(g/cm^2)); ε – electric field (~0.5 kV/cm); ρ - the liquid argon density (~1.4 g/cm^3).

For a long muon track escaping the detector, the momentum is determined exploiting the multiple scattering along the track, studying the displacements with respect to a straight line. The procedure, implemented as a Kalman filter technique and validated on cosmic ray stopping muons, allows a resolution $\Delta p/p$ that can be as good as 10%, depending mainly on the track length [5].

3.3. Particle identification

Successful particle identification depends largely on the efficiency of the spatial and calorimetric reconstruction. In case of particles stopping in the detector the energy deposition per track length unit as a function of the deposited energy of the residual range allows to identify particles species (muon, pion, kaon, proton). The information from 3D track reconstruction is the input to a neural network based algorithm for particle identification [6]. A high efficiency has been achieved, above 85%, for particles with deposited energy around 30 MeV and higher (which corresponds to the length of at least 6 cm).

Electrons are fully identified by the characteristic electromagnetic showering; they are well separated from π^0 via dE/dx signal around the conversion point, conversion distance and π^0 invariant mass measurement at the level of 10^{-3}. This feature guarantees a powerful identification of the CC electron neutrino interactions, while rejecting NC interactions to a negligible level.

4. Data analysis

The ICARUS analysis framework allows a complete kinematical reconstruction of the neutrino interactions. The analysis of events collected during 2010 is on-going. They are very important in order to develop and test tools for the automatized reconstruction procedure. Analysis of the muon momentum in the CC interaction events, the study of the muon directions, and also measurement of the ν_μ energy show good agreement with Monte Carlo simulation performed in FLUKA [7].

One simple example of the fully reconstructed NC neutrino event with production of η particle is presented in figure 2. There are two electromagnetic cascades pointing to the primary vertex and tracks which were reconstructed and analysed. The conversion distances of cascades are measured to be 26 cm and 12 cm respectively. The associated invariant mass $m_{\gamma\gamma}^* = 512 \pm 48$ MeV is compatible with the η mass. The ionization signal in 2.5 cm of the initial part of the cascade amounts to be 4.3 and 5.1 m.i.p. (minimum ionizing particle) on average. This is clear signature of a pair conversion confirming the expected e/γ identification capabilities of the detector. The trajectory visible in both projections, Collection and Induction-2, is the result of the chain of particle decays. Starting from primary vertex they are following: pion, muon and electron/positron. Those tracks were reconstructed

Journal of Physics: Conference Series **408** (2013) 012078 doi:10.1088/1742-6596/408/1/012078

in 3D and precise energy measurement was provided. The total momentum of the event is around 2.2 GeV.

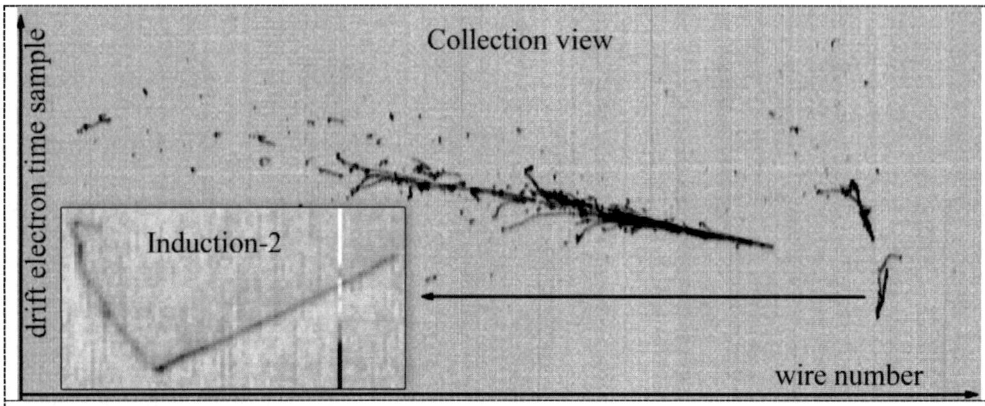

Figure 2. An example of ν_μ NC interaction from the CNGS beam. The whole event is shown in the Collection wire plane. Trajectories of the chain of decaying particles are also shown in the Induction-2 wire plane on the zoomed image.

5. Summary

The ICARUS detector is so far the biggest LAr detector ever built. It has been successfully installed in the Gran Sasso underground laboratory and it is presently collecting data. Thanks to very good spatial-calorimetric resolution and its underground operation a LAr detector is a perfect tool in research of neutrino physics and proton decay. Therefore the experience gained from the ICARUS experiment will be very valuable for the future research in these fields. As far as reconstruction and data analysis are concerned, one should stress that collected neutrino interaction events are used in order to develop and test automatic event reconstruction, allowing batch analysis of the data collected in long run periods.

References

[1] Rubbia C 1977 *CERN-EP*/**77-08**
[2] Rubbia C *et al.* 2011 *JINST* **6** P07011
[3] Kegl B, Krzyżak A, Linder T and Zeger K 2000 *IEEE Transactions on Pattern Analysis and Machine Intelligence* **22** 281
[4] Amoruso S *et al.* 2004 *Nucl. Instr. Meth.* A **523** 275
[5] Ankowski A *et al.* 2006 *Eur. Phys. J.* C **48** 667
[6] http://www.ire.pw.edu.pl/rsulej/NetMaker/
[7] Battistoni G, Muraro S, Sala P R, Cerutti F, Ferrari A, Roesler S, Fasso A and Ranft J *Proc. of the Hadronic Shower Simulation Workshop* 2007 **896** 31
Fasso A, Ferrari A, Ranft J and Sala P R 2005 *CERN-20050-10 INFN/TC_05/1 SLAC-R-773*

Journal of Physics: Conference Series **408** (2013) 012079 doi:10.1088/1742-6596/408/1/012079

First measurements of muon production rate using a novel pion capture system at MuSIC

S Cook[1], R D'Arcy[1], M Fukuda[2], K Hatanaka[2], Y Hino[3], Y Kuno[3], M Lancaster[1], Y Mori[4], TH Nam[3], T Ogitsu[5], H Sakamoto[3], A Sato[3], NM Truong[3], A Yamamoto[5], M Yoshida[5] and M Wing[1]

[1]Department of Physics and Astronomy, University College London, Gower Street, London WC1E 6BT, UK

[2]Research Centre for Nuclear Physics, 10-1 Mihogaoka, Ibaraki, Osaka 567-0047, Japan

[3]Department of Physics, Graduate School of Science, Osaka University, 1-1 Machikane, Toyonaka, Osaka 590-0043 Japan

[4]Kyoto Research Reactor Institute, 2 Asashiro-Nishi, Kumatori-cho, Sennangu, Osaka 590-0494, Japan

[5]High Energy Accelerator Research Organisation, KEK, 1-1 Oho, Tsukuba, Ibaragi 305-0801, Japan

E-mail: scook@hep.ucl.ac.uk

Abstract. The MuSIC (Muon Science Innovative Channel) beam line at RCNP (Research Centre for Nuclear Physics), Osaka will be the most intense source of muons in the world. A proton beam is incident on a target and, by using a novel capture solenoid, guides the produced pions into the beam line where they subsequently decay to muons. This increased muon flux will allow more precise measurements of cLFV (charged Lepton Flavour Violation) as well as making muon beams more economically feasible. Currently the first 36° of solenoid beam pipe have been completed and installed for testing with low proton current of 1 nA. Measurements of the total particle flux and the muon life time were made. The measurements were taken using thin plastic scintillators coupled to MPPCs (Multi-Pixel Photon Counter) that surrounded a magnesium or copper stopping target. The scintillators were used to record which particles stopped and their subsequent decay times giving a muon yield of 8.5×10^5 muons $W^{-1}_{\text{proton beam}}$ or 3×10^8 muons s^{-1} when using the RCNP's full power (400 W).

1. Introduction

The Muon Science Innovative Channel (MuSIC) aims to generate the most intense muon beam in the world. More importantly it aims to have the highest proton to muon yield in the world. Muons are increasingly becoming of interest for those wanting to probe the energy and precision frontiers; MuSIC hopes to provide the technology to make it possible to pursue these aims.

Creating a high intensity muon beam will open up many areas of research. Planned programs include: precision measurements of $\mu \to eee$; research and development into using FFAGs as muon storage rings; nuclear muon capture; and material science use of the muon beam such as muon spin rotation. Further in the future, high intensity muon beams can be used to drive neutrino factories and muon colliders.

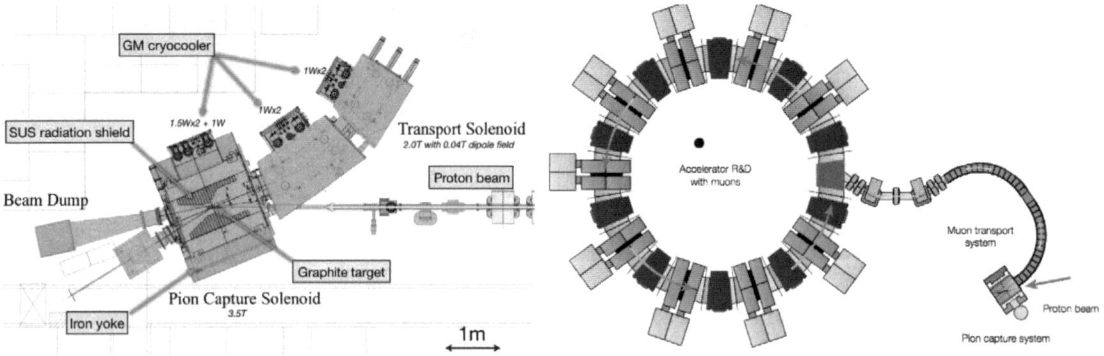

Figure 1. Left, a schematic of the current status of the MuSIC pion production system. Right, the proposed, fully completed, facility.

2. Overview

MuSIC is a muon beam line being built at the Research Centre for Nuclear Physics (RCNP) in Osaka, Japan. In addition to using the majority of the proton beam in pion production it also employs a novel pion capture solenoid to vastly increase its proton to muon efficiency.

The pion capture solenoid provides nearly 2π Sr acceptance in the production region. This captures backwards travelling pions using a 3.5 T superconducting magnet that completely surrounds the production target. The target is a 20 cm long graphite cylinder with a radius of 2 cm.

Stainless steel shielding is used to protect the superconducting solenoids from excessive heating. Cooling for the solenoid at 4 K is provided by 3 GM cryocoolers (two with a power of 1.5 W each and one of 1 W). The entire apparatus is enclosed in an iron yoke.

MuSIC produces muons from the decay of pions; to allow time for this to occur there is a transport solenoid after the capture solenoid, this also provides momentum selection. The transport solenoid has a 3 m diameter and extends for 180°. It has a static 2 T magnetic field as well as an additional 0.04 T magnetic field that can be used for fine adjustments and charge selection.

The proton beam is supplied by the RCNP cyclotron, it is a 400 W proton beam with maximum current of 1 μA (with a planned upgrade to 5 μA) and energy of 392 MeV. It is a continuous proton beam.

Commissioning of MuSIC began in 2009 and consisted of the installation of the pion capture solenoid and the first 36° of the transport solenoid. There have been three successful runs, the most recent of which was in July 2011 when the muon component of the beam was measured. The initial runs have been performed using a lower, 1 nA, current pending measurements of the neutron flux which will allow use of the full 1 μA current.

3. Measurement of muon flux

3.1. Measurement of the muon lifetime

In order to establish both the presence of muons in the beam and their flux the muon lifetime was measured. The experimental set up consisted of two plastic scintillators (both $400\times50\times3.5$ mm^3) on either side of a copper stopping target ($370 \times 80 \times 6$ mm^3). This arrangement was held in place using an aluminium surround positioned at the end of the beam pipe.

To measure the lifetime a simple trigger system was devised to detect stopped muons and then measure the time taken for the decay electron to be emitted. A gate of 20 μs was used to look for the decay electron. The logic for detection of a muon was a hit in the front scintillator

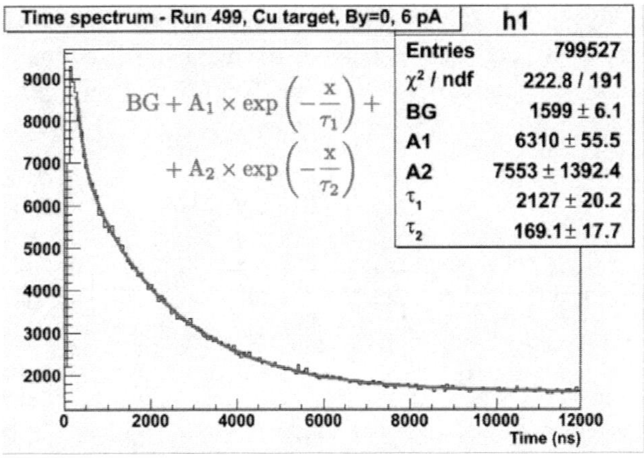

Figure 2. Decay time for stopped muons. Fitted with two exponentials, corresponding to the copper stopping time and the combination of free muon decay times with muons stopping in the scintillator (see table 3.1).

Table 1. Muonic decay times. The measured values for plastic scintillator and a free muon are the same as we did not have sufficient data to resolve these two components.

Material	Measured (ns)	Expected (ns)	Ref
Carbon (scintillator)	2127 ± 20	2026.3 ± 1.5	[1]
Free Muon		2197.034 ± 0.021	[2]
Copper	169 ± 17	164.0 ± 1.6	[1]

but not the rear whilst an electron was taken as a single hit in either the front or the rear scintillator.

A multi-hit TDC was used to time every hit that triggered only one of the scintillators after the initial muon. Readout from the scintillators was done using MPPCs which have an output signal proportional to the number of photons that strike their surface. MPPCs were used instead of traditional PMTs as they work well in strong magnetic fields and can be directly mounted to the scintillator without light guides, which reduces the cost and complexity of the system.

The beam supplied during the measurement was a 6 pA proton beam at 392 MeV. This exceeds the pion production threshold.

A preliminary analysis has been carried out in which two exponential curves and a constant background have been fitted to the data (figure 2). The two exponentials accounted for the decay of negative muons in Cu; and a combined term for decay of free muons or muons captured in the plastic scintillator (carbon capture). The approximation of the capture of muons in carbon for capture in plastic is discussed in [1]. The measurement of the muon in copper was 169±17 ns whilst the measured combined value for a free muon and muon in carbon is 2127±20 ns given in table 3.1. As muon capture only occurs for negative muons we can use the ratio of the deviations from the predicted values as a measure of the ratio of positive to negative muons (R) using the equation:

$$R = \frac{\tau_{\text{measured}} - \tau_C}{\tau_{\text{free}} - \tau_C} = \frac{2127-2026}{2197-2026} \approx 0.6 \quad (1)$$

Where τ_{measured} our measurement of the muon lifetime and τ_C and τ_{free} are the lifetimes given in

[1] and [2] respectively. Finally using the rate at which the system was triggered with stopped muons a measure of the muon rate gives a beam current of 2000 muons s^{-1} this implies a rate of 3×10^8 muon s^{-1} should be achievable when using the full proton beam current of 1 μA.

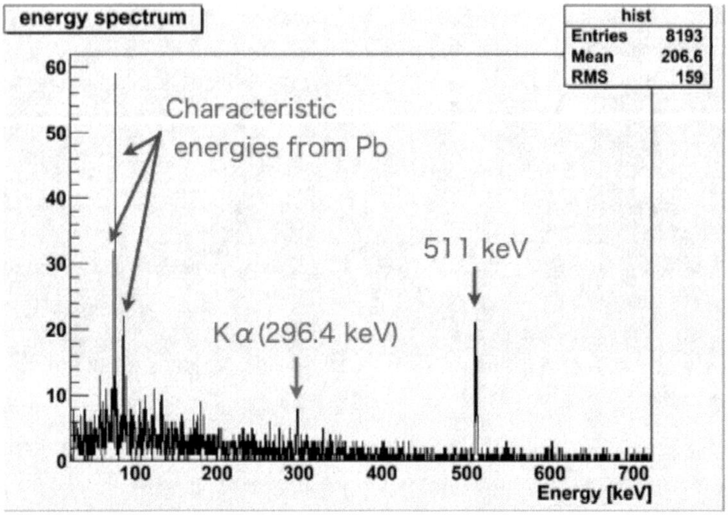

Figure 3. Energy spectrum of X-rays detected using the Ge detector. The three low energy peaks are due to emissions from the surrounding lead, the peak 511 KeV is due to electron-positron annihilation, the Kα peak is produced by muonic magnesium; 26 entries were counted at the Kα peak, this corresponds to a muon flux of 10^8 muons s^{-1} given a 1 μA proton beam.

3.2. Measurement of muon X-rays

As well as measuring the muon lifetime in copper the emission of X-rays from muonic magnesium was also measured. This was done using a Ge detector.

The detector was set slightly back from the end of the beam pipe and shielded from stray radiation using lead, paraffin and cadmium. Using the scintillators to reject non-stopping muons an energy spectrum of the detected X-rays was constructed and the Kα muonic peak identified (figure 3). Preliminary estimates accounting for acceptance of the Ge detector imply that the number of muonic X-rays seen correspond to a flux of 10^8 muons s^{-1} when using a 1μA proton beam.

4. The prospects

At time of writing a fourth beam time is being planned to measure the background neutron flux in order to establish whether further shielding is required. Assuming a successful run then beam time is scheduled for early 2012 in order to begin using higher beam currents with the ultimate aim of measuring the highest intensity muon beam in the world. Beyond that it is hoped that the commissioning of the complete facility can be completed in the next 4 years.

References

[1] Suzuki T, Measday D F and Roalsvig J P 1987 *Phys. Rev. C* **35**(6) 2212–2224
[2] Nakamura K *et al.* (Particle Data Group) 2010 *J. Phys. G* **37** 075021

Recent studies on the PRISM FFAG ring

J Pasternak[1,2], on behalf of the PRISM Task Force

[1]Imperial College London, London, UK
[2]ISIS, Rutherford Appleton Laboratory, STFC, OX11 0QX, UK

E-mail: j.pasternak@imperial.ac.uk

Abstract. Next generation lepton flavour violation experiments require high intensity and high quality muon beams. The PRISM Task Force focuses on accelerator R&D for realizing such beams using an FFAG ring. The scaling and non-scaling designs for the PRISM FFAG ring are discussed. The recent studies on injection/extraction systems and matching to the solenoid channel are outlined. The future plans for the study are presented.

1. Introduction

Although the Standard Model (SM) has proven to be extremely successful describing elementary particles and their interactions, it is believed to be only an approximation to a more fundamental theory, which yet needs to be discovered. Currently many experiments are intensively searching for any new phenomena setting new limits on various SM extensions. It is believed, that charged lepton flavour violating processes and in particular muon to electron conversion, are expected to be a fruitful tool to search for physics beyond the SM. The COMET and Mu2e experiments, which are capable of measuring muon to electron conversion with a sensitivity of $<10^{-16}$, are now in the advanced stage of the approval. In order to achieve even greater sensitivity ($<10^{-18}$) beyond the current generation experiments, the PRISM (Phase Rotated Intense Source of Muons) experiment was proposed, which will use an FFAG ring for longitudinal phase-space rotation. This will not only allow for reduction of the muon beam energy spread, but will also purify muons from any unwanted contamination like pions. Substantial progress in defining the PRISM system was achieved, including the baseline ring design based on 10 identical cells with triplet scaling FFAG configuration, together with its large aperture magnets construction and the development of an RF system with Magnetic Alloy (MA) cores. A test ring accelerator with six cells equipped with the prototype RF system was successfully assembled at RCNP in Osaka and phase rotation was demonstrated using alpha particles [1].

However, several technological challenges remains to be solved before a feasibility of the PRISM/PRIME system (PRIME denotes the dedicated detector system to be used with the PRISM muon beam) could be established and the experiment proposed. The PRISM Task Force currently addresses [2, 3] these issues utilizing synergies with other muon accelerator projects such as the Neutrino Factory and the Muon Collider. The activities of the Task Force include the detailed design of the injection and extraction system and the transfer line between the solenoidal pion decay channel and the FFAG ring; search for an alternative FFAG ring designs; developments for high gradient RF systems [4]; and the studies of kicker and septum magnets. This paper reports briefly on the progress obtained by the PRISM Task Force.

2. The baseline design

The PRISM/PRIME system requires a proton driver capable to produce an ultra short proton bunch length (~10ns) at the pion production target in order to efficiently use the phase rotation. Several options to produce such a proton beam have been considered including using the Main Ring at J-PARC or the Booster at Fermilab but it is rather clear, that in order to produce the high power required future new facilities are needed with the specification similar to the Neutrino Factory proton driver [5].

For the pion capture the backward collection in the high field solenoid is preferred, which allows to deliver sufficient pion intensity at low energy reducing the unwanted high intensity pion contamination. The pions will decay producing a muon beam in the solenoidal transport channel, which may contain the bend solenoids for further background reduction. The muon beam will then be injected into the FFAG ring, where the beam energy spread will be reduced (by a factor of ~10) using RF phase rotation and muons will be cleaned from any contamination in several turns. The high quality muon beam would be then extracted and sent to the stopping target, where the muon to electron conversion may occur. Downstream the stopping target the PRIME detector will be located, which will be optimized for detection of the electron signal from muon to electron conversion events. The main accelerator parameters for the PRISM are collected in Table 1.

Table 1. Main parameters for PRISM.

Parameter	Value
Proton beam power	1-2 MW
Target type	solid
Pion capture field	5-10 T
Momentum acceptance	±20 %
Reference μ^- momentum	40-68 MeV/c
Harmonic number	1
Minimal acceptance (H/V)	3.8/0.5 π cm rad
Repetition rate	100-1000 Hz
RF voltage per turn	3-5.5 MV
RF frequency	3-6 MHz

3. Front end and injection

The muon transport and injection for PRISM are particularly demanding as beam with not only large transverse emittances but also with a large momentum spread needs to be simultaneously matched into the FFAG ring acceptance. The beam optics of the superconducting solenoidal channel and the FFAG accelerator are quite different, which requires a careful adjustment of the beam conditions between the two systems to minimize the losses. One of the key points is the necessary dispersion matching: as injection is foreseen in the vertical direction, the unavoidable vertical dispersion created by the septum needs to be corrected to zero and the horizontal dispersion (small or zero in the solenoidal channel) needs to be adjusted to a correct value in the ring.

A dedicated front end system has been designed to meet these difficult requirements. It consists of several modules, each playing an important role in the beam transport. Firstly, a solenoidal adiabatic matching section is used to increase the beta function from the low value in the superconducting decay solenoid to the value in the alternating gradient (AG) part and the FFAG ring. It is followed by a quadrupole matching section (currently quadruplet, but more can be added). Downstream the horizontal dispersion creator is located consisting of a pair of equal, but opposite strength rectangular dipole magnets, followed by two circular FFAG cells with total π horizontal phase advance. Once the horizontal dispersion is introduced, it needs to be kept constant up to the FFAG ring. This is achieved using the straight FFAG sections [6], which are also used for the betatron matching. Finally, two vertical deflection sections (one incorporating the injection septum), which independently match the vertical dispersion and adjust the direction of the incoming beam line are used. The vertical and horizontal layouts of the AG part of the front end are shown in Fig 1. and Fig 2., respectively. The performance of the beam transport in the front end is currently being addressed via the particle tracking.

Several injection geometries were studied. The initial design used the same kickers for both injection and extraction. The current solution separates injection and extraction systems. The injection will use 1 vertical septum followed by either 1 or preferentially 2 kickers each located in consecutive cells. The kicker magnet design has been studied assuming the existing pulsed magnet technology to be used. No show-stopper has been found so far, but due to a high repetition rate and a short rise time dedicated prototyping is mandatory.

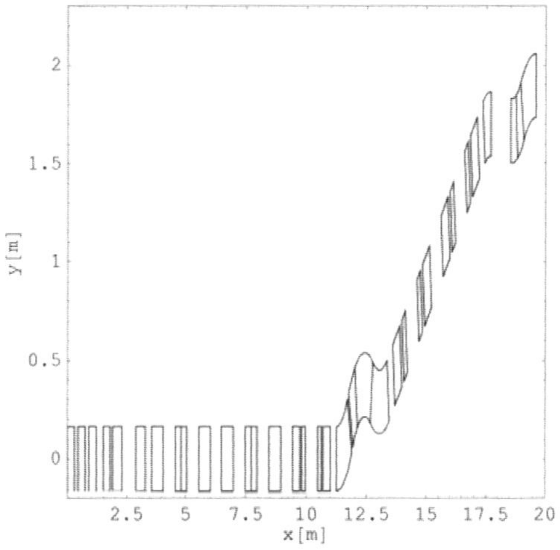

Figure 1. The vertical layout of the PRISM front end (AG part). The decay solenoid, located on the left would be on the different level than the FFAG injection septum located on the right.

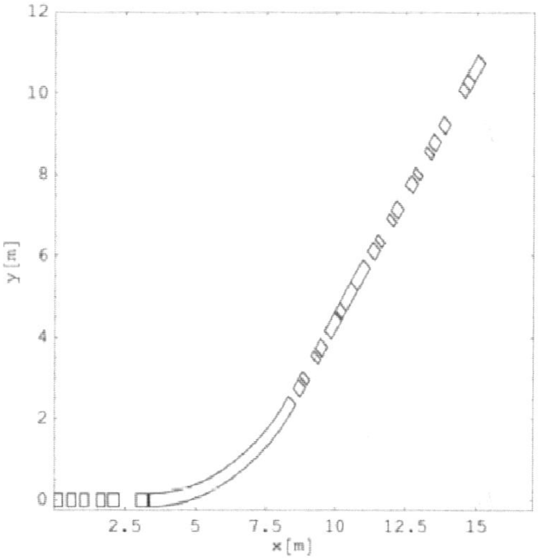

Figure 2. The horizontal layout of the PRISM front end (AG part). The big curved section is the FFAG dispersion creator. The injection septum is the last element on the upper right.

4. Alternative FFAG ring designs

The PRISM Task Force not only continues the development of the baseline ring, but also undertakes studies of alternative options. The goal of this task is to maximize the ring acceptance, facilitating the injection/extraction requirements and space constraints. Several alternative ring designs have been produced and various options include.

- New symmetric scaling FFAG ring, in particular based on FDF triplet arrangements.
- Advanced scaling FFAG [7] using the straight FFAG cells in the "straight sections" with zero net bending and the circular FFAG cells in the compact arcs. The layout of such a racetrack ring is shown in Fig 3.
- Advanced scaling FFAG incorporating arc sections with different radii ("egg-shaped", as shown in Fig. 4).
- Scaling FFAGs with superperiods introduced by rearranging the geometry in circular cells.
- Non-scaling FFAG utilizing the combined function magnets with only dipole and quadrupole magnetic field components.

The acceptance and performance tracking studies for the proposed alternative rings are under way using alternative accelerator codes and the first results look promising.

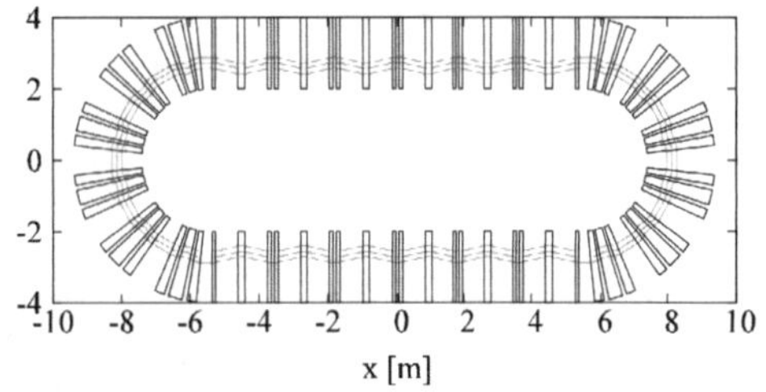

Figure 1. The layout of the advanced racetrack scaling FFAG ring for PRISM.

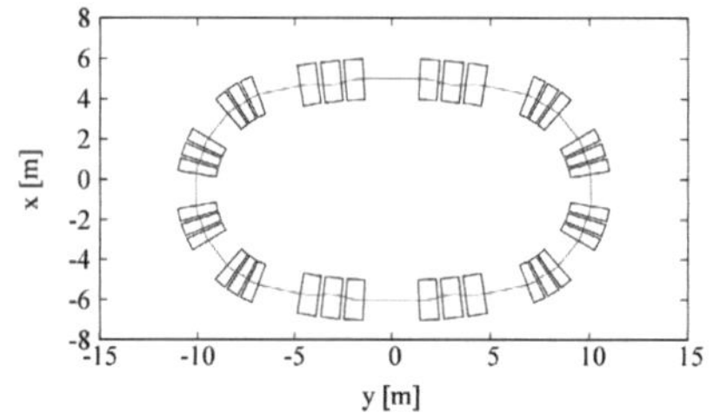

Figure 2. : The layout of the advanced scaling FFAG using arcs with different radii ("the egg-shaped ring").

5. Summary and future plans

The PRISM Task Force continues to work on the challenges, which need to be overcome before the feasibility of the PRISM system can be established. Substantial progress has been achieved in many areas such as: the design of the muon front end, studies of the injection system geometry, in development of the high gradient RF system (see [3, 4]). In addition several alternative ring designs, based on recently developed advanced concepts in the domain of FFAG accelerators, have been proposed. Future studies will focus on further optimization of the front end and the injection system, systematic comparison of the performance achieved by alternative ring designs and continuation of the hardware studies including the RF system, the kicker and septum magnets. The PRISM Task Force aims to publish the feasibility report in 2012. If it turns successful, it may start working on the possible tests of muon phase rotation and various PRISM subsystems using the MuSIC muon beam at RCNP in Osaka [8], which opens a possibility for experimental verifications of the developed solutions.

References

[1] Sato A 2010 *Demonstration of phase rotation using alpha particles in the six-sector PRISM-FFAG* in Proc. of IPAC'10 Kyoto
[2] Pasternak J *et al.* 2010 WEPE056 in Proc. of IPAC'10 Kyoto
[3] Pasternak J *et al.* 2011 MOPZ004 in Proc. of IPAC'11 San Sebastian
[4] Ohmori C *et al.* 2011 *High Gradient Magnetic Alloy Cavities for J-PARC Upgrade* in Proc. of IPAC'11 San Sebastian
[5] Abrams R J *et al.* 2011 *Interim Design Report (IDR)* IDS-NF-020 *Preprint* arXiv:1112.2853v1 [hep-ex]
[6] Lagrange J-B *et al.* 2009 *Zero-chromatic FFAG straight* section in Proc. of FFAG'09 Chicago
[7] Lagrange J-B *et al.* 2009 *Zero-chromatic FFAG straight section* in Proc. of FFAG'09 Chicago
[8] Sato A *et al.* 2011 *MuSIC the World's Highest Intensity DC Muon Beam using a Pion Capture System* in Proc. of IPAC'11 San Sebastian

Journal of Physics: Conference Series **408** (2013) 012081 doi:10.1088/1742-6596/408/1/012081

Modeling neutrino-nucleus interactions in the few-GeV regime.

N Jachowicz, P Vancraeyveld

Ghent University, Department of Physics and Astronomy,
Proeftuinstraat 86, B-9000 Gent, Belgium.

E-mail: `natalie.jachowicz@ugent.be`

Abstract.

Detecting neutrinos and extracting the information they bring along is an ambitious task that requires a detailed understanding of neutrino-nucleus interactions over a broad energy range. We present calculations for quasi-elastic neutrino-induced nucleon knockout reactions on atomic nuclei and neutrino-induced pion production reactions. In our models, final-state interactions are introduced using a relativistic multiple-scattering Glauber approximation (RMSGA) approach. For interactions at low incoming neutrino energies, long-range correlations are implemented by means of a continuum random phase approximation (CRPA) approach. As neutrinos are the only particles interacting solely by means of the weak interaction, they can reveal information about e.g. the structure of nuclei or the strange quark content of the nucleon that is difficult to obtain otherwise. We investigated these effects and present results for the sensitivity of neutrino interactions to the influence of the nucleon's strange quark sea.

1. Introduction

The experimental confirmation of neutrino oscillations raised an enormous experimental and theoretical interest in the oscillation properties of these particles. For an accurate interpretation of the experimental data, a thorough understanding of the interactions involved in oscillation experiments is indispensable. Cross sections for neutrino-nucleus interactions play an important role. Whereas the experimental observable is often an inclusive cross section, a detailed study of exclusive cross sections is indispensable for a thorough understanding of these processes.

2. Quasi-elastic neutrino scattering

The one-fold differential cross section for the scattering process $A(\nu, lN)$ is given by

$$\frac{d\sigma}{dT_N} = \frac{M_N M_{A-1}}{(2\pi)^3 M_A} 4\pi^2 \int \sin\theta_l d\theta_l \int \sin\theta_N d\theta_N k_N f_{rec}^{-1} \sigma_M \left[v_L R_L + v_T R_T + h v_{T'} R_{T'} \right], \qquad (1)$$

with M_N, T_N and $\vec{k}_N$ the mass, kinetic energy and momentum of the ejectile, M_A and M_{A-1} the mass of the target and residual nuclei. The direction of the outgoing lepton and nucleon is determined by $\Omega_l(\theta_l, \phi_l)$ and $\Omega_N(\theta_N, \phi_N)$. The recoil factor is denoted by f_{rec}. The quantity σ_M is the weak variant of the Mott cross section.

In Eq.(1), v_L, v_T and $v_{T'}$ are the longitudinal, transverse and interference kinematic factors and R_L, R_T and $R_{T'}$ the accompanying structure functions, reflecting the influence of nuclear

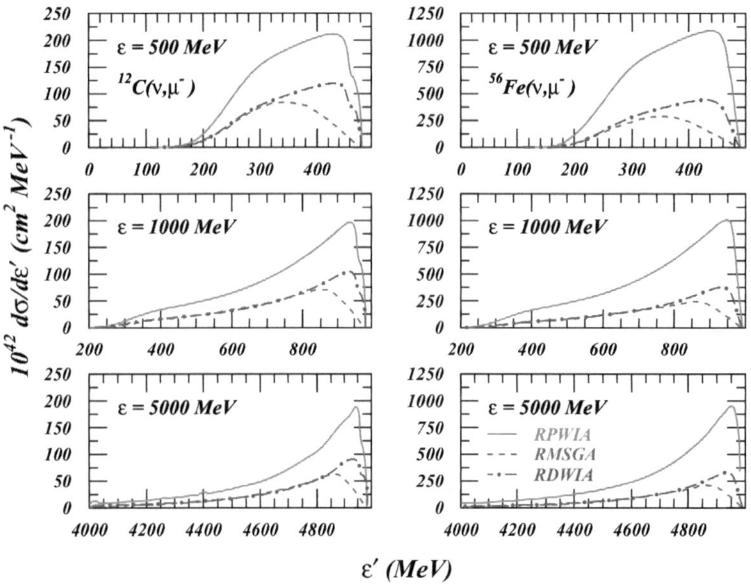

Figure 1. Cross sections for the charged-current reactions ^{12}C(ν_μ, μ^-) and ^{56}Fe(ν_μ, μ^-) as a function of the energy of the outgoing muon ε'. The different curves compare the RMSGA results (dashed) with the RDWIA (dash-dotted) and the RPWIA limit (full line).

dynamics on the scattering process [1]. The helicity of the incoming neutrino is denoted by h. In our numerical calculations, bound-state wave-functions are obtained within the Hartree approximation to the $\sigma - \omega$ model, adopting the W1 parametrization for the different field strengths.

We introduce final-state interactions adopting a relativistic multiple-scattering Glauber approximation (RMSGA) [1, 2]. As a semi-classical approach, this technique exploits the advantages of the kinematics conditions reigning at sufficiently high energies, where high momentum transfers strongly favor forward rescattering of the outgoing nucleon. The Glauber technique assumes linear trajectories for the ejectile and frozen spectator nucleons in the residual system. The influence of the nuclear medium on the outgoing nucleon's wave function are condensed in the eikonal phase $\mathcal{G}[\vec{b}(x,y),z]$, that summarizes the effects of the scattering reactions the ejectile undergoes. This results in a scattering wave function that can be written as $\phi_F(\vec{r}) = \mathcal{G}[\vec{b}(x,y),z]\ \phi_{k_N,s_N}(\vec{r})$, with $\phi_{k_N,s_N}(\vec{r})$ a relativistic plane wave. In the limit of vanishing final-state interactions ($\mathcal{G} = 1$) , the formalism becomes equivalent to the relativistic plane wave impulse approximation (RPWIA) [1]. Figure 1 shows the influence of final state interactions on cross sections for charged-current processes on ^{12}C and ^{56}Fe, and compares Glauber and RDWIA calculations. In the region where both approaches are valid, their results are in excellent agreement.

3. Strangeness

The strangeness content of the nucleon influences neutrino-nucleus cross sections and has an important impact on several cross-section ratios. We compared the influence of axial as well as vector strangeness on ν and $\overline{\nu}$ cross-section ratios [3]. We compared the impact of the weak strangeness form factors on the ratio of proton-to-neutron knockout, the ratio of neutral-to-charged current cross sections, on the Paschos-Wolfenstein relation for protons and neutrons, and on the longitudinal helicity asymmetry for neutrinos and antineutrinos. Figure 2 summarizes the main results. The longitudinal helicity asymmetry for antineutrinos is most sensitive to strangeness effects. In general, antineutrino-induced processes exhibit a more outspoken strangeness sensitivity than their neutrino counterparts. The overall sensitivity of $R_{NC/CC}$ ratios to strangeness effects is considerably smaller than that of $R_{p/n}$.

Whereas in PVES the smallness of the axial strangeness effects impedes the determination of g_A^s, in neutrino scattering a thorough understanding of vector strangeness effects is a prerequisite for extracting information on the axial strangeness. Hence a combined analysis of parity-violating electron scattering and neutrino-induced processes would offer the best prospects for

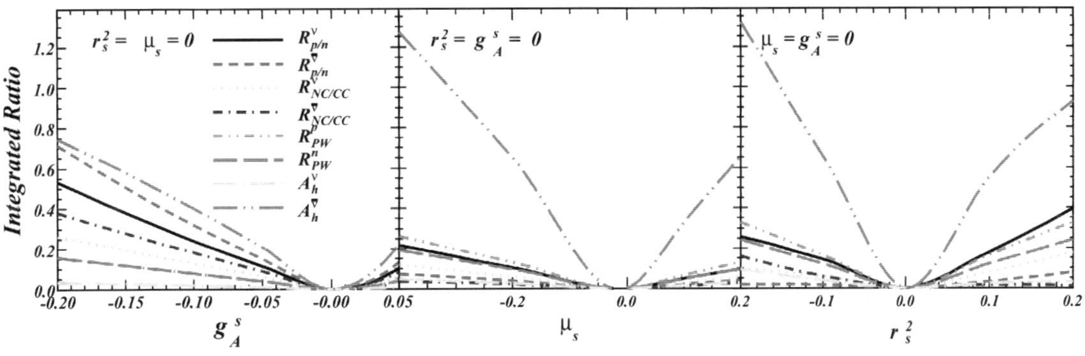

Figure 2. Comparison between the strangeness influence on various ratios of total cross-sections in terms of the relative sensitivity $\left|\frac{R(s=0)-R(s)}{R(s=0)}\right|$, as a function of the strangeness form factors g_A^s, μ_s and r_s^2 for 1 GeV neutrino scattering off ^{12}C.

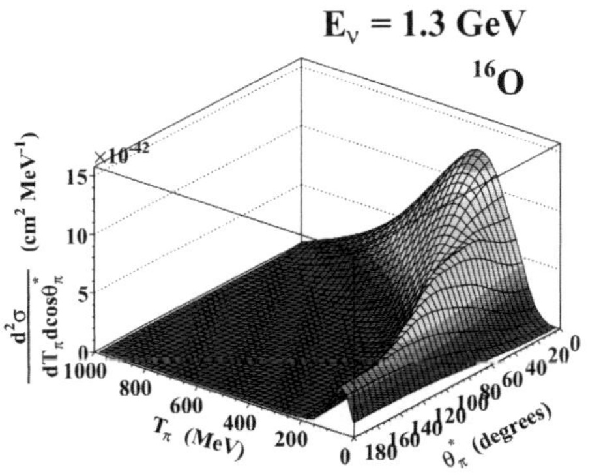

Figure 3. Resonant pion production off ^{16}O for 1.3 GeV incoming neutrinos.

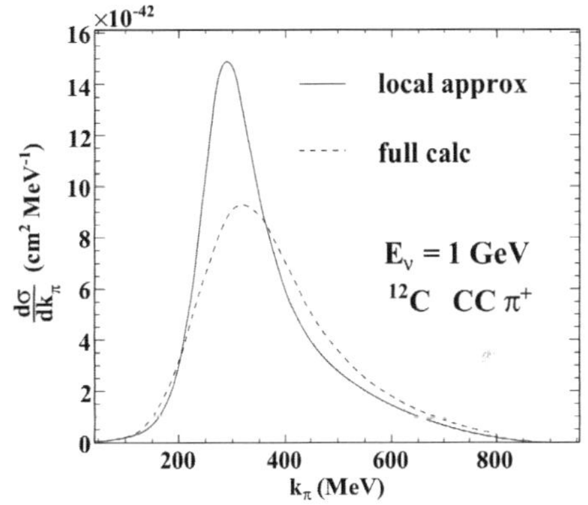

Figure 4. Comparison of full and local approximation calculations for coherent pion production.

a thorough understanding of the influence of the nucleon's strange quark sea on electroweak processes.

4. Weak one-pion production on a nucleus

At slightly higher energies, resonant and coherent pion production become important reaction channels. The Δ-mediated pion-production cross section is evaluated adopting the relativistic plane-wave impulse approximation for the calculation of the matrix element. The single particle wave functions are the same as those used in section 2. Medium modifications for the Δ were implemented as discussed in Ref. [4]. Figure 3 shows the differential cross section for neutrino-induced pion-production off ^{16}O.

Coherent pion production leaves the nuclear target intact by transferring only small four-momenta. Figure 4 compares full- and local-approximation calculations for charged-current coherent pion production on ^{12}C.

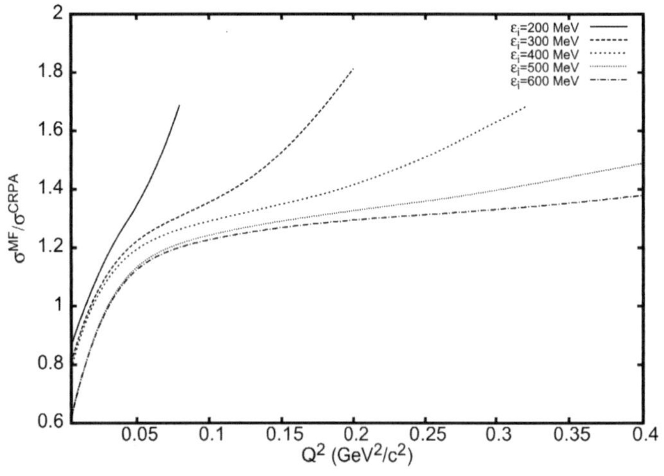

Figure 5. Ratio of cross sections obtained with mean field (MF) wave functions to cross sections including continuum RPA correlations (CRPA) as a function of Q^2 for incoming neutrino-energies ranging from 200 to 600 MeV and with ^{12}C as target nucleus.

5. Long-range CRPA correlations

For inclusive cross sections, especially at higher energies, the Fermi Gas model is remarkably accurate. At lower energies however, nuclear effects become preponderant. In a Random Phase Approximation (RPA) approach, long-range correlations between the nucleons in the nucleus are introduced. Whereas in a mean-field calculation a nucleon experiences the presence of the others only through the mean-field generated by their mutual interaction, the random phase approximation additionally allows the particles to interact by means of the residual two-body force. The random phase approximation describes a nuclear state as the coherent superposition of particle-hole contributions [5, 6].

$$|\Psi_{RPA}\rangle = \sum_{C} \left\{ X_{(\Psi,C)} \left|ph^{-1}\right\rangle - Y_{(\Psi,C)} \left|hp^{-1}\right\rangle \right\}. \tag{2}$$

The summation index C stands for all quantum numbers defining a reaction channel unambiguously. In our model, the continuum RPA equations are solved using a Green's function approach in which the polarization propagator is approximated by an iteration of the first-order contribution. The unperturbed wave functions are generated using either a Woods-Saxon potential or a HF-calculation using a Skyrme force. The latter approach makes self-consistent HF-RPA calculations possible. As is shown in figure 5, long-range Random Phase Approximation correlations account for a considerable reduction of cross sections at low incoming neutrino-energy.

References

[1] Martínez M C, Lava P, Jachowicz N, Ryckebusch J, Vantournhout K and Udías J M 2006 *Phys. Rev.* C **73** 024607
[2] Ryckebusch J, Debruyne D, Lava P, Janssen S, Van Overmeire B and Van Cauteren T 2003 *Nucl. Phys.* A **728** 226
[3] Jachowicz N, Vancraeyveld P, Lava P, Praet C and Ryckebusch J 2008 *Phys. Rev.* C **76** 055501
[4] Praet C, Lalakulich O, Jachowicz N and Ryckebusch J 2009 *Phys. Rev.* C **79** 044603
[5] Jachowicz N, Heyde K, Ryckebusch J and Rombouts S 1999 *Phys. Rev.* C **59** 3246
[6] Jachowicz N, Heyde K, Ryckebusch J and Rombouts S 2002 *Phys. Rev.* C **65** 025501

Study of neutrino interaction with T2K on-axis neutrino detector Proton Module

T Kikawa for the T2K Collaboration

Department of Physics, Kyoto University, Kyoto 606-8502, Japan

E-mail: kikawa@scphys.kyoto-u.ac.jp

Abstract.
T2K is a long baseline neutrino oscillation experiment. We have constructed a new neutrino detector, "Proton Module" to measure the neutrino cross section precisely using the T2K neutrino beam. We report the status of the data acquisition and the Monte Carlo study.

1. Introduction

T2K (Tokai-to-Kamioka) [1] is a long baseline neutrino oscillation experiment aiming to search for the electron neutrino appearance and to measure the oscillation parameters associated with the muon neutrino disappearance precisely. A high intensity neutrino beam from J-PARC (Japan Proton Accelerator Research Complex) is measured with the 280 m near detector complex (ND280) and the 295 km far detector (Super-Kamiokande). For the oscillation analysis of T2K, there is a large systematic error attributed to the uncertainty of the neutrino cross section. The charged current quasi-elastic (CCQE) interaction ($\nu_\ell + n \rightarrow \ell^- + p$) events are used as signal in T2K. The ν_μ CCQE cross section on ^{12}C was measured by MiniBooNE [2], NOMAD [3] and LSND [4]. However, they cannot be consistently described with a currrent neutrino interaction model (Fig. 1). The T2K on-axis neutrino beam covers the energy region of 1.5$\sim$4 GeV, which is not covered with the past experiments (Fig. 2). We have constructed a new neutrino detector, "Proton Module" (Fig. 3) to measure the neutrino cross section precisely with the T2K on-axis neutrino beam. The Proton Module consists of 34 tracking planes surrounded by 6 veto planes

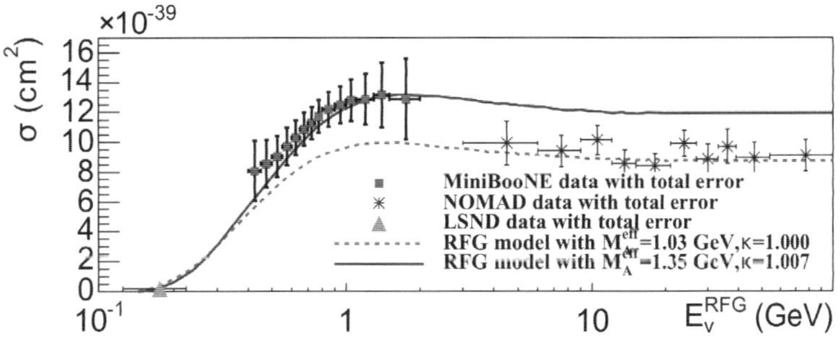

Figure 1. Cross section of the ν_μ CCQE interaction as a function of the neutrino energy [2]

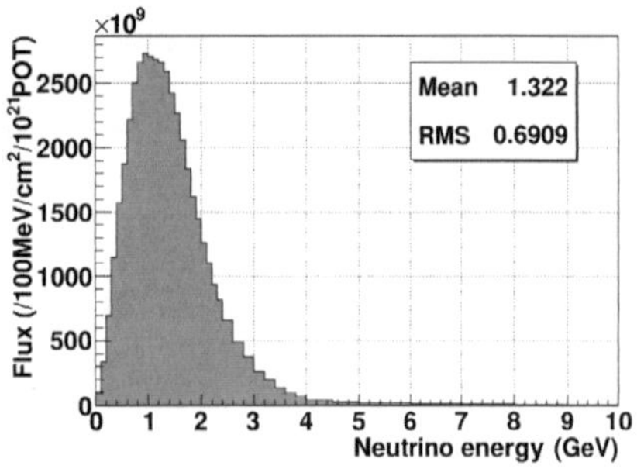

Figure 2. The T2K on-axis neutrino beam flux

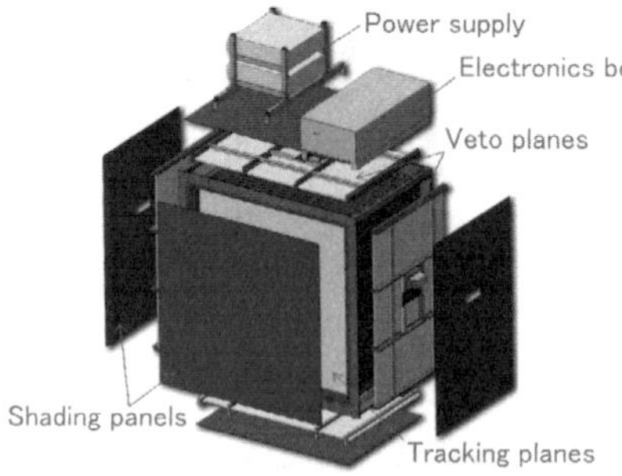

Figure 3. Schematic view of the Proton Module

(Fig. 3). Each tracking plane consists of 32 scintillator bars. Scintillation lights are collected and transported with a wavelength shifting fiber which is inserted in a hole at the center of the scintillator bar. Then the lights are read out by a Multi-Pixel Photon Counter (MPPC) attached to one end of the fiber. The dimension of the entire module is $1.42 \times 1.42 \times 0.96$ m^3, total target mass is 556 kg, and total number of channels is 1204. Since the Proton Module is a fully-active detector, it can reconstruct tracks of various kinds of particles from the interaction point. The Proton Module is placed in the upstream of seven identical detectors called INGRID. INGRID consist of a sandwich structure of iron target plates and tracking scintillator planes. Although their main purpose is to monitor the neutrino beam direction and intensity, they are used as muon counters in this study.

2. Data acquisition

The Proton Module started the beam data acquisition on November 16, 2010 and suspended it on March 11, 2011 because of the Tohoku earthquake. The Proton Module suffered no damage from the earthquake. During this time period, 1.1×10^{20} protons on target (POT) of physics data was accumulated. Gains of all MPPCs were monitored and they are stable as shown in Fig. 4. Figure 5 shows the daily neutrino event rate normalized by POT. It is stable within the margin of statistical error.

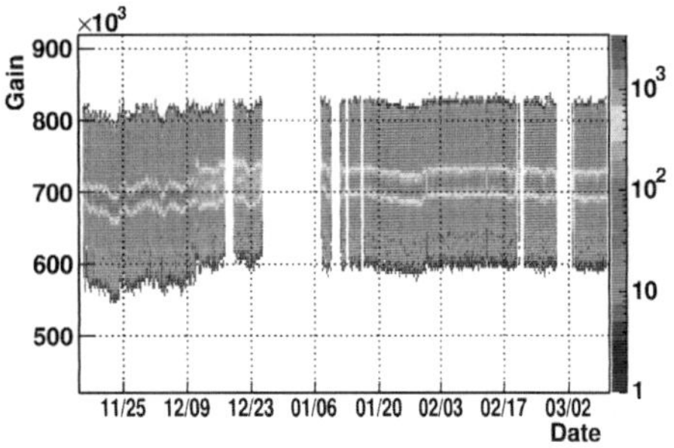

Figure 4. MPPC gain history

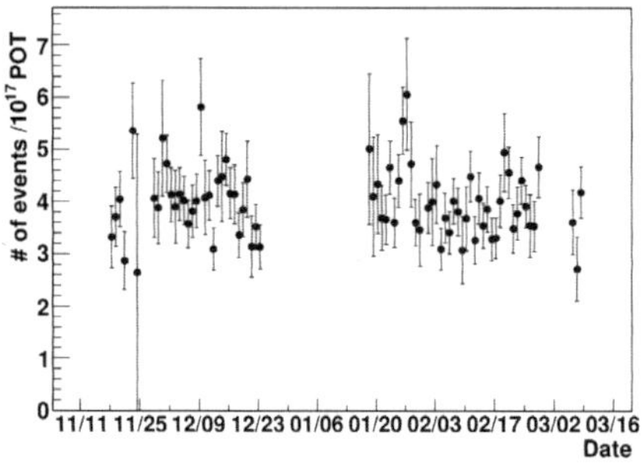

Figure 5. Daily neutrino event rate

3. Monte Carlo study

The analysis method is being studied with Monte Carlo simulations. To simulate the detection of the beam neutrinos by the Proton Module, three Monte Carlo simulation programs are utilized: JNUBEAM [1] for a simulation of neutrino beam line, NEUT [5] for a simulation of neutrino interactions and GEANT4 [6] for a simulation of the detector response.

3.1. Neutrino event selection

At the beginning, tracks are reconstructed using hit information of each channnel. As a result of the simulations, it turned out that most of the tracks reconstructed in the Proton Module are caused by the incoming particles generated by the neutrino interactions in the wall of the detector hall. Four selections are applied to reject these background events. First, events which have a hit at a veto plane in the upstream extrapolated position from reconstructed tracks are rejected. Second, events whose reconstructed vertex is outside of fiducial volume is rejected. The fiducial volume is defined as a volume other than edge scintillator bars. Third, events which have only one track from the vertex are rejected. This is a temporary selection and one track events will be used for the CCQE study in the future. Forth, events which have no tracks in INGRID located just behind the Proton Module are rejected. After these selections, the purity of the neutrino events becomes 99.7 %. Result of the neutrino event selection is summarized in Table 1.

Table 1. Summary of the neutrino event selection (Monte Calro simulation per 1.1×10^{20} POT)

Selection criteria	Neutrino events	Background events	Efficiency	Purity
Tracks in the Proton Module	21337	477585	79%	4.2%
Veto plane cut	18320	61016	67%	23.1%
Fiducial volume cut	14478	23627	53%	38.0%
# of tracks from the vertex ≥ 2	7897	2464	29%	76.2%
Tracks in INGRID	4109	12	15%	99.7%

3.2. CCQE event selection

After the neutrino event selection, additional three selections are applied to enhance the CCQE events. First, events which have two tracks from the vertex are selected (Fig. 6) because final state particles of CCQE are two particles (a muon and a proton). Second, events which have one track matching with tracks in INGRID are selected (Fig. 7) because a muon track is expected to be reconstructed in INGRID but a proton track is not. Third, likelihood selection is applied; the likelihood is defined by the opening angle and the coplanarity angle (Fig. 8, 9). The opening angle is the angle between the two reconstructed tracks and the coplanarity angle is the angle between $\vec{\nu} \times \vec{\mu}$ and $\vec{\nu} \times \vec{p}$. Events whose likelihood is more than 0.6 are selected (Fig. 10). After these selections, the purity of the CCQE events becomes 74%. Result of the CCQE event selection is summarized in Table 2 and an event display of the selected CCQE event is shown in Fig. 11.

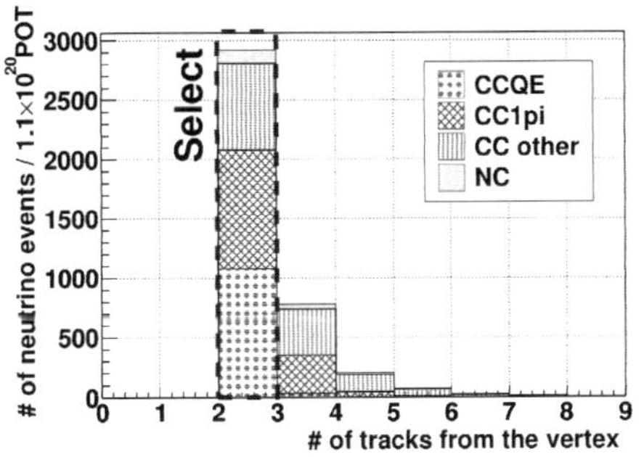

Figure 6. Number of the tracks from the vertex

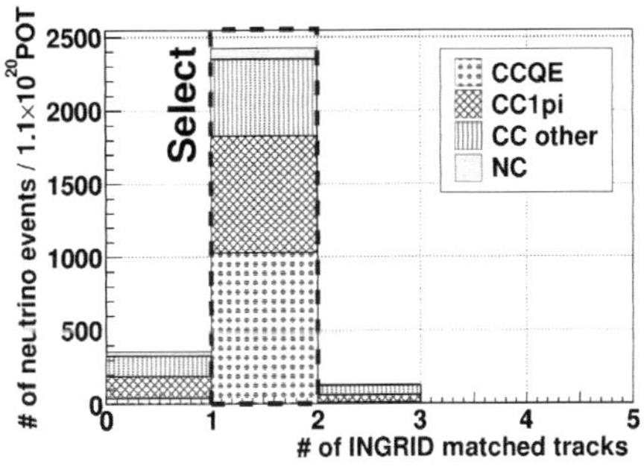

Figure 7. Number of the INGRID matched tracks

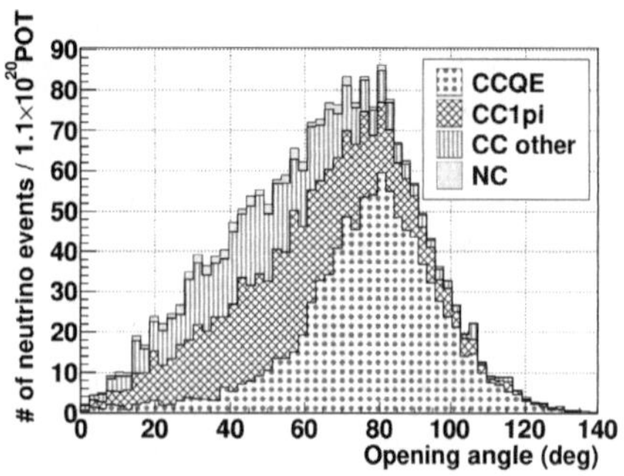

Figure 8. Opening angle between the two tracks

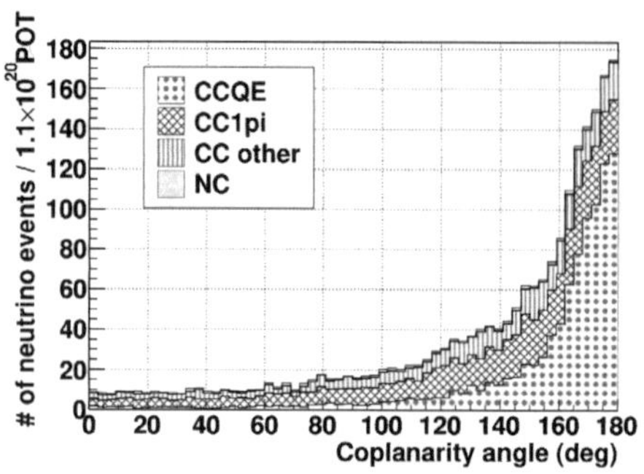

Figure 9. Coplanarity angle

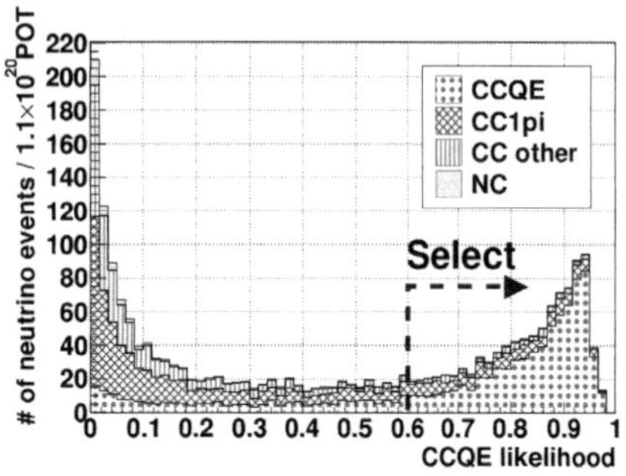

Figure 10. CCQE likelihood

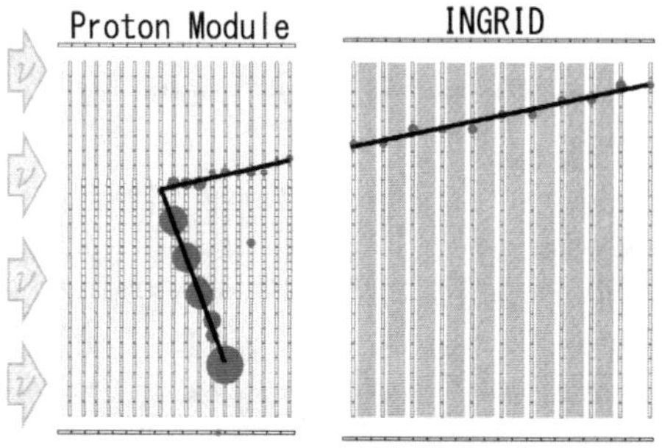

Figure 11. Event display of a selected CCQE event

Table 2. Summary of the CCQE event selection (Monte Calro simulation per 1.1×10^{20} POT)

Selection criteria	CCQE events	Non-QE events	Efficiency	Purity
Neutrino event selection	1108	3013	12.9%	27%
# of tracks from the vertex = 2	1094	1817	12.7%	38%
# of INGRID matched tracks = 1	1029	1391	12.0%	43%
CCQE likelihood > 0.6	871	313	10.2%	74%

4. Conclusion

The Proton Module acquired the beam data successfully. Based on the Monte Carlo simulation, the method of the CCQE selection is optimized. Our study is going ahead toward the first physics result with the Proton Module.

References

[1] K Abe *et al.* (T2K Collaboration), 2011, *Nucl. Instrum. Meth. A*, **659**
[2] A A Aguilar-Arevalo *et al.* (MiniBooNE Collaboration), 2010, *Phys. Rev. D*, **81**, 092005
[3] V Lyubushkin *et al.* (NOMAD Collaboration), 2009, *Eur. Phys. J. C*, **63**, 355
[4] L B Auerbach *et al.* (LSND Collaboration), 2002, *Phys. Rev. C*, **66**, 015501
[5] Y Hayato, 2002, *Nucl. Phys. Proc. Suppl.*, **112**, 171
[6] S Agostinelli *et al.* (GEANT4 Collaboration), 2003, *Nucl. Instrum. Meth. B*, **506**, 250

Journal of Physics: Conference Series **408** (2013) 012083 doi:10.1088/1742-6596/408/1/012083

Study of multiple scattering in high magnetic fields

Daniel M Kaplan[1] **and Thomas J Roberts**[2]

[1] Illinois Institute of Technology, Chicago, IL 60616, USA
[2] Muons, Inc., Batavia, IL 60510, USA

E-mail: kaplan@iit.edu

Abstract. Muon cooling for a neutrino factory or muon collider can be achieved using low-Z absorbers in strong focusing fields. Proposed cooling lattices place absorbers in solenoidal fields ranging up to 30 to 40 T. The cooling performance of these lattices is determined by the interplay of ionization energy loss and Molière scattering, but Bethe's classic treatment of Moliere scattering ignores the helical motion of charged particles in solenoidal fields. When this motion is taken into account, the performance of these lattices can be better than predicted by simulations using the standard treatment.

1. Introduction

The question of the effect of strong magnetic fields on multiple scattering was raised by Lebrun in 1999 [1]. He realized that Bethe's derivation [2] of the Molière scattering distribution implicitly assumed straight-line motion between scatters. In contrast, in a strong magnetic field, a charged particle executes circular or helical motion, and in the limit of infinitely strong field, such a particle's path is confined to a magnetic-field line. Clearly, this must reduce the multiple-scattering-induced spread in position as a charged particle traverses a scattering medium, as Lebrun demonstrated using Geant 3. Since this is exactly the configuration proposed in many implementations of muon cooling, simulations of muon-cooling channels risk underestimating their effectiveness, especially as the beam emittance shrinks and the field strength grows to 20 T and beyond [3].

In order to understand the degree of underestimation that may occur in such simulations, we undertook to compare the treatment of scattering in magnetic fields in the principal codes used in recent MAP and NFMCC cooling studies: G4beamline [4] and ICOOL [5]. To provide a simple and well-understood baseline, we also evaluated the effect using a Gaussian approximation to the scattering distribution.[1]

2. Gaussian Model

In our Gaussian study, we let each muon start at one end ($z = 0$) of a 30-cm-long liquid-hydrogen scatterer (a typical muon-cooling-channel absorber length), with longitudinal momentum $p_z = 200\,\mathrm{MeV}/c$ and transverse momenta $p_x = p_y = 0$. We subdivided the absorber into thinner and thinner longitudinal slices, tracing each muon's path as a straight

[1] For simplicity, we disabled dE/dx energy loss in all three calculations.

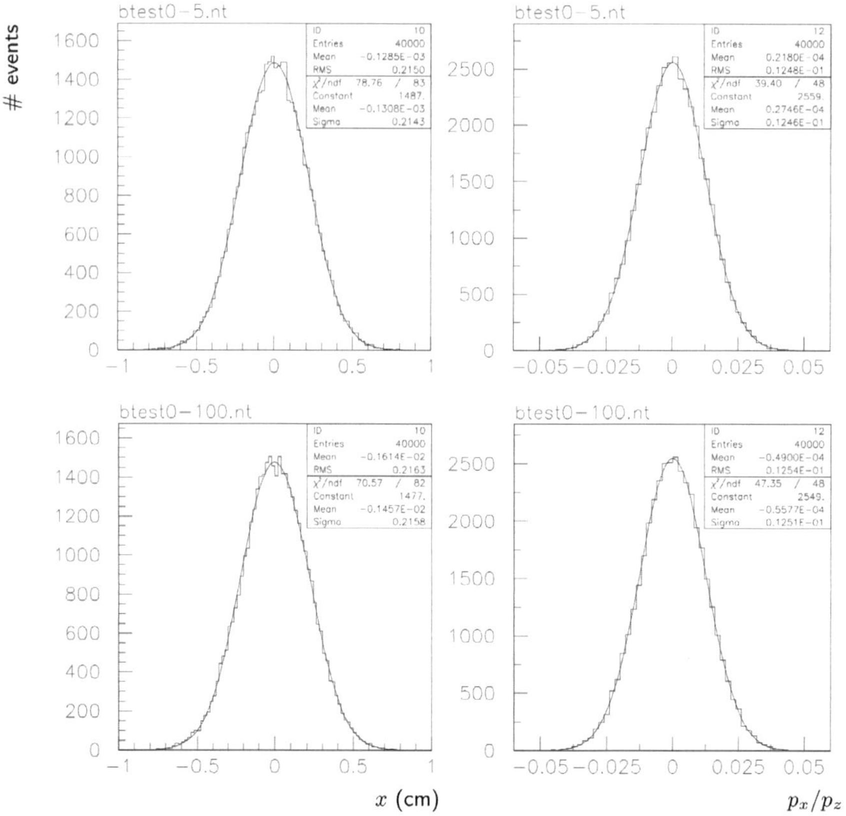

Figure 1. Distributions in (left) x and (right) $\tan\theta_x$ at the end of a 30 cm LH_2 absorber in Gaussian multiple-scattering model, for (top) 5 and (bottom) 100 absorber slices.

line (if $B = 0$) or helix (if $B \neq 0$) between the centers of successive slices. At the center of each slice, we allowed θ_x and θ_y to change by random amounts, with r.m.s. variation [6]

$$\sigma_{\theta_x} = \sigma_{\theta_y} = \frac{13.6\,\mathrm{MeV}}{\beta c p}\sqrt{\frac{\Delta z}{X_0}}\left[1 + 0.038\ln\left(\frac{30\,\mathrm{cm}}{X_0}\right)\right], \tag{1}$$

where Δz is the width of a slice and $X_0 = 866\,\mathrm{cm}$ is the radiation length of liquid hydrogen [6]. Figure 1 shows an example of the transverse spatial and angular distributions at the end of the absorber for the cases $B = 0$ and $n_{\mathrm{slices}} = 5$ or 100. As expected, the number of slices makes no difference if $B = 0$.

We then allowed the longitudinal magnetic field to increase up to 100 T and studied the number of slices required for the r.m.s. x value at the end of the absorber to stabilize, finding that 10 slices suffice (Fig. 2). We observe a dramatic suppression of spatial (but not angular) spread as the magnetic field increases, by a factor of ≈ 5 at 30 T and ≈ 20 at (the admittedly impractical) 100 T.

3. G4beamline

We next performed a similar study with the G4beamline simulation package. The "knob" comparable to absorber subdivision is the choice of the maxStep parameter, which is 100 mm by default. It represents the largest step along the particle trajectory that Geant 4 is

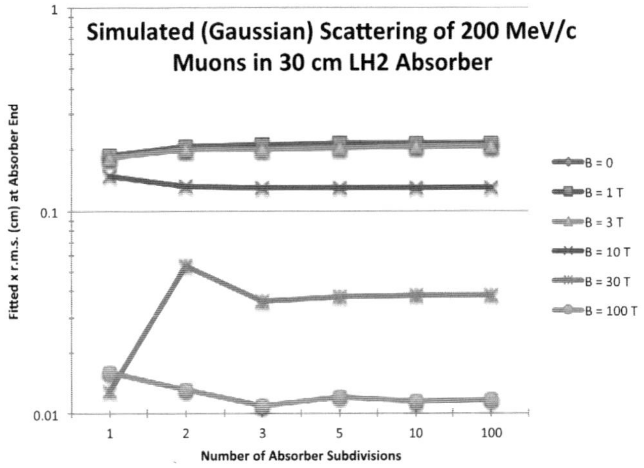

Figure 2. Root-mean-square value of x after 30 cm LH_2 absorber vs number of slices, in Gaussian model, for longitudinal B fields indicated.

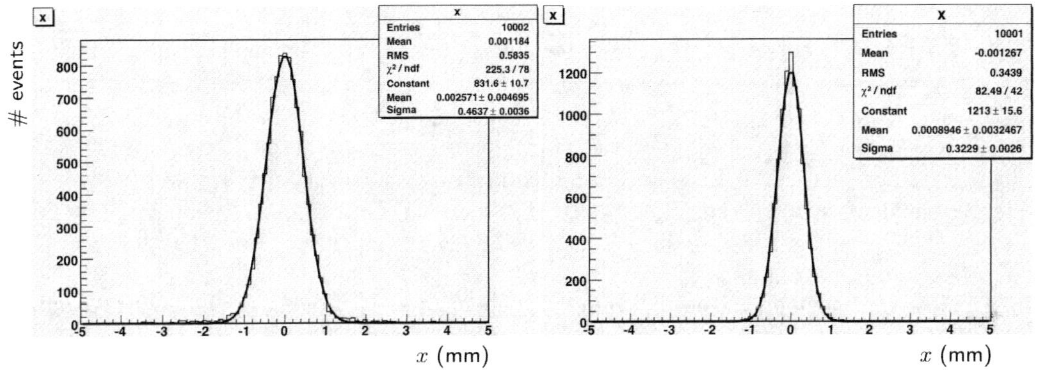

Figure 3. Distributions in x after 30 cm LH_2 absorber with $B = 30\,\mathrm{T}$, in G4beamline simulations, with (left) the default value maxStep = 100 mm and (right) maxStep = 1 mm.

permitted to take. Figure 3 compares the r.m.s. x distributions in a 30 T longitudinal field after traversal of a 30 cm LH_2 absorber by 200 MeV/c muons with maxStep = 100 or 1 mm. The default value of maxStep leads to an overestimate of the r.m.s. spread by about 50%.

4. ICOOL

Finally we studied the same situation using ICOOL. The relevant knob here is the choice of the ZSTEP parameter, which we took as 10^{-1} or 10^{-3} m. In Fig. 4 we see that the spatial spread is independent of the value of ZSTEP.

5. Discussion and Conclusions

We have verified Lebrun's conclusion that high magnetic fields suppress some of the effects of multiple scattering and can thus produce smaller beam emittances in ionization cooling than one might naively expect. Because of this effect, G4beamline cooling results at high fields are not reliable unless the maxStep parameter is chosen carefully. We have discussed this with Yonehara, who used G4beamline to simulate helical cooling channels at fields up to $\approx 15\,\mathrm{T}$ [7]. We found that he performed a maxStep sensitivity study and so was not

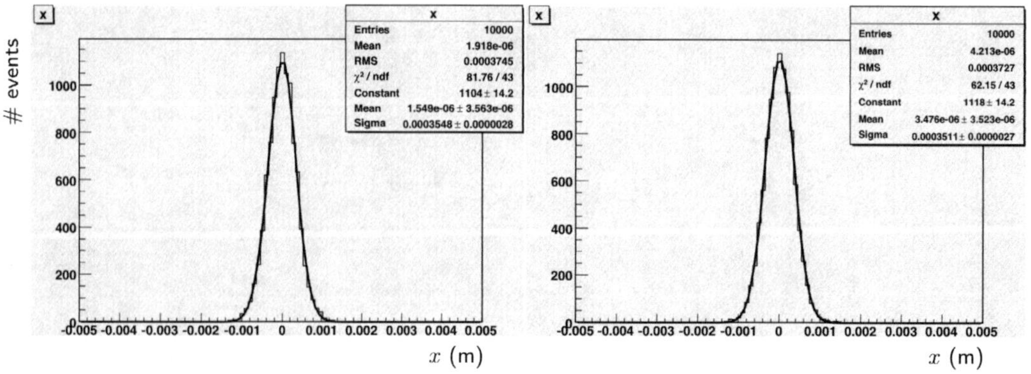

Figure 4. Distributions in x after 30 cm LH_2 absorber with $B = 30\,\text{T}$, in ICOOL simulations, with (left) ZSTEP $= 10^{-1}$ m and (right) ZSTEP $= 10^{-3}$ m.

misled by this effect. In contrast, ICOOL uses a variable-stepping algorithm by default, and so is largely insensitive to the user's choice of ZSTEP. Thus Palmer's simulations of 30–40 T "final-cooling" channels [3] are also insensitive to the effect. Modification of the Geant 4 variable-stepping agorithm to take this physics into account is under study. Until it is successfully implemented, G4beamline users are warned to include a study of stability with respect to variation of maxStep if they are using fields of $\sim 5\,\text{T}$ and above.

It is worth noting that the three simulations disagree at the $\pm 10\%$ level on the amount of scattering in 30 cm of LH_2, with (for example) $x_{r.m.s.} = 0.38$, 0.32, and 0.35 mm (with negligible statistical uncertainty) for $B = 30\,\text{T}$ in the Gaussian, G4beamline, and ICOOL cases. This is not surprising, since the Bethe calculation [2] and the PDG formula (Eq. 1) are based on approximate (Thomas–Fermi) atomic potentials, whose use is unnecessary for hydrogen [8]. G4beamline and ICOOL use more recent scattering models based on exact hydrogen wavefunctions [9]. The residual 10% disagreement between G4beamline and ICOOL was not investigated.

Acknowledgments

We thank R. Fernow for providing the ICOOL example deck used in this study and for useful conversations. Work supported by U.S. D.O.E.

References

[1] Lebrun P 1999 Modeling of multiple scattering effects in the presence of a strong magnetic field *NFMCC Note 30* `http://nfmcc-docdb.fnal.gov/cgi-bin/ShowDocument?docid=30`
[2] Bethe H A 1953 *Phys. Rev.* **89** 1256
[3] Palmer R B, Fernow R C and Lederman J L 2011 Muon collider final cooling in 30–50 T solenoids *PAC-2011-THOBN2*
[4] G4beamline website: `http://g4beamline.muonsinc.com`
[5] Fernow R C 2001 ICOOL – a tool for muon collider simulations, in *Muon Colliders at the High Energy Frontier*, ed A C Caldwell and B King (Paramus, NJ: Rinton); see also `https://pubweb.bnl.gov/~fernow/icool/readme.html`
[6] Nakamura K *et al.* (Particle Data Group) 2010 *J. Phys. G* **37** 075021
[7] Yonehara K, Derbenev Y S and Johnson R P 2010 Helical channel design and technology for cooling of muon beams *AIP Conf. Proc.* **1299** 658
[8] Tollestrup A V 1999 Moliere scattering for 186 MeV/c muons in hydrogen *NFMCC Note 20* `http://nfmcc-docdb.fnal.gov/cgi-bin/ShowDocument?docid=20`
[9] Fernow R C 2006 Scattering in ICOOL *NFMCC Note 336* `http://nfmcc-docdb.fnal.gov/cgi-bin/ShowDocument?docid=336`

The MICE luminosity monitor

A Dobbs[1], D Forrest[2] , F J P Soler[2]

1 Physics Department, Blackett Laboratory, Imperial College London, SW7 2BW, UK
2 School of Physics and Astronomy, University of Glasgow, Glasgow, G12 8QQ, UK

Abstract. The MICE experiment will provide the first measurement of ionisation cooling, a technique suitable for reducing the transverse emittance of a tertiary muon beam in a future neutrino factory accelerator facility. MICE is presently in the final stages of commissioning its beam line. The MICE luminosity monitor has proved an invaluable tool throughout this process, providing independent measurements of particle rate from the MICE target, normalisation for beam line detectors and verification of simulation codes.

1. Introduction

Substantial research and development work has justified the neutrino factory as the facility of choice for cataloguing the parameter space of neutrino oscillations [1]. No other proposed facility provides a more promising venue for the discovery of leptonic CP violation over as great a range of values of the θ_{13} oscillation angle, or greater precision in measuring the oscillation parameters than the neutrino factory.

A significant cost driver of such an accelerator facility is the cooling channel, which must reduce the phase space volume of the muon beam to within the acceptance of the downstream acceleration system. In an ionisation cooling channel, muons undergo energy loss in low density absorbers and the longitudinal momentum is restored in accelerating Radio Frequency (RF) cavities.

The Muon Ionisation Cooling Experiment (MICE) seeks to provide a first ever demonstration of ionisation cooling [2]. Whilst the physics of the process is not in doubt, significant engineering challenges exist in realising a cooling channel in practice. The goal of MICE is to measure a fractional change in emittance of order $\sim 10\%$ to an error of 1% with two scintillating fibre trackers at the entrance and exit of the channel (see figure 1). MICE has reached its first major milestone: a high purity muon beam line capable of providing sufficient muons for precision measurement of emittance.

2. Luminosity monitor and beam loss

The MICE target operates parasitically to the host proton synchrotron ISIS, dipping into the halo of the beam in the last few milliseconds of the acceleration cycle. This produces beam loss, which is measured by a series of 3 and 4 metre argon gas ionisation tubes along the synchrotron, approximately 2-3 m from the beam. The beam loss experienced by ISIS, and consequently the muon rate in MICE, will vary depending upon beam conditions and target control parameters, such as dip depth and the time of injection in the beam. In order to characterise the beam line ahead of future analyses, it is essential that MICE can both measure and predict particle rate.

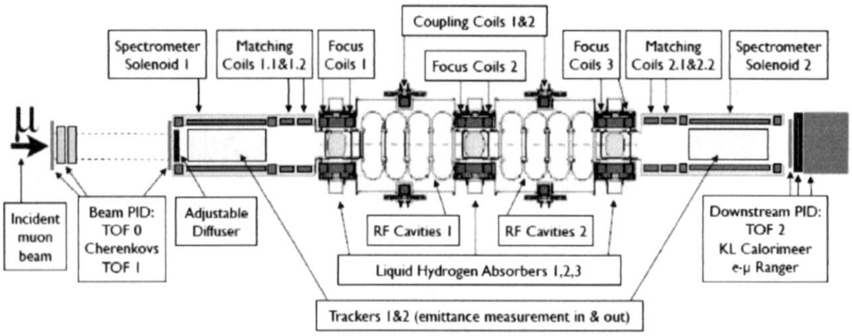

Figure 1. The full MICE cooling channel with absorbers and RF cavities braced by scintillating fibre trackers and particle identification detectors.

The luminosity monitor (LM), shown in figure 2, provides a means to measure particle rate independently of the host accelerator beam loss monitors. Two pairs of scintillators are read out through low noise photomultiplier tubes and the coincidence of hits in each pair (C12 and C34), as well as the four-fold coincidence (C1234), are recorded in data. Placed 10 m from the target at a similar polar angle to the MICE beam line, the LM allows for normalisation of all detectors along the beam line as well as verification of physics models and simulation codes [3].

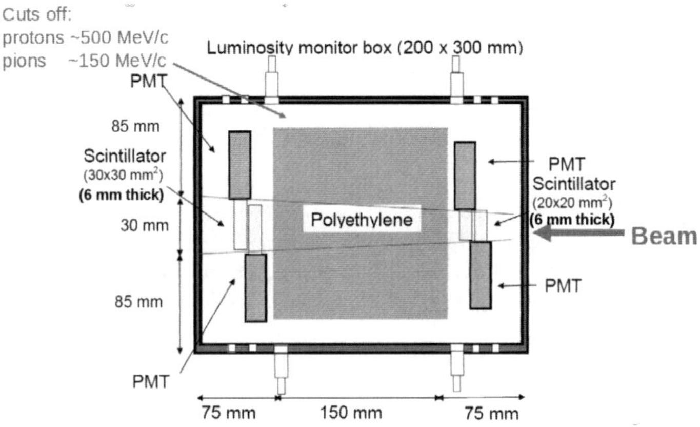

Figure 2. The luminosity monitor consists of two pairs of scintillators read out through low noise photomultiplier tubes. The second set is shielded by 15 cm of polyethylene which acts as a filter stopping low energy protons and pions.

LM commissioning (figure 3) has demonstrated an expected linear relationship between counts in each of the LM channels and beam loss. Predefined beam loss conversions can then be applied to convert counts to protons on target.

3. Simulations and results

The geometry of the luminosity monitor and target were modelled in G4Beamline, a general purpose simulation program based on GEANT4 [4]. In order to increase statistical accuracy, the area of the scintillator planes and plastic were increased by a factor of 100 and 2×10^9 protons on target were simulated using Grid resources. This simulation was successful in producing a reasonable rate at the first set of scintillators (C12). A second set of simulations with the actual experimental geometry was created to study the particle transmission rate and to determine the rates in the downstream scintillators (C34 and C1234) with respect to C12.

The number of protons on target that interact with the target and create beam loss in the ISIS ring depend upon the target thickness and geometry. At present, MICE employs a titanium

Journal of Physics: Conference Series **408** (2013) 012084 doi:10.1088/1742-6596/408/1/012084

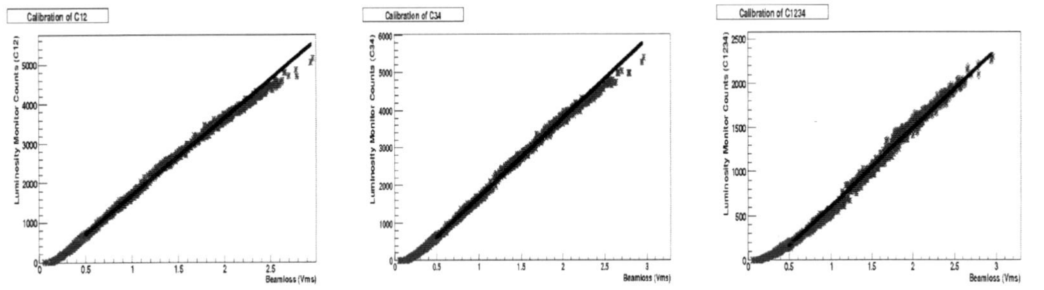

(a) (b) (c)

Figure 3. First data from the luminosity monitor showing a direct relationship between luminosity and beam loss on all three scaler output channels. Graphs show counts for coincidences of signals for PMTs (a) 1,2 (b) 3,4 and (c) 1,2,3 and 4. The straight lines indicate a linear fit to the data between 0.5 Vms and 3 Vms, since the method for calculating beam loss is unreliable below 0.5 Vms.

cylindrical target with inner radius 2.3 mm and outer radius 3 mm. As a consequence of the geometry, protons encounter a variable thickness of titanium. To study this effect, a rectangular target with cross-section 1×1 mm^2 and length 10 mm was also simulated. This target is similar to that previously employed by MICE in early data taking. The ratio of the counts produced from the cylindrical target to the long and thin rectangular target (the geometry factor) was found to be $\eta = 0.169 \pm 0.001$.

The geometry factor can be understood qualitatively by considering the mechanisms in which a proton is lost in the beam when it encounters the target. This was studied with the Objective Ring Beam and Injection Tracking code (ORBIT) [5], which is a particle tracking software package for accelerator rings, including space-charge effects. ORBIT simulations with the correct dimensions of the target were compared to other simulations in which the target was 10 mm long and 1 mm wide (long and thin). Preliminary results from these simulations show that the ratio of protons from the ISIS beam lost in the target area compared to those lost in the downstream collimators, is 5.4 for the long and thin target and 2.1 for the cylindrical MICE target [6]. This suggests that this ratio is 2.5 times larger in the long thin target than in the MICE cylindrical target. In the G4beamline simulation it was found that there is a factor of $1/\eta = 5.9$ times more beam loss in the long thin target than the cylindrical target. While the two independent calculations do not agree numerically, it does suggest that targets with a greater average material thickness do suffer more local beam loss than thinner targets. The beam loss mechanisms along the ISIS ring can be due to a mixture of scattering, which deflect the protons from their ideal trajectory, and from energy loss in the target.

Five hadronic physics models in GEANT4 were studied: QGSP_BIC, the default physics model in MICE, QGSP_BERT, QGSC, QGSC_CHIPS and LHEP [7]. Momentum distributions for QGSP_BIC are shown in figure 4. For each model, the yield (number of particles per proton-on-target, POT, per cm^2) was calculated as follows:

$$\mathrm{Y} = \frac{\text{LM Counts}}{\text{POT} \times \text{Area}} \times \frac{1}{\eta} \tag{1}$$

The results are shown in table 1, finding close agreement with data. The errors in the data are dominated by the unknown uncertainty in the calibration of the host accelerator beam loss monitors from which we infer the number of protons interacting with the target.

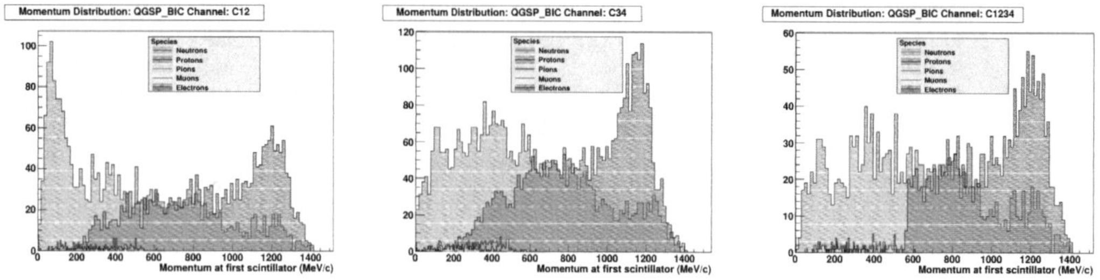

(a) (b) (c)

Figure 4. The momentum distributions at (a) the first, (b) second and (c) both sets of scintillators using the QGSP_BIC hadronic model.

4. Conclusions

A luminosity monitor has been designed, built and installed at the ISIS synchrotron, performing well as a beam line tool, as MICE reaches the final stages of commissioning its beam line. It has also provided an excellent device for studying the validity of simulation codes, with QGSP_BIC, QGSP_BERT and QGSC_BERT the most consistent with MICE data. As the experiment moves forward, the LM will be useful for normalising particle rate in all the MICE detectors and to independently measure beam loss from ISIS. Simulations of proton interactions in the MICE target compared to ORBIT simulations show some differences in the expected local beam loss, probably since ORBIT takes into account disruptions in the proton trajectory in the ISIS ring. The authors acknowledge the support of the Science and Technology Facilities Council (STFC).

Table 1. Response of the luminosity monitor as a function of hadronic model. Each simulation has an identical geometry description and is given the same set of functional parameters, including 2×10^9 protons on target. Errors in data are dominated by the 30% estimated error attributed to the beam loss calibration whilst errors in the simulation are statistical.

	Yield $\left(10^{-8}\frac{\text{counts}}{\text{POT cm}^2}\right)$		
Model	C12	C34	C1234
Data	1.72 ± 0.52	0.82 ± 0.25	0.79 ± 0.24
QGSP_BIC	1.67 ± 0.01	0.84 ± 0.03	1.06 ± 0.04
QGSP_BERT	1.80 ± 0.02	0.81 ± 0.03	1.05 ± 0.04
LHEP	1.03 ± 0.01	0.25 ± 0.01	0.29 ± 0.02
QGSC_CHIPS	0.74 ± 0.01	0.37 ± 0.02	0.49 ± 0.03
QGSC_BERT	1.80 ± 0.02	0.84 ± 0.03	1.06 ± 0.04

References

[1] Choubey S *et al.* (IDS-NF Collaboration) 2011 (*Preprint* `arXiv:1112.2853 [hep-ex]`)
[2] Bogomilov M *et al.* (MICE Collaboration) 2012 *JINST* **7** P05009 (*Preprint* `arXiv: 1203.4089 [physics-acc-ph]`)
[3] Forrest D 2011 *PhD Thesis (University of Glasgow)*
[4] Muons Inc G4Beamline Simulation Software http://g4beamline.muonsinc.com
[5] Galambos J, Holmes J, Olsen D, Luccio A and Beebee-Wang J 1999 ORBIT user manual, version 1.10 Tech. rep. http://www.docstoc.com/docs/29691669/ORBIT-User-Manual
[6] Dobbs A J 2011 *PhD Thesis (Imperial College, London)*
[7] CERN Geant4 http://geant4.web.cern.ch/geant4/

Journal of Physics: Conference Series **408** (2013) 012085 doi:10.1088/1742-6596/408/1/012085

NUFACT11 Round Table Discussion – Questions & Answers

A Blondel[1] and I Efthymiopoulos[2] (on behalf of the NUFACT11 participants)
[1]DPNC University of Geneva, 1205 Geneva, Switzerland
[2] CERN, 1211 Geneva, Switzerland

E-mail: alain.blondel@cern.ch , Ilias.Efthymiopoulos@cern.ch

Abstract. This presents the response to questions formulated by a round table discussion panel to the NUFACT11 participants. The main points made were as follows.
Neutrino mass *is* physics beyond the Standard Model; it presents a very deep puzzle to solve, with the answers to several fundamental questions as potential reward. Precision measurement of oscillation parameters is one essential way to access information needed to solve this puzzle and there is a relatively clear (but not easy) way forward. Large θ_{13} makes the appearance signals larger, but does not allow to relax the requirements on high beam intensity and large detector masses; it creates a difficult challenge on systematic errors, which will require dedicated ancillary experiments. Of particular interest is a low intensity muon storage ring for required cross-section measurements.
There exist already extended bottom-up international collaboration in both physics and R&D experiments. All three main regions have plans for upgraded super-beams and associated detectors, with a variety of (mostly) complementary baselines, proceeding in an incremental way. The ultimate precision and verification of the neutrino mixing picture requires new types of neutrino beams based on storage rings. For low energy neutrinos, the beta-beam is a possible (but substantial) intermediate step. The Neutrino Factory offers the best precision and best sensitivity to deviations from the standard picture. Coordinated international R&D is pursued. International collaboration at the top level would be beneficial in acknowledging the process, so as make sure that intermediate steps are effectively building up in complementary fashion towards the ultimate goals.

1. The round table and the questions.

The round table discussion took place on 1st August in the CERN main auditorium. The invited members were: S. Bertolucci (CERN), K. Nishikawa (KEK), J. Strait (Fermilab), J. Womersley (STFC), T. Nakada (ECFA), S. Myers (CERN), S. Komamya (ICFA). The discussion was chaired by John Ellis (CERN). A number of questions had been prepared by the NUFACT11 scientific program committee and were used as a starting point for the discussion (see appendix). Somewhat unconventionally this round table discussion took place on the first day of the meeting. As an outcome the panel members addressed the participants with a number of questions.

Sergio Bertolucci
I hear that international collaboration on coordinated R&D is lacking. Could the community propose a realistic plan for common and staged neutrino R&D?

Journal of Physics: Conference Series **408** (2013) 012085 doi:10.1088/1742-6596/408/1/012085

Sachio Komamiya
Physics question: θ_{13} & δ are similar to V_{ub} and CP phase in the quark sector. In heavy flavors, angles are not really the interesting thing – and CP violation there is found to be insufficient to generate Baryogenesis. How about neutrinos? We know about Lepto-genesis, is there further fundamental physics in neutrinos beyond the numerical values of the angle and phase?
Steve Myers
As an accelerator physicist, there are too many options – we need to kill some to make progress. Need to make a choice, define the next steps, define critical R&D and move on. What is it that the community really wants to build?

Tatsuya Nakada
In the incremental approach, where does a Beta Beam fit? This question is in regard both to the physics case and the R&D needs.

Koichiro Nishikawa
In the near term, the international framework should be bottom up and not top down? At what point should an international framework be forced from the top?

Jim Strait
A. If you didn't have a Neutrino Factory, how precise a measurement of parameters can be done with Superbeams before reaching their limitations?
B. What "external" measurements, e.g. particle production, neutrino cross sections, etc., can be done to improve the current systematic error limits on Superbeam experiments, and what are the ultimately limiting systematic errors?

John Womersley
A. Do I need to worry whether neutrino and antineutrino oscillation parameters are the same, or worry about the LSND anomaly?
B. Do neutrinos play a role in dark matter, especially if there is no light neutralino?
C. Since STFC and funding agencies from other countries in Europe, the US, and Japan are investing substantially in MICE, please consider, given the new indications that θ_{13} may be large, the importance of continuing to support R&D for the Neutrino Factory (and ionization cooling in particular).

2. The Answers
These questions were written up by the organizers and distributed to the participants, then debated during dedicated Working Group sessions at the workshop (which thus really did some *work*!) – or as usual during informal discussions at coffee breaks or workshop dinner. The answers were collected through the NUFACT working group conveners or directly by messages to the editors. These questions can be grouped in three sets, which are not completely disconnected:
2.1 Fundamental physics questions (Komamiya, Womersley A and B)
2.2 Technical physics questions regarding neutrino oscillation experiments (Strait A and B, Nakada)
2.3 Strategic questions regarding staging, international collaborations and R&D programs

2.1. Fundamental Physics questions
The information we can obtain from neutrinos is complementary to the one from the quark sector. We stress that the study of quark flavor physics led to many more results than just the determination of the CP phase and of the mixing angles, it led to a precise experimental verification of the CKM

mechanism and is very sensitive to new physics at scales around and above the electro-weak scale and couplings commensurate to those of the Standard Model particles, which would contribute to observables via direct or loop processes. In that way quark flavor physics leads to stringent constraints on this type of physics beyond the Standard Model.

In contrast, neutrino mass is *direct* manifestation of Physics Beyond the Standard Model. The relevant energy scale can be anywhere from very low scale (sub-eV) up to the GUT scale (10^{16} GeV).

We do not know the mechanism for the generation of neutrino masses; many models have been proposed, invoking such concepts as Grand Unification, String Theories or compositeness. It is fair to say that so far none of these has proven very predictive, the neutrino masses and mixing themselves having been subject to a large number of incorrect predictions.

In this context, the precise measurement of mixing angles and CP phase (the neutrino "PMNS" matrix) is one essential element (a "first step"?) in deciphering this fascinating puzzle. The values of the parameters of the PMNS mixing matrix and of the neutrino mass matrix (or matrices?) are either fundamental constants of nature or are derivable from some currently-unknown higher theory. In either case we must determine these numbers experimentally, and there is no other way to do it than by dedicated oscillation experiments of the type that were discussed at NuFact11. We must explore neutrino mixing experimentally to the full before we will be in any position to create or test any higher scale model. A guideline recommendation (in absence of a more quantitative argument) is to test the neutrino mixing paradigm with the same level of accuracy as that of the quarks.

The puzzle receives complexity with respect to charged quarks and leptons because Majorana mass terms are allowed in addition to the more standard Dirac mass terms. In many models it arises naturally that there should exist in addition to the three known (left-handed) neutrino species, an additional set of, at least two and generally three, right-handed neutrinos. These 'sterile' neutrinos may have the same masses as the known left-handed ones in the situation where neutrinos receive their mass only from Dirac mass terms, e.g. from the Higgs Yukawa couplings. If there exist, in addition, Majorana mass terms, right-handed neutrinos acquire different masses from the left-handed ones by the "see-saw mechanism". Although most Grand Unified Theory scenarios send these right-handed neutrinos to very high masses, this needs not be the case and one or more of the 'sterile' neutrinos could exist at low masses, providing a possible candidate for dark matter, or providing visible signals at the LHC. Some could be massive enough to provide a natural mechanism for the Baryon-Antibaryon asymmetry of the Universe via Leptogenesis. The presentations by Shaposhnikov, Mavromatos, Mohapatra and Senjanovic among others have highlighted the depth of the neutrino questions and the variety of solutions neutrinos have to propose to well known cosmological and particle physics questions.

More specifically for Dark Matter (DM): we know that the left-handed neutrinos are a small part of the DM (hot DM); it is known that keV right-handed (sterile) neutrinos can provide all of the dark matter (warm DM). This possibility is maybe even slightly favored over cold DM by structure formation arguments.
Furthermore, if DM is made of WIMPs, neutrinos do play a role in the search for it: DM annihilates into neutrinos, solar neutrinos are a background in DM direct detection experiments, NC neutrino scattering experiments are needed to get information on DM-nucleus interactions, and finally there are experimental similarities between neutrino and DM detector requirements.

Measuring the masses of neutrinos and establishing the existence or otherwise of Majorana mass terms are further aspects of this detective work which may become accessible in the upcoming decades. Yet these important steps will not answer all questions. We still do not know for instance how the

Journal of Physics: Conference Series **408** (2013) 012085 doi:10.1088/1742-6596/408/1/012085

Majorana phases could be measured should neutrinos be Majorana particles. We certainly do not have the slightest idea yet how to perform measurements on sterile neutrinos and if or how they can ever be discovered. For these reasons, any hints, however experimentally disputed they may be, like the LSND anomaly – or the recent OPERA neutrino time-of-flight anomaly-- must be followed up carefully because sterile neutrinos may quite naturally exist -- or because some other anomaly may exist that we have not actually thought about theoretically.

To conclude: we know enough about neutrinos to predict that they offer an extremely rich and deep field of investigation, and we know of a reasonably well defined experimental track to follow. However it is fair to expect surprises along the way. Experiments should be ready for that.

2.2. Technical questions regarding neutrino oscillation experiments.

The physics reach of future neutrino oscillation experiments in CP violation and Mass Hierarchy is intimately associated with issues of systematic errors and backgrounds in a way that is connected to the exact value of the mixing angle θ_{13}. This is a rapidly evolving field and the following will provide a snapshot of the present situation. More definite answers will certainly be available by NUFACT12! A summary of present sensitivity estimates can be found in the report by W. Winter at the workshop.

Several studies have shown that unless θ_{13} is very large (above $\sin^2 2\,\theta_{13} \gtrsim 0.11$) it will be very difficult for the combination of presently approved experiments (T2K, NOvA, and the reactor experiments Double Chooz, Reno, Daya Bay) to make a *definitive* determination of the neutrino mass hierarchy, sign(Δm^2_{13}). (see e.g. S. Prakash et al in these proceedings and the article of Patrick Huber, Manfred Lindner, Thomas Schwetz, Walter Winter, JHEP 0911:044,2009, arXiv:0907.1896v1 [hep-ph]). Superbeam experiments with a longer baseline than NOvA will be needed to answer that question.

Concerning the CP violation, the evidence from T2K, MINOS (and in November 2011, Double Chooz) indicates a value of $\sin^2 2\,\theta_{13}$ in the vicinity of 0.08±0.03; the error is asymmetric so the value is non-zero with a significance in excess of more than 3σ. The situation will evolve rapidly in 2012. This large value has the advantage that the signal reaction $\nu_\mu \rightarrow \nu_e$ is larger and easier to separate from backgrounds. The counter-intuitive effect that it reduces to less than 20% the CP asymmetry:

$$A_{CP} = (P(\nu_\mu \rightarrow \nu_e) - P(\bar{\nu}_\mu \rightarrow \bar{\nu}_e)) / (P(\nu_\mu \rightarrow \nu_e) + P(\bar{\nu}_\mu \rightarrow \bar{\nu}_e)).$$

The reason is that, while the numerator is proportional to $\sin\theta_{13}.\sin\delta$, the denominator is driven by the appearance probability itself which scales with $\sin^2 2\theta_{13}$. As a consequence the relative statistical accuracy on the asymmetry itself does not improve. *The large value of θ_{13} does not allow to relax the need for high intensity and high detector mass.*

Furthermore, the measurement becomes very sensitive to systematic uncertainties on the signal itself, i.e. on cross-sections within experimental cuts, and to the knowledge of the matter effect. One should bear in mind that ν_e and $\bar{\nu}_e$ cross sections can be different from those of ν_μ and $\bar{\nu}_\mu$ of the same energy because of the interplay of muon mass effects with i) nuclear effects and ii) pion production thresholds; they should be measured.

The estimates of sensitivity to δ that can be found in the literature are often relatively vague about the assumptions made on systematic uncertainties; their dedicated understanding will be a major subject of study towards NUFACT12. In this context several aspects of neutrino experiments will be emphasized:

1. The near detector design

Journal of Physics: Conference Series **408** (2013) 012085 doi:10.1088/1742-6596/408/1/012085

2. The ancillary experiments to improve predictions of neutrino flux (hadroproduction experiments) for the Superbeam experiments
3. Ancillary measurements of appearance signal cross-sections

The best (2-3% precision) measurements of neutrino cross-sections were made with muon (anti)neutrino beams in the early 80's above 40 GeV neutrino energy, on heavy isoscalar targets with dedicated exposures to a narrow band beam (at CERN) or dichromatic quad-triplet beam (at Fermilab). In the energy range of interest (200 MeV to 5 GeV) for the determination of the neutrino mixing angles in beam experiments, the best measurements are rather in the 7-15% range. To measure a CP asymmetry, the maximum value of which is at most about 20%, when one or two of the four relevant neutrino cross-sections is measured to $\pm 7\%$, and the others not at all, is a difficult starting point for a 5σ discovery. In addition to the high intensity neutrino beams and large detector masses required, a demonstration that the experiment can be designed to control systematic errors at the few % level will be mandatory before a clear conclusion on the chance of a future project to "discover Leptonic CP" can be assessed.

The Neutrino Factory design studies have developed at this point the most complete discussion of systematic uncertainties – with a stored beam of muons, the flux can be predicted from muon decay to better than a percent, so a well designed near detector can measure all relevant cross-sections (ν_μ , ν_e, $\bar{\nu}_\mu$ $\bar{\nu}_e$) to this accuracy.

Does this imply that a definitive measurement of CP violation must await construction of a Neutrino Factory? The answer is: perhaps not. A mini-Neutrino Factory may provide the Superbeam experiments with the cross-section measurements they need. One of the highlights of the NUFACT10 and NUFACT11 workshops was the presentation, (A. Bross et al), of the 'mini-neutrino-factory' or Low-energy Low-intensity Low-cost Neutrino Factory (L^3NF), in which a storage ring of medium energy is used to store muons directly in a way similar to the antiproton production systems of CERN and FNAL. The quality of cross-section measurements which are trademarks of the Neutrino Factory should be preserved, but the required muon intensity is considerably reduced. This experiment can also pursue search for sterile neutrinos in its own right and constitutes an interesting R&D step towards the full size Neutrino Factory. The experiment is under evaluation but seems to be promising.

It remains very likely that the Neutrino Factory will offer the best precision and there exist a significant fraction of phase space where only the Neutrino Factory can make a definitive measurement. In addition, measurements specific to high energy electron neutrinos such as $\nu_e \rightarrow \nu_\tau$ cannot be made anywhere else. However, as was emphasized in the round table discussion, the Neutrino Factory is a large step to climb, and there is considerable interest from the community to proceed in an incremental fashion, building up expertise and community, ascertaining the limitations of the Superbeam experiments along the way.

What is the role of the beta-beam in an incremental approach? The Working Group WG1 (neutrino oscillations) could not reach a consensus on this issue; we summarize here the pros and cons.

By construction the Beta Beam provides a low energy (up to 1 GeV for the CERN-SPS baseline scenario) beam of pure ν_e or $\bar{\nu}_e$ depending on the stored ion. There has been considerable progress on understanding how to produce sufficient amounts of the baseline ions, ^{6}He and more critically ^{18}Ne. The perceived virtue of beta-beam lies in the fact that it can be deployed on the CERN site using the PS and SPS as ion accelerators; it remains true that the ion source and pre-acceleration, as well as the storage ring, equivalent in magnetic rigidity to a 300 GeV/c proton storage ring, are non-trivial. Moreover, the issue of irradiation by ion losses and decay products of the CERN accelerators and of the storage ring is considerable. The baseline scenario Beta Beam matches well the Water Cherenkov

Journal of Physics: Conference Series **408** (2013) 012085 doi:10.1088/1742-6596/408/1/012085

technology and a distance of the order of 300-500 km. Here the physics aim would be essentially the search and measurement of the CP violation. The baseline to the Frejus site at 130 km is acknowledged to be on the short side of optimal. At large values of θ_{13} the backgrounds are not critical, so the duty cycle requirements can be somewhat relaxed.

From the point of view of flux control, the Beta Beam should offer many of the advantages offered by the Neutrino Factory: the number of ions can be directly measured and the neutrino flux prediction should be accurate to better than 1%. Beta Beams will be able to measure the relevant ν_e, ($\bar{\nu}_e$) cross-sections with an accuracy of this order. However since the signal is the appearance of a ν_μ, ($\bar{\nu}_\mu$), these cross-sections need to be known, especially in the ~250 MeV region of the first oscillation maximum for the Frejus experiment, where nuclear effects and the muon mass effect are important. The Beta Beam will rely on measurements that have to be performed in a low energy pion decay beam (or mini-Neutrino Factory of low energy), in a rather unfavorable region – it remains to be demonstrated that the desired accuracy can be reached. There is at the moment no proposal worked out for a near detector station in the beta-beam and it is difficult to judge what precision can be achieved on a CP asymmetry. The sensitivity curves shown so far indicate that a Beta Beam to Frejus would be of similar accuracy (and nicely complementary) to a Superbeam in measuring the CP phase. Both the Superbeam in question, which involves a 4MW Superconducting Proton Linac to be built, and the Beta Beam, are seen are rather high steps for an 'incremental' path. A more complete assessment of these issues will be available when EUROnu and LAGUNA-LBNO design studies will be completed, in 2012 and 2014 respectively.

2.3 Strategic questions regarding staging, international collaborations and R&D programs

There are presently four neutrino beams in operation performing neutrino oscillation experiments. The CNGS beam at CERN serves the experiments OPERA and ICARUS. The NUMI beam at Fermilab serves MINOS, MINERvA, ArgoNeuT and will serve NOvA; the Booster neutrino beam at Fermilab serves MiniBooNE, SciBooNE, and will serve the MicroBooNE experiment. The T2K beam serves the T2K experiment.

There exist already a number of ***bottom-up*** international collaborations to study future projects. Several regular international meetings occur yearly: the NBI workshops for neutrino beams and instrumentation, the NUFACT workshops, dedicated to Superbeam, Beta Beam and neutrino factories; the NNN workshops dedicated to the large underground detectors for nucleon decay and neutrino experiments; the NUINT workshops for studies of neutrino interaction cross-sections. The ongoing oscillation experiments T2K, MINOS, OPERA, Double Chooz are international collaborations involving physicists originating from two or all of the three world regions. The ongoing design studies for DUSEL (USA), T2HK (Japan), EUROnu and LAGUNA-LBNO are more regional and address a more concrete localized scenario – a necessary step to identify and solve concrete issues. The International Design Study for the Neutrino Factory is inter-regional, but it is not addressing a localized scenario. The R&D experiments such as MERIT (at CERN), MICE (at RAL) and EMMA (at Darsbury lab) for the Neutrino Factory, or 60GHz ion source (LPSC, Grenoble, and IAP, Nizhny Novgorod), isotope production (ANL, INFN Legnaro) for Beta Beams, are run by international collaborations.

The question of top-down coordination was already raised after the first NUFACT99 workshop and while it did not materialize by a program of coordinated oscillation experiments, it led to the international collaboration for the Neutrino Factory R&D experiments mentioned above.

Has time come for a ***top-down*** coordination?

The number of questions to be answered in neutrino physics is large: CP violation, mass hierarchy, precise measurements of parameters, tests of unitarity, searches for sterile neutrinos and search for other more exotic scenarios. While a Neutrino Factory is claimed to be able to solve a large fraction of these questions in a definitive way, no large laboratory has proposed to host it, and it is hoped that one can have a first look at these questions by a number of incremental and complementary steps with advantages in terms of funding profile and continuity of activity, while the most critical R&Ds for the long term are being pursued.

There appears to be three next-generation Superbeam experiments being discussed or proposed – one in Japan with T2HK (295 km) or Tokai to Okinoshima (658km), on in the the USA with DUSEL (1300 km), and, one in Europe with either the CERN to Frejus (130km) project which is also suitable to a Beta Beam, or from CERN to Pyhasalmi (2285km), with an improved replica of the CNGS beam, which is a good baseline for a Neutrino Factory. (The DUSEL baseline itself would be suitable for a lower energy Neutrino Factory). One can see from the variety of baselines (and detectors) that there are complementarities in these approaches, both for physics and for the detector technologies. As mentioned above it is almost certain that once choices have been made international collaborations will be put in place for each of the successful projects. Note that care is sometimes taken to make sure that the project is suitable for a long term vision with more ambitious beams.

The question is the following. Is there space and need for collaboration and focusing of efforts to realize sooner a more ambitious project? Or is the variety and wealth of physics such that regional projects can be carried out in such a way that different experiments situated in different places provide each their piece of the puzzle or the necessary verifications while maintaining a deliberate level of competition? In the longer term we will no doubt pursue even more ambitious facilities, however it would appear premature to decide on the exact makeup and location of such projects now. Enhanced high-level support for performing the R&D necessary to position ourselves to be able to build such facilities constitutes a much valued investment for the future to ensure that progress in this important field continues in a healthy way. The working group were unanimous that e.g. the Muon Cooling R&D should be pursued actively and efficiently, as it will certainly be needed for the Neutrino Factory or a Muon Collider.

Acknowledgements

We would like to thank wholeheartedly the round table participants for their enthusiastic response to our invitation and of their careful preparation of the questions asked by the program committee. That the committee replied by formulating questions of their own was a tremendous stimulus to the whole workshop. The program committee prepared the initial questions and contributed substantially to phrasing and correcting the answers. The answers themselves originated from discussions in the working groups. The working group conveners animated these discussions, worked hard and late in concatenating and formulating the answers in their conclusions; the success of NUFACT is greatly due to their dedication and skill.

Journal of Physics: Conference Series **408** (2013) 012085 doi:10.1088/1742-6596/408/1/012085

APPENDIX: Questions asked to the panel by the NUFACT11 Program Committee

21 May 2011

Future Neutrino Facilities in the Global Physics Environment

Proposed Questions for NuFact11 Round Table
by the NUFACT11 Scientific Program Committee

(NB emphasis is placed on neutrino oscillation experiments)

1. Is the "incremental" approach to experimental neutrino oscillation physics the best way for the international community to proceed, or should a big step to a Neutrino Factory be advocated? If a big step is deemed appropriate, when should it be envisioned?

2. Up to what levels do you think the regional neutrino programs should work, and when and how do you envision that a fully international framework of neutrino projects is needed and should be formed? What do you think is "missing" that prevents a decision being made on a large international neutrino project: a determination of θ_{13}, a community consensus, a political decision at ICFA level, technological issues…?

3. What room do you see for accelerator-based neutrino projects besides the other particle physics projects (e.g. in Europe besides the LHC)?

4. What might be the scientific impact of LHC results on the neutrino program?

5. What do you think are the most important goals in neutrino oscillation physics and what do you think are the best ways to get there? What relative importance/organization should be given to the mainline program (unraveling 3x3 mixing including CP violation) with respect to search for exotics such as sterile neutrinos, Non-standard interactions, etc

6. How important is it that proposed giant neutrino detectors serve additional physics needs – how bad is it if they don't?

7. What importance do you attach to the Neutrino Factory being on the upgrade path toward a Muon Collider?

AUTHOR INDEX

AUTHOR INDEX

Institute of Physics
Dirac House, Temple Back
Bristol BS1 6BE UK

CURRAN ASSOCIATES INC
proceedings.com